Environment and Aquaculture in Developing Countries

Edited by

R.S.V. Pullin
H. Rosenthal
J.L. Maclean

INTERNA... LI... RESOU... ...FÜR ...BEIT

Environment and Aquaculture in Developing Countries

Edited by

R.S.V. Pullin
H. Rosenthal
J.L. Maclean

1993

Published by the International Center for Living Aquatic Resources Management, MCPO Box 2631, 0718 Makati, Metro Manila, Philippines and Deutsche Gesellschaft für Technische Zusammenarbeit (GTZ) GmbH, Postfach 5180, D-6236 Eschborn, Federal Republic of Germany

Printed in Manila, Philippines

Pullin, R.S.V., H. Rosenthal and J.L. Maclean, Editors. 1993. Environment and aquaculture in developing countries. ICLARM Conf. Proc. 31, 359 p.

ISSN 0115-4435
ISBN 971-8709-05-3

Cover: Khao Sam Roi Yot (Mountain of Three Hundred Peaks) Natural Park is located in Prachuab Khiri Khan province in Thailand. The 98-km^2 park contains a 40-km^2 marsh, an important sanctuary for birds which is listed in the Asian Wetlands Directory of the IUCN as a site of global conservation importance. Vast areas of marsh in the park have been converted into shrimp ponds. The prestigious Siam Society, a national NGO has urged action by the government. Although many of the shrimp ponds have ceased operation (because of disease caused by intensive culture and limited water supply), new ponds were still being constructed in December 1992. Photo courtesy of Peter Edwards.

Available at US$15 (surface mail), $22 (airmail), P350 from: • ICLARM, MCPO Box 2631, 0718 Makati, Metro Manila, Philippines • International Specialized Book Services, 5804 N.E. Hassalo St., Portland, Oregon 97213-3644 USA (use airmail price) • Ernst S. Toeche-Mittler GmbH, Versandbuchhandlung, Hindenburgstrasse 33, D-6100 Darmstadt, Germany (use airmail price). Please make payments payable to ICLARM in US$ by international money order or bankdraft. We can accept US$ checks only if from a US-based bank due to high clearance fees of other banks.

ICLARM Contribution No. 941

Contents

Foreword

The resources available to ensure the continuance of life on earth are finite. Any resource can only serve a limited number of purposes at the same time and place. This is particularly true of water which is a fundamental requirement not only for aquatic but also for terrestrial organisms. It similarly applies to nutrients and energy.

With an increasing demand for food, energy and space by growing population, the pressure of exploitation is reaching alarming levels on an increasing number of species and over an expanding area. To avoid overexploitation and loss, the resources essential for human survival must be used efficiently and wisely. This requires channeling their utilization in ways that fulfill multiple and complementary objectives wherever possible.

Modern aquaculture appeared at a time when many claims for use of the resources had been made and competition was growing for those niches still available. Labor was becoming increasingly expensive, leading to intensification in terms of rationalization and mechanization to reduce costs. This meant higher stocking densities and higher demand for feed and energy. Among the most immediate environmental consequences were overloading of the waters with nutrients, contamination with chemicals for the treatment of diseases and pests, and ecological damage through the installation of voluminous infrastructure. The demand for feed increased the pressure on other living resources such as small pelagic fish utilized as fishmeal.

Most of the more conspicuous mistakes made so far were committed by developed countries. Some at least could have been avoided through more awareness, foresight and readiness to renounce fast profits which were both questionable and harmful in the long term. The most important lesson to be learnt from the past is more consideration for the need to understand better the environmental and social context in which aquaculture is being developed. Such better understanding should then lead to the establishment of a general policy to guide development action in the most promising directions and to keep negative side effects to a minimum.

In the majority of developing countries, intensification is of less immediate concern, though on a mid- and long-term basis related problems will gain in importance. The more urgent question is how to make the best possible use of the productivity of natural systems without radical environmental changes and at low levels of costly inputs. What is needed for the future is an approach which makes use of the experience available, adds to the existing know-how through continued research efforts, elaborates and refines guidelines, and creates appropriate frameworks for further development. Aquaculture production is in great demand, but it must not be achieved without due regard to safeguarding our basis of survival.

This proceedings volume presents detailed reviews of pertinent environmental issues and the conclusions and recommendations of an international conference convened by the International Center for Living Aquatic Resources Management (ICLARM) and the Deutsche Gesellschaft für Technische Zusammenarbeit (GTZ),

GmbH at the Bellagio Conference and Study Centre of the Rockefeller Foundation in September 1990. Only for a few of the issues are clear solutions becoming apparent. Much remains to be done and only intensive collaboration among all parties concerned will bring us closer to success. The results of this conference should be seen as a step in this direction.

Dr. Martin Bilio
Senior Adviser for Living
Aquatic Resources Utilization
Deutsche Gesselschaft für Technische
Zusammenarbeit (GTZ), GmbH
Federal Republic of Germany

Acknowledgements

This conference was made possible by the Rockefeller Foundation which offered the use of its magnificent Bellagio Study and Conference Center as the venue and by the Deutsche Gesellschaft für Technische Zusammenarbeit (GTZ) GmbH, which provided a generous grant to cover organizational and publishing costs.

An Overview of Environmental Issues in Developing-Country Aquaculture*

ROGER S.V. PULLIN
International Center for Living Aquatic Resources Management
MCPO Box 2631, 0718 Makati
Metro Manila, Philippines

PULLIN, R.S.V. 1993. An overview of environmental issues in developing-country aquaculture, p. 1-19. *In* R.S.V. Pullin, H. Rosenthal and J.L. Maclean (eds.) Environment and aquaculture in developing countries. ICLARM Conf. Proc. 31, 359 p.

Abstract

Aquaculture, like all interventions by humans to exploit or manage natural resources for food production, has the potential for causing environmental harm as well as for improving livelihood and nutrition. Aquaculture development must be undertaken in a broad intersectoral context, considering especially its interactions with agriculture, forestry and capture fisheries and its environmental consequences. This paper examines types of aquaculture development and discusses the concept of sustainability and demographic, political and economic factors before giving examples of recent developments and criteria for assessing others.

Introduction

Aquaculture, like all food production by farming, has large effects on the environment, many of which can be negative: occupation and fragmentation of former natural habitats; reduction of the abundance and diversity of wildlife and changes in soil, water and landscape quality. The same applies to agriculture (Simons 1988, 1989). Because farming will remain the mainstay of most developing-country economies for the foreseeable future and will cause much environmental change, it is essential that the potential negative effects of further development of aquaculture be thoroughly appraised. Environmental protection and nature conservation now have much higher profiles in the political arena, mass media and public awareness than before. Environmental impacts at the relatively new frontier of aquaculture need very careful attention.

This paper gives working definitions of terms (aquaculture, developing countries, environment, sustainability and agroecology) and discusses broad concepts, summarizes the status of developing-country aquaculture and considers the future of aquaculture in developing countries, emphasizing the search for sustainability in the face of rapid change.

Aquaculture

Aquaculture is defined here as a modification of the definition proposed by FAO (1990a), omitting FAO's criterion that produce can be considered as derived

*ICLARM Contribution No. 737

from aquaculture only if raised under individual or corporate ownership. Aquaculture is the farming of aquatic organisms, including fish, molluscs, crustaceans and aquatic plants. Farming implies some form of intervention in the rearing process to enhance production, such as regular stocking, feeding, protection from predators, etc.

This definition includes enhanced fisheries (stock enhancement, aquatic ranching and management of natural aquatic environments) within the scope of production systems considered. FAO (1990b) includes in aquaculture statistics those "culture-based" fisheries that are stocked annually with propagated juveniles, but regards fisheries that are established through single or intermittent introductions as contributing to capture fisheries production.

Aquaculture can be broadly classified as extensive, having no feed or fertilizer inputs; semi-intensive, having some fertilizer and/or feed inputs; and intensive, largely reliant on feed inputs (Edwards et al. 1988a; Pullin 1989). Enhanced fisheries resemble extensive aquaculture with low levels of human intervention. Classification of aquaculture according to the economic goals or status of culturists - for example as 'subsistence', 'commercial' and 'entrepreneurial' - has also been attempted but is usually confusing. In much of Asia and Africa, fish is 'the other staple' (other than grains), the main animal protein source of the people. *All* farmers who try something new and profitable can be considered 'entrepreneurs' whatever the scale of their operations. Subsistence aquaculture barely exists. Virtually all aquaculture has a profit motive in cash or in kind.

Developing Countries and Development

A developing country is defined here largely as in a UN (1989a) report: all of Africa, Asia (excluding Japan), Latin America and Oceania (excluding Australia and New Zealand). This UN report referred back to a 1963 UN distinction between 'developed' and 'developing' countries based on population growth and pronounced it still valid:

> No other criterion, be it per capita (sic) income, urbanization, literacy, industrialization, etc., defines this dichotomy so sharply as the level of fertility. With exceedingly few exceptions, it can be said that where the gross reproduction rate is greater than 2.0, the country is a 'developing' one, and where it is less, the country is 'developed'.

Singapore, the Republic of Korea and Taiwan are here excluded from the definition of developing countries.

The Club of Rome recognized the limits to development.

> We are further convinced that demographic pressure in the world has already attained such a high level, and is moreover so unequally distributed, that this alone must compel mankind to seek a state of equilibrium on our planet (Meadows et al. 1972).

So, would 'transformation' be a better term than development? Probably not, as human 'states of equilibrium' are always highly dynamic. Development can be defined simply as the betterment of living standards for the disadvantaged. Betterment implies improved quality of life in, for example, health, education and recreation.

Environment

The term 'environment' is defined here broadly as the whole ecosystem and its nonliving and living resources, including human beings.

Sustainability

Sustainability has become a fundamental consideration for all development that involves the use of natural resources, particularly agriculture. The concept of sustainability is examined here because of the parallels that can be drawn between aquaculture and agriculture.

The Concept of Sustainability

"Productivity without sustainability is mining" (Dover and Talbot 1987). They pointed to various viewpoints on sustainable agriculture: supplying enough food for all - the food sufficiency/productivity viewpoint; maintaining average output indefinitely without depleting renewable resources - the ecological/stewardship viewpoint; and conserving the sociocultural aspects of rural society - the community viewpoint.

The Brundtland Report (WCED 1987) stated:

> The concept of sustainable development does imply limits - not absolute limits but limitations imposed by the present state of technology and social organization on environmental resources and by the ability of the biosphere to absorb the effects of human activities.
>
> .... sustainable development can only be achieved if population size and growth are in harmony with the changing productive potential of the ecosystem.

The Consultative Group on International Agricultural Research (CGIAR) has a Sustainability Committee which has stated:

> .... all Centers [International Agricultural Research Centers (IARCs)] should view the concept of sustainability as a guidepost to the development and introduction of agricultural techniques and technologies. IARC research and other activities should seek to maximize output and increase efficiency in the use of inputs, while minimizing the extraction of nutrients and organic matter from the soil and the contamination of the environment.... Production of good quality genetic materials through basic scientific work on the physical, chemical, and biological processes involved in plant and animal growth remains the critical contribution that the CGIAR Centers can make to the objective that must underlie sustainable agriculture: achieving more production per unit of land and input at less total energy and environmental cost (CGIAR 1989).

Dixon and Fallon (1989) have provided an excellent review of the concept of sustainability and the difficulties of defining it. They favor:

> a socioeconomic definition.... one that revolves around social and economic well-being for the present generation and retention of future options for our children.

A recent IUCN publication (McNeely et al. 1990) defined sustainable development as "a pattern of social and structural transformations (i.e., development) that optimizes the economic and other societal benefits available in the present without jeopardizing the likely potential of similar benefits in the future".

Is Sustainability a Realistic Goal?

Development policymakers now tend to use three criteria for assessing the efficacy of change: sustainability, equitability and environmental soundness. Of these, sustainability is the hardest to apply.

Ekins (1989) praised the Brundtland Report but also saw problems in reconciling economic growth and sustainability. He referred to Mrs. Brundtland's call for economic growth that is forceful and at the same time socially and environmentally sustainable; to the Report's call for growth rates of at least 5% in developing countries and 3-4% in industrialized countries and to its statement that:

> such growth rates could be environmentally sustainable if

> industrialized nations can continue the recent shifts in the contents of their growth towards less material- and energy-intensive activities and the improvement of their efficiency in many materials and energy (WCED 1987).

Ekins (1989) preferred routes to sustainability based on improved human welfare for the disadvantaged, not just increases in production and consumption and added the following:

> In poverty-stricken countries one can imagine a high weighting being given to production even at some environmental cost. In richer countries, with long-term survival at stake, as the [Brundtland] Commission notes, the rational choice would tend to favour environment and safety of the future, while of course not ruling out the possibility that some of these choices might also produce some production growth.

Against such complexities, is 'sustainability' a realistic concept? The word itself fails to combine a sense of durability *and* adaptability - for which it is difficult to find a single English word. Development should comprise evolutionary improvements in human welfare for both rural areas and rapidly growing cities and must make sense in terms of changing internal and foreign relations. "Evolvability" would be an appropriate, though clumsy, term. Indeed the fitness of organisms to their ecological niches and the evolution of species are useful analogies: capacity to adapt and prosper depends upon diversity. Diversity and scope for change in the technologies implemented for aquaculture development will be their best guarantee of lasting long enough to make substantial, environmentally compatible contributions to development and of leading to further positive changes. No system with human interventions lasts unchanged indefinitely.

Altieri (1989) presented the following conclusion on sustainability that can be applied to aquaculture as well as to agriculture:

> Current efforts aimed at soil and water conservation, improved food security, germplasm conservation, etc. will serve to counteract hunger or loss of resources temporarily.... However, ultimate sustainability will be reached as farmers: .... increase their access to land, resources and a suitable technology that allows them to manage these resources ecologically.

Agroecology and the Interaction of Aquaculture with Other Sectors

Agroecology is an approach advocated for devising sustainable agricultural systems and has been called a new paradigm for world agriculture (Altieri 1989). It emphasizes 'strength in diversity' of farming systems. This approach ignored, however, the potential contributions of aquatic production as did those of Charlton (1987), Dover and Talbot (1987), Tivy (1987), Edwards (1989) and Pimentel et al. (1989). In these and many other, otherwise excellent publications, the authors' vision was limited to terrestrial food production: largely a crop sector viewpoint. Edwards et al. (1988a) provided a framework for integrated agriculture and inland aquaculture but did not mention agroforestry. Agroforestry scientists are considering integration with crop and livestock production, but not yet with aquaculture (Hart 1987; Raintree and Torres 1987). In other words, hardly anyone has yet taken a truly holistic, all sectors approach.

Aquaculturists must themselves avoid similar narrowness of vision and recognize the very broad sectorally interdependent context in which aquaculture development takes place (Smith 1986). A broad intersectoral approach, based on ecosystem management, is essential for developing-country aquaculture development. Hopefully the realization will spread that an alliance of aquaculture with agroecology greatly strengthens this new paradigm (Lightfoot 1990). Similarly, aquaculture in the coastal zone, capture fisheries, industry, tourism and shipping all interact. Land-based activities (agriculture, forestry and industry) have profound effects on aquatic ecosystems (Chua et al. 1989a, 1989b). Moreover, most coastal aquaculture depends upon terrestrial resources for construction materials and other inputs, particularly feeds.

The Status of Developing-Country Aquaculture

General Considerations

Aquaculture is underdeveloped throughout the developing world. Much has been written about Asia being a region of strength and various countries being advanced in aquaculture. In comparative terms this is true. As a statement on the status of aquaculture vis-à-vis other sectors, it is highly misleading. Even in Asia, the number of farmers and coastal dwellers involved in part-time or full-time aquaculture is almost certainly below 1% of food producers. Most developing-country aquaculture is still the farming of undomesticated organisms in poorly understood systems. The supportive research base for developing-country aquaculture is weak (Pullin and Neal 1984). For enhanced fisheries it is very weak.

Recent Statistics

Table 1 summarizes aquaculture production in the developing regions in 1990. The predominance of Asia (especially China) is obvious. Table 2 gives the global picture in terms of aquatic environments. It is complex and dynamic (Csavas 1988) and there is little or no correlation between the growth of aquaculture and GDP (Table 3).

FAO (1989b) concluded that recent growth of aquaculture had been less dramatic than had been forecast in the mid-1970s: especially in developing countries where, apart from the spectacular rise of the shrimp culture industry, there had been little change in seaweed or mollusc production and a decline in African aquaculture production. A further conclusion was that growth in production of high value produce had been greatest and that "rural small-scale integrated aquaculture" had grown more slowly than had been anticipated.

Shrimp culture has indeed expanded in Asia and Latin America, largely because technology was available and there were sites to be occupied, albeit often at high environmental and social costs (see below). Seaweed culture has also prospered. Such operations have validated to some extent the philosophy that emerged from the 1976 World Conference on Aquaculture in Kyoto, that aquaculture could, and should, be developed through production-orientated research, parallel to which more basic research would also expand. However, the result is a developing-country aquaculture sector in which a few subsectors have prospered but the majority still lack reliable technology likely to attract new entrants, especially those of low-income groups.

Statistics currently available for 1988 (FAO 1990b) focused on a global increase of 19% in the value of aquaculture produce over 1987. Much of this was due to a 16% increase in production in China. Elsewhere, where research and development have been well supported, progress has sometimes been rapid in relative terms though still limited in absolute terms. For example, in Malaŵi, small-scale pond culture has increased from a very low level pre-1980 to several hundred ponds at present and expansion is continuing (ICLARM-GTZ 1991). However, the total national annual

Table 1. Aquaculture production (t) of important commodity groups in developing countries of four regions. (Source: FAO data (1992) except for India where production of carps is estimated at 200,000 t·year^{-1} and Taiwan Fisheries Bureau 1986, 1991).

	Year	Asia (excl. China) (12 countries)	China	SubSaharan Africa (29 countries)	West Asia/ North Africa (11 countries)	Latin America/ Caribbean (21 countries)	Totals
FINFISH							
Carps and other cyprinids	1984	691,469	1,766,158	515	14,888	1,837	2,474,867
	1985	545,916	2,383,786	389	40,293	1,147	2,971,531
	1988	633,372	3,858,500	647	17,846	4,155	4,514,520
	1990	711,441	4,124,478	1,126	48,011	4,588	4,889,644
Tilapias and other cichlids	1984	71,768	18,100	4,141	3,328	26,311	123,648
	1985	105,325	77,120	7,006	26,745	17,976	234,172
	1988	136,091	39,000	7,726	1,469	31,736	216,022
	1990	154,716	160,369	8,799	30,555	26,622	381,061
Misc. spp.[a]	1984	459,259	9,359	4,430	26,589	7,422	507,059
	1985	410,380	118,683	3,185	2,117	19,605	553,970
	1988	508,938	32,671	2,843	60,098	7,500	612,050
	1990	661,049	204,434	4,251	15,691	51,864	937,289
Bivalve molluscs							
Oysters	1984	21,768	40,688	37	119	41,795	104,407
	1985	21,077	76,866	38	137	39,917	138,035
	1988	73,954	73,954	170	214	53,641	201,933
	1990	17,936	111,326	39	157	49,100	178,558
Mussels	1984	48,131	136,582	-	99	1,111	185,923
	1985	48,586	128,860	-	114	1,136	178,696
	1988	77,981	429,675	-	316	2,230	510,202
	1990	75,097	495,895	-	179	2,431	573,602
Misc. spp.[b]	1984	70,093	165,521	-	-	947	236,561
	1985	57,497	177,438	-	-	5,917	240,852
	1988	46,641	441,197	-	-	966	488,804
	1990	47,981	367,421	-	-	1,312	416,714
Seaweeds	1984	61,000	1,640,756	-	-	6,804	1,708,560
	1985	244,446	1,693,258	-	-	4,924	1,942,628
	1988	79,162	1,581,370	-	-	23,113	1,683,645
	1990	373,176	1,662,087	-	-	38,017	2,073,280
Crustaceans[c]	1984	94,646	22,021	36	25	6,520	123,248
	1985	124,418	51,027	42	26	9,515	185,028
	1988	168,393	202,319	80	7	18,186	388,985
	1990	337,169	226,385	299	63	25,132	589,048
Totals	1984	1,518,134	3,799,185	9,159	45,048	92,747	5,464,273
	1985	1,557,645	4,707,038	10,660	69,432	100,137	6,444,912
	1988	1,724,532	6,658,686	11,466	79,950	141,527	8,616,161
	1990	2,378,565	7,352,395	14,514	94,656	199,066	10,039,196

[a]Includes all other finfish, such as catfish, milkfish, mullets and a wide range of freshwater, brackishwater and marine species.
[b]Includes all other bivalves, such as ark shells, clams, cockles, etc.
[c]Mainly marine and brackishwater shrimp species, but also includes freshwater prawns.

Table 2. Global aquaculture production (t) by environment in 1987 (FAO 1989a).

Major groups cultured	Marine	Freshwater	Brackishwater	Total	World total (%)
Finfish	403,571	6,005,630	384,240	6,793,441	51.4
Crustaceans	34,806	61,592	478,508	574,906	4.4
Molluscs	2,572,395	9,473	90,526	2,672,394	20.2
Seaweeds	3,133,981	13	5,479	3,139,473	23.8
Others	26,715	839	148	27,702	0.2
Total	6,171,468	6,077,547	958,901	13,207,916	
Percentage of world total	46.7	46.0	7.3		

Table 3. Average annual growth rates of GDP and aquaculture production in Asian countries during 1983-87.

Country	Aquaculture[a] (% growth)	GDP[b] (% growth)
China	30.66	11.02
Japan	0.94	3.80
Korea, Republic of	9.24	9.54
Philippines	6.30	-0.66
Indonesia	10.59	4.06
Taiwan	6.89	8.94
Vietnam	9.20	5.32
Bangladesh	9.29	4.04
Thailand	14.16	6.16
Malaysia	1.92	3.90

[a]Based on ADCP Aquaculture Minutes (various issues); FAO Fish. Circ. 815, 1989.
[b]Based on Asian Dev. Outlook 1989; ADB 1989; World Development Report 1988.

Table 4. Population estimates (millions) for world, developed countries and the main developing regions of the world as assessed in 1984 (UN 1989b).

	1960	1980	2025
World	3,019	4,450	8,206
Developed countries	945	1,137	1,396
Developing countries	2,074	3,313	6,809
Africa	280	479	1,617
Latin America/Caribbean	217	361	779
China	657	996	1,475
South Asia	595	949	1,855
Southeast Asia	226	361	688
Oceania	16	23	38

aquaculture production is still only about 150 t. In Nepal, cultured finfish production (carps) increased from 150 t in 1982 to 5,175 t in 1988 - a huge relative increase, but a small absolute tonnage.

Further Development

Demographic Considerations

The population of developing countries increased from 1.7 billion in 1950 to 3.7 billion in 1985 and a further increase to 6.8 billion by 2025 is estimated (UN 1989b). Table 4 summarizes the picture. It is not all discouraging. Feeding growing populations poses a major, but not insuperable, challenge if more productive and profitable farming systems are developed and implemented quickly. However, demographic changes towards a better balance between population and resources are sorely needed.

Political and Economic Considerations

In campaigns against poverty and for environmental conservation, most decisions and actions are taken by nation states. However, natural habitats and their biota do not conform to national boundaries. Many waterbodies and catchments are shared and one nation's actions affect others downstream or across the water. Effective environmental conservation requires transnational cooperation. Rhodes (1986) found this possible, cited its proven success in the eradication of smallpox and expected the same in the avoidance of nuclear conflict. Concerning the latter, he stated:

> The preeminent transnational community in our culture is science. With the release of nuclear energy in the first half of the twentieth century that *model commonwealth* [present author's emphasis] decisively challenged the power of the nation-state.

A leading article in The Economist (Anon. 1990), titled "Goodbye to the nation state?", emphasized moves towards federation in Europe, but found nationalism and tribalism to be highly durable and forecast more emergent nation states over the next 50 years, including perhaps a quite different map of Africa. It also recognized a "Commonwealth model" as a future mechanism for action on transnational issues in which nation states retain sovereignty and cooperate on "foreign policy, defence and some aspects of trade". Environmental issues were not mentioned, but clearly the time is ripe for increasing transnational cooperation in balancing development and conservation strategies. The World Conservation Union (IUCN) has an Ecology Commission, a Sustainable Development Commission and a new strategic plan based on sustainability (IUCN 1991), through which some of these efforts could perhaps be coordinated.

Aquaculture scientists can also help to set environmental issues in aquaculture development in such a transnational context, despite the enduring background of regional, national and intranational vested interests. A well-publicized nonaquaculture example that adversely affects developing countries is the Common Agricultural Policy of the EEC. It is, however, rather unfair to single out developed countries and regions with such examples, when in the developing countries themselves the poor are usually kept in poverty because the *status quo* protects the interests of the rich. Inequity pervades human endeavor, irrespective of the stage of development of national economies.

Grigg (1985) recognized the differing situations of the main developing regions and found that in Africa the natural environment presents still unresolved problems for crop and livestock production which together with civil strife, wars and lack of skilled personnel have caused a decline in agricultural production per caput and a shortfall in local supplies in a majority of countries. His view of Asia was that the small size of farms in most countries need not be a hindrance to increased food production and there could be much higher farm yields and incomes in South and Southeast Asia, as has happened in East Asia. For Latin America, he found the problem of reducing malnutrition to be mainly a matter of income distribution and land reform rather than increasing agricultural production.

Ruddle and Rondinelli (1983) provided a framework for development and identified the failure of the 'trickle down' approach, the need for close partnership with target beneficiaries, and posed the following questions, relevant to the present underdeveloped status of developing-country aquaculture:

> In what ways could production of a major resource be increased, made more efficient or of better quality, without affecting renewability of the resource? What are the probable ecosystemic and sociocultural costs of recommended changes? How can these be ameliorated? Are there other major renewable resources in the area *not* being utilized at present that could be

> developed without major changes, by existing resource system structures? What are the probable ecosystemic and sociocultural costs involved in this?

Another important consideration, mainly for small-scale operations (but also relevant to larger-scale aquaculture and always location specific) is the target fish yield set by development agencies for aquaculture systems. This is often set too high in the belief that the high yields obtained by research institutions and resource-rich farmers (for example, 10 $t \cdot ha^{-1} \cdot year^{-1}$ or more for semi-intensive manured pond aquaculture) must also be approached on farms, or else the development is not worthwhile. For existing fish farmers and potential new entrants, especially in the poorest areas, this is a counterproductive approach. Modest yields from aquaculture can give great nutritional and income improvements in developing-country rural areas. For example fishpond yields of 2 to 3 $t \cdot ha \cdot year^{-1}$ would be very attractive in much of rural Africa.

What Kinds of Aquaculture Make Sense in Developing Countries?

The Brundtland Report gave these encouraging words on aquaculture development:

> Aquaculture can be undertaken in paddy fields, abandoned mining excavations, small ponds and many other areas with water, as well as on various commercial scales: individual, family, cooperative or corporate. The expansion of aquaculture should be given high priority in developing and developed countries.

This is correct, but such aquaculture development must be socially equitable, environmentally compatible, and have sufficient diversity and scope for change to adapt to changing circumstances.

Intensive aquaculture (in effect, using the feedlot principle) usually poses much greater threats to the environment than does extensive or semi-intensive aquaculture. Intensive fish farms are often heavy users of antibiotics and disinfectants and their operators need to be aware of the dangers of release of such chemicals to the natural environment including the possibilities of producing drug-resistant pathogens (see Austin, this vol.). Pollution by intensive aquaculture is well known. Fish fecal wastes and uneaten food in effluents from fish farms and in settlement from cages have high biological oxygen demands (BODs) and contain large quantities of particulate matter and nutrients.

Such impacts greatly threaten sustainability, but some less intensive aquaculture systems can also be short-lived (nonsustainable). The following examples illustrate some of the relevant issues.

Milkfish *(Chanos chanos)* pen aquaculture in Laguna de Bay, a shallow 90,000 ha eutrophic lake adjacent to Metropolitan Manila, Philippines, grew from a single experimental pen in 1970 to about 7,000 ha of pens in 1974 producing a mean yield of about 7 $t \cdot ha^{-1}$ $year^{-1}$ (Pullin 1981). They resembled extensive fish 'ranches' with some individual units as large as 50 ha or more. At its peak, the total pen area may have been as much as 34,000 ha. It is now about 2,800 ha and the average annual yield is about 3.8 $t \cdot ha^{-1}$. Its expansion phase was a 'goldrush' in which the pen owners (mainly upper-class citizens) got richer and the lake's small-scale fishers and aquaculturists suffered greatly. The decline has been because of conflicts, losses due to typhoons and a reduction in the lake's productivity attributed to multiple causes (pollution, turbidity due to catchment erosion, and altered flushing patterns due to flood control structures). This was a nonsustainable, socially inequitable development and one not to emulate.

Parallel to the milkfish pens' rise and fall, there was a rise (and has recently been a fall, because of deteriorating water quality) in small-scale tilapia cage culture in Laguna de Bay. As it developed, this tilapia growout operation generated a large demand for tilapia fry and fingerlings which was met by small-scale hatchery operators in the villages of lakeside provinces. There were probably over 600 hatcheries in 1983. They also sold tilapia seed to other farmers countrywide. The benefits that accrued to hatchery operators, their families and communities were great (Gaite et al. 1985; Yater and Smith 1985).

The subsequent decline of these hatcheries because of the decline of the Laguna de Bay tilapia cage growout operations and competition for markets elsewhere from new tilapia seed suppliers does not negate the very real benefits in terms of improved housing, purchase of household appliances, education, etc., that this tilapia hatchery development gave to some lakeside villages for a short but significant period. Since in most aquaculture worldwide hatchery/nursery and growout operations are separate, such examples of interdependence, competition and change are common. This development was not sustainable, but still helped some villages for a significant period.

In both these examples, environmental degradation, in this case of the lake water, was a contributing factor to their nonsustainability.

The environmental problems caused by the expansion of shrimp farming in Latin America and Asia (another aquaculture goldrush) are also becoming serious. These include destruction of mangroves, salinization of inland areas (soils and aquifers used for domestic water supply), land subsidence and watertable changes due to excessive pumping, pollution of adjacent areas by farm effluents, poor hygiene favoring the spread of diseases, misuse of antibiotics and other chemicals, and social disruption (New 1990).

Small-scale hatchery and growout operations are socially and environmentally desirable. Small-scale is a term synonymous with the household/village-level, i.e., operations run by an individual smallholder and family or a village community group. The bulk of the food production in developing nations will be by small-scale producers for the foreseeable future. The Brundtland Report stated:

> Most developing nations need more effective incentive systems to encourage production, especially of food crops. In short, the 'terms of trade' need to be turned in favour of the small farmer.

This applies to aquaculture as well as to agriculture. It would be socially, politically and environmentally undesirable to promote new technologies for aquaculture development that would lead to this food production role being substantially transferred from small-scale producers to larger concerns. The aquaculture systems best suited to small-scale producers are low-input systems, particularly small ponds, cages and pens. It has been previously argued (Pullin 1989) that small-scale, semi-intensive aquaculture systems, particularly those integrated with agriculture, are less environmentally disruptive than larger or more intensive systems. Nature conservation organizations worldwide have recognized that small-scale, diverse farming systems permit much better coexistence of agriculture and wildlife, in terms of the latter's abundance and diversity, than larger-scale monocrop or factory farming systems.

Nevertheless, the long-term survival and improvement of small-scale farms, while advantageous for the needs of rural peoples and for environmental conservation, may not suffice for the needs of all who will depend upon the food production sector. Urbanization is proceeding rapidly and affordable produce will have to be available to maintain the social fabric of developing-country cities. This will probably require the development of some large-scale aquaculture enterprises (run by estates and corporations) to increase fish supply and lower prices.

Bimbao and Smith (1988) reviewed the economics of tilapia production in the Philippines and found that the purchasing power of the average Filipino had declined by about 30% from 1983-85 (due mainly to high inflation) and that tilapia, like most fish, was becoming beyond the purchasing power of the majority of low-income consumers; this in a country where rice and *fish* form the staple diet. This illustrates the difficulty in fostering development that will provide a balance between benefits to low-income producers and low-income consumers. Tilapia retails at about US$3.00/kg in Metro Manila markets. Bimbao and Ahmed (1990) found that an expansion of tilapia supply of up to 40% would not depress prices significantly. Thereafter, prices would fall with further increases in supply but would still be attractive to producers. The situation is probably very different elsewhere; for example, concurrent tilapia prices in Bangkok were only about US$0.60/kg.

Opportunities for and needs of rural producers and consumers, urban fringe producers and consumers and city dweller consumers must all be considered. Therefore *all* kinds of aquaculture, small- to large-scale and extensive to intensive, including enhanced fisheries, can make sense in developing-country aquaculture development, depending upon the needs of different sections of the community.

It is also most important to view aquaculture development as only one of several options for meeting livelihood and nutritional needs. One must avoid the naive assumption that aquaculture 'must' be able to fill the fish supply or protein gap. Fish is only *one* protein source. Expansion of fish supply through aquaculture must be weighed against the pros and cons of increased supply of vegetable and other animal proteins.

Experience Gained in Developed Countries

In the pursuit of environmentally compatible aquaculture development for developing countries, experience gained in developed countries is useful. However, this must be applied with a realistic appreciation of developing-country needs and constraints. Environmental conservation and human needs must be balanced. Where pristine habitats are disappearing there should be all possible efforts to conserve their remnants but developing countries need realistic policies and legislation to suit their circumstances. The contrast between Philippine coastal waters and Scandinavian fjords is an example. The former support the needs of millions of people, have virtually no pristine habitats and have suffered massive loss of, and damage to, coral reef and mangrove ecosystems. The latter include many pristine and near-pristine habitats and support very low human populations. Clearly, achievable environmental targets are different for these two contrasting locations.

This also applies to aquaculture development in inland waters. Costa-Pierce and Roem (1990) studied waste production and efficiency of feed utilization in cage culture in a tropical eutrophic reservoir (Saguling, near Bandung, West Java, Indonesia) in which 1,300 cage units produced 2,550 t of common carp (*Cyprinus carpio*) in 1988. The percentage loss of feed nutrients in Saguling was low (C, 5.4%; N, 3.5%; P, undetectable) and the sedimentation rate (2.0-25.0 $g{\cdot}m^{-2}{\cdot}day^{-1}$; mean, 13.3 $g{\cdot}m^{-2}{\cdot}day^{-1}$) lower than that reported for temperate zone cage culture: e.g., 150 $g{\cdot}m^{-2}{\cdot}day^{-1}$ (Merican and Phillips 1985); 17-26 $g{\cdot}m^{-2}{\cdot}day^{-1}$, (Enell and Löf 1983). The reservoir also received about 150,000 $m^{3}{\cdot}day^{-1}$ of organic wastes from the Bandung conurbation and the natural sedimentation rate was 1.7 ± 1.2 $g{\cdot}m^{-2}{\cdot}day^{-1}$. This aquaculture development now supplies about 20% of the freshwater fish supply for a conurbation of 3 million people.

In examining the environmental impact of, and in setting limits to, the density and siting of cages based on the carrying capacity of lakes, it is essential to balance benefits against any additional polluting effect of the cages and to consider exactly what environmental and social targets are achievable. Costa-Pierce (1990) made the general point that the environmental impact of cage culture in the Saguling Reservoir was insignificant compared to the impacts of raw sewage and of water level fluctuations, which confound attempts to estimate absorptive capacity.

A further example concerns the use of chemicals in developing-country aquaculture. The misuse of antibiotics and pesticides should be prevented everywhere. Steroid hormones, however, against which there is a blanket ban in food production in some developed countries, can be used to good effect in developing-country aquaculture to produce monosex male tilapias (Guerrero 1982). There is as yet no comparable practical method for avoiding the problems of uncontrolled tilapia breeding. Alternative methods, such as hybridization, have many disadvantages, e.g., use of additional (sometimes exotic) species and strict management requirements. The benefits from hormonal production of monosex male tilapia fry (by short-duration treatment of fry) are clear and the technique is beginning to be adopted in some developing countries (Macintosh et al. 1985). All available evidence suggests that hormones are eliminated in a few days and no residues could possibly be left in fish sold to consumers (Johnstone et al. 1983). Rothbard et al. (1990) even found that an androgen fed to tilapia for 11 weeks as a growth promoter was also eliminated in less than a week upon cessation of treatment. There are some risks to hatchery workers if they mishandle androgenic hormones and there are possibilities of contamination of the fish farm and adjacent environment (Rothbard et al. 1990), but clearly it would be unwise to prohibit entirely techniques, that are valuable and that can be applied safely, involving the use of hormones in tilapia culture in developing countries.

Sewage-fed fish culture affords another example. The risks may be insignificant, provided that postharvest handling is hygienic and produce is well-cooked (Edwards 1985). In China, India, Indonesia and Vietnam, large quantities of fish and vegetables are raised on human excreta. Excreta reuse through aquaculture may be one of the least environmentally harmful disposal methods. It should also be more feasible in developing countries where sewage wastes and wastewater are less mixed with detergents, heavy metals and other chemicals than in industrialized countries, though this is a worldwide constraint.

All the above examples show that technical advice and policy formulation for aquaculture development must be attuned to specific needs and opportunities, rather than being constrained by foreign cultural biases. Experience gained in and technical expertise from developed countries can assist aquaculture development with environmental protection in the developing regions provided that this requirement is met.

Criteria for Assessing Environmental Impact and Benefits

Sets of summary criteria, impacts and benefits have been published for appraising developing-country aquaculture development (McAllister 1988; Pullin 1989; Tables 5 and 6). Table 5 highlights social and environmental criteria and touches on international equity issues. Its message is that the route to the greatest good for the greatest number is fraught with complex issues and side effects. The table merely notes these and makes no explicit judgments.

Table 6 takes a more structured approach to the social and environmental pros and cons of different types of aquaculture. Here the judgments are more explicit and clearly favor the development of semi-intensive systems.

Both tables identify only the broad categories of impacts and benefits. More detailed frameworks are required for specific situations.

Table 5. Social, environmental and esthetic criteria for and against aquaculture development projects (McAllister 1988).

Criteria	<u>Yes</u> Favorable project	<u>No</u> Less favorable or unfavorable project
SOCIAL		
Whose income does it benefit?	Poor	Middle class, rich
Capital needed	Low capital	Capital intensive
Return to worker/ family	Self-employment	Low wages
Operated by	Individual, family co-op or community	Company
Gender	Benefits men, women & children	Exploits/neglects women and children
Disturbance to culture, customs	None	Some, much
Working conditions ~capture fishery or gleaning	High quality	Low quality
Nutritional quality ~natural food	Equal to or greater than	Lower than
Food for	Poor	Wealthy
Effect on public health (drinking water, mosquitos, parasites, etc.)	Low	High
Who made the decision?	Local community after mature debate & discussion	NGO, Washington, London, Ottawa, consulting company
ENVIRONMENTAL		
Culture method	Polyculture	Monoculture

continued

Table 5 (continued)

Criteria	Yes Favorable project	No Less favorable or unfavorable project
Relation to natural environment	Displaces none or little	Replaces one with the other
Used as an excuse not to restore natural environment or uses restoration funding	Not so	Is used
Uses few artificial genetic strains	Not so	Is used
Risks escape for selected strains into nature	No	Yes
Cultured stock	Native	Exotic
Risks to native species - disease. hybridization, extinctions, etc.	None	Some, much
Disease, predators competitors controlled by	Biological means	Chemical means or by drugs
Fertilizers	Organic	Chemical
Output into natural environment - chemical, organic and physical	Low	High
Culture subjectivity to disease/stress	Low	High
Facility design effect on wildlife predators	Naturally excludes birds & mammals	Fish eating ducks, herons, cormorants otters, seals, etc. controlled by gun
ESTHETIC		
Culture area	Beautiful	Ugly
WORLD ORDER		
Produces food for	Developing countries	Developed countries
Profit flow ratio	High for Developing countries	High for developed countries
Needed supplies and most goods from	Developing countries	Developed countries
Import of foreign technology	None	Some, much
Needed 'seed' stock from	Developing countries	Developed countries

Table 6. Developing-country aquaculture systems: environmental impact and benefits for producers. Extensive systems are defined as having no feed or fertilizer inputs; semi-intensive systems as having some feed and/or fertilizer inputs; and intensive systems as being mainly reliant on external feed inputs. The possible consequences of exotic breed transfers apply to all systems listed here (Pullin 1989). Enhanced fisheries are not included here because of the general lack of developing-country examples.

System	Environmental Impact	Benefits
EXTENSIVE		
1. Seaweed culture	May occupy formerly pristine reefs; rough weather losses; market competition; conflicts/failures, social disruption	Income; employment; foreign exchange
2. Coastal bivalve culture (mussels, oysters, clams, cockles)	Public health risks and consumer resistance (microbial diseases, red tides, industrial pollution); rough weather losses; seed shortages; market competition especially for export produce; failures, social disruption	Income; employment; foreign exchange; directly improved nutrition
3. Coastal fishponds (mullets, milkfish, shrimps, tilapias)	Destruction of ecosystems, especially mangroves; increasingly noncompetitive with more intensive systems; non-sustainable with high population growth; conflicts/failures, social disruption	Income, employment, foreign exchange (shrimps); directly improved nutrition
4. Pen and cage culture in eutrophic waters and/or on rich benthos (carps, catfish, milkfish, tilapias)	Exclusion of traditional fishers; navigational hazards; conflicts, social disruption; management difficulties; wood consumption	Income; employment; directly improved nutrition
SEMI-INTENSIVE		
1. Fresh- and brackishwater ponds (shrimps and prawns; carps, catfish, milkfish, mullets, tilapias)	Freshwater: health risks to farm workers from waterborne diseases. Brackishwater: salinization/acidification of soils/aquifers. Both: market competition, especially for export produce; feed and fertilizer availability/prices; conflicts/failures, social disruption	Income; employment; foreign exchange (shrimps and prawns) directly improved nutrition
2. Integrated agriculture-aquaculture (rice-fish; livestock/poultry-fish; vegetables-fish and all combinations of these)	As freshwater above, plus possible consumer resistance to excreta-fed produce; competition from other users of inputs such as livestock excreta and cereal brans; toxic substances in livestock feeds (e.g., heavy metals) may accumulate in pond sediments and fish; pesticides may accumulate in fish	Income; employment; directly improved nutrition; synergistic interactions between crop, livestock, vegetable and fish components; recycles on-farm residues and other cheap resources
3. Sewage-fish culture (waste treatment ponds; latrine wastes and septage used as pond inputs; fish cages in wastewater channels)	Possible health risks to farm workers and consumers; consumer resistance to produce	Income; employment; directly improved nutrition; turns waste disposal liabilities into productive assets
4. Cage and pen culture, especially in eutrophic waters or on rich benthos (carps, catfish, milkfish, tilapias)	As extensive cage and pen systems above	Income; employment; directly improved nutrition

continued

Table 6 (continued)

System	Environmental Impact	Benefits
INTENSIVE		
1. Freshwater, brackishwater and marine ponds (shrimps and prawns; fish, especially carnivores - catfish, snakeheads, groupers, seabass, etc.	Effluents/drainage high in BOD and suspended solids; market competition, especially for export product; conflicts/failures, social disruption	Income; employment; foreign exchange
2. Freshwater, brackishwater and marine cage and pen culture (finfish, especially carnivores - groupers, seabass, etc. - but also some omnivores, such as common carp)	Accumulation of anoxic sediments below cages due to fecal and waste feed build-up; market competition, especially for export produce; conflicts/failures, social disruption; consumption of wood and other materials	Income; foreign exchange (high priced carnivores); a little employment
3. Other - raceways, silos, tanks, etc.	Effluents/drainage high in BOD and suspended solids; many location-specific problems	Income; foreign exchange; a little employment

Conclusions

Developing-country aquaculture development is needed to help alleviate poverty and increase protein food supply. Poverty and effective environmental conservation cannot coexist. Pauly et al. (1989) illustrate this well for the issue of prevention of dynamite fishing. Development must complement environmental conservation, not compete with it. Therefore, developing-country aquaculture development must be pursued in harmony with realistic environmental conservation objectives, with transnational cooperation and with effective legislation.

This will require much more reliable information on the environmental impact of developing-country aquaculture than is available at present. This in turn will require much more research on existing and evolving developing-country aquaculture systems, not just extrapolations from developed-country experience.

Above all, developing-country aquaculture development and its environmental aspects must be considered in a broad intersectoral context so that the use of natural resources to meet evolving human needs (whether in agriculture, fisheries, forestry or aquaculture) can be optimized with respect to environmental conservation.

Finally, because aquaculture is a relatively new and underdeveloped sector in most developing countries, it will come under increasingly close scrutiny with respect to its environmental impact, perhaps even unfairly so in comparison with the safeguards demanded for better-known sectors, especially agriculture. This can lead to incomplete and unbalanced commentaries. For example, several agricultural serials publicized the paper by Scholtissek and Naylor (1988) on the possibility of new flu viruses from pig-duck-fish zoonoses in Chinese integrated farming but omitted to summarize the rejoinders to this paper published by aquaculturists (for example, Edwards et al. 1988b), who pointed out the improbability of this in most integrated farming systems.

For these and other similar warnings against the possible environmental hazards of aquaculture development, what is needed is a balanced view - not underestimating the environmental concerns associated with developing-country aquaculture development but placing these in a broad rural development context in which human needs and all development options and environmental issues receive full consideration.

References

ADB. 1989. Asian development outlook 1989. Asian Development Bank, Manila, Philippines. 207 p.

ADB. 1990. Annual report 1989. Asian Development Bank, Manila, Philippines. 199 p.

Altieri, M.A. 1989. Agroecology: a new research and development paradigm for world agriculture. Agriculture, Ecosystems and Environment 27: 37-46.

Anon. 1990. Goodbye to the nation state? The Economist, June 23-29, 1990: 9-10.

Bimbao, M.A.P. and M. Ahmed. 1990. Market potential of tilapia in the Philippines: an assessment. Paper presented at the Asian Regional Workshop on Tilapia Genetics, 29-31 August 1990, Central Luzon State University, Muñoz, Nueva Ecija, Philippines.

Bimbao, M.A.P. and I.R. Smith. 1988. Philippine tilapia economics: industry growth and potential, p. 539-551. *In* R.S.V. Pullin, T. Bhukaswan, K. Tonguthai and J.L. Maclean (eds.) The Second International Symposium on Tilapia in Aquaculture. ICLARM Conf. Proc. 15, 623 p.

CGIAR. 1989. Sustainable agricultural production: report of the CGIAR Sustainability Committee, MT/89/14. Consultative Group on International Agricultural Research, Washington, D.C. 28 p.

Charlton, C.A. 1987. Problems and prospects for sustainable agricultural systems in the humid tropics. Appl. Geogr. 7: 153-174.

Chua, T.E., J.N. Paw and E. Tech. 1989a. Coastal aquaculture development in ASEAN: the need for planning and environmental management, p. 57-70. *In* T.-E. Chua and D. Pauly (eds.) Coastal area management in Southeast Asia: policies, management strategies and case studies. ICLARM Conf. Proc. 19, 254 p.

Chua, T.E., J.N. Paw and F.Y. Guarin. 1989b. The environmental impact of aquaculture and the effects of pollution on coastal aquaculture development in Southeast Asia. Mar. Pollut. Bull. 20(7): 335-343.

Costa-Pierce, B.A. 1990. Multiple regression analysis of plankton and water quality relationships as affected by sewage inputs and cage aquaculture in a eutrophic tropical reservoir. Paper presented at the Second Asian Reservoir Fisheries Workshop, 15-19 October 1990, Hangzhou, China.

Costa-Pierce, B.A. and C.M. Roem. 1990. Waste production and efficiency of feed use in floating net cages in a eutrophic tropical reservoir, p 257-271. *In* B.A. Costa-Pierce and O. Soemarwoto (eds.) Reservoir fisheries and aquaculture development for resettlement in Indonesia. ICLARM Tech. Rep. 23, 378 p.

Csavas, I. 1988. Problems of inland fisheries and aquaculture, p. 137-180. *In* Fishing industry in Asia and the Pacific. Asian Productivity Organization, Tokyo.

Dixon, J.A. and L.A. Fallon. 1989. The concept of sustainability: origins, extensions and usefulness for policy. Soc. Nat. Resour. 2:73-84.

Dover, M.J. and L.M. Talbot. 1987. To feed the earth: agroecology for sustainable development. World Resources Institute, Washington, DC. 88 p.

Edwards, C.A. 1989. The importance of integration in sustainable agricultural systems. Agric. Ecosyst. Environ. 27: 25-35.

Edwards, P. 1985. Aquaculture: a component of low cost sanitation technology. World Bank Tech. Pap. No. 36. The International Bank for Reconstruction and Development, Washington, DC.

Edwards, P., R.S.V. Pullin and J.A. Gartner. 1988a. Research and education for the development of integrated crop-livestock-fish farming systems in the tropics. ICLARM Stud. Rev. 16, 53 p.

Edwards, P., C. Kwei-Lin, D.J. Macintosh, K.L. Wee, D. Little and N.J. James-Taylor. 1988b. Fish farming and aquaculture. Nature (Lond.) 333:505-506.

Ekins, P. 1987. Beyond growth: the real priorities of sustainable development. Environ. Conserv. 16: 5-6 and 12.

Enell, M. and J. Löf. 1983. Environmental impact of aquaculture sedimentation and nutrient loadings from fish cage culture farming. Vatten/Water 39 (4):364-375.

FAO. 1989a. Aquaculture production (1984-1987). FAO Fish. Circ. 815 Rev. 1. FAO, Rome.

FAO. 1989b. Review of recent developments in aquaculture. COFI/89/Inf. 5. Committee on Fisheries, FAO, Rome.

FAO. 1990a. The definition of aquaculture and collection of statistics. Aquaculture Minutes No. 7. Inland Water Resources and Aquaculture Service, Fisheries Department, FAO, Rome.

FAO. 1990b. Global aquaculture value increases by another 19%. Followup of aquaculture statistics. Aquaculture Minutes No. 8. Inland Water Resources and Aquaculture Service, Fisheries Department, FAO, Rome.

FAO. 1992. Aquaculture production (1984-1990). Fishery Information, Data and Statistics Service. FAO Fish. Circ. 815 (Rev. 4), 260 p.

Gaite, M.C., J.N.A. Morales, O.C.R. Orilla and B.B. Pili. 1985. The adoption of tilapia farming and its impact on the community of Sto. Domingo, Bay, Laguna, Philippines, p. 44-49. *In* I.R. Smith, E.B. Torres and E.O. Tan (eds.) Philippine tilapia economics. ICLARM Conf. Proc. 12, 261 p.

Grigg, D. 1985. The world food problem, 1950-1980. Blackwell, Oxford.

Guerrero, R.D. III. 1982. Control of tilapia reproduction, p. 309-316. *In* R.S.V. Pullin and R.H. Lowe-McConnell (eds.) The biology and culture of tilapias. ICLARM Conf. Proc. 7, 432 p.

Hart, R.D. 1987. Research and development strategies to improve integrated crop, livestock and tree systems, p. 29-95. *In* Proceedings of the Workshop on Farming Systems Research, 17-21 February 1986. International Crops Research Institute for the Semi-Arid Tropics, Patancheru, India.

ICLARM-GTZ. 1991. The context of small-scale integrated agriculture-aquaculture systems in Africa: a case study of Malawi. ICLARM Stud. Rev. 18, 302 p.

IUCN. 1991. Caring for the world. A strategy for sustainable living. World Conservation Union, United Nations Environmental Programme and World Wildlife Fund, Gland, Switzerland.

Johnstone, R., D.J. Macintosh and R.S. Wright. 1983. Elimination of orally-administered 17α-methyltestosterone by *Oreochromis mossambicus* (Tilapia) and *Salmo gairdneri* (rainbow trout) juveniles. Aquaculture 35:249-257.

Lightfoot, C. 1990. Integration of aquaculture and agriculture: a route to sustainable farming systems. Naga, ICLARM Q. 13(1):9-12.

Macintosh, D.J., T.J. Varghese and G.P. Satyanarayana Rao. 1985. Hormonal sex-reversal of wild-spawned tilapia in India. J. Fish Biol. 26:87-94.

McAllister, D.E. 1988. Aquaculture, yes, no, maybe? Sea Wind 3(1):13-18.

McNeely, J.A., K.R. Miller, W.V. Reid, R.A. Mittermeier and T.B. Werner. 1990. Conserving the world's biological diversity. International Union for the Conservation of Nature, Gland, Switzerland; World Resources Institute, Conservation International, World Wildlife Fund, US; and the World Bank, Washington, DC.

Meadows, D.H., D.L. Meadows, J. Randers and W.W. Behrens III. 1972. The limits to growth. A report for the Club of Rome's project on the predicament of mankind. Universe Books, New York.

Merican, Z.O. and M. Phillips. 1985. Solid waste production from rainbow trout, *Salmo gairdneri* Richardson, cage culture. Aquacult. Fish. Manage. 1:55-69.

New, M.B. 1990. Shrimp farming experiences environmental problems. EC Cooperational Bulletin, 3 p.

Pauly, D., G. Silvestre and I.R. Smith. 1989. On development, fisheries and dynamite: a brief review of tropical fisheries management. Nat. Resour. Modelling 3:307-329.

Pimentel, D., T.W. Culliney, I.W. Butler, D.J. Reinemann and K.B. Beckman. 1989. Low-input sustainable agriculture using ecological management practices. Agric. Ecosyst. Environ. 27: 3-24.

Pullin, R.S.V. 1981. Fish pens of Laguna de Bay, Philippines. ICLARM Newsl. 4(4): 11-13.

Pullin, R.S.V. 1989. Third-world aquaculture and the environment. Naga, ICLARM Q. 12(1):10-13.

Pullin, R.S.V. and R.A. Neal. 1984. Tropical aquaculture. Need for a strong research base. Mar. Policy 8:217-228.

Raintree, J.B. and F. Torres. 1987. Agroforestry research in a farming systems perspective: the ICRAF approach, p. 95-106. *In* Proceedings of the Workshop on Farming Systems Research, 17-21 February 1986. International Crops Research Institute for the Semi-Arid Tropics, Patancheru, India.

Rhodes, R. 1986. The making of the atomic bomb. Simon and Schuster Inc., New York.

Rothbard, S., Y. Zohar, N. Zamora, B. Levavi-Sivan, B. Moav and Z. Yaron. 1990. Clearance of 17-α-ethnyltestosterone from muscle of sex-inversed tilapia hybrids for growth enhancement with two doses of the androgen. Aquaculture 89:365-376.

Ruddle, K. and D.A. Rondinelli. 1983. Transforming natural resources for human development: a resource systems framework for development policy. Resource Systems Theory and Methodology Series No. 1. United Nations University, Tokyo.

Scholtissek, C. and E.R. Naylor. 1988. Fish farming and influenza pandemics. Nature (Lond.) 331:215.

Simons, I.G. 1988. The environmental impact of pre-industrial agriculture. Outlook on Agriculture 17:90-96.

Simons, I.G. 1989. Land transformations during the period of pre-industrial agriculture. Outlook on Agriculture 18:85-91.

Smith, I.R. 1986. Aquaculture development: it's a matter of broader context, p. 303-317. *In* Proceedings of the SADCC Workshop on the Development of Village Level and Commercial Aquaculture in the SADCC Region, 29 September-3 October 1986, Kariba, Zimbabwe. Southern African Development Co-ordination Conference, Lilongwe, Malaŵi.

Taiwan Fisheries Bureau. 1986. Fisheries Yearbook. Taiwan Area 1985: 28-74.

Taiwan Fisheries Forum Bureau. 1991. Fisheries Yearbook. Taiwan Area 1990. 169-203.

Tivy, J. 1987. Nutrient cycling in agro-ecosystems. Appl. Geogr. 7:93-113.

UN. 1989a. World population prospects 1988. Department of International Economic and Social Affairs. Population Studies No. 106. ST/ESA/SER.A/106. United Nations, New York.

UN. 1989b. World population at the turn of the century. Department of International Economics and Social Affairs. Population Studies No. 111. ST/ESA/SER.A/111. United Nations, New York.

WCED (World Commission on Environment and Development). 1987. Our common future. Oxford University Press, Oxford.

World development report 1988. Published for the World Bank by Oxford University Press, New York. 307 p.

Yater, L.R. and I.R. Smith. 1985. Economics of private tilapia hatcheries in Laguna and Rizal Provinces, Philippines, p. 15-32. *In* I.R. Smith, E.B. Torres and E.O. Tan (eds.) Philippine tilapia economics. ICLARM Conf. Proc. 12, 261 p.

Discussion

BILIO: The statistics from all types of brackishwater aquaculture need to be closely examined.

CSAVAS: The distinction between marine and brackishwater aquaculture is vague; for example, milkfish production comes from both marine and brackishwater. The distinction should really be between freshwater and coastal aquaculture.

BILIO: This distinction is not ideal but it is much better.

EDWARDS: The paper draws attention to the low level of aquaculture development in general, even in Asia. This could be easily seen from comparison of protein and energy (calorie) production of aquaculture and agriculture for human food.

PULLIN: On statistics, I believe that FAO is now planning to separate statistics for hatchery and growout operations and for the former to differentiate between hatchery operations used to stock open waters once or infrequently (in which case these statistics would be considered part of culture-based or enhanced fisheries) or frequently, in which case the hatchery operations are classified as aquaculture production. Is this correct? If so, will it work?

MARTÍNEZ-ESPINOSA: Using such a modified definition of aquaculture, about half the current Latin American aquaculture production would disappear and be considered as culture-based fisheries. For Cuba it would be almost 100%. The questions of ownership are important. The treatment of hatchery data needs more discussion. This is a very important point for Latin America.

The Impacts of Aquaculture Development on Socioeconomic Environments in Developing Countries: Toward a Paradigm for Assessment

KENNETH RUDDLE*
Centre for Development Studies
University of Bergen
Strφmgaten 54, N-5007 Bergen
Norway

RUDDLE, K. 1993. The impacts of aquaculture development on socioeconomic environments in developing countries: toward a paradigm for assessment, p. 20-41. *In* R.S.V. Pullin, H. Rosenthal and J.L. Maclean (eds.) Environment and aquaculture in developing countries. ICLARM Conf. Proc. 31, 359 p.

Abstract

The principal components of a resource systems paradigm for analyzing the human ecological consequences of aquaculture development in developing countries are presented, specifically for freshwater pond aquaculture, based on examples from eastern Central Africa, southern China and Panamá. Aquaculture as an innovative food production technology is examined in terms of the main perceptions in forming policy design, and the innovation adoption process by small-scale farm households. The principal social characteristics that influence the manner in which any innovation is received are discussed and, in terms of those, the impacts at the household, community and governmental-international agency level of the adoption of aquaculture, as well as major administrative changes within an old-established system, are examined.

Introduction

Mirroring the great complexity of many tropical agroecosystems are the tasks of inducing development in them, whether it be transforming traditional systems (Ruddle and Grandstaff 1978) or introducing new ones. Not only must the complex interactions of the biological and physical components of the systems be understood and accounted for, but, equally, so must the complex characteristics of the human managers of and consumers from these systems, as well as complicating factors introduced by the larger regional, national and international society, which impinge on and often constrain local managerial options.

Thus, the problems associated with any natural resource development are not just technical and agronomic, not ecological and not socioeconomic. They are essentially problems in human ecology, which embraces all these factors and much more.

The Human Ecological Perspective

The "natural" environment that forms the context in which any individual, community or nation exists and functions,

*Present address: Matsugaoka-cho 11-20, Nishinomiya-shi, Hyogo-ken 662, Japan.

must, by definition, include physical, biological and social phenomena. "Other people" and institutions influence social behavior and must be coped with, as with any other component of the environment. Thus, in any ecological examination of the environmental impact of aquaculture development, the "sociocultural environment" or "human environment" must be given equal weight with the "biophysical" components.

Human ecology is not a distinct science with specialized practitioners. Like aquaculture, it is rather a distinctive point of integration of several disciplines. For present purposes, human ecology may be defined as the study of how and for what purposes humans use resources and incorporate them into society and, by so doing, how humans, resources and society become transformed (c.f. Bennett 1976). Such studies should, therefore, be holistic, use a systems methodology, in which human behavioral (sociocultural) factors and environmental (biophysical) factors interact reciprocally and, wherever possible, be quantified in terms of energy, materials, information and cash or cash equivalent flows, to facilitate both the analysis of individual systems and meaningful comparison among systems.

Human ecology is concerned essentially with human adaptation; i.e., "the rational or purposive manipulation of social and biophysical environments" (Bennett 1976), and assessing the performance of adaptation by measuring the rate(s) and analyzing consequences of sustained yield of the resource(s) in use. Understanding the form(s) that adaptation will take from such strategic behavior is the key issue in policy-oriented human ecology, which focuses on the joint objectives of "environmental integrity" and human survival at reasonable levels of security, based on the sustained-yield use of natural resources. Human ecology as a policy science of sustained yield and resource use processes must, therefore, deal with power and control over resources and, in the process, over society.

The practical application of human ecology to address development issues has been retarded by lack of suitable paradigms. Attempts have been made to overcome that by focusing specifically on resource systems (Fig. 1) (Ruddle and Grandstaff 1978; Grandstaff et al. 1980; Ruddle and Rondinelli 1983; ICLARM and GTZ 1991), emphasizing the flows of energy, materials and information.

Based on the resource system approach, in this paper I outline a paradigm for analyzing the impact of the development of small-scale pond aquaculture on the social and economic domain of developing-country environments. This is not without difficulties, because aquaculture remains essentially a localized and innovative human adaptation. This is no less true of aquaculture as a field of scientific endeavor which, despite established institutes, is only now gaining recognition as a "multidiscipline". As a consequence, holistic, human ecological studies in the field are rare and socioeconomic information is scant, fragmented and of extremely limited time depth.

Thus, here I treat aquaculture as any other agrotechnological innovation in the generalized terms of external influences, attributes of society relative to innovation, innovation adoption process, and impacts on society (Fig. 2). The paper is based mainly on my field research in southern China and Malaŵi, as well as on secondary sources for Panamá.

External Influences

The principal objectives of aquaculture development in developing countries are to enhance the production of fish as human food, and thereby to improve the livelihood of farm families, by upgrading household nutritional status and/or increasing cash

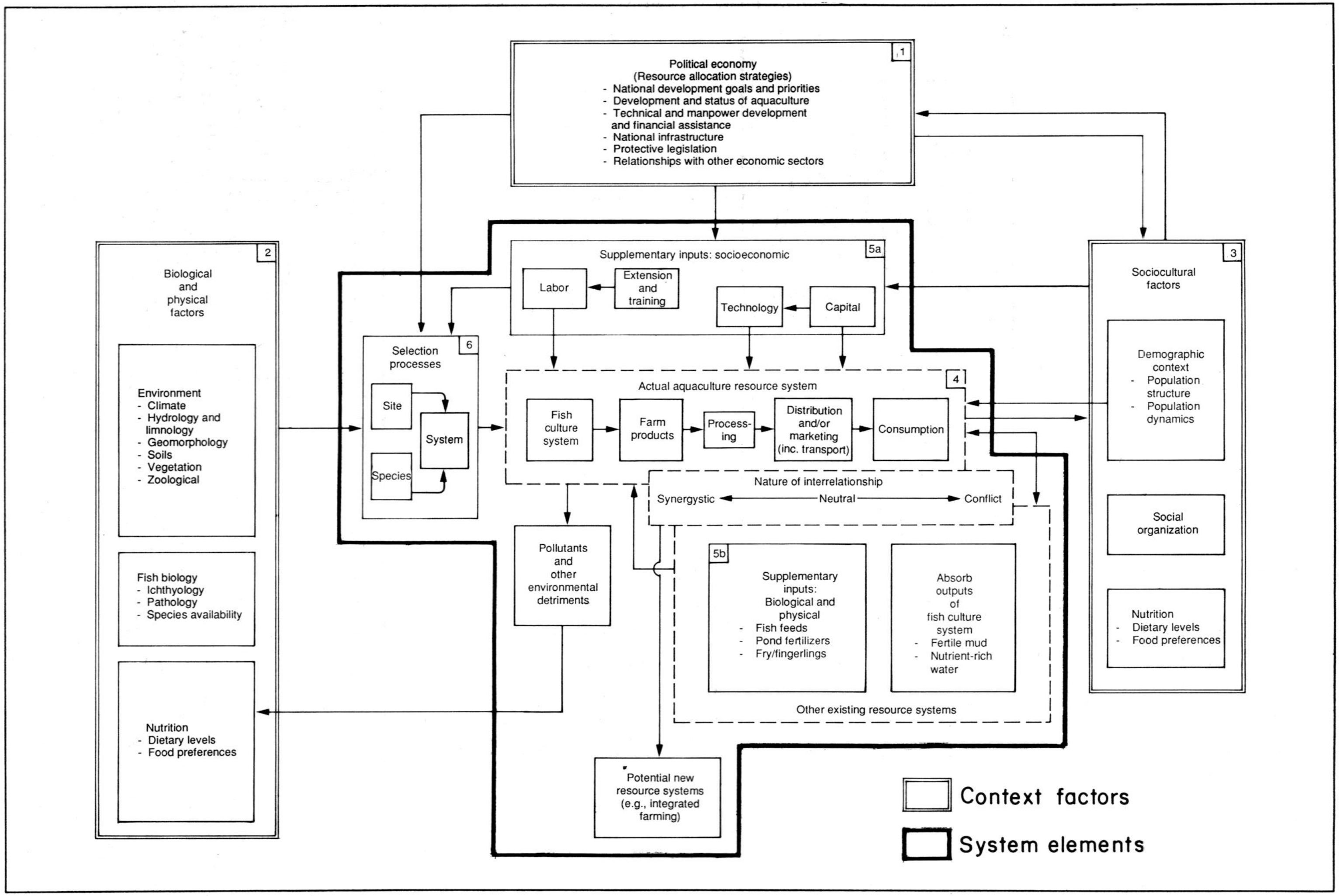

Fig. 1. A resource systems model of aquaculture in human ecological context.

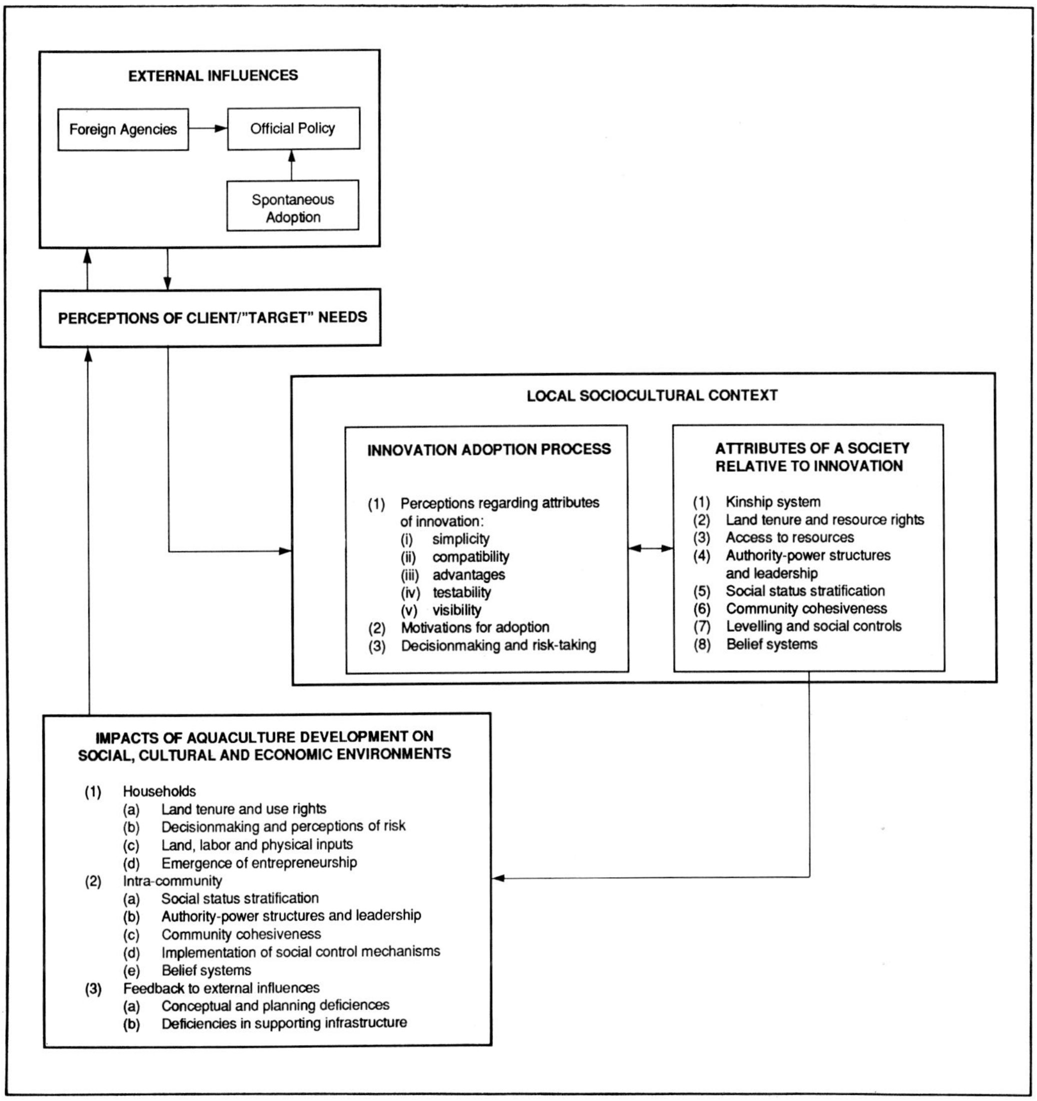

Fig. 2. Paradigm of potential impacts of aquaculture innovation on social environments.

or in-kind income. There are other relatively minor objectives.

But aquaculture is a new and not uncommonly strange technology to many organizations involved in development, and most members of assistance organizations, versed mainly in agricultural development, are ill-equipped to assess either the merits of aquatic food production, the difficulty of sustaining aquaculture where there has been no continuous tradition of it, or the potential negative impact on environments and societies of inappropriate aquaculture development. On the other hand, aquaculture is often promoted vigorously as a panacea by specialists with vested interests in it. Thus given the precarious nutritional circumstances under which vast numbers of people in developing countries exist, coupled with the incidence of hunger and the occurrence of outright famine, there is a natural tendency to react with incautious optimism to any promising means, such as aquaculture, of raising food production levels.

Essentially, development of aquaculture has been severely retarded in developing countries by the failure of agencies, governments and farmers to appreciate its basic requirements. In particular it is not well appreciated that aquaculture development must be integrated within overall and comprehensive rural development programs, and also that it must be supported by an appropriate range of economic, physical, institutional, structural and other investments.

Further, in many nations aquaculture may have little or no role to play in the national development process. In some instances, "aquaculture" may just have become a fashionable term entered into national policy documents merely to satisfy the policies of donors or assistance agencies. In some nations or regions there may be no justification other than this, and efforts and funds might be better directed to developing other sources of animal and vegetable protein, lest effort and funds be squandered.

Commonly, too, the human ecological implications of biological and physical environmental constraints to aquaculture development are not fully appreciated. In particular, high elevation above sea level and aridity or drought-proneness are fundamental and severe constraints on the development of aquaculture, because successful adaptation to them adds greatly to the risk burden of small-scale farm households and communities. Since rates of fish growth correlate positively with temperature, many parts of developing countries are suboptimal for fish growth. This also limits the species that can be cultured. Obviously, where a reliable water supply is not available for at least most of the year, aquaculture is infeasible. Paradoxically, therefore, many areas best suited to aquaculture development are naturally fish-rich and are already exploited by freshwater capture fisheries. This is not to say that aquaculture development in such areas would be pointless, rather it highlights the need for a parallel development of distribution and marketing systems for fish products.

Aquaculture development must conform to the overall development goals and policies of a country and be adapted to local biological, physical and sociocultural norms that will be critical in determining its success. Many otherwise excellent aquaculture programs have failed because these two fundamental factors were not fully considered (FAO 1985). This has led to the frustrated expectations and negative perceptions of client populations with respect to official competence and sincerity in attempts to promote development and thus to the emergence of new behavior patterns among client populations that can only retard the acceptance of later innovations.

Innovation Adoption Process

As is now well established, the principal factors in the process which impinge on the incorporation of any innovation into rural society are: (1) perceptions with respect to the innovation and the requirements for its successful adoption, together with motives for adoption; and (2) decisionmaking, principally regarding the perceived risks inherent in its adoption.

Perceptions Regarding an Innovation

Most official perceptions, including those of donors, underlying the introduction of aquaculture are that it can: (1) improve local nutritional levels and variety, by both the direct production of food and through the increase of household incomes; (2) increase local self-reliance in food supply, particularly in remote areas; (3) supplement the yield of often declining capture fisheries; and (4) generate employment opportunities. In some cases the perceptions of change agents go beyond these basic aspirations, as in Panamá, where the introduction of community aquaculture is viewed by the government as an instrument of rural social development that facilitates the introduction of new organizational and managerial structures (Molnar et al. 1985).

Perceptions of a target population influence the degree of success of efforts to introduce and diffuse an innovation. In particular, major determinants of success are: (1) the attractiveness of perceived benefits as related to costs of adoption; (2) compatibility of the objectives, demands, opportunity costs, and lifestyle of an innovation with existing resource systems and its potential for integration with them; and (3) the degree of perceived complexity of the innovation and its relationship with existing skills or the ability and means to acquire additional skills.

Voluminous research on the acceptance of innovation in rural societies demonstrates five basic and generic attributes that characterize any new technology and affect the way in which a target population perceives it (Rogers and Shoemaker 1971). These perceptions will largely determine the way in which communities respond to the proposed changes that the technology heralds. These attributes are simplicity, compatibility, advantageousness, testability and visibility (Table 1).

Table 1. Attributes of an aquaculture innovation affecting its adoption.

Attribute	Characteristics of attribute	Examples	Reference
(1) Simplicity	Aquaculture system and related institutions is simple, divisible and easily mastered without special skills-training by many people.	Panamá: Small-scale farmers perceive the techniques, skills and vocabulary of aquaculture to be the same as those of traditional farming, and therefore easily mastered.	Molnar et al. (1985)
		Central Malaŵi: aquaculture perceived as not complex and can be learned by children.	Likongwe (1989)
		Southern Malaŵi: aquaculture skills learned from parents*.	Mills (1989)
(2) Compatibility	Aquaculture system is compatible with existing farming systems, in terms of land, capital, labor, risk and opportunity cost, as well as with community behavioral norms and social roles.	Malaŵi: Adoption of aquaculture not perceived as disrupting pre-existing labor allocation, and as having only a low labor demand.	Likongwe (1989) Mills (1989)
(3) Advantageous	Aquaculture makes better use of on-farm resources and provides greater benefits than do customary practices alone.	Southern Malaŵi: advantages perceived (in rank order) are: (1) additional cash income, (2) enhanced social status within community, (3) provision of additional items for reciprocal exchange, (4) improved household nutrition. Central Malaŵi: (1) improved household nutrition, (2) increased income. (However, in Dedza area of Central Malaŵi as well as in Cameroon, aquaculture is adopted so as to accrue prestige.)	Likongwe (1989) Mills (1989) Nji (1986) Molnar et al. (1985)
(4) Testability	Aquaculture must be capable of being tested by a representative sample of community members, so as to be potentially available to all households.	In Central and Southern Malaŵi, adopted by a range of households representing various economic strata and not generally viewed as being limited to better-off families.	Banda (1989) Likongwe (1989) Mills (1989)
(5) Appraisability	The essential qualitative and other results of an aquaculture system are quickly appraised by potential adopters through casual observation and conversation with peers.	This has characterized the adoption process among small-scale fish farmers in Malaŵi.	Banda (1989) Likongwe (1989) Mills (1989)

*However, the knowledge and skills required to construct and manage a fishpond are viewed with "considerable interest and awe", and the social skills required to deal with the Fisheries Department and the Fish Farmers' Club are seen as complex (Mills 1989).

Table 2. Supply of inputs to household fishponds.

Input	Extrapolated application rate (t·ha·year^{-1})	Actual application rates: Produced by household (t)	($)	(%)	Supplied externally (t)	($)	(%)
		Household 1					
Elephant grass	7.58	2.50	50.76	100.0	0.00	0.00	0.0
Pig excrement	151.50	42.00	127.92	84.0	8.00	24.36	16.0
Human excrement	10.60	1.84	24.24	52.5	1.66	21.94	47.5
Kitchen and field waste	13.60	2.25	0.76	50.0	2.25	0.76	50.0
Sugar cane waste	60.60	0.00	0.00	0.0	20.00	253.80	100.0
Concentrates	0.27	0.00	0.00	0.0	0.09	13.70	100.0
Fingerlings	-	0.00	0.00	0.0	-	338.41	100.0
Totals	-	-	203.68	-	-	652.97	-
		Household 2					
Elephant grass	12.60	2.50	50.76	100.0	0.00	0.00	0.0
Pig excrement	113.60	22.50	101.52	100.0	0.00	0.00	0.0
Human excrement	25.60	5.07	66.98	100.0	0.00	0.00	0.0
Silkworm waste	8.3	1.66	42.26	100.0	0.00	0.00	0.0
Kitchen and field waste	0.00	0.00	0.00	0.0	0.00	0.00	0.0
Sugar cane waste	25.20	0.00	0.00	0.0	5.00	50.76	100.0
Concentrates	8.83	0.00	0.00	0.0	1.75	507.61	100.0
Fingerlings	-	0.00	0.00	0.0	-	203.05	100.0
Totals	-	-	261.52	-	-	761.42	-
		Household 3					
Elephant grass	25.25	2.50	50.76	100.0	0.00	0.00	0.0
Pig excrement	229.50	22.72	45.68	100.0	0.00	0.00	0.0
Human excrement	30.10	2.98	39.26	100.0	0.00	0.00	0.0
Concentrates	10.10	0.00	0.00	0.0	1.00	152.28	100.0
Fingerlings	-	0.00	0.00	0.0	-	101.52	100.0
Totals	-	-	135.71	-	-	253.80	-
		Household 4					
Elephant grass	28.40	3.75	76.14	100.0	0.00	0.00	0.0
Pig excrement	34.09	4.50	13.71	100.0	0.00	0.00	0.0
Human excrement	34.84	4.60	61.04	100.0	0.00	0.00	0.0
Fingerlings	-	0.00	0.00	0.0	-	135.36	100.0
Dipterex	15.15 (kg)	0.00	0.00	0.0	2.00 (kg)	3.55	100.0
Teaseed cake	60.60 (kg)	0.00	0.00	0.0	8.00 (kg)	0.73	100.0
Totals	-	-	150.89	-	-	139.64	-

Source: Ruddle and Zhong (1988)

Risk-Taking and Decisionmaking

Patterns of resource use within developing-country communities usually reflect the decisions of large numbers of households and small groups, in addition to individuals, as individual needs are generally satisfied through small group relationships. However, there is generally considerable variation in microeconomic behavior, which is caused in large part by the differential use of and access to renewable resources, as well as by perceptions that influence risk-taking and decisionmaking in social and economic activities (Johnson 1972; Rutz 1977; Ruddle 1984).

Most small-scale farmers plant a combination of crops, or may opt to include a fishpond within their operation, to spread risk. But, as in tropical agroecosystems, combinations of crops, livestock and fish also ensure the exploitation of a range of available ecological niches on a farm, as well as providing a balanced household food supply, thereby also enhancing household food security. Thus, distinguishing a single motive with respect to farming systems and the decision to diversify operations by adding new elements is a complex problem.

The productive activities of most developing-country rural households consist of several complementary economic activities that together provide a balance of subsistence goods. Commonly, small-scale fishing, animal husbandry, hunting and collecting of forest products are the economic complements of cultivation. In integrated systems of agriculture-aquaculture, some of these components are tightly fitted within a single system. Thus an evaluation of risks is made with respect to multiple activities and not just a dominant one. An innovation, such as aquaculture, will not be adopted if it is perceived as unduly risky to household basal subsistence.

Diversification of sources of household income has also been a traditional risk-spreading device. Thus, for example, small-scale farmers are not uncommonly either part-time or seasonal fishers. Opportunities to earn complementary, supplementary or totally alternative sources of income have been enhanced by urbanization and by the industrialization and commercialization of economies. The relative stability of such non-agricultural sources of income adds greatly to the opportunity cost of small-scale farming and, ironically, exacerbates risk by destabilizing the age and gender balance of rural labor supply. This is offset, however, where incomes earned in those other sectors are used to finance small-scale farming or to mitigate the risks inherent in adopting an innovation such as aquaculture (Laan 1984).

Where seasonality of rainfall is pronounced most agricultural labor inputs are concentrated within a typically short four-month growing season. Thus agricultural labor productivity is limited by the amount of indispensible tasks that can be performed during that limited time period. Risk is introduced by the physical ability of available labor to undertake the tasks - the incidence of illness and infirmity, age and gender, among other factors - and by both the capacity to hire labor and the availability of additional labor. Aquaculture development schemes that threaten this delicate balance by virtue of either labor demands or increased opportunity cost of labor are unacceptable to most rural communities in developing countries.

Farm household decisionmaking ranges from very deliberate problem-solving to automatic, subconscious behavior, based in large part on traditional teaching and community norms. In developing nations, most tend toward the latter type (Simon 1957). Moreover, most decisionmakers apparently prefer to follow the precedents of established patterns, rather than to process data themselves. This may go far toward explaining "tradition", "conservatism", and "habit" in resource use, and may help to explain the considerable uniformity observed in decisionmaking among small-scale farmers (Simon 1957).

All aspects of decisionmaking are closely tied to local culture, technological levels, and peer group pressure. Small-scale farmers rarely make massive decisions, but they do make small, incremental decisions that in aggregate might lead to large-scale changes. Thus they gradually adjust existing patterns so that an optimum pattern might evolve over a longer time period. This is closely related to the availability of information; the completeness of an individual's "decision environment" being a function of both the customary and formal levels of education achieved, the communication system, and the motivation and willingness to make an effort to obtain information.

There are major differences between community and individual decisions. The former appear to be more deliberate, explicit and often better publicized. Generally, too, they are of greater importance in establishing and maintaining patterns of resource use within communities and regions. The efficiency with which decisions are made, together with the distribution of benefits, are the two major determinants of success in the adoption of an innovation, especially a community endeavor.

In agricultural households in Central Malaŵi, for example, most household decisions are made after consultation between both conjugal partners and, despite a matrilineal social organization, the wife's brother(s) have no say in the decisions (Mthinda 1980; Phiri 1981), even in female-headed households. The traditional division of agricultural responsibilities and associated decisionmaking regarding resource allocation has important implications for the development of aquaculture and, in particular, for the development of integrated farming. Of the fish farmers surveyed in Zomba District, Southern Malaŵi, 31% of respondents reported disagreements with their wives over agricultural resource allocation. Since in a matrilineal society women perceive themselves as being principally responsible for household subsistence, which is seen as being quite different from aquaculture, the decision about the use of potential pond inputs, such as maize bran, devolves on the women rather than on the men, who are those primarily responsible for aquaculture decisionmaking. Five per cent of fish farmers reported that their wives had refused to permit the use of maize bran as a pond input, claiming that there were more pressing uses for it (Mills 1989).

It has been frequently averred that many subsistence farmers in developing countries are concerned to produce only enough to ensure their household subsistence base, since they value "leisure" more than a cash income obtained from sale of agricultural surpluses. However, in Malaŵi, for example, surveys demonstrate the western economic rationality of small-scale farmers who live above a subsistence base. For example, Minford and Ohs (1970) concluded that there is a significant positive response by farmers to higher producer prices and a negative response to higher consumer prices; Brown (n.d.), in a nationwide survey, found that planting decisions depend more on yield risk than on price or income factors, that increased income promoted increased labor input, and that higher prices and incomes led to increased consumption, higher output and greater use of improved technology; and Gordon (1971) calculated the supply response function and a production function for each major crop and found that a 1% price increase evoked a 2.3% increase in cultivated area.

Subsistence households in Zomba District, in contrast, generally base their farming decisions more on social considerations than on economic factors. On the other hand, small-scale farmers in transition from household subsistence to an incipient commercial orientation, or those with a holding large enough to produce a saleable surplus, base their decisionmaking more on economic than social factors (G.A. Banda, pers. comm.).

Impacts of Aquaculture Development

Without long-term evaluation exercises or other longitudinal studies, it is impossible to assess the human ecological impact *per se* of aquaculture development. This is a task for the future. At this stage in the development of a paradigm for that purpose, potential impacts of aquaculture development on households, its intra-community consequences and, from these levels, feedback to and impacts on external influences, as manifested in modifications to policy, programs and projects, or in the provision of supplementary inputs to aquaculture or complementary physical and institutional infrastructures, are examined. These can all be regarded as indicators, although not amenable to precise measurement, of the human ecological consequences of aquaculture development.

The Impacts of Aquaculture Development on Households

LAND TENURE AND USE RIGHTS

Innovations such as aquaculture may be most easily introduced and have the least social environmental impact, other things being equal, into communities where land and improvements to it, as well as access to water resources and the like, are vested in the individual. The degree to which an individual will be permitted use of land owned by the extended kin group or the community varies according to tradition and particularly to conditions within ethnic groups brought about by recent or ongoing socioeconomic change. Members of most developing-country rural societies are still bound by traditional kinship obligations, which they can ignore only at their peril. The fundamental sociological element in many developing-country societies is the extended family, that forms a village or hamlet, commonly based on a founding lineage or its descendants.

A special and widespread case in developing countries is presented by matrilineal kinship societies, since descent and inheritance, with associated primary rights to resources, is through the female. In many matrilineal societies an area occupied by a matrilineage is recognized as being owned by that lineage as a corporate group. The lineage has the general usufruct right to that area, whereas the individual families which compose that lineage have the recognized usufruct. Under matrilineal kinship systems land and resource rights are inherited through the female line, and decisions are made by a woman's brothers; i.e., a man seeking permission to build a fishpond would usually have to petition his wife's brother(s). Traditionally, residence rules have been uxorilocal (i.e., on marriage a man takes up residence in his wife's household).

In such societies a male is in an anomalous position with respect to resource rights, since although the head of his household, as a consequence of matrilineality and uxorilocality, respectively, he obtains access to land through the female line and resides in his wife's village. A husband works the land of his wife or wives. Further, although heading his own household, a man must generally submit to the authority of his wife's brothers, with respect to his wife, his children, and her land. Not uncommonly, therefore, husbands experience insecurity of tenure and divorce rates might be high. On divorce, or on the death of his wife, a man returns to his mother's village and his former wife, or her descendants, retains all the usufruct rights to the lands that formerly they cultivated jointly.

Many traditional subsistence communities are bound by close interpersonal relationships via institutions for sharing, which demonstrate that all people in a community are linked together in one way or another, and that everybody has access rights to a common property resource, as well as rights to demand a share in the product of resources to which specific households hold exclusive rights. Commonly, subsistence activities are not viewed as a business undertaking, rather as operations on which depends the welfare of all members of a community. Widespread are sets of related concepts which stress that more fundamental than increasing profits, spreading risks and gaining assistance, is the individual's right to survive; that nobody should go hungry and that everybody should share in the results of an economic activity, however meagre they might be. Such obligations have an important bearing on the decision to adopt or not innovations leading to increased productivity, as a concomitant increase is the requirement to share.

In many such societies, people who ignore or neglect their duties toward kin are criticized and, finally, deserted by them, thereby depriving themselves of social and economic security in time of need (e.g., Mitchell 1951). In traditional communities in parts of Southern Malaŵi, for example, a person who ignores his kinship obligations to such an extent that he is deserted by his relatives is considered to be the target of sorcery by them. Further, he himself was likely to be suspected of being a sorcerer (Mitchell 1951).

It has been widely claimed that the insecurity of male tenure under matrilineal systems of social organization is a common disincentive to make long-term investment and improvement in a land unit, the usufruct right of which belongs to a wife (e.g., Lamport-Stokes 1970). This varies greatly, however, and depends in large part on the personality characteristics of the individual male.

Land scarcity coupled with the disincentives inherent in the traditional social structure are an increasing cause of divergence from cultural norms in any matrilineal society. Men living uxorilocally frequently seek to establish their own farms and particularly those for cash crops or other long-term investments.

As a consequence of such highly complex interaction of personal usufruct rights to land, especially in areas of matrilineal societies and high population density, either those obtained through village membership or those acquired individually, patterns of cultivation rights in many developing countries are extremely complicated. It is almost meaningless to attempt to resolve into a checklist of simple, generalizable principles, particularly since its inherent complexity is now being additionally ramified by population growth and resultant pressures on land and land fragmentation, and as a result of increasing divergences from nominally prescribed cultural forms attributable to general "modernization".

Access to Resources

In Zomba District of Southern Malaŵi, 76% of fish farming households have a total land holding of more than 2 ha, and only 16 have 1.0 ha or less (Banda 1989). Most fish farmers therefore have relatively large land holdings. Those with less than 0.8 ha who practice fish farming usually do so because they have a piece of waterlogged land not suitable for crop production. Such people are usually young and lack experience to obtain additional land of better quality. Thus few fish farmers are in the subsistence household category (defined as having less than 1.6 ha), since they are unwilling to risk adoption of an as yet unproven innovation and only 10.5% of those sampled are in the "below subsistence category" (defined as having less than 0.8 ha).

Together with land, family labor is the principal household resource in rural Malaŵi. Extra labor is obtained by hiring, the role of the formerly important reciprocal labor having declined with the rapid socioeconomic change that has occurred in Southern Malaŵi (Banda 1989).

The Emergence of Entrepreneurship

A rare case of household entrepreneurship on a large scale in an old-established aquaculture system is well illustrated by changes that have occurred during the last decade in the dike-pond system of integrated farming in the Zhujiang (Pearl River) Delta of Guangdong Province, southern China (Ruddle and Zhong 1988).

From late 1978 there occurred in rural China a progressive repudiation of the notion of a highly collectivized and egalitarian society and the concomitant beginnings of a mixed economy. In essence, the reforms decollectivized many agricultural practices, transformed from *de facto* to *de jure* the status of the individual family as the fundamental rural economic unit, and removed the controls that prevented households from freely marketing surplus production. Farm households obtained considerable freedom in deciding how to allocate their own capital, labor and management resources. As a result, productivity and household incomes rose dramatically and household activity schedules and the allocation of labor also became more flexible (Ruddle and Zhong 1988). Thus variations emerged among households in terms, among other things, of the allocation of working capital and labor to the system, management strategies and levels of productivity, the energy efficiency of household ponds, and household economics. This was, of course, a response to individual household circumstances that affected their physical and financial capacity to supply different inputs (such as excrements and elephant grass on the one hand; and purchased fingerlings and concentrated feed on the other) at different rates, as well as differing perceptions with regard to the comparative worth of traditional and modern inputs.

By manipulating energy inputs to actualize opportunity costs of on-farm sourced inputs, and thereby offset costs of purchased energy sources, modelling demonstrated that household cash incomes could be raised 3-12% (Ruddle and Zhong 1988). In practice, rates would probably be higher, since substituting concentrated feed for traditional inputs would improve water quality, thus, all things being equal, improving fish yields.

However, to make these simple changes at the household level requires important changes at higher levels in the system. Among the most important of these are the regular and sufficient supply of concentrated feed to local pond operators from the newly opened factory in the county; the adoption of concentrated feeds as a consequence of repeated and successful demonstration effects; a continued market for household produced and surplus excrements that would absorb these surpluses as they continued to increase with the wider adoption of concentrates; the ability to divert sugar cane waste to other productive industrial uses (e.g., pelletized feeds); and the industrial use of silkworm waste, among others.

Impacts of Aquaculture Development on Communities

Social Status Stratification

This may correspond to property ownership patterns and especially to land tenure. Large landowners, local officials and relatively better educated persons often become leaders and exert disproportionate influence on the adoption of innovations.

Social stratification with respect to wealth and authority can either impede or facilitate the adoption and sustainment of innovation, depending on local circumstances. In some cases, other factors being equal, the more homogeneous a community the more likely is an innovation to be widely adopted (Oxby 1983).

On the other hand, participation or sponsorship by widely respected elites may legitimize an undertaking and encourage broad participation, as in parts of Malaŵi, where adoption of aquaculture by better-off households is perceived to reduce the element of risk for the poorer ones. Further, elites may provide technical leadership, ensure good management, and monitor the distribution of benefits (Molnar et al. 1985). In other cases this may evolve into or reinforce a dependency relationship, rather than fostering self-reliant community development. Further, depending on the nature of the client-patron relationship and the existence or not of paternalism and/or personalism, households may participate in an aquaculture project for political reasons rather than from true commitment.

Whereas elites generally support innovations that they perceive as offering additional opportunities for them to fulfill their duties (and thereby enhance their social status) as generous insurers of general community welfare, as in aquaculture developments in Zomba District, Southern Malaŵi (Mills 1989), they can also undermine projects to thwart perceived diminishment of their own status. Any innovation may be perceived by large landowners as a threat to a pool of cheap and readily available labor, for example, and so might be undermined to remove competition, to monopolize access to a new resource, or to achieve other personal ends.

Authority-Power Structures and Leadership

Many developing-country societies are characterized by hierarchical social organization, which ranges from a regional leader/tribal chief through an individual family member. Commonly, this is reflected in the spatial organization of the territories with which each social unit is associated. In general, the highest level commonly has three main categories of duty toward his followers: land allocation, judicial matters, and ritual and religious responsibilities. The second level is frequently that of the village headman, whose various social roles mirror at the village level those of the regional leader/tribal chief. Dispute settlement is invariably a major duty, particularly concerning rights to and conflict over the use of land, water and other resources. Village headmen are also responsible for representing the interests of their villagers in dealings with other groups.

The structure of authority and leadership in matrilineal societies is relatively complex, because while resources are controlled through the female line, villages are governed according to principles of male leadership. This leads to several inherent paradoxes and tensions within the social system. Among the most important of these are: the competition between a husband and his spouse's brother for control of both his wife and children; the ambivalent position of a man in his wife's village; the conflict between the principle of uxorilocality and that of male governance; the structural opposition between proximate generations as a result of emphasis on the unity of the sibling group; the division of a mother's loyalties between her sibling group and her children; and the political competition among uterine brothers for their sisters' loyalty. These inherent conflicts among principles of social organization may cause serious social tensions within small communities and can be easily exacerbated by economically beneficial innovations, such as aquaculture.

As would be anticipated from changing political, economic, and social conditions, as well as from the personality traits of individuals, in few if any societies do everyday social relationships and patterns of behavior correspond exactly to such nominal ideals of social structure. Monetization, urbanization, industrialization and the resultant migration for wage labor have been among the principal factors that have caused major changes in developing-country social organization.

The principal function of community leadership with respect to the adoption of innovation is the ability to mobilize and then sustain participation, by generating labor contributions, organizing infrastructure, interfacing with external authorities, effecting consensual decisionmaking and ensuring the equitable distribution of benefits, such that community harmony is maintained.

Good leadership can speed the adoption process and ensure its continuity. On the negative side, however, strong leadership can sap the initiative of other participants to develop managerial skills and overdependence on a single strong individual leader can imperil long-term continuity of development. Poor Panamánian farmers, for example, interact best when the leader is a respected member of the community but still a peer; i.e., he is neither so wealthy nor so powerful that he is no longer regarded as being a member of the local group (Molnar et al. 1985).

Community Cohesiveness

Factionalism within a community stemming from social and/or economic status, caste, or ethnicity, among other causes, can be a major obstacle to the successful adoption and dissemination of an innovation, particularly when access to common property resources, such as water, is required. In addition to analyzing the cost-benefits to themselves, individual households also monitor those for their group as a whole. Whereas costs and benefits need not balance exactly for individual households, a net gain must be guaranteed by a group (Kikuchi et al. 1978). Further inequities may be interpreted in the context of former existing intra-community conflicts and resultant factions (Peterson 1982), which tend to become exacerbated as a consequence. A positive aspect of intra-community conflict and factionalism, on the other hand, is its possible indication of a healthy level of social energy which might be harnessable in support of a well-designed development project. In contrast, its lack may indicate lethargic resignation and pessimism toward innovation, among other negative traits (Whyte and Albert 1976).

In Panamá, for example, although community aquaculture projects did not engender any new intra-community conflicts, some became enmeshed in earlier ones between members of collective farms and independent farmers. The latter, resentful over their exclusion from the aquaculture development project, attempted to undermine it. However, rather than achieving their object they strengthened the cohesiveness of the communal farmers, since aquaculture became a further rallying point in their ongoing conflict with the independents (Molnar et al. 1985). Evidence of the success of innovations in general demonstrates that commitment to an undertaking, regardless of what stimulates it, is more important to the success of a project than skills and competence, which can be learned (Leonard and Marshall 1982).

Implementation of Social Control Mechanisms

In many societies worldwide, levelling mechanisms are fundamental in controlling the individual and in functioning to maintain community social order and social status ranking. An individual is prevented by a variety of social pressures, obligations, proscriptions and punishments from advancing economically beyond his or her defined social role. On the contrary, people are commonly enmeshed by sets of reciprocal rewards for conduct appropriate to their social status. As a consequence, in many developing-country societies, an individual who decides to devote time to economically productive activities, as opposed to socially productive activities, is commonly regarded as a deviant who must bear heavy social costs.

Individuals and households incur additional risks by participating in a collective aquaculture enterprise that they would not face in private undertakings. These are no different from the additional risks encountered in any collective innovation. Especially important is the individual's perception of the balance of costs and benefits (rewards), or "distributive justice", among all participants (Popkin 1979). Participants tend to avoid situations perceived as being unfair and to seek out those with a clear demonstration of equity (Homans 1961). Perceptions that a group endeavor has obvious winners and losers can torpedo the undertaking in the short term and, worse, lead to long-term and profound social upheaval within a community.

Social controls, levelling mechanisms, or sanctions are widely applied to either households or individuals within communities when a majority perceives that wealth accumulation by a few members is to the detriment of the group. To prevent such an occurrence from causing deep discord, some objective criterion of balance or distributive justice is generally recognized.

Thus, the rules of access to a collective or common property resource must depend on some invariable criteria. In aquaculture this might be, for example, distribution of harvest share proportionate to units of labor supplied. However, a system of benefits strictly proportionate to inputs does not recognize inherent inequalities of individual or household energy, talent, or motivation, or the ability of a deprived household to supply labor. The enforcement of strict equity principles could therefore lead to resentment. This may be managed, as on community managed aquaculture projects in Panamá (Molnar et al. 1985), for example, by informal compensatory mechanisms, which, despite the normatively mandated equal rewards, distribute the harvest by size and quality classes according to individual inputs. In other communal aquaculture projects in Panamá, records are kept to apportion accurately the harvest benefit by labor input (Molnar et al. 1985). Also in Panamá, conforming to the rules of the rural culture, which are enforced by peer pressure, a greater share of the harvest than is mandated by input is obtained by impoverished families, the infirm, or the otherwise unfortunate (Molnar et al. 1985).

Such mechanisms tend to be blurred where an enterprise is operated within a kinship or fictive kinship network and resource use is characterized by sharing, pooling, generalized exchange and nonreciprocal giving, rather than by the reciprocity or commercialism characteristic of distant kin and nonkin (Sahlins 1965). In nonkinship-based enterprises, the degree of participation correlates closely with anticipated direct benefits. If costs continuously exceed benefits, individuals will not contribute to collective action, unless coerced by peer pressure, punishment, or threat of sorcery, for example. Further, innovation is likely to be resisted by those who benefit least from existing situations, since they may perceive it as likely to perpetuate current inequities (Alexander 1975) and, hence, power.

Levelling mechanisms may be implemented when activities such as the accumulation of wealth by a particular individual or family within a community are perceived by traditional leaders as a threat to either their power or status, or both, and to the unity of the community. In Malaŵi, for example, individual advancement is often perceived by guardians as a threat to the unity of the female corporate group (Mitchell 1956) and is restrained by accusations of sorcery (Lawry 1981), the frequency of which commonly leads to the fissioning of villages. For this reason successful farmers will occasionally have difficulty in obtaining hired labor (Lawry 1981).

Social levelling mechanisms are strongest where there is little to share (Humphrey 1971). In the Zomba District of Southern Malaŵi belief in witchcraft is so strong that small-scale farmers, including fish farmers, dare not produce beyond a certain level, for fear of being bewitched by their peers (G.A. Banda, pers. comm.).

Banda (1989) noted that jealous members of communities in Mwanza District deliberately damage the fishponds of youngsters, of whom they have no fear. The ponds of successful farmers are also damaged or the stocked fish stolen (G.A. Banda, pers. comm.).

One of the negative consequences of the moral requirement to share resources is that some farmers might elect not to improve their economic level, for example, by incorporating a fishpond into their holding. The commonly stated rationale for this is that there is little point in working harder - and possibly thereby also incurring social sanctions - if one will be pressured to give away a large portion of the fruits of the extra labor to relatives or other members of the community (Mitchell 1951; Lawry 1981).

With the increasing commercialization of the rural economy in Zomba District, Southern Malaŵi, for example, there are emerging signs of a corresponding decline in the social role of pond-cultured fish as both giftgiving and subsidized sales items to relatives and friends. Of the subsistence fish farmers interviewed, 37% revealed an increasing inclination toward purely commercial sales of their fish, since they were nowadays no longer receiving equal value in reciprocal exchanges for their fish, and that "...some neighbors were taking advantage of their fish farming activity" (Mills 1989). Similarly, 36% of the commercially oriented fish farmers claimed that they had been motivated toward commercial activities owing to the decline beyond an acceptable level in the economic value of return reciprocity. Two commercial fish farmers in the area characterized their commercial efforts as a deliberate attempt to escape potential reciprocal obligations. One such farmer started to sell more of his product in a local market rather than using the conventional pondside site venue in order to escape the increasing and, as he perceived, unneccessary economic burden of subsidizing below-market-price sales to friends and relatives. Further, he was able to obtain prices 10% higher than nonsubsidized rates in the conventional venue (Mills 1989). However, in so doing a farmer might incur social sanctions, or be subjected to levelling mechanisms, since the amount of fish available for subsidized sales to friends and relatives resident in the producing community is reduced.

Belief Systems

Throughout the developing world, belief systems have a major impact on resource use. Local constraints embedded in traditional magico-religious systems may be the most widespread community-wide sociocultural factor either impeding the development of aquaculture in Africa (Grove et al. 1980), or, in contrast, being overwhelmed by it in combination with other elements of change. Such phenomena are not generalizable since they occur as a plethora of local details.

Commonplace throughout Africa, for example, is the animistic belief that ancestral and guardian spirits of a household or community reside in a wide variety of natural or manmade landscape features. For this reason the development of aquaculture may be either retarded or precluded entirely in a particular locality by a refusal to modify the environment, such as to excavate fishponds, or even to modify those constructed by ancestors (Grove et al. 1980). In Malaŵi, such beliefs have retarded the development of small-scale aquaculture in parts of Lilongwe District (D.H. Ng'ong'ola, pers. comm.).

Impact on External Influences

CONCEPTUAL AND PLANNING DEFICIENCIES

As the history of past failures in developing countries demonstrates, the development of sustainable aquaculture requires sound policy, well-conceived planning, and proper implementation via biotechnical and socioeconomic research that works in tandem with a dedicated extension service. These are indispensable for ensuring successful and sustained development. Ill-conceived projects coupled with contradictory objectives have been highlighted as a major source of prior failures in aquaculture development projects (FAO 1975). For example, in one appraisal of the status of aquaculture in developing countries it was observed that "most failures of aquaculture development programs in Africa so far can be explained by the lack of qualified technicians and of an adequate infrastructure, as well as by the absence of government policy specifically aimed at this form of development" (Coche 1983). This feedback has revealed itself in the widespread acknowledgement of (a) conceptual and planning deficiencies, and (b) in infrastructural deficiencies in systems.

Among the principal impacts on external influences of developing-country aquaculture development programs is the realization that failure has stemmed fundamentally from not having viewed aquaculture as a system, the success of which depends on the parallel development of a physical and institutional support infrastructure. The long-term objectives of alleviating malnutrition and poverty in developing countries do not depend just on increasing the production of a balanced mix of foodstuffs, as is commonly considered, as much as on increasing their distribution for either direct food use or of the distribution of the benefits of the commodities, when they are not used for direct local consumption. For example, throughout southern Africa transportation costs are perhaps the principal constraint on supplying inexpensive fish to consumers; as in Zambia, where they can sometimes be three or four times the value of the fish transported (Subramaniam 1986).

The provision of appropriate institutions is of perhaps greater importance. Most evaluations of integrated rural development programs assume either the prior existence of, or the ability to create quickly, an institutional structure appropriate to local needs and capable of distributing the resources for decentralized investment and production to far-flung and diverse rural regions (Rondinelli and Ruddle 1977). This is particularly true for a new idea, such as aquaculture. Most reports on the status of aquaculture development in developing countries stress the overwhelming institutional deficiences that preclude rapid growth of the sector.

A combination of at least four basic institutional deficiencies commonly occurs in programs for small-scale aquaculture development:

(i) most organizations that provide technical inputs and/or services are either absent or exist in only their traditional forms or surrogates and the latter are usually inadequate for promoting and sustaining aquaculture development directly;

(ii) such institutions are rarely linked into a hierarchy of supporting institutions so as to provide a reliable flow of inputs and their resultant unreliability makes adoption of their innovations, services and techniques by small-scale farmers unnecessarily risky;

(iii) owing to a combination of scarce finances, ineffective linkages, lack of skilled manpower and weak political support, among other things, existing institutions generally have a low administrative capacity to deal with the complex problems and procedures of aquaculture and overall rural development; and

(iv) newly introduced governmental institutions are commonly incompatible with the traditions, behavior and cultural patterns of local "target" societies. As such they may be a further source of alienation and increased impoverishment.

Like appropriate technology, appropriate institutions to serve aquaculture and other sectors of the economy should be adaptable to the wide and complex variety of problems and conditions characteristic of developing countries. The development and transfer of supporting institutions, like aquaculture technology transfer and development, must blend adaptation, innovation and creativity with an intimate knowledge of local capabilities and constraints.

DEFICIENCIES IN THE SUPPORTING INFRASTRUCTURE FOR AQUACULTURE

Extension and information. The lack of a well organized aquaculture extension service - that indispensable link between researcher, administrator and producer - is usually a consequence of the general scarcity of trained specialist personnel and is a factor contributing to low levels of development, particularly of the small-scale rural sector.

As a consequence of deficiencies in infrastructure, the acquisition and reinforcement of initial aquaculture skills from earlier adopters is of major importance. In Zomba District, Malaŵi, for example, 24% of fish farmers and former fish farmers obtained them from neighboring farmers, 26% from distant farmers, and 26% from personal observation of the activity by other farmers (Banda 1989). Farmers appear to prefer to consult with other local, earlier-adopter farmers of approximately similar social and economic levels, since it is perceived as giving a better evaluation of the risks involved than would consultation with extension personnel from an experiment station backed by the resources of government (Banda 1989).

Generally, early adopters of an innovation have higher social status, such as village headmen, that appear to make them better evaluators of a new activity, like aquaculture, at least in the early stages of an introduction. If they are successful, other farmers with lower status tend to emulate them (Banda 1989). Thus in Zomba District, the role of one key farmer, an early adopter, was cited by 75% of established fish farmers as the principal means by which initial fish farming skills were reinforced and enhanced (Mills 1989). This farmer, a respected member of the community, is locally renowned for running a successful commercially oriented aquaculture operation. Once initial activities have developed, reinforcement and enhancement of skills takes place through contacts with key farmers, the aquaculture and agricultural extension services, and fish farmer clubs.

The situation is similar in Central Malaŵi, where the principal skills learned interpersonally from other farmers are pond site selection (19% of respondents), pond construction (14%) and fish-feeding (21%) (Likongwe 1989). The informal learning of aquaculture skills correlates inversely with the frequency of extension service contacts (Likongwe 1989). Such skill acquisition appears to be systematic, in that informal skill acquisition for pond selection correlates positively with that for pond site selection, arranging water supply and fish feeding.

In Zomba District, knowledge about pond construction and management is not given freely, except to family and close friends. Rather, it is perceived as being difficult to acquire and thus an extremely valuable asset that enhances social status within the community. As such, it is a commodity the giving of which requires some form of recompense (Mills 1989).

Fish Seed Supply. Among the major constraints to the more rapid development of small-scale aquaculture in developing countries is seed fish supply. A response to this problem in Malaŵi, for example, was the institution of the "fingerling debt" system, whereby farmers are supplied with free seed for their start-up crop from the extension center. In return they incur the obligation to provide free seed to another farmer when he begins operations (O.V. Msiska, pers. comm.).

Credit. In most instances, the provision of credit is essential to the establishment of a new enterprise. However, few potential small-scale fish farmers in developing countries have suitable assets for use as collateral. Further, given the risks inherent in adopting a generally locally unproven innovation, especially in the absence of a perceived adequate supporting infrastructure, small-scale farmers are usually reluctant to raise capital by selling other farm produce to start aquaculture, and thus to divert investment in such proven agricultural inputs as field crop fertilizers, until they are confident in the profitability of aquaculture, as in Southern Malaŵi, for example (Banda 1989).

Marketing. The commercial distribution of the products of a new enterprise is commonly problematical in developing countries. As a consequence, it has been suggested that aquaculture development be planned to take advantage of existing distribution and marketing systems for the products of capture fisheries (Pillay 1977).

However, this is unnecessary in many areas, particularly where fresh fish is a preferred food and in short supply, at least seasonally. Thus in the Zomba District of Southern Malaŵi, for example, the sale of cultured fish, regardless of species, is a simple task, since demand always far outstrips supply. Indeed, some farmers ration the amount of fish sold to customers attending a harvest to ensure reasonably equitable sales (Mills 1989). Further, some enterprising farmers have developed specialized markets. Most farmers simply notify the community of an impending harvest, and customers arrive on the appointed day. In Zomba District, 96% of commercially oriented fish farmers sell their fish on site. This obviates the needs for specialized marketing strategies, the role of fish traders, and processing, thereby side stepping potential constraints caused by infrastructural deficiencies.

Prices of cash sales are tempered by reciprocity and urban demand. Although there is a close-to-current-market price that fish farmers could seek, many sales to friends and relatives are "socially priced" at a lower rate. Nevertheless, farmers are mostly able to obtain close to current market prices (Mills 1989).

Protective legislation. Aquaculture developments in developing countries are commonly unprotected by formal legislation, and customary law does not usually apply to the practice. On the contrary, legislation governing water abstraction, river pollution, public health, fish handling, and water or coastal and lakeshore rights all hinder the development of aquaculture, especially by small-scale farmers. Thus one impact of aquaculture development, as in Southern Malawi, is a widely perceived need to formalize the processes for securing sites and water, and obtaining various licenses and obtaining legal protection against theft and the like. Legislation is also required to protect the water rights of fish farmers against upstream pollution and damage to watersheds. Insurance against accidental loss is also required by small-scale operators, particularly to mitigate perceived risk that retards dissemination of aquaculture.

IMPACTS ON COMMUNITY WELFARE

Public health. There is little evidence that the development of small-scale aquaculture anywhere has had an adverse impact on public health, by spreading disease. Regardless of this, however, public health measures must be planned as an integral part of aquaculture development, since if the development of aquaculture is perceived locally to be responsible for increasing morbidity rates of schistosomiasis and other debilitating diseases, the innovation will be critically evaluated by adopters and nonadopters with respect to the trade-off between nutritional and economic benefits compared with health implications. Aquaculture development has the potential to increase hazards to human health, primarily from waterborne diseases (especially schistosomiasis, malaria and guinea worm) and fish parasites. Schistosomiasis is perhaps the most prevalent debilitating disease, with an incidence of 70-90%.

Conclusions

Aquaculture development in developing countries has had a chequered history, at best. This is particularly true of Africa, but also, to a lesser extent, of Asia. In large part, failures have stemmed from social, cultural and economic causes, and, in particular, from the lack of appreciation among policymakers and planners that aquaculture is just one dimension of a much larger human ecological system.

Above all, aquaculture must be seen in context. In particular, but not exclusively, it must be planned for and assessed in the context of (1) national development policy and goals and existing levels of national development and the constraints that these impose on sustainability; (2) its socioeconomic and other relationships to alternative sources of both animal and vegetable protein; (3) the physical and biological environments into which its introduction is proposed; (4) the socioeconomic environment into which its introduction is proposed; and (5) its capacity for integration with existing or potential resource use practices, especially farming systems.

A thorough understanding of the sociocultural environmental context into which the introduction of small-scale aquaculture is proposed has been outlined in this paper. The tailoring of systems to fit that context is an absolutely essential prerequisite to any development project, since congruence or not with often complex sociocultural variables guarantees either the success or failure of the introduction of any innovation, all other variables being equal. It is axiomatic therefore, although not often appreciated, that aquaculture must be adapted to society; the converse is not worthy of consideration.

References

Alexander, P. 1975. Innovation in a cultural vacuum: the mechanization of Sri Lanka fisheries. Human Organization 34(3):333-344.

Banda, G.A. 1989. An investigation of factors for the adoption of fish farming in a matrilineal social system: the case of Zomba District. 22 p. (incomplete). Institute of Social Research, University of Malaŵi, Zomba.

Bennett, J.W. 1976. The ecological transition. Pergamon, New York.

Brown, C.P. n.d. The marketing of primary produce. Department of Economics, University of Malaŵi, Zomba. (MS).

Coche, A.G. 1983. Aquaculture development. CIFA/83/5. FAO, Rome.

FAO. 1975. Aquaculture planning in Africa. 12 p. FAO, Rome.

FAO. 1985. Progress in the development of aquaculture in Africa. Committee for Inland Fisheries of Africa, 6th Session, October 1985. CIFA/85/12; 18 p. FAO, Rome.

Gordon, J.G. 1971. A model for estimating future agricultural acreage and production in Malaŵi. Agrarian Development Studies Report No. 4. Wye College, University of London, London.

Grandstaff, T.B., K. Ruddle, J.N. Hawkins, T.C. Foin and W.G. Davis. 1980. Implementing the resource systems approach to transformational development. Resour. Manage. Optim. 1:145-160.

Grove, J.H., D.R. Street and P.D. Starr. 1980. Review of aquaculture development activities in Central and West Africa. Research and Development Series No. 28, 29 p. Auburn University International Center for Aquaculture, Alabama.

Homans, G.C. 1961. Social behavior: its elementary forms. Hartcourt, Brace and World, New York.

Humphrey, D.H. 1971. Preliminary report of the smallholder innovation research project. Chancellor College, University of Malaŵi, Zomba. (mimeo.).

ICLARM-GTZ. 1991. The context of small-scale integrated agriculture-aquaculture systems in Africa: a case study of Malaŵi. ICLARM Stud. Rev. 18, 302 p.

Johnson, A.W. 1972. Individuality and experimentation in traditional agriculture. Human Ecol. 1:149-159.

Kikuchi, M., G. Dozina and Y. Hayami. 1978. Economics of community work programs: a communal irrigation project in the Philippines. Econ. Dev. Cult. Change 16(2):211-226.

Laan, van der E. 1984. Factors contributing to the rapid development of small-scale irrigation schemes along the Sénégal river. Agric. Admin. 17:203-213.

Lamport-Stokes, H.J. 1970. Land tenure in Malaŵi. Soc. Malaŵi J. 23(2):59-87.

Lawry, J.M. 1981. Group formation and dynamics and household decision making in Liwonde Agricultural Development Division. 111 p. LWADD, Liwonde, Malaŵi. (MS).

Leonard, D.K. and D.R. Marshall, Editors. 1982. Institutions of rural development for the poor: decentralization and organizational linkages. University of California, Institute for International Studies, Berkeley.

Likongwe, J.S. 1989. Mechanisms of informal extension and the rate of innovation diffusion among small-scale farmers and nonfish farmers in Lilongwe, Dedza and Ntcheu districts in Central Malaŵi. Bunda College of Agriculture, University of Malaŵi, Lilongwe. 80 p.

Mills, G.G. 1989. Entrepreneurship in small-scale fish farming in the Zomba District, Malaŵi. 60 p. Department of Sociology, Chancellor College, University of Malaŵi, Zomba.

Minford, P. and P. Ohs. 1970. Supply response of Malaŵi labor. (mimeo.).

Mitchell, J.C. 1951. An outline of the social structure of Malemia area. Nyasaland J. 4(2):15-47.

Mitchell, J.C. 1956. The Yao village. Manchester University Press, Manchester.

Molnar, J.J., N.B. Schwartz, and L.L. Lovshin. 1985. Integrated aquacultural development: sociological issues in the cooperative management of community fishponds. Sociologia Ruralis 15(1):61-80.

Mthinda, C. 1980. Decision making in smallholder farms within the Lilongwe Land Development Programme Area. Lilongwe, Bunda College of Agriculture, University of Malaŵi, Lilongwe. 15 p. B.Sc. research paper.

Nji, A. 1986. Social, cultural and economic determinants of the adoption of fish pond culture in Menoua, Mezam and Momo divisions of

Cameroon. Dschang University Center, Dschang, Cameroon.

Oxby, C. 1983. Farmers' groups in rural areas of the third world. Community Dev. J. 18(1):50-59.

Peterson, S. 1982. Allocation of aquaculture resources, p. 21-30. *In* L.J. Smith and S. Peterson (eds.) Aquaculture development in less developed countries: social, economic and political problems. Westview Press, Boulder, Colorado.

Phiri, C.D. 1981. Decision making in smallholder farms within the Lilongwe Land Development Programme Area. Bunda College of Agriculture, University of Malaŵi, Lilongwe. 24 p. B.Sc. research paper.

Pillay, T.V.R. 1977. Planning of aquaculture development. Fishing News Books, Farnham, Surrey.

Popkin, S.L. 1979. The rational peasant. University of California Press, Berkeley.

Rogers, E.M. and F. Shoemaker. 1971. Communication of innovations. The Free Press, New York.

Rondinelli, D.A. and K. Ruddle. 1977. Appropriate institutions for rural development: organizing services and technology in developing countries. Philipp. J. Public Adm. 21 (1):35-52.

Ruddle, K. 1984. Normative models and human behavior: some theoretical issues in household resource use. Bull. Natl. Mus. Ethnol. 5(3):801-822.

Ruddle, K. and T.B. Grandstaff. 1978. The international potential of traditional resource systems in marginal areas. Technol. Forecast. Soc. Change 11:119-131.

Ruddle, K. and D.A. Rondinelli. 1983. Transforming natural resources for human development: a resource systems framework for development policy. United Nations University, Tokyo.

Ruddle, K. and G.F. Zhong. 1988. Integrated agriculture-aquaculture in south China: the dike-pond system of the Zhujiang Delta. Cambridge University Press, Cambridge.

Rutz, H.J. 1977. Individual decisions and function systems: economic rationality and environmental adaptation. Am. Ethnol. 4:156-174.

Sahlins, M. 1965. On the sociology of primitive exchange, p. 207-223. *In* M. Banton (ed.) The relevance of models for social anthropology. Tavistock, London.

Simon, H.A. 1957. Rational choice and the structure of the environment. Psychol. Rev. 63:129-138.

Subramaniam, S.P. 1986. Recent trends in fish utilization and marketing in Zambia, p. 424-436. *In* Fish processing in Africa: proceedings of the FAO Expert Consultation on Fish Technology in Africa, 21-25 January 1985, Lusaka, Zambia. FAO, Rome.

Whyte, W.F. and G. Albert. 1976. Power, politics and progress: social change in rural Peru. Elsevier Scientific Publishing Co., New York.

Aquaculture and Management of Freshwater Environments, with Emphasis on Latin America

MANUEL MARTÍNEZ-ESPINOSA and UWE BARG
Inland Water Resources and Aquaculture Service
Fishery Resources and Environment Division
FAO, Via delle Terme di Caracalla
00100 Rome, Italy

MARTINEZ-ESPINOSA, M. and U. BARG. 1993. Aquaculture and management of freshwater environments, with emphasis on Latin America, p. 42-59. *In* R.S.V. Pullin, H. Rosenthal and J.L. Maclean (eds.) Environment and aquaculture in developing countries. ICLARM Conf. Proc. 31, 359 p.

Abstract

The development of freshwater aquaculture and culture-based fisheries in most developing countries faces socioeconomic, environmental, biotechnical, institutional and legislative constraints. In view of the increasing need and pressure for environmental conservation and control, emphasis is given to difficulties already experienced in the implementation of development plans, management schemes and regulatory measures, as related to the utilization of aquaculture and fisheries resources.

Socially oriented aquaculture development projects require long-term public sector support and improved project formulation and implementation to be successful in terms of socioeconomic feasibility and ecological compatibility. Culture-based fisheries in lakes and reservoirs need careful assessment as to the potential ecological and genetic risks of exotic fish introductions and restocking measures. Intensive aquaculture carries risks of polluting the aquatic environment with nutrient- and chemical-laden effluents, spreading fish diseases and causing disruption of wild stocks through fish escapes.

Multisectoral efforts should be continued on the integration of aquaculture with agriculture, forestry and inland fisheries as well as its inclusion in river basin planning and management schemes. Regulatory measures to prevent and to limit negative environmental impact of aquaculture should be enforceable and cost-effective. Constructive formulation of such measures can benefit from the scientific environmental capacity concept, using modern hazard assessment methodologies.

Introduction

When referring to aquaculture and environmental issues it is very important to classify types of aquaculture. We refer to extensive, semi-intensive and intensive aquaculture and culture-based fisheries. In terms of sustainable and environmentally sound development, however, it is fundamental for these classifications that people involved in the activity are also considered (e.g., producers, consumers, retailers, etc.). Socially oriented aquaculture can be clearly differentiated from cash-crop oriented aquaculture. In this context, aquaculture should be classified, when required, making use of both technical and socioeconomic criteria. In many cases, it may prove useful to refer to "rural aquaculture" within the general framework of agricultural development. Rural aquaculture can be practised at subsistence level as well as at semi-intensive level with varying degrees of cash-crop orientation.

This paper concentrates on Latin America, as representative of the developing world where semi-intensive (shrimp) and intensive aquaculture (salmon) have experienced a rapid development, whereas culture-based fisheries are just beginning to develop. For much of rural aquaculture, results have been well below the expectations and resources involved.

Aquaculture operations can have both beneficial and adverse effects on the environment. Aquaculture can provide a way to use agriculture wastes to make marginal lands more productive. Fish convert plant and animal wastes into high quality protein and enrich pond mud for subsequent use on crop land. Aquaculture can be a major source of fish supply where inland fisheries have been virtually eliminated because of damming, canalization and other modifications of the ecosystem. On the other hand, some freshwater aquaculture can have negative impacts on natural habitats and their biota, through the culture process itself, and on human health.

Rural Aquaculture

Rural freshwater aquaculture in developing countries is not normally a source of environmental pollution but may have some negative impacts; for example, the spread of exotic fish species and consequent effects on native species and habitats, loss of genetic diversity, the spread of fish diseases and public health problems due to unsafe produce, poor working conditions, waterborne diseases and parasites.

Another important negative impact of rural aquaculture derives from the numerous failures of projects aimed at developing the sector with the purpose of contributing to a rise in the standard of living of small-scale fish farmers in developing countries. Considering that rural aquaculture is just one activity carried out by farmers, the negative impact in this case is not on the environment alone, but also on the farmer who is also part of the ecosystem. It is in this context that anything that threatens the sustainability of aquaculture can be considered as harmful to the environment. The objective of developing "environment-compatible" sustainable aquaculture will not be achieved if most of the development projects fail.

Analysis of social-oriented rural aquaculture projects (for example, Smith and Peterson 1982; Engle 1986; Martínez-Espinosa 1986; Wijkström 1986; FAO-SIDA 1987; FAO-UNDP-Norwegian Ministry of Development Cooperation 1987; Martínez-Espinosa 1990) concluded that international assistance had not produced the expected results at farm level to establish a self-supporting aquaculture sector. A change of strategy has been proposed towards institution building, planning, economics, credit facilitation, etc., thereby addressing a range of social factors.

Rural aquaculture must be considered as an additional component of agriculture and should therefore benefit from the wider experience accumulated in this field. Special efforts must be made in the identification of the potential target groups in aquaculture development projects, as well as in the appraisal of the socioeconomic characteristics. Project formulation based on wrong assumptions about social, economic and political factors are the primary cause of failures. The main reason for the high desertion rate has been established to be the lack of motivation in farmers due to poor returns with no surplus to commercialize. It is only in very specific cases that the subsistence type of aquaculture, designed for self-consumption, is likely to be pursued after the completion of external assistance. Not enough attention has been given to special credit facilities for small-scale aquaculturists on terms and conditions which address their weak economic conditions and particular needs.

Integrated agriculture-aquaculture is still evolving as a technology applicable in many developing and industrialized countries. Environmentally friendly, model integrated farming systems, developed on research stations are rarely adopted because of their complexity. Very few Farming Systems Research and Extension (FSRE) initiatives attract sufficient adopters to show any real impact.

Why are FSRE and agroecosystem tools so little used in systems that incorporate aquaculture? The first and most important reason is the institutional structure in which agricultural research and development is conducted. FSRE requires social scientists to work alongside biologists. Such interdisciplinary teams require all members to have a working knowledge of the other disciplines but educational programs rarely offer appropriate courses (Edwards et al. 1988). Agroecosystem analysis and the new farming systems it inspires require integration of crops, livestock, fish and forestry but these commodities are often separated into different departments, ministries and research institutes at national and international levels. Institutional barriers inhibit the growth of expertise, the flow of funds and the use of FSRE and agroecosystem tools (Lightfoot 1990).

Integrated aquaculture, if developed, can become a very useful tool for water management. Run-off from the catchment area and the nutrients it carries can be stored in fishponds and used for other purposes such as livestock, irrigation and domestic consumption. Ponds can raise or sustain the local water table thus facilitating the excavation of shallow wells. Nutrients in the run-off from small basins can be trapped in fishponds where animal excreta and plant residues can also be recycled into proteins (Pretto 1989).

Aquaculture in Reservoirs and Lakes

Freshwater aquaculture in lakes and reservoirs can have all the abovementioned negative impacts, including eutrophication through the organic fallout from fish cages, impairment of esthetic qualities and disruption of resident fish stocks.

The main constraint to the development of this sector is the lack of fisheries management plans for reservoirs. For example, in Venezuela, "the management of these water bodies constitutes a conflict area among different government organizations which share legal competence" (translated from Novoa, in press). Moreover, many such waterbodies do not have even the minimum required infrastructure for exploitation and management of their fisheries, and physical and biological limitations prevent, in many cases, any development. Very often, long distances to markets and a lack of functional roads discourage attempts to develop activities. Despite these limitations, culture-based fisheries in reservoirs and lakes have a significant potential, especially in regions such as Latin America. This potential has been only partially realized but affords examples of successful experiences of relevance to other parts of the world.

Virtually all the large drainage basins in Latin America are progressively becoming regulated by dams and will ultimately form chains of reservoirs if existing plans are completed. Out of a total of 183 dams scheduled for the Paraná River sub-basin (Argentina, Paraguay, Brazil), 121 are already in operation or under construction. Thirty-two dams are being built in the Magdalena River basin which will also become a chain of reservoirs and others are envisaged for its upper and middle sections. There are, however relatively few dams in the large Amazon and Orinoco basins at present. This is why appropriate measures from properly planned stepwise construction should be formulated now to avoid the problems prevailing in the other basins mentioned (Quirós 1989).

The measures adopted to mitigate the loss of ichthyofauna due to the construction of dams and other obstructions include fish passes in the dams and periodic stocking of waterbodies with fingerlings produced in hatcheries. In Brazil, there is a law providing for the installation of fish passes (Quirós 1989). In practice, however, very few have been installed in Brazilian dams or in the rest of Latin America, and their effectiveness and economic feasibility are controversial. Introduction of exotic species to lakes and reservoirs also has important impacts. In Cuba, production of fishes in reservoirs derives entirely from exotic species: 90% from tilapias (*Oreochromis aureus*, *O. mossambicus*) and the rest from cyprinids (*Cyprinus carpio*, *Ctenopharyngodon idella*, *Aristichthys nobilis*, *Hypophthalmichthys molitrix*). In Mexico, six main groups of fishes are produced in reservoirs, four native (*Chirostoma* spp., *Chirostoma estor*, *Ictalurus punctatus*, *Micropterus salmoides*), and two exotics (tilapias and cyprinids) represent 80% of the total catch (Olmos, in press). In the northeast of Brazil, two-thirds of the production through culture-based fisheries in reservoirs is from introduced species, both nonindigenous from other basins (*Cichla ocellaris*, *Cichla temensis*, *Plagioscion squamossisimus*, *Plagioscion surinamensis*, *Arapaima gigas*, *Colossoma macropomum*, *Astronotus ocelatus*, *Macrobrachium amazonicus*), and exotics (tilapias and cyprinids) (Studart Gurgel, in press).

Fish introductions to the region and fish transfers among basins have been reviewed by Welcomme (1988). There is no rational policy on introductions and follow-up control, but further progress in the development of extensive aquaculture activities in these waterbodies will obviously improve awareness of the need for this. Less than 10% of the estimated potential area for culture-based fisheries of Latin American lakes and reservoirs is being exploited at present. Some successful examples are Cuba, 16,000 t (1986); Brazil, 19,000 t (1987) from over 100 reservoirs in the northeast; and Mexico 109,000 t (1987), which constitutes almost half the total freshwater aquaculture production in the region (Juarez, in press).

In these successful cases, different organizational strategies were adopted. For Cuba, there is a government-owned company (Empresa Nacional de Acuicultura), that integrates all the activities from research to development. This has avoided any problems of legal competence among organizations, and has allowed efforts to be concentrated on technical and social issues. Three components of the development process which are often neglected in other countries receive in this case the attention they deserve: previous studies of resident fauna and limnology, fisheries resources management, and general follow-up procedures. Studies are carried out to establish when, which species and how many fish to stock and how best to regulate fishing effort, organize groups of fishers and guarantee sales of the produce to another government-owned company which commercializes it.

This integrated form of management, which includes the operation of a network of hatcheries to produce the seed (mostly for tilapia), has led to a wide variety of basic studies. For example, studies of fish parasites suggesting optimal times for fish stocking (Vinjoy 1988); studies of abundance of fish predators, some of them introduced species, and work on their control with selective fishing gear; studies on aquatic weeds and their control, including the use of herbivorous carps and mechanical methods; monitoring of pollution, particularly from sugar and paper factories, as well as other industrial wastes, fertilizers and pesticides; absolute control of species and quantities of fish stocked; and a genetic improvement program with tilapia at a research center in Pavón, Santa Clara Province (Marí-Diaz, in press).

In the semiarid zone of the Brazilian northeast, the Departamento Nacional de Obras Contra las Secas (DNOCS) has been responsible for the last 40 years for the fisheries exploitation of 104 reservoirs, with a total water surface of 140,000 ha and a mean annual production of 17,000 t. DNOCS also carries out periodic stocking of native and exotic species produced in their hatcheries. These culture-based fisheries activities are based on studies of fish population dynamics and on effort regulations to limit entry.

There is still controversy on cleaning of the reservoir bottom before it is flooded. The policy that prevails is to leave it untouched in order to offer more fish refuges and feeding areas, but submerged vegetation can hamper fishing. For Brazil, Godoy (1985) reports 35 structures to assist fish migration after dam construction of which 21 are located in the DNOCS area. With respect to the introduction of species from other parts of the country or exotic to the Region, DNOCS's experience is, in general terms, positive because it has increased the yield of commercially important species (Studart Gurgel, in press).

In Chile, the rapidly growing salmonid culture industry has already attracted criticism with respect to its impact on the aquatic environment, particularly freshwater lakes, concerning mainly the effects of wastes and escapes of fishes. Some of this criticism also questions the esthetic impact of the activity. Alternative uses of the environment such as recreation and tourism do not seem to mix well with fish farming (Munro 1990).

Much has still to be learned about the long-term effects of aquaculture. Existing regulations have been established empirically based on the quantities of harvested fish per unit of area. Criteria for the maximum permitted amounts of solids and chemical compounds leaving and entering aquaculture installations are still being discussed (see below: Aquaculture legislation and conservation of freshwater ecosystems). When rivers carry farm wastes into lakes, this must be included in the overall calculations of their carrying capacity for aquaculture. It is very difficult to collect all the necessary information for this, particularly in developing countries.

Intensive Freshwater Aquaculture

Intensive aquaculture in the majority of developing countries is largely aimed at export markets, although tourism-related consumption may generate some seasonal demand. Intensive freshwater aquaculture in many developing countries is only just starting to expand; for example, trout and smolt production in salmonid culture (*Oncorhynchus mykiss, O. kisutch, O. tschawytscha, O. masou, Salmo salar*) in Chile (Munro 1990) and tilapia culture (*Oreochromis aureus* x *O. hornorum*) in Costa Rica (Martínez-Palacios et al. 1989), grass carp (*Ctenopharyngodon idella*) and eel (*Anguilla japonica*) in China (FAO 1983), snakehead (*Channa striatus*), walking catfish (*Clarias* spp.) and giant freshwater prawn (*Macrobrachium rosenbergii*) in Thailand (Boonyaratpalin and Akiyama 1989).

The nature and extent of the environmental impact of intensive freshwater aquaculture, drawn mainly from experience by industrialized countries (Alabaster

1982; EIFAC 1988; Pursiainen 1988; EIFAC 1990), depend on site selection and engineering design of aquaculture installations; selection of appropriate species and breeds; intensity of culture systems; quality of farm management and husbandry (disease precautions, stress avoidance, feed formulation, feeding strategies and water quality); disposal and treatment effluents; uses of waters affected by farm effluents; and the sensitivity of these recipient waters to farm effluents.

The freshwater environment can be affected by the release from fish farms of uneaten food; feces and dissolved excretory products; high microbial loads, parasites, disease organisms and vectors; and aquaculture chemicals, such as anesthetics, disinfectants, biocides, food additives, drugs for disease prophylaxis/treatment and inorganic and organic fertilizers. The chemistry of the recipient waterbodies and their bottom sediments can be changed by an increase in suspended solids, by biochemical and chemical oxygen demands, and by increased nutrient loading, particularly nitrogen and phosphorus. Thus, quantitative and qualitative changes in the biota (bacteria, protozoa, plankton, benthos and fish) of recipient waterbodies are very likely. Nutrient and organic enrichment may lead to local eutrophication and hypoxia or anoxia (see also Beveridge 1984), although the fertilization of some oligotrophic waters may increase fish production. Wild fish populations may be threatened by diseases and parasites emanating from high levels of infection in aquaculture installations. Aquaculture chemicals and their residues may cause sublethal and lethal effects on wild aquatic organisms, depending on their potential for bioaccumulation, their toxicity and characteristics of physicochemical persistence. The self-purification capacity of recipient waters may be diminished by antibiotics that inhibit microbial growth. Also, the excessive use of drugs may generate drug-resistant pathogens.

Such impacts on rivers and freshwater lakes also depend on the residence time and amount of water flowing through the aquaculture installations and on the kinetics of the recipient waters. Here, for instance, seasonal changes in river flow, lake flushing and water space around cages may well influence the dilution and distribution of contaminated waters.

Inland fisheries can be affected by intensive freshwater aquaculture. Abstraction of water, diversion of watercourses, dam and pond construction, and setting up of fishpens and cages in openwater areas, can have serious implications for aquatic wild life. The feeding and breeding habitats of many species could be disturbed. With aquatic life cycles being disrupted and recruitment reduced, overall productivity may be lowered (Dunn 1989).

Fish species with good characteristics for farming and stocking and good marketability have been and will continue to be introduced to new habitats. However, introduced exotic species and genetically modified breeds may alter and impoverish local aquatic biodiversity and genetic resources, as they may affect endemic species via competition, predation, destruction of habitats, transmission of parasites and diseases, and interspecific breeding (Welcomme 1988).

Related concerns and recommendations on the conservation and utilization of aquatic genetic resources have been formulated at various expert consultations dealing with genetic resources of fish, stock enhancement and aquaculture genetics (FAO/UNEP 1981; EIFAC 1982; Chevassus and Coche 1986). Joint efforts by working parties on stock enhancement and introductions of the European Inland Fisheries Advisory Commission of FAO (EIFAC) and the working group on introductions and transfers of marine organisms of the

International Council for the Exploration of the Sea (ICES) led to the formulation of related Codes of Practice and protocols (EIFAC 1984; FAO 1986; Turner 1988), which have been adopted by EIFAC member countries and which have been reviewed subsequently by FAO regional fisheries commissions such as the Commission for Inland Fisheries of Latin America and the Caribbean (COPESCAL), the Committee for Inland Fisheries of Africa (CIFA) and the Indo-Pacific Fishery Commission (IPFC) (see also IPFC 1988a, 1988b; FAO 1990; FAO 1991). Many countries have enacted legislation to regulate and control the movement of eggs, larvae/juveniles and adult stages of exotic fish species, often combined with compulsory certification of stocks to be free of certain diseases and banning of all movements of diseased stocks (Van Houtte et al. 1989).

Good husbandry, particularly efficient use of feeds and fertilizers, is essential to minimize the harmful effects of effluents from fish farms. Feed formulation is of great importance. Feeding rates can be optimized in order to avoid overfeeding (New 1987). Poor processing and storage of feed results in losses and deteriorative changes, including microbial contamination and content of harmful substances, e.g., solvent residues, aflatoxin, botulinum toxin, therapeutic drugs, etc. (Tacon 1987). Extruded diets may help to improve feed quality and digestibility thus reducing pollution, but the extrusion technology is costly and complex (Clarke 1990).

Fish farm effluents can also be treated to reduce their impact. However, high flow rates and dilute concentrations of pollutants in such effluent pose problems. Treatment facilities must be efficient, yet economically feasible to install and to operate. Treatment technologies for intensive aquaculture are being developed in industrialized countries based on sedimentation, decantation, biological oxidation and filtration (Petit and Maurel 1983). Often these techniques are designed for "high-tech" systems (Mäkinen et al. 1988). Economic constraints in production and operating costs often make the treatment of fish farm wastes difficult to support (Muir 1982), particularly in developing countries.

Integration of aquaculture with other activities (agriculture, industrial and urban use of water) is likely to be the most effective means of development, by sharing water use or enhancing its value sufficiently to allow investment in improved water supply or treatment (Muir and Beveridge 1987).

Aquaculture Legislation and Conservation of Freshwater Ecosystems

According to Howarth (1990), aquaculture ought to be environmentally regulated for its own interest. Aquaculture depends upon a good aquatic environment. It is particularly vulnerable to excessive abstraction and water contamination from a range of industrial, agricultural and domestic sources. However, aquaculture is also susceptible to risks of self-pollution. Hence, its own interests justify measures directed towards the regulation of water abstraction, and the prevention of unacceptable contamination, including pollution and emission of potentially harmful substances from fish farms. Likewise, the restriction of fish movements is sometimes required to avoid the spread of diseases between farmed populations, and from farmed populations into the wild and vice versa.

Legislative and administrative measures aiming at the environmental compatibility of the various aquaculture practices should be considered within the broader legislative context governing aquaculture. Most developing countries have little or no aquaculture-specific legislation purposely designed to protect or allow this activity. However, many aquaculturists must cope with complex laws and regulations on land tenure, waste use, environment protection, pollution prevention, public health and fisheries in general. Few of these are specifically drafted to promote or regulate aquaculture, and confusion, conflicts and overlapping exist (Van Houtte et al. 1989).

Constructive and adaptive concepts on regulations are needed in order to avoid obstacles to aquaculture development and, equally, to ensure that the aquatic environment is adequately protected. Apparent over-regulation and legal uncertainties can, however, hamper aquaculture development, by creating significant barriers to the establishment or the continued operation of aquafarms.

Various preventative and remedial measures for controlling and managing the environmental impact of aquaculture have been developed and applied (Van Houtte et al. 1989; see also McCoy 1989; Bye 1990; Quiney 1990; EIFAC 1992):

1) When planning the use of land and water resources the zoning of areas for aquaculture purposes should be included. Also, once protected areas and waterbodies such as parks and nature reserves are established, it should be clearly stated where and under which conditions aquaculture practices would be permitted, if at all.
2) In some countries, e.g., Venezuela, Mexico, Philippines and France, environmental impact assessment studies on the potential effects of the proposed aquaculture operation are required prior to the authorization for the installation of a fish farm.
3) In France, activities/installations may be termed and classified as environmentally critical undertakings, being subject to special declaration or authorization procedures.
4) The installation of effluent quality control equipment or water discharge treatment facilities is being promoted through fiscal incentives like direct subsidies or tax deductions and exemptions. Charges or taxes on polluting effluents also exist, e.g., in Poland, Hungary and France.
5) The introduction of pollutants into freshwater ecosystems through aquaculture or other industries may be regulated by setting quantitative and qualitative limits to the waste waters discharged (FWPCA 1968). Also, in order to meet these limits, the treatment of effluents prior to discharge is often required.

In general, regulatory standards on the water quality of effluents and the recipient waterbodies have been adopted to meet scientifically derived water quality criteria (FWPCA 1968; Sprague 1976; EEC 1978; Alabaster and Lloyd 1982). Both standards and criteria are determined according to the choice of water quality objectives formulated. The main advantage of the water quality objectives approach is that standards can be set appropriate to particular uses of the water resources (GESAMP 1986), but the application and usefulness of water quality standards have also been increasingly questioned by scientists and decisionmakers (see also GESAMP 1991). This approach also neglects the effects of aquatic sediments, bioaccumulation in the food chain, the impact and interactions of multiple sources and the overall load and the persistence characteristics of potentially harmful substances.

There are now modern conceptual frameworks for environmental management and pollution control such as the Environmental Capacity Approach (GESAMP 1986, 1991) including Hazard Assessment methodology (Bro-Rasmussen and Christiansen 1984; Landner 1988, 1989). The concept of environmental capacity, originally proposed by Cairns (1977), is based on the definition of the assimilative capacity, which is defined as the "ability of receiving system or ecosystem to cope with certain concentrations or levels of waste discharges without suffering any significant deleterious effects" (see also Cairns 1989). The environmental capacity approach works well as an interactive environmental management strategy. Other traditionally used complex strategies, based on environmental quality objectives or simple but readily enforceable strategies, such as those based on uniform emission standards, maximum allowable concentrations in effluent, the black/grey/white lists (Hellawell 1986) or the application of principles of best practicable means available, are considered as simple components of this adaptive, interactive strategy (GESAMP 1986, 1991).

This scientific approach requires technical and socioeconomic inputs as parallel, interactive and complementary activities in decisionmaking in integrated, environmentally compatible, development planning. It emphasizes the objectivity and independence of technical inputs and their influences on decisions related to socioeconomic feasibility. It also emphasizes that the acceptability of environmental impact rests on much more than political considerations. Such acceptability can be determined scientifically, assuming that the environmental capacity can be quantified. Once the environmental capacity of a given substance is determined, it can be apportioned for various resource uses and needs. It is also important to recognize that many ecosystems do have the potential to recover from pollution, provided that corrective or remedial measures are implemented.

The methodology for the assessment of the environmental capacity, which is site- and contaminant-specific, uses critical pathway analysis for both conservative and nonconservative contaminants and establishment of environmental quality objectives, criteria and standards. Faced with the inevitability of several sources of uncertainties in real situations, a probabilistic approach is used as an alternative to deterministic analysis. The methodology recommended (GESAMP 1986) consists of three decision stages. Socioeconomic goals (priorities and objectives) are assessed in the planning stage, considering present and future use of resources. In the preliminary scientific assessment stage, the environmental capacity is derived and quantified, resulting in the setup of allowable inputs. Finally, monitoring provides a continuous test of whether the environmental capacity is balanced, exceeded or underutilized. Consequently, adaptation measures may be required.

Within the environmental capacity approach, hazard assessment is a key scientific tool for predicting possible adverse effects of the discharge of pollutants. It is based on the relationship of the expected environmental concentration of a chemical substance (to which target organisms are potentially exposed) and the toxicological properties of the substance, i.e., the predicted concentrations with potential/possible adverse biological effect (Cairns et al. 1978). The prediction of the environmental concentration starts with the determination of exposure-related data (Landner 1988), which refer to the rate of chemical substance input, the properties of the substance and the environment. The persistence and the distribution of the substance is evaluated from data on physicochemical characteristics, biogeochemical behavior, biodegradability, bioaccumulation potential and bioavailability. Biological effects are predicted on the basis of acute and chronic toxicity studies or are calculated on the basis of quantitative stucture activity relationships (QSARs) (Könemann 1981; Boudou and Ribeyre 1989; Halfon 1989).

Regulatory measures may then be adopted upon comparison of the predicted environmental concentration (in water, sediments and organisms) and the information on lowest concentration where adverse biological effects can be expected.

As for pollution control of African waters, for which ecotoxicological data are scarce, Biney et al. (1987) list the following strategy options for the management of polluting discharges:

1) Limitation of the effluent by means of *rigid* effluent standards, both with chemical concentration limits and/or with a toxicological limit derived by simple acute toxicity tests on effluent. However, the specific characteristics of the receiving waterbody are not considered.
2) Limitation of the effluent by *flexible* standards. Here, the limits are calculated in order to maintain water quality criteria in a *specific* waterbody. Also, in this case, the limit can be defined as a threshold of the chemical and/or its toxic effects.
3) For some chemical substances it is scientifically unsound and insufficient for environmental protection to set up objectives, criteria and regulations for water alone. The classical case is mercury (Moore and Ramamoorthy 1984). In such cases, it is necessary to indicate objectives or criteria for another environmental compartment (e.g., sediments and/or fish).
4) In some cases, where a species is shown to be particularly sensitive to certain substances (e.g., crustaceans to pesticides), an "indicator species"-oriented management strategy has to be preferred to the water quality criteria approach (Hellawell 1986).
5) The classification of chemical substances in use in a country into "black", "grey" and "white" lists can be of help (Hellawell 1986), especially in the framework of a hazard assessment approach to water quality control, i.e., the comparison of predicted environmental exposure with available toxicity data. This does not necessarily mean that blacklisted chemicals are to be totally banned, but that they should be used only under certain conditions and strict controls.

In summarizing, it is emphasized that the particular role of aquaculture in utilizing land and water resources for food production calls for an integrative and flexible environmental legislation which is enforceable, effective and adaptive to the socioeconomic conditions and development needs in local communities, particularly as prevailing in developing countries. Modern scientific methodologies and conceptual frameworks for the assessment and prediction of environmental impact are being developed and should be applied in environmental management and regulation of human activities, including aquaculture.

Institutional Aspects

During the last decades there has been an increasing awareness that environmental protection cannot succeed as an isolated activity. Biswas (1978) stressed that planning and management of water development projects should be such that the economic benefits are maximized without causing serious impacts on the environment. In addressing preventative control of water pollution in developing countries, Diamant (1978) emphasized the need for simultaneous planning of water supply and wastewater disposal projects, for water pollution control legislation and for river basin authorities to be established. Prioritizing community water supply and water for agriculture, the 1977 UN Water Conference recommended integrated planning of water management (Falkenmark 1977). At the 1981 UN Interregional Meeting of International River Organizations, there was general agreement that environmental considerations have to be included in the development of shared water resources, but not on the degree of detail, nor on the weight that should be given to environmental factors; and, there was a clear impatience among participants from developing countries to get on with development while identifying positive as well as negative aspects of projects (United Nations 1983). Effective environmental management must be inseparable from land and water management and pursued in harmony with socioeconomic interests in the catchment zone (Eren 1977; Singh 1977; Lundqvist et al. 1985).

Can one local government organization represent all interests when planning development in a catchment area or river basin (Reynolds 1985) and how can resource and environmental conflicts be managed (Bateld 1985)? What kind of planning methodology is best applied (see also Biswas 1985; Pantulu 1985; United Nations 1988)? In attempting answers to these interrelated questions, it is clear that two conditions must be met if planning is to be useful: there must be the need for desirable changes or for actions to prevent undesirable changes, and there must be the political will and ability, including financial capacity, to put the plan into effect.

Political will in developing countries may be weaker than in developed countries. Indicative planning is conditional on the total or partial approval of the different social, economic and political actors involved. Each of these groups has a different rationality and, additionally, in most developing countries, social and economic inequities constitute serious obstacles to participation and consensus in planning and decisionmaking. Moreover, some of these actors may represent foreign interests with little concern for the long-term consequences of their actions. For example, Latin America has seen, during the last three decades, the formulation by governments of sound plans which, however, were never applied. Likewise, Satia (1986) states that "many plans bear no evidence of internal initiative within the country but are rather fruits of external pressure". Taking the case of overfishing and competition between capture and culture fisheries in Laguna de Bay, Philippines, Smith (1982) also advocates a participatory and more decentralized fisheries management approach.

In short, it is important to recognize that existing planning deficiencies, lack of coordination and difficulties in the implementation of management plans constitute severe constraints to many sectoral and multisectoral development efforts, which undoubtedly need to be considered when addressing the environmental implications of aquaculture development.

Legal and administrative tools to create or enforce rational systems for water management, land use or fisheries and aquaculture development are frequently proposed. Experience has shown that in some cases such regulatory measures can act as incentives or disincentives. In many other cases, however, these measures have been difficult to apply or to control by relevant authorities, often due to inappropriate institutional infrastructure. Moreover, these measures are often not specific and therefore not effective in eliminating undesirable uses of land and water resources. It is therefore emphasized that there is a growing need to integrate this experience on both the formulation of adequate regulatory measures and adaptive institutionalization into ongoing efforts of establishing rules and institutions aiming (sometimes exclusively) at environmental conservation.

There are guidelines available, e.g., on the development of inland fisheries and aquaculture in the context of multiple use of resources (Bernacsek 1984; Alabaster 1985; Petr 1985; Scudder and Conelly 1985; Vanderpuye 1985; Welcomme 1985; Baluyut 1986; Sreenivasan 1986; Dunn 1989), on land use planning (FAO 1989), and on rural area development planning (Bendavid-Val 1990). However, it is stressed that solutions and approaches will be required that are country- and basin area-specific and, in particular, also site-specific when it comes to aquaculture development, depending on the environmental and socioeconomic circumstances encountered.

Last not least, it is recalled that often it is the discrepancy in distribution of economic power among the various social groups in a country which leads to environmentally degrading practices. On one side, it is poverty and marginalization which forces subsistence farmers to unsustainable land use (see also Lundqvist et al. 1985). On the other side, it is considerable short-term gains from trading in international markets which make national and foreign entrepreneurs embark on environmentally unacceptable land and water usage. Relevant national policies need to be adjusted accordingly.

Conclusions

Freshwater aquaculture development in Latin America, Africa and Asia needs the benefit of internal and external experience for preventing environmental damage and for avoiding harmful effects of aquatic pollution and physical degradation on aquaculture resources. Environmental compatibility, both in socioeconomic/sociocultural and biophysical terms, needs to be emphasized in aquaculture development efforts by public and private initiatives.

Institutional, economic, social and political factors in the aquaculture planning process are as important or more important than biotechnical factors, particularly in developing countries. This is especially true for those types of aquaculture that have a social orientation and therefore a strong governmental intervention. Sustainability is also of particular importance for socially oriented aquaculture development projects, and increased efforts have to be made to improve efficiency in the implementation of such projects and to reduce successfully their negative impacts on humans and the environment.

Small-scale aquaculture practices need to be better integrated with other rural activities in agriculture, forestry and capture fisheries. Integrated agriculture-aquaculture farming systems show a good potential in this respect. However, these systems still have to be studied in greater depth in terms of financial and economic viability, diversity and site-specificity.

Similarly, it is important to guarantee the success of culture-based fisheries development in lakes and reservoirs. Strategies to compensate for the loss of aquatic fauna due to physical obstructions of rivers are directly linked to important environmental issues, such as the introduction of exotic species, the spread of diseases and the loss of genetic diversity.

Furthermore, catchment area management approaches should also be considered, if not promoted, where appropriate, when formulating policy recommendations covering institutional and regulatory frameworks for both inland fisheries and aquaculture development as well as for environmental conservation and rehabilitation of freshwater ecosystems.

Only few developing countries are

enacting legislation supporting or even dealing with aquaculture development, which, instead, is often being constrained by regulations on land tenure, water use and public health. Additional legislative measures for prevention and control of aquatic pollution will have to be considered by aquaculturists, also for their own interests. Constructive and adaptive aquaculture-related regulations need to be implemented in order to avoid obstacles in socioeconomic growth in rural communities and to ensure that resources in the aquatic environment are properly managed and protected.

The potential of intensive freshwater aquaculture to degrade the natural environment is considerable, particularly in view of current trends of intensification and rapid expansion of the industry. In this respect, environmental management options will include improvements in farming performance (especially related to feeds and feeding, stocking densities and water quality management, disease prevention and chemical usage) and in the selection of sites and species, installation of effluent treatment facilities and, where required, strict enforcement of environmental regulations specific to intensive freshwater aquaculture.

Acknowledgements

The authors would like to sincerely thank Dr. M. Beveridge (Institute of Aquaculture, Stirling), Dr. R.S.V. Pullin (ICLARM), Dr. T. Petr and Dr. H. Naeve (FAO) for their support through fruitful conceptual discussions, technical comments and editorial amendments.

References

Alabaster, J.S. 1982. Report of the EIFAC workshop on Fish-Farm Effluents. Silkeborg, Denmark, 26-28 May 1981. EIFAC Tech. Pap. 41, 166 p. FAO, Rome.

Alabaster, J.S., Editor. 1985. Habitat modification and freshwater fisheries. Proceedings of a Symposium of the European Inland Fisheries Advisory Commission (EIFAC). FAO, Rome, and Butterworths, London.

Alabaster, J.S. and R. Lloyd. 1982. Water quality criteria for freshwater fish. 2nd edition. FAO, Rome, and Butterworths, London.

Baluyut, E.A. 1986. Planning for inland fisheries under constraints from other uses of land and water resources: general considerations and the Philippines. FAO Fish. Circ. 798, 30 p.

Bateld, K. 1985. Environmental conflicts: structure and management, p. 209-218. *In* J. Lundqvist, U. Lohm and M. Falkenmark (eds.) Strategies for river basin management (environmental integration of land and water in a river basin). D. Reidel Publishing Company, Dordrecht, Holland.

Bendavid-Val, A. 1990. Rural area development planning: a review and synthesis of approaches. FAO/ESP/TMAP 21, 289 p. FAO, Rome.

Bernacsek, G.M. 1984. Guidelines for dam design and operation to optimize fish production in impounded river basins (based on a review of the ecological effects of large dams in Africa). CIFA Tech. Pap. 11, 98 p. FAO, Rome.

Beveridge, M.C.M. 1984. Cage and pen fish farming. Carrying capacity models and environmental impact. FAO Fish. Tech. Pap. 255, 131 p.

Biney, C., D. Calamari, H. Naeve, T.W. Maembe, B. Nyagakeni and M.A.H. Saad. 1987. Scientific bases for pollution control in African inland waters. Chem. Ecol. 3:49-74.

Biswas, A.K. 1978. Water development and environment, p. 21-32. *In* E.A.R. Ouano, B.N. Lohano and N.C. Thanh (eds.) Proceedings of the AIT/CIDA International Conference on Water Pollution Control in Developing Countries, Bangkok, Thailand. 21-25 February 1978, Asian Institute of Technology, Bangkok. Vol. II, 264 p.

Biswas, A.K. 1985. Methodology for monitoring and evaluation of integrated land and water development, p. 49-61. *In* J. Lundqvist, U. Lohm and M. Falkenmark (eds.) Strategies for river basin management (environmental integration of land and water in a river basin). D. Reidel Publishing Company, Dordrecht, Holland.

Boonyaratpalin, M. and D.M. Akiyama. 1989. The aquaculture industry in South-East Asia, p. 41-56. *In* Proceedings of the Third International Symposium on Feeding and Nutrition in Fish. Toba, Japan.

Boudou, A. and F. Ribeyre. 1989. Fundamental concepts in aquatic ecotoxicology, p. 35-76. *In* A. Boudou and F. Ribeyre (eds.) Aquatic ecotoxicology: fundamental concepts and methodologies. Vol. I. CRC Press, Florida.

Bro-Rasmussen, F. and K. Christiansen. 1984. Hazard assessment - a summary of analysis and integrated evaluation of exposure and potential effects from toxic environmental chemicals. Ecol. Modelling 22(1-4): 67-84.

Bye, V.J. 1990. Legal, political and social constraints in aquaculture, p. 123-144. *In* N. De Pauw and R. Billard (eds.) Aquaculture Europe 1989 - business joins science. Spec. Publ. No. 12. European Aquaculture Society, Bredene, Belgium.

Cairns, J. 1977. Aquatic ecosystem assimilative capacity. Fisheries 2(2):5-7.

Cairns, J. 1989. Applied ecotoxiology and methodology, p. 275-290. *In* A. Boudou and F. Ribeyre (eds.) Aquatic ecotoxicology: fundamental concepts and methodologies. Vol. II. CRC Press, Florida.

Cairns, J., K.L. Dickson and A.W. Maki, Editors. 1978. Estimating the hazard of chemical substances to aquatic life. Summary and conclusions. STP 657. American Society for Testing and Materials, Philadelphia.

Chevassus, B. and A.G. Coche, Editors. 1986. Report of the Symposium on Selection, Hybridization and Genetic Engineering in Aquaculture of Fish and Shellfish for Consumption and Stocking. Bordeaux, France, 27-30 May 1986. EIFAC Tech. Pap. 50, 52 p. FAO, Rome.

Clarke, S. 1990. Extruded diets and pollution reduction. Fish Farmer 4(1):95.

Diamant, B.Z. 1978. Preventive control of water pollution in developing countries, p. 9-15. *In* E.A.R. Ouano, B.N. Lohano and N.C. Thanh (eds.) Proceedings of the AIT/CIDA International Conference on Water Pollution Control in Developing Countries, 21-25 February 1978, Bangkok, Thailand. Asian Institute of Technology, Bangkok. Vol. I, 744 p.

Dunn, I.G. 1989. Development of inland fisheries under constraints from other uses of land and water resources: guidelines for planners. FAO Fish. Circ. 826, 53 p. FAO, Rome.

Edwards, P., R.S.V. Pullin and J.A. Gartner. 1988. Research and education for the development of integrated crop-livestock fish systems in the tropics. ICLARM Stud. Rev. 16, 53 p.

EEC. 1978. European Economic Community Council Directive of 18 July 1978 on the quality of freshwaters needing protection or improvement in order to protect fish life. Official J. Europ. Comm. 21(L222): 1-10.

EIFAC. 1982. Report of the Symposium on Stock Enhancement in the Management of Freshwater Fisheries. Budapest, Hungary, 31 May-2 June 1982 in conjunction with the Twelfth Session of the European Inland Fisheries Advisory Commission. EIFAC Tech. Pap. 42, 43 p. FAO, Rome.

EIFAC. 1984. Report of the EIFAC Working Party on Stock Enhancement, 16-19 May 1984, Hamburg, Germany. EIFAC Tech. Pap. 44, 22 p. FAO, Rome.

EIFAC. 1988. Report of the First Session of the EIFAC Working Party on Fish-Farm Effluents, 29 May-1 June 1987, The Hague, The Netherlands. EIFAC/XV/88/Inf. 19. FAO, Rome.

EIFAC. 1990. Report of the Second Session of the EIFAC Working Party on Fish-Farm Effluents, 10-12 October 1988, Verona, Italy, p. 69-77. *In* Report of the Sixteenth Session of the European Inland Fisheries Advisory Commission, 15-22 May 1990, Prague, Czechoslovakia. FAO Fish. Rep. 446.

EIFAC. 1992. Report of the Third Session of the EIFAC Working Party on Fish-Farm Effluents, 28-31 May 1991, Tihany, Hungary. FAO, Rome.

Engle, C.R. 1986. Criterios para la preparación y evaluación de proyectos de acuicultura de subsistencia. Serie RLAC/85/44-PES-8. 79 p. FAO Oficina Regional para América Latina y el Caribe. Santiago, Chile. (In Spanish).

Eren, T. 1977. The integrated watershed approach for development project formulation, p. 9-14. *In* S.H. Kunkle and J.L.Thames (eds.) Guidelines for watershed management. FAO Conservation Guide 1, 293 p. FAO, Rome.

Falkenmark, M. 1977. UN Water Conference: Agreement on Goals and Action Plan. Ambio 6(4): 222-227.

FAO. 1983. Freshwater aquaculture development in China. Report of the FAO/UNDP study tour organized for French-speaking African countries. FAO Fish. Tech. Pap. 215, 125 p.

FAO. 1986. Report of the Fourteenth Session of the European Inland Fisheries Advisory Commission, 27 May-3 June 1986, Bordeaux, France. FAO Fish. Rep. 364, 71 p.

FAO. 1989. Guidelines for land use planning. (By the FAO inter-departmental working group on land use planning). 121 p. FAO, Rome.

FAO. 1990. Informe de la quinta reunión del grupo de trabajo sobre recursos pesqueros de la Comisión de Pesca Continental para América Latina y el Caribe (COPESCAL), 24-28 septiembre 1990, Santiago, Chile. FAO Fish. Rep. 442, 16 p. (In Spanish).

FAO. 1991. Report of the Eighth Session of the Committee on Inland Fisheries of Africa (CIFA), 21-25 October 1990, Cairo, Egypt. FAO Fish. Rep. 448.

FAO-SIDA. 1987. Technical consultation on aquaculture in rural development. Project GCP/INT/436/SWE/REP/2, 84 p. Aquaculture for Local Community Development Programme, 27-30 October 1987, Lusaka, Zambia. FAO, Rome.

FAO-UNDP-Norwegian Ministry of Development Cooperation. 1987. Thematic evaluation of aquaculture. FAO, Rome. 129 p.

FAO/UNEP. 1981. Conservation of the genetic resources of fish: problems and recommendations.

Report of the Expert Consultation on the Genetic Resources of Fish, 9-13 June 1980, Rome. FAO Fish. Tech. Pap. 217, 43 p.

FWPCA. 1968. Water quality criteria. Report of the National Technical Advisory Committee to the Secretary of the Interior. Federal Water Pollution Control Administration (FWPCA), Washington, DC.

GESAMP. 1986. Environmental capacity. An approach to marine pollution prevention. Rep. Stud. GESAMP 30, 49 p. IMO/FAO/UNESCO/WMO/WHO/IAEA/UN/UNEP Group of Experts on the Scientific Aspects of Marine Pollution. FAO, Rome.

GESAMP. 1991. Global strategies for marine environmental protection. Rep. Stud. GESAMP 45, 36 p. IMO/FAO/UNESCO/WMO/WHO/IAEA/UN/UNEP Joint Group of Experts on the Scientific Aspects of Marine Pollution. IMO, London.

Godoy, M.P. 1985. Aquicultura actividade multidisciplinar: escadas o outras facilidades para passagens de peixes, estaçoes de piscicultura. Florianópolis, Brazil. (In Portuguese).

Halfon, E. 1989. Mathematical models for predicting the fate of contaminants in freshwater-ecosystems, p. 257-274. *In* A. Boudou and F.Ribeyre (eds.) Aquatic ecotoxicology: fundamental concepts and methodologies. Vol. II. CRC Press, Florida.

Hellawell, J.M. 1986. Biological indicators of freshwater pollution and environmental management. (Pollution monitoring series). Elsevier Applied Science Publishers, England.

Howarth, W. 1990. The law of aquaculture: the law relating to the farming of fish and shellfish in Britain. Fishing News Books Ltd., Farnham, Surrey.

IPFC. 1988a. Report of the Seventh Session of the Indo-Pacific Fishery Commission Working Party of Experts on Aquaculture, 1-6 August 1988, Bangkok, Thailand. FAO Fish. Rep. 411, 37 p.

IPFC. 1988b. Report of the Fourth Session of the Indo-Pacific Fishery Commission Working Party of Experts on Inland Fisheries, 8-14 September 1988, Kathmandu, Nepal, and Report of the Workshop on the Use of Cyprinids in the Fisheries Management of Larger Inland Water Bodies of the Indo-Pacific, 8-10 September 1988, Kathmandu, Nepal. FAO Fish. Rep. 405, 51 p.

Juarez, R. El aprovechamiento de los embalses en América Latina a través de la práctica de la acuícultura de repoblación, p. 3-31. *In* Manejo y explotación acuicola de embalses de agua dulce en América Latina. Documento de Campo No. 19. Proyecto Regional AQUILA GCP/RLA-075/ITA. Brasilia, Brazil. (In Spanish). (In press).

Könemann, H. 1981. Quantitative structure-activity relationships in fish toxicity studies. Part I: Relationship for 50 industrial pollutants. Toxicology (19):209-221.

Landner, L. 1988. Hazardous chemicals in the environment - some new approaches to advanced assessment. Ambio 17(6):360-366.

Landner, L., Editor. 1989. Chemicals in the aquatic environment: advanced hazard assessment. Springer, Berlin.

Lightfoot, C. 1990. Integration of aquaculture and agriculture: a route to sustainable farming systems. Naga, ICLARM Q. 13(1):9-12.

Lundqvist, J., U. Lohm and M. Falkenmark 1985. River basin strategy for coordinated land and water conservation: synthesis and conclusions, p. 5-17. *In* J. Lundqvist, U. Lohm and M. Falkenmark (eds.) Strategies for river basin management environmental integration of land and water in a river basin). D. Reidel Publishing Company, Dordrecht, Holland.

Mäkinen, T., S. Lindgren and P. Eskelinen. 1988. Sieving as an effluent treatment method for aquaculture. Aquacult. Eng. 7:367-377.

Marí-Diaz, A. Pesquerias derivadas de la acuicultura en aguas interiores de Cuba, p. 71-104. *In* Manejo y explotación acuícola de embalses de agua dulce en América Latina. Documento de Campo No. 19. Proyecto Regional AQUILA GCP/RLA-075/ITA. Brasilia, Brazil. (In Spanish). (In press).

Martínez-Espinosa, M. 1986. La acuícultura rural en América Latina: estudio de casos en Colombia, Costa Rica y El Salvador. Serie RLAC/87/PES-15, 61 p. FAO Oficina Regional para América Latina y el Caribe, Santiago, Chile. (In Spanish).

Martínez-Espinosa, M. 1990. Informe de la Reunión sobre acuicultura rural en América Latina y el Caribe, 24-26 Octubre 1989, Maracaibo, Venezuela. FAO Fish. Circ. 829, 39 p.

Martínez-Palacios, C.A., M.S. Chavez de Martinez and M.A. Olvera Novoa. 1989. La nutrición y la alimentación en la acuicultura de América Latina: una diagnosis. GCP/RLA/075/ITA. Documento de Campo No. 17. FAO, Rome (In Spanish).

McCoy, H.D. 1989. Commercial aquaculture zones. A legislative proposal. Aquacult. Mag. (Nov./Dec. 1989):39-46.

Moore, J.W. and S. Ramamoorthy. 1984. Heavy metals in natural waters: applied monitoring and impact assessment. Springer, New York.

Muir, J.F. 1982. Economic aspects of waste treatment in fish culture, p. 123-135. *In* J.S. Alabaster (ed.) Report of the EIFAC Workshop on Fish-Farm Effluents, 26-28 May 1981, Silkeborg, Denmark. EIFAC Tech. Pap. 41, 166 p. FAO, Rome.

Muir, J.F. and M.C.M. Beveridge. 1987. Water resources and aquaculture development, p. 321-324. *In* R.P. Lim, A.B. Viner, L.H.S. Lim and J.I. Furtado (eds.) Sustainable clean water. Proceedings of the Regional Workshop on Limnology and Water Resources Management in the Developing Countries of Asia and the Pacific, 29

November-5 December 1982, University of Malaya, Kuala Lumpur, Malaysia. Ergeb. Limnol./Adv. Limnol. 28.

Munro, A.L.S. 1990. Report to FAO on the management of salmonid resources of Chile. November 1989-January 1990. Project TCP/CHI/8851. FAO, Rome.

New, M.B. 1987. Feed and feeding of fish and shrimp - a manual on the preparation and presentation of compound feeds for shrimp and fish in aquaculture. ADCP/REP/87/26, 275 p. FAO, Rome.

Novoa, D. Manejo de embalses con fines piscicolas: informe Venezuela, p. 163-193. *In* Manejo y explotación acuícola de embalses de agua dulce en América Latina. Documento de Campo No. 19. Proyecto Regional AQUILA GCP/RLA-075/ITA. Brasilia, Brazil. (In Spanish). (In press).

Olmos, E. Situación actual y perspectivas de las pesquerias derivadas de la acuicultura en México, p. 105-162. *In* Manejo y explotación acuícola de embalses de agua dulce en América Latina. Documento de Campo No. 19. Proyecto Regional AQUILA GCP/RLA-075/ITA, Brasilia, Brazil. (In Spanish). (In press).

Pantulu, V.R. 1985. Ecosystem modelling of a river basin, p. 31-40. *In* J. Lundqvist, U. Lohm and M. Falkenmark (eds.) Strategies for river basin management (environmental integration of land and water in a river basin). D. Reidel Publishing Company, Dordrecht, Holland.

Petit, J. and P. Maurel. 1983. L'épuration des eaux en pisciculture d'eau douce: ses méthodes à travers le monde. Piscic. Fr. 72:21-30

Petr, T., Editor. 1985. Inland fisheries in multi-purpose river basin planning and development in tropical countries: three case studies. FAO Fish. Tech. Pap. 265, 166 p.

Pretto, R. 1989. La agro-acuicultura en Latinoamérica: antecedentes y perspectivas. Trabajo presentado en la Reunión sobre la Planificación de la Acuicultura en América Latina. FAO, Proyecto Regional AQUILA, 8-12 Mayo 1989, Caracas, Venezuela. 12 p. (In Spanish).

Pursiainen, M., Editor. 1988. National contributions on suspended solids from land-based fish farms. Paper presented at the First Session of the EIFAC Working Party on Fish-Farm Effluents, 29 May-1 June 1987, The Hague, The Netherlands. (Riista - ja kalataloudcn tutkimuslaitos, kalantutkimusosasto) Monistettuja julkasuja No. 74, 93 p. Helsinki, Finland.

Quiney, D. 1990. Réglementation et pollution piscicole. Le vent du Nord? Aqua Revue 32(Août-Septembre 1990):7-9.

Quirós, R. 1989. Structures assisting the migrations of non-salmonid fish: Latin America. COPESCAL Tech. Pap. 5, 41 p. FAO, Rome.

Reynolds, P.J. 1985. Ecosystem approaches to river basin planning, p. 41-48. *In* J. Lundqvist, U. Lohm and M. Falkenmark (eds.) Strategies for river basin management (environmental integration of land and water in a river basin). D. Reidel Publishing Company, Dordrecht, Holland.

Satia, N.P.B. 1986. Planning for aquaculture in Africa, p. 17-42. *In* J.L. Gaudet and D. Parker (eds.) Summary of proceedings and selected papers. Symposium on the Planning and Implementation of Fisheries Management and Development Programmes in Africa, 7-11 October 1985, Lusaka, Zambia. FAO Fish. Rep. 360, 155 p.

Scudder, T. and T. Conelly. 1985. Management systems for riverine fisheries. FAO Fish. Tech. Pap. 263, 85 p.

Singh, G. 1977. Watershed organizations and socio-economic factors, p. 263-270. *In* S.H. Kunkle and J.L. Thames (eds.) Guidelines for watershed management. FAO Conservation Guide 1. FAO, Rome.

Smith, I.R. 1982. Mismanagement of inland fisheries and some corrective measures, p. 88-100. *In* T. Petr (ed.) Summary report and selected papers presented at the IPFC Workshop on Inland Fisheries for Planners, 2-6 August 1982, Manila, Philippines. FAO Fish. Rep. 288, 191 p.

Smith, L.J. and S. Peterson, Editors. 1982. Aquaculture development in less developed countries: social, economic and political problems. Westview Press, Boulder, Colorado.

Sprague, J.B. 1976. Current status of sublethal tests of pollutants on aquatic organisms. J. Fish. Res. Board Can. (33):1988-1992.

Sreenivasan, A. 1986. Inland fisheries under constraints from other uses of land and water resources: Indian subcontinent and Sri Lanka. FAO Fish. Circ. 797, 68 p.

Studart Gurgel, J.J. Sobre el aprovechamiento pesquero de embalses construidos en el Brasil, especialmente en la región del semi-árido, p. 33-70. *In* Manejo y explotación acuícola de embalses de agua dulce en América Latina. Documento de Campo No. 19. Proyecto Regional AQUILA GCP/RLA-075/ITA. Brasilia, Brazil. (In Spanish). (In press).

Tacon, A.G.J. 1987. The nutrition and feeding of farmed fish and shrimp - a training manual. 2. Nutrient sources and composition. GCP/RLA/075/ITA Field Document 5. FAO, Brazil.

Turner, G.E. 1988. Codes of practice and manual of procedures for consideration of introductions and transfers of marine and freshwater organisms. EIFAC Occas. Pap. 23, 44 p. FAO, Rome.

United Nations. 1983. Experiences in the development and management of international river and lake basins. Proceedings of the United Nations Interregional Meeting of International River Organizations, 5-14 May 1981, Dakar, Sénégal. Natural Resources/Water Series No.

10. United Nations, New York. 424 p.
United Nations. 1988. Assessment of multiple objective water resources projects: approaches for developing countries. TCD/SEM.88/3; INT/84/R36, 125 p. United Nations, New York.
Van Houtte, A.R., N. Bonucci and W.R. Edeson. 1989. A preliminary review of selected legislation governing aquaculture. ADCP/REP/89/42, 81 p. FAO, Rome.
Vanderpuye, C.J. 1985. Evaluation guidelines for rational planning and management of tropical and subtropical inland fisheries under constraints from other uses of land and water resources: Africa. FAO Fish. Circ. 789, 44 p.
Vinjoy, M. 1988. Análisis cuantitativo del parasitismo de peces en ambalses. Boletín Técnico de Acuicultura No. 15. Cuba.
Welcomme, R.L. 1985. River fisheries. FAO Fish. Tech. Pap. 262, 333 p.
Welcomme, R.L. 1988. International introductions of inland aquatic species. FAO Fish. Tech. Pap. 294, 318 p.
Wijkström, U.N. 1986. Acuícultura y desarollo rural. Trabajo presentado en la 19a Conferencia Regional de la FAO para América Latina y el Caribe, 5-3 Agosto 1986, Barbados. 24 p. (In Spanish).

Discussion

ROSENTHAL: This paper clearly highlights three areas of concern: 1. the resource systems in which aquaculture takes place and the environmental quality therein; 2. the environmental conditions within the farming system itself; and 3. the effect that aquaculture activities can have on surrounding environments.

The concept of assimilative capacity is interesting but can be risky. For example, industrial operations often try to push waste receiving capacity to its limits and we are hardly able to set safe limits. For this reason, the various conventions concerned with the North Sea have controversially debated the options to define its assimilative capacity. ICES at present feels unable to adopt this approach because of the many interactions among environmental factors (natural and from pollution) about which we know little. Scientists can therefore best help society and its environmental regulatory authorities by documenting environmental change while politicians will have to decide what changes are acceptable to society. Politicians usually want scientists to set benchmark numbers, but this can be difficult and is often dangerous.

PULLIN: The paper is concerned almost exclusively with the freshwater environment. Freshwater is becoming a very limited resource. It is used for irrigation and domestic supply and wastewater disposal. Freshwater aquaculture may become less and less viable if it competes with rather than complements these uses, especially close to the huge and rapidly growing cities of some developing countries. For example, Laguna de Bay, the 90,000 ha freshwater lake adjacent to Metropolitan Manila and famous for its cage and pen aquaculture, will probably have to supply some of the city's drinking water in future.

ROSENTHAL: So again, we must consider the dynamic nature of these situations and the multiple use options involved when we assess sustainability.

BILIO: In African reservoirs that are used for drinking water supply, aquaculture is generally prohibited. Only fisheries are allowed. If the purpose of such reservoirs is for irrigation or power generation, however, both aquaculture and culture-based fisheries are possible. One prerequisite for such culture-based fisheries is to know in advance the characteristics of the catchment area (geology and limnology).

MARTÍNEZ-ESPINOSA: This assimilative capacity approach, although not yet fully formulated has a flexibility that no other approach has ever had. It is based on the situations of various cases. For example, if a trout farm effluent causes eutrophication in a freshwater body used for drinking purposes, this is serious. If, however, it goes almost directly into the sea, it is less serious. This is a site-specific approach. This illustrates why we are recommending it.

EDWARDS: You mentioned integration of intensive aquaculture with agriculture. We can also conceive using the effluents from intensive aquaculture in less intensive aquaculture, such as semi-intensive ponds containing filter-feeding fish - bearing in mind Dr. Pullin's comment about water scarcity. Of course, the other resources, especially land, and the farmers' and consumers' interest in these species have also to be there, but at least this seems theoretically feasible.

MARTÍNEZ-ESPINOSA: We must be careful though not to attempt to introduce too many new and demanding technologies into rural areas when trying to develop 'socially-oriented' aquaculture. Too many projects ignore this and bring in, for example, biogas and wind power technologies.

EDWARDS: I totally agree. Such technologies rarely get beyond the drawing board or prototype development in most rural development projects. I was referring specifically to the productive reuse of effluents from intensive farming systems. There are examples in Thailand of the reuse of effluents from intensive culture of walking catfish (*Clarias* spp.) in semi-intensive fishponds.

ROSENTHAL: It is clear that planning for fisheries and aquaculture in reservoirs must be done from the inception of the planning of the reservoir development, not as an afterthought. By analogy, the use of waste-heat from power stations must be planned as part of the whole development.

PULLIN: This planning for fisheries and aquaculture development in new reservoirs is now becoming more commonplace. A good example is the World Bank-funded project for the Saguling and Cirata reservoirs near Bandung, Indonesia[1]. The World Bank is now making this a matter of policy for new developments and ICLARM is advising the Bank on another development in Orissa, India.

---

[1]Costa-Pierce, B.A. and O. Soemarwoto, Editors. 1990. Reservoir fisheries and aquaculture for resettlement in Indonesia. ICLARM Tech. Rep. 23, 278 p.

Aquaculture and Conservation of Genetic Diversity

Stefano Cataudella
Dipartimento di Biologia
Università di Roma "Tor Vergata"
Via E. Carnevale, 00173 Rome, Italy

Donatella Crosetti
Central Institute for Marine Research (ICRAM)
Respighi 5, 00197 Rome, Italy

CATAUDELLA, S. and D. CROSETTI. 1993. Aquaculture and conservation of genetic diversity, p. 60-73. *In* R.S.V. Pullin, H. Rosenthal and J.L. Maclean (eds.) Environment and aquaculture in developing countries. ICLARM Conf. Proc. 31, 359 p.

Abstract

The history of aquaculture genetics and the effects of aquaculture on natural aquatic genetic resources are summarized. Modern approaches to fish genetic characterization are discussed. Fish gene pools and genetic impoverishment in fish populations are reviewed with reference to environmental change, the effects of capture fisheries and enhanced fisheries, introduction of exotic species and hybridization. The conservation of fish genetic resources is considered extremely important and approaches to *ex situ* conservation are discussed, together with examples of international, regional and national efforts.

Introduction

Agriculture and animal husbandry have been developed and intensified by sustained manipulation and modification of natural ecological balances. Wide knowledge of living organisms and terrestrial ecosystems has brought understanding of the effects of thousands of years of such manipulations and awareness that they define the place of humans in nature. During this process, however, genetic resources and potential have been irretrievably lost. These losses have ethical and esthetic aspects as well as causing an actual diminution of production potential.

Living aquatic resources, although exploited by man since ancient times, have so far suffered less than terrestrial resources in this context. However, the difficulty of access to aquatic environments and the size of the hydrosphere have not prevented humans from overexploiting fish stocks and degrading many aquatic environments through the impact of activities on land.

Conservation of genetic resources in relation to aquaculture must take into account the two main approaches to aquaculture: zootechnical and ecological. The former follows the principles of intensive rearing, using breeds adapted to the aims of farmers, and invites genetic manipulation to develop better breeds. The latter uses seed similar to wildtypes to enhance fisheries. For both, the maintenance of good performance requires thorough knowledge of the characteristics of the original populations and of the changes induced by human intervention (choice of broodstock, techniques used for captive breeding and larval rearing, and genetic changes due to stock management, e.g., inbreeding). Without this knowledge there could be reduced performance due to genetic deterioration from the original populations. Moreover, the availability of the original populations of selected and manipulated lines is vital to permit the recovery of genes lost by intentional or unintentional manipulations.

The objective of aquaculture is to farm fish for food and income and to keep such farming options open for future generations. A good understanding of genetic variation within farmed aquatic populations is a basic prerequisite for efficient long-term management.

The application of genetics to aquaculture and realization of the need for conservation of aquatic genetic resources (which are largely nondomesticated) are very recent (Ryman 1991). Maintenance of intraspecific genetic variability has been long neglected in the management of natural populations. Indeed, farmed aquatic species were formerly considered to be more or less genetically homogeneous, with a large amount of variation caused by environmental factors. The existence of high levels of inter- and intrapopulation genetic variation in fish was not widely recognized until the 1960s, with the introduction of biochemical analyses in population genetics (mainly electrophoresis of enzymes) which led to the detection of previously unrecognized genetic variation. These techniques are now used very widely (e.g., Ferguson and Thorpe 1991; Utter 1991; Whitmore 1991). It is very important to investigate the consequences of breeding practices on the genetic structure and performance of cultured stocks. Macaranas and Fujio (1990) have suggested investigations on genetic variability and differences at loci controlling biochemical traits, genetic differences at loci controlling performance traits and the correlation between these.

Parallel to the development of aquaculture has emerged the need to preserve natural living aquatic resources (inter- and intrapopulation variation) and to establish conservation programs at local, national, regional and international levels. The widespread transfers of aquatic species and the enormous impact of human interventions on aquatic genetic resources (from increasing fishing efforts and intensification of aquaculture practices) are now becoming widely discussed in relation to conservation (Hindar et al. 1991). However, understanding of how to conserve aquatic genetic resources remains weak. Most assumptions derive from experience with other organisms and theoretical speculation (Frankel and Soulé 1981; Nelson and Soulé 1987). The conservation of cultured fish genetic resources can use similar approaches to those applied for the conservation of genetic resources of farm animals (Siler et al. 1984). There has been increasing interest in conservation of fish genetic resources in the last decade as evidenced by numerous national and international meetings (e.g., FAO/UNEP 1981; Ryman 1981; STOCS 1981; Pullin 1988; Das and Jhingran 1989; Prace Vurh Vodnany 1989; Billington and Hebert 1991; ICLARM 1992). An expert consultation on "Utilization and conservation of aquatic genetic resources" was organized by FAO in November 1992 in Rome.

There are at present two main fields in aquaculture genetic research: genetic characterization of farmed stocks and biotechnology. Although there is an extensive literature on the theory of breeding plans and their application to livestock and plant production, little attention has yet been given to this in aquaculture (Gall 1990).

Manipulation of genotypes for production purposes and conservation of genetic resources in nondomesticated populations should be complementary. Human activities have strong impacts on aquatic genetic resources. Artificial selection may be intentional (through fisheries management and by selection of broodstock) or unintentional (where no specific breeding goals exist, but the culturist must still choose broodstock). For example, the larger individuals may be removed from a pond and sent to market and the smaller fish used for restocking.

Genetic Characterization

General Considerations

A prerequisite to gene pool conservation is the characterization of the genetic structure of populations (FAO/UNEP 1981). Wild and farmed populations should be monitored to determine their genetic structure. Indeed, to manipulate better the variation associated with quantitative genetic traits in captive fish populations, aquaculture geneticists should understand the source of the variation of such traits (Robinson and Doyle 1990). Moreover, ecological and taxonomical studies are required, particularly in tropical regions where most ecosystems are poorly understood and many species still undescribed (FAO/UNEP 1981). In freshwaters, data should be collected within localities, between localities within a drainage system, and between drainage systems (Meffe 1986). Documentation of aquatic genetic resources includes database development and circulation of literature.

Genetic analyses can provide important information on the taxonomic status of populations. Some landlocked forms of *Salmo salar* in North America were formerly given subspecific status *(S. salar sebago)*. This was not confirmed by protein polymorphism and mtDNA analyses (Stahl 1983, 1987; Birt et al. 1986). This species is better considered as comprising thousands of reproductively isolated breeding groups, the adults having strong homing behavior, returning to spawn in their natal rivers (Davidson et al. 1989).

Many techniques are used in the characterization of fish genetic resources. In the last decade, new techniques from molecular biology have been introduced. Although any one technique can provide interesting data, a multitechnique approach is better (Chevassus and Coche 1986; Cataudella et al. 1987). Comparative analyses of electrophoretic, meristic and morphological variation should be carried out to ensure unbiased estimates of genetic variation (FAO/UNEP 1981). These techniques facilitate detection of genetic markers that are characteristic of some stocks, in order to monitor quantitative and qualitative genetic changes. Genetic markers can be used to label specific hatchery strains or natural populations unambiguously, or even as identification systems for patenting strains (Gyllensten and Wilson 1987). The main genetic markers of fish are morphological characteristics, isozymes and mitochondrial DNA.

Multivariate Analysis of Morphometric and Meristic Data

Multivariate analysis of morphometric and meristic data is less expensive and laborious than electrophoresis and is useful for defining taxa and forms. However, it must be carefully analyzed, as the expression of gene action on morphological characters is greatly influenced by environmental conditions (phenotypic plasticity). Sometimes, morphological analyses enable the detection of variation in fish shape caused by selection practices (Corti et al. 1988; Eknath et al. 1991), or provide information on the origin of populations present in mixed stock fisheries: as for Atlantic salmon off the Greenland coast, where different stocks have been delineated according to scale morphology and growth patterns (Thorpe, in press).

Morphometry has been recently defined as "traditional (multivariate) morphometrics" or "geometrical morphometrics" (Reyment 1991). The former is the formal codification of morphometrics as we generally know it, an extension of the univariate and multivariate statistics applied to biological characters that are generally represented by meristic characters (in fish, for instance, number of vertebrae, fin rays, gill rakers, etc.) or by distance characters (total length, maximum height, fin length, etc.). Numerical techniques (principal components and derivates, canonical analysis, etc.) have been developed and are widely used to study morphological variation in fish.

Geometrical morphometrics (Bookstein 1991) represents a completely new approach that analyzes the form described in space by a set of landmarks through superimposition methods, an extension of D'Arcy Thomson's (1917) intuition: differences in the form are geometrically described by the deformation of the plane given by the superimposition of one form over the other. Given a formal description of the algebra (Bookstein 1991), it is evident that this approach is particularly suited also to study how an individual changes in shape during growth, relative to differences in the environment. This has particular interest in aquatic animals.

Starch-gel Electrophoresis of Enzymes

Among biochemical techniques, starch-gel electrophoresis of proteins is the predominant tool for population genetics, and databases of proteins are constantly expanding. Its success is based on being a quick and simple technique that provides high resolution. Indeed, the combination of gel electrophoresis and histochemical staining is the fastest and most economical method for surveys of variation at a large number of loci. Since their introduction in the early 1960s, electrophoretic techniques have been widely used by population geneticists to clarify the status of various taxa, and to document the status of wild and cultured stocks.

Many questions in fisheries and aquaculture, like stock assessment, analysis of mixed stock fisheries or recognition of hybrid populations, can be addressed readily through the study of genetic variability within and among populations (e.g., Ryman and Utter 1987; Ferguson and Thorpe 1991; Gauldie 1991; Whitmore 1991). Although most information is available on temperate species, studies are spreading to tropical species as well, especially those with economic interest. Indeed, stock identification is essential to both fish conservation and fisheries and aquaculture management.

Correlation between the level of individual multiple locus heterozygosity and performance traits (such as growth rate or growth-related traits, etc.) has been suggested for different cultured organisms (Leary et al. 1984; Macaranas and Fujio 1986; Sbordoni et al. 1987; Kohen 1991).

Mitochondrial DNA

Restriction enzyme analyses of mitochondrial DNA (mtDNA) have greater resolving power and require fewer samples than allozyme studies. The relative ease of applying this method to fish and the rapid rate of mtDNA nucleotide divergence have made mtDNA analysis a very powerful method, now more and more used in fish biology and fisheries management (Ferris and Berg 1987; Bermingham et al. 1991; Ferguson and Thorpe 1991). In particular, it is an extremely useful tool in the detection of intraspecific variability (Avise et al. 1990; Seyoum and Kornfield 1992; Crosetti et al., in press).

Cytogenetics and Karyotyping

Fish chromosomes are small and homologies are difficult to detect without refined techniques to show specific staining and banding. However, there are groups which differ markedly in their karyotypes and cytogenetic studies can provide significant markers. The Atlantic salmon, for instance, has a diploid number varying from 55 to 60, and differences in chromosome arm numbers have been found among its European populations (Davidson et al. 1989). Polymorphisms in chromosome number have been detected in different populations and among individuals of the same population in *Seriola dumerili* (Vitturi et al. 1986).

Fish Gene Pools

The gene pool of a given population or species is the set of genotypes of individuals that form that particular population or species (Rab 1989). Being a dynamic open system, it varies with time and can be easily disturbed. At a given time, a particular gene pool has evolved to become adapted to the local physical, chemical and biotic conditions of its environment. The gene pools of today are the result of evolutionary processes that have affected those taxa over thousands or millions of years. For domesticated organisms, however, humans have bypassed natural selection with rapid and very strong selection for specific traits.

Many fish species have a complex structure of subpopulations more or less genetically differentiated. Some are composed of populations which are spatially and genetically isolated and cannot mix their gene pools. This applies particularly to freshwater fish species that extend across different watersheds.

The concept of a fish stock has been well discussed to define its different meanings in fisheries and aquaculture (STOCS 1981). Particular care has been devoted to discriminating between fish stocks for fishing purposes (more precisely defined as mixed fisheries stocks), and stocks in the genetic sense, i.e., subpopulations more or less reproductively isolated by time or spawning locations (Nelson and Soulé 1987).

Genetic Impoverishment in Fish Populations

There are many degrees of genetic impoverishment, as evident by the scale of danger criteria established by IUCN for taxa threatened with extinction. Very often, only the threat of species extinction is strong enough to induce some kind of human reaction towards protection. However, in the long term, reduced genetic variability or the extinction of a single stock could be very serious.

Human activities are the primary cause of genetic impoverishment in many fish stocks and species and trends towards their extinction. Aquatic genetic resources can be affected by environmental changes (particularly pollution); fisheries; artificial selection and domestication in aquaculture; and transfers and introduction of species (which may result in hybridization, introgression and founder and bottleneck effects) (FAO/UNEP 1981; Wohlfarth 1986).

In natural fish stocks, loss of diversity is most evident in changes in species composition in intensive and selective fisheries, although these changes may be confounded with losses brought about by eutrophication, other forms of pollution, or changes in the environment caused by humans (such as siltation of spawning sites or introduction of exotic species) which may render a stock more vulnerable to the stress of overexploitation (Smith 1968; Regier 1973). Selective elimination of subpopulations or stocks has attracted the most attention (STOCS 1981). Within a stock, Nelson and Soulé (1987) distinguish between undirected loss of genetic variation (inbreeding, genetic drift) and directional changes (selection). These are distinct problems. Genetic erosion is the loss of genes within a population/species, resulting in a much smaller gene pool in the individuals surviving from the original population/species, which had formerly much higher genetic diversity.

Environmental Change

Human activities cause severe habitat alterations and produce different forms of pollution, e.g., eutrophication, toxins from industrial wastes and thermal pollution. In some regions, water quality has been severely affected by precipitation acidified by combustion of fossil fuels. Some fish species, especially their early life history stages, are extremely sensitive to water acidification (FAO/UNEP 1981). In southern Norway, this has caused the total disappearance of fish from hundreds of lakes, and of salmon in many river systems (Gjedrem 1981). Hydroelectric power development has also affected fish populations through the building of dams which created barriers to spawning migrations and destroyed spawning sites and nurseries. New dams have led to the extinction of many salmon stocks (Saunders 1981).

Capture Fisheries

Capture fisheries can cause the loss of aquatic genetic resources as a consequence of overexploitation or from selective fishing practices. In fisheries, the first stocks to disappear are those with properties most desirable to the fisheries or to future enhancement or aquaculture efforts, e.g., rapid growth and high catchability (Thorpe and Koonce 1981). Recent technological improvement in fishing gear has resulted in increasing harvesting efficiency, which in some cases has caused the depletion of some fish stocks and species. Overexploitation may also cause genetic drift, where only a limited gene pool survives harvesting. Fisheries management may also have significant but involuntary effects on genetic resources, by setting up restrictions for time and location of capture and type of gear used. In particular, mesh size can be very selective, affecting a particular size of the population.

In the salmon fisheries of West Greenland, different genetic stocks are pooled together in fished stocks, as they gather in common pelagic feeding areas (Stahl 1987). These "mixed-stock fisheries" are therefore difficult to protect, but harvesting may cause overexploitation of the numerically smaller stocks. Overfishing of Baltic salmon has resulted in slower growth, smaller size at maturity and earlier homing. In Lake George, fishing pressure reduced the size and maturation size of *Oreochromis niloticus* over an eleven-year period (Gwahaba 1973; Lowe-McConnell 1982).

Overfishing can also affect an entire fish community. In Lake Malaŵi, intensive fishing of cichlids has caused the decrease of large species, which dominated before, and now small species predominate (Turner 1976).

Enhanced Fisheries

There is a growing need to enhance fisheries, particularly inland fisheries, by stocking of hatchery-raised juveniles followed by harvesting through fishing. Fisheries enhancement has three purposes: to create new fisheries (often introducing exotic species into a waterbody); to supplement natural production to increase harvests; and to mitigate fish losses resulting from human activities (Lannan et al. 1989).

There may be insufficient local broodstock for the required mass propagation and other strains are often imported. The growing practice of enhancing fisheries formerly based on wild populations by introductions of hatchery-raised stocks can cause severe genetic damage. Hatchery-raised stocks are commonly much less variable genetically than wild stocks. For example, hatchery-raised salmon are used for stocking rivers flowing into the Baltic. Originally, these rivers produced about 10 million smolts annually, but the building of hydropower stations in the 1940-1960s made natural spawning impossible. In Sweden, the Swedish Water Law obliged the generating companies to build hatcheries to compensate for this loss. In Norway, 10-15 million Atlantic salmon fry are produced annually for restocking purposes (Hansen et al. 1987a) and are released into rivers regardless of their origin. Moreover, escapees from the sea cages of salmon farms are 'homeless', and reproduce in a wide range of rivers (Hansen et al. 1987b). It is highly probable that such escapees from cages and land-based farms interact with local native populations (Skaala et al. 1990).

Introduced populations can also cause direct genetic changes by interbreeding or have more indirect effects through predation, competition for food, mates, spawning sites, or introduction of pathogens and parasites. The breeding of hatchery-raised fish with wild populations can produce hybrid or introgressed progeny that may be less adapted to the local environment, on the assumption that local populations have evolved to be well adapted.

Over much of Europe, rivers and streams have been and are regularly restocked with non-native fry. Often, hatchery programs work against the aim of preserving natural populations, and may cause a complete replacement of natural populations with hatchery fish. Natural populations of brown trout *(Salmo trutta)*, each showing specific allele frequencies and at least one allele at high frequency not previously reported, have been identified in Spain (Garcia-Marin et al. 1991). Sea ranching is well developed in Japan, where seed from shore-based hatcheries are released to the wild where they grow and are subsequently harvested by the existing fisheries (Davy 1991).

Introduction of Exotic Species

Waterbodies have often been stocked with exotic fish (Billington and Hebert 1991) without evaluating the genetic and ecological impacts on local aquatic, genetic resources and habitats. There are many cases of successful colonizations by exotic species [for example, common carp *(Cyprinus carpio)* and brown trout *(Salmo trutta)* in North America, or rainbow trout *(Oncorhynchus mykiss)* in Europe and Latin America]. However, little is known of the genetic consequences. There are some examples of introductions which have caused severe damage to local populations either by hybridizing with them, or by outcompeting against them, leading eventually to extinction. Moreover, introductions may fail in their objectives and may lead to genetic degradation of natural populations; for example, whitefish in Czechoslovakia (Barus 1989). Introductions and transfers usually comprise small numbers of fish. This can cause bottlenecks and founder effects and reduced genetic variation in future generations.

Introductions have increased with demands from hatcheries and improved communications. However, many introductions and transfers are uncontrolled and undocumented. It is often difficult to determine the origin of introduced stocks.

Pullin (1988) cites many cases where tilapias have been found in countries to which they are exotic and were previously unrecorded. The earliest transfers of tilapias in North Africa were probably by the French Foreign Legion from one well to another (Thys van den Audenaerde 1988). *Oreochromis mossambicus* was first reported in Asia in 1938, when two females and three males of unknown origin were accidentally discovered in Java (Schuster 1952).

Other problems are the accuracy of identification of the original stock and its possible admixtures of closely related species. There are often mistakes in the nomenclature of introduced species, creating much confusion. The Kafue strain of *O. macrochir* was introduced to fish culture ponds at Kipopo (Katanga, Zaïre) under the name of *andersonii*, and *O. mortimeri* to Katanga under the name *"mossambica"* (Lowe-McConnell 1988).

As an example of these introductions, *Barbotes* spp., endemic to Lake Lanao (Philippines) have become rare following the intentional introductions of *Clarias batrachus*, *Channa striata*, *O. mossambicus*, together with the accidental introduction of *Glossogobius giurus*. The native fish were formerly the principal sources of livelihood and foodfish for the local people (Frey 1969).

In Queensland (Australia), tilapias are considered as pests (Bluhdorn and Arthington 1990). Negative impact on indigenous fish can be considered general for tropical America, where piscivorous fish were introduced predominantly (Fernando 1991).

Introductions can produce hybrids with indigenous species. Introgressive hybridization has occurred in Lake Itasy in Madagascar, where *O. niloticus* crossed with *O. macrochir* has produced the so-called "tilapia trois-quarts" (Vincke 1971; Daget and Moreau 1981), and in the Philippines with *O. mossambicus* and *O. niloticus* (Taniguchi et al. 1985; Macaranas et al. 1986).

O. niloticus was introduced into Lake Victoria in a shipment of *Tilapia zillii* from Lake Albert, or spread through the drainage system, having being used in culture trials carried out at Kajansi (Uganda) (Lowe-McConnell 1988). In the same lake, *T. zillii* has largely displaced the endemic *O. variabilis*.

In Lake Titicaca, endemic species of *Orestia* suffered a rapid population decline from infection with sporozoan parasites introduced with trout (FAO/UNEP 1981). Introductions of non-native salmonids have spread diseases into new habitats, sometimes to the detriment of native fish (Allendorf and Leary 1988).

Hybridization

Hybridization can be carried out in breeding programs to obtain heterosis and required characteristics in the hybrid progeny, but may also be an unplanned consequence of the introduction of one population (or species) in the distribution area of another. Hybrids between local and domesticated stocks may affect neighboring ecosystems if their genes and parasites spread.

Although hybridization is often used to increase the amount of genetic variation in a population, it can be a mode of genetic diversity (Nelson and Soulé 1987). The effect of loss of gene exchange between subpopulations is to increase the variance within groups, decrease the variance between groups and decrease the total variance (Crow and Kimura 1970). However, as intraspecific hybridization breaks co-adapted gene complexes and can cause an outbred depression (which has been often empirically evidenced, although numerical data are scarce), the fitness of hybrid generations usually declines with time.

The change in genotypic variance from hybridization between small neighboring subpopulations may provide beneficial new alleles without harming local adaptation (Allendorf 1983).

Intraspecific hybridization mixes gene pools. The present 'Ivory Coast' or 'Bouaké' strain of *O. niloticus* is actually a mixture of Volta and Nile strains, which became mixed in captivity. Its genetic identity depends on the degree of hybridization between strains which is not well documented (Nugent 1988).

Over one hundred combinations of distant hybridizations have been recently carried out in China, including crosses between families, subfamilies, genera and species. Five crosses between common carp strains showed hybrid vigor with potential application in aquaculture (Wu 1990). The Kurst strain of common carp is produced through hybridization to obtain higher cold resistance, achieved through excessive accumulation of body fat (Gall 1990).

Conservation of Fish Genetic Resources

Ex situ conservation

Ex situ fish gene banks offer an alternative to *in situ* conservation of genetic resources, but are difficult and costly to establish and to maintain. They include techniques such as rotational line crossing and sperm cryopreservation. These are particularly valuable where only a few hatchery populations survive from formerly abundant stocks. Although requiring expensive skilled labor and high technology, they may be in some cases the only remaining means to prevent extinction of some species and strains.

Fish gene banks should, as far as is practical, represent the total gene pool. In species where the spawning season is protracted, material must be collected and maintained to represent the entire spawning season. An adequate number of individuals must be kept to maintain high genetic diversity. FAO recommends effective population size, Ne, of at least 50 for short-term conservation, and 500 for long-term conservation (FAO/UNEP 1981). These figures vary slightly with authors (Tave 1986; Smitherman and Tave 1987). Fish farms and public aquaria can play important roles in the breeding and management of endangered species.

In addition to stock collections, gamete cryopreservation is a useful technique in the conservation of genetic variability and has the advantage of maintaining high levels of genetic variability, without maintaining large numbers of breeders (Harvey 1987). Spermatozoa can be collected from a range of known strains and stored. Sperm banks should be managed following the Codes of Practice already established for livestock sperm and embryo banks. Cryopreservation of fish sperm constitutes a haploid gene bank: only half the genome is stored. Moreover, although techniques for sperm cryopreservation are established for many fish species, no sperm collection has yet been established for commercial purposes. Finfish eggs and embryos cannot yet be cryopreserved.

Examples of International, Regional and National Efforts

The European Inland Fisheries Advisory Commission (EIFAC), working with FAO, has developed Codes of Practice for fish introductions and transfers (Turner 1987). Recommendations have been formulated and addressed to different audiences: international organizations, governments, aquaculturists, fisheries managers, conservationists and research scientists (FAO/UNEP 1981).

The Working Group on Genetics of the International Council for Exploration of the Sea (ICES) is promoting the development of an International Register of Available Strains of Fish and Shellfish.

ICLARM and FAO are coordinating a new international database (FishBase) that will incorporate a tilapia strain registry and museum data, the assemblage of which is the responsibility of the Zoologisches Institut und Museum, University of Hamburg (Pullin 1990).

ICLARM, with Philippine and Norwegian collaborators, established the GIFT (Genetic Improvement of Farmed Tilapias) project with the aim of developing more productive stocks of tilapias by selection for high growth rate and other economically important traits (Pullin et al. 1991). This project includes a study of tilapia genetic resources in Asia and Africa, and the establishment of a collection of promising strains from new importations from Africa and from existing Asian cultured stocks.

A major component of the AADCP (ASEAN-EEC Aquaculture and Coordination Programme) focuses on the role of genetic manipulation and population genetic techniques in aquaculture and fisheries management in the ASEAN region, twinning the National Aquaculture Institute (NAGRI) in Thailand with the Institute of Aquaculture of the University of Stirling, Scotland.

IDRC has promoted the AGNA (Aquaculture Genetics Network in Asia) project, as a link between aquaculture genetics projects in Asian countries with each other and with Dalhousie University in Canada. The objectives are to develop new superior strains of fish by efficient artificial selection and hybridization, to maximize the rate of domestication and to minimize the inbreeding of present stocks.

A Nordic Symposium on Gene Banks (1978) held in Helsinki recommended the establishment of gene banks in each of the Nordic countries with contact between groups to exchange information, and research results, particularly for gamete storage (Gjedrem 1981).

Nyman and Norman (1987) suggested a national strategy for the conservation of Atlantic salmon genetic resources in Sweden, where most rivers have been modified by the building of dams for hydroelectric power production. The strategy relies on restocking with river-specific stocks bred from at least 25 pairs of broodstock and requires that wild populations be carefully protected from hybridization with hatchery stocks, and used as *in situ* gene-banks.

In Hungary, a government-sponsored program maintains 18 "landraces" of common carp, including nine native races and nine exotic races imported from elsewhere in Europe and Asia (Lannan et al. 1989), and coordinates activities to meet the needs of aquaculture and natural resource management.

Genetics is one of the most important fields of scientific research in Czech ichthyology and applied fisheries science (Barus 1989) and three specialized meetings on fish genetics have been organized in Czechoslovakia. Endangered species have been surveyed with analyses of the causes of threats, present status, future prospects and conservation efforts (Lusk 1989). A specific program has been launched for the preservation of the wild Danube carp, *Cyprinus carpio carpio* (Krupka et al. 1989).

Despite their richness in endemic fish genetic resources, most African countries have given low priority to aquaculture and conservation of fish genetic resources. Countries with sites of special significance for aquatic genetic resources generally lack funds to invest in their conservation. However, Malaŵi has recognized the importance of genetic conservation and prohibits introductions of exotic fish to protect the native species, ecology and fisheries of Lake Malaŵi. Many African waterbodies are shared among countries and conservation programs should be established at the regional level.

The International Study on Artemia (ISA) was created in 1978 to establish an interdisciplinary approach to the characterization of Artemia strains (chemical composition, value as food for aquaculture species, general biology, and genetics). In a recent review of the genetics of Artemia, Abreu-Grobois (1987) stressed the importance of protecting natural populations of Artemia from extinction due to habitat destruction or by the indiscriminate introductions of more competitive species when inoculating brine-shrimp-free salterns.

Conclusions

Aquaculture production totalled about 15.3 million tonnes in 1990, using more than 140 species: a wide range, for different environmental conditions and markets. Constant progress is being made in improving the methodologies for characterization of aquatic genetic resources. The evidence of damage to natural populations continues to accumulate. Basic research provides improved techniques and methodologies for multidisciplinary approaches. Analyses of these complex problems must include environmental and economic costs and benefits so that participants are fully aware of their importance in development programs and do not view them as merely of academic interest.

Conservation of aquatic genetic resources concerns the whole world and requires a global approach and global awareness. Financial support by donors is essential to initiate well-coordinated programs that both support fundamental research and facilitate stock conservation in specialized centers and natural waters.

References

Abreu-Grobois, F.A. 1987. A review of the genetics of Artemia, p. 61-100. *In* P. Sorgeloos, D.A. Bengtson, W. Decler and E. Jaspers (eds.) Artemia: research and its application. Universa Press, Wetteren, Belgium.

Allendoıf, F.W. 1983. Isolation, geneflow, and genetic differentiation among populations, p. 51-65. *In* C.M. Schonewald-Cox et al. (eds.) Genetics and conservation. Benjamin-Cummings, Menlo Park, California.

Allendorf, F.W. and R.F. Leary. 1988. Conservation and distribution of genetic variation in a polytypic species, the cutthroat trout. Conserv. Biol. 2:170-184.

Avise, J.C., W.S. Nelson, J. Arnold, R.K. Koehn, G.C. Williams and V. Thorsteinsson. 1990. The evolutionary genetic status of icelandic eels. Evolution 44:1254-1262.

Barus, V. 1989. Tropical tasks of Czechoslovak genetics in fisheries and in fish gene pool conservation. Prace Vurh Vodnany 18:5-8.

Bermingham, E., S.H. Forbes, K. Fridland and C. Pla. 1991. Discriminations between Atlantic salmon *(Salmo salar)* of North American and European origin using restriction analyses of mitochondrial DNA. Can. J. Fish. Aquat. Sci. 48:884-893.

Billington, N. and P.D.N. Hebert, Editors. 1991. International Symposium on "The Ecological and Genetic Implications of Fish Introductions (FIN)". Can. J. Fish Aquat. Sci. 48 (suppl. 1), 181 p.

Birt, T.P., J.M. Green and W.S. Davidson. 1986. Analysis of mitochondrial DNA in allopatric anadromous and nonanadromous Atlantic salmon, *Salmo salar*. Can. J. Zool. 64: 118-120.

Bluhdorn, D.R. and A.H. Arthington. 1990. Somatic characteristics of an Australian population of *Oreochromis mossambicus* (Pisces: Cichlidae). Environ. Biol. Fish. 29:277-291.

Bookstein, F. 1991. Morphometric tools for landmark data. Geometry and biology. Cambridge University Press, Cambridge. 435 p.

Cataudella, S., L. Sola, M. Corti, R. Arcangeli, G. La Rosa, M. Mattoccia, M. Cobolli-Sbordoni and V. Sbordoni. 1987. Cytogenetic, genic and morphometric characterization of groups of common carp, *Cyprinus carpio*, p. 113-129. *In* K. Tiews (ed.) Selection, hybridization and genetic engineering in aquaculture. Vol. 1. Springer-Verlag, Berlin.

Chevassus, B. and A.G. Coche, Editors. 1986. Report on the "Symposium on Selection, Hybridization and Genetic Engineering in Aquaculture of Fish and Shellfish for Consumption and Stocking", 27-30 May 1986, Bordeaux, France. EIFAC Tech. Pap. 50, 52 p.

Corti, M., R.S. Thorpe, L. Sola, S. Cataudella and V. Sbordoni. 1988. Multivariate morphometrics in aquaculture: a case study of six stocks of the common carp, *Cyprinus carpio* from Italy. Can. J. Fish Aquat. Sci. 45:1-7.

Crosetti, D., J.C. Avise, F. Placidi, A.R. Rossi and L. Sola. Geographic variability in the grey mullet *Mugil cephalus*: preliminary results of mtDNA and chromosomes analyses. Aquaculture. (In press).

Crow, J.F. and M. Kimura. 1970. An introduction to population genetics theory. Harper and Row, New York. 591 p.

Daget, J. and J. Moreau. 1981. Hybridation progressive entre deux espèces de *Sarotherodon* (Pisces, Cichlidae) dans un lac de Madagascar. Bull. Mus. Hist. Nat. (France). 4E se., A. Zool. 3:689-703.

Das, P. and A.G. Jhingran. 1989. Fish genetics in India. Proceedings of the Symposium on "Conservation and Management of Fish Genetic Resources of India". Today and Tomorrow's Printers and Publishers, New Delhi. 266 p.

Davidson, W.S., T.P. Birt and J.M. Green. 1989. A review of genetic variation in Atlantic salmon, *Salmo salar* L., and its importance for stock identification, enhancement programmes and aquaculture. J. Fish Biol. 34:547-560.

Davy, F.B. 1991. Mariculture research and development in Japan. IDRC-MR 298e, 97 p.

Eknath, A.E., J.M. Macaranas, L.Q. Agustin, R.R. Velasco, M.C.A. Ablan, M.J.R. Pante and R.S.V. Pullin. 1991. Biochemical and morphometric approaches to characterize farmed tilapias. Naga, ICLARM Q. 14(2):7-9.

FAO/UNEP. 1981. Conservation of the genetic resources of fish: problems and recommendations. FAO Fish. Tech. Pap. 217, 43 p.

Ferguson, A. and J.E. Thorpe, Editors. 1991. Biochemical genetics and taxonomy of fish. J. Fish Biol. 39 (suppl. A), 357 p.

Fernando, C.H. 1991. Impacts of fish introductions in tropical Asia and America. Can. J. Fish Aquat. Sci. 48 (suppl. 1):24-32.

Ferris, S.D. and W.J. Berg. 1987. The utility of mitochondrial DNA in fish genetics and fisheries management, p. 277-300. *In* N. Ryman and F. Utter (eds.) Population genetics and fishery management. Washington Sea Grant Program, University of Washington, Seattle.

Frankel, D.H. and M.E. Soulé. 1981. Conservation and evolution. Cambridge University Press, Cambridge. 327 p.

Frey, G.D. 1969. A limnological reconnaissance of Lake Lanao. Verh. Int. Ver. Theor. Angew. Limnol. 17:1090-1102.

Gall, G.A.E. 1990. Basis for evaluating breeding plans. Aquaculture 85:125-142.

Garcia-Marin, J.L., P.E. Jorde, N. Ryman, F. Utter and C. Pla. 1991. Management implications of genetic differentiation between native and hatchery populations of brown trout *(Salmo trutta)* in Spain. Aquaculture 95:235-249.

Gauldie, R.W. 1991. Taking stock of genetic concept in fisheries management. Can. J. Fish Aquat. Sci. 48:722-731.

Gjedrem, T. 1981. Conservation of fish populations in Norway, p. 33-36. *In* N. Ryman (ed.) Fish gene pools. Ecol. Bull., Stockholm, Vol. 34.

Gwahaba, J.J. 1973. Effects of fishing on the *Tilapia nilotica* (Linné, 1857) population in Lake George, Uganda, over the past 20 years. E. Afr. Wildl. J. 11:317-328.

Gyllensten, U. and A.C. Wilson. 1987. Mitochondrial DNA of salmonids: inter- and intraspecific variability detected with restriction enzymes, p. 301-319. *In* N. Ryman and F. Utter (eds.) Population genetics and fishery management. Washington Sea Grant Program, University of Washington, Seattle.

Hansen, L-P., R.A. Lund and K. Hindar. 1987a. Possible interaction between wild and reared Atlantic salmon in Norway. ICES C.M. 1987/ M:14.

Hansen, L-P., K.B. Doving and B. Jonsonn. 1987b. Migration of farmed adult Atlantic salmon with and without olfactory sense, released on the Norwegian coast. J. Fish Biol. 30:713-721.

Harvey, B.J. 1987. Gamete banking and applied genetics in aquaculture, p. 257-264. *In* K. Tiews (ed.) Selection, hybridization and genetic engineering in aquaculture. Vol. 1. Springer-Verlag, Berlin.

Hindar, K., N. Ryman and F. Utter. 1991. Genetic effects of cultured fish on natural fish populations. Can. J. Fish. Aquat. Sci. 48:945-957.

ICLARM. 1992. Meeting on "International concerns in the use of aquatic germplasm", June 1992, Cavite, Philippines.

Kohen, R.K. 1991. The genetics and taxonomy of species in the genus *Mytilus*. Aquaculture 94:125-145.

Krupka, I., J. Meszaros, P. Rab and V. Slechtova. 1989. Toward the preservation of endangered fish gene pools - example of wild Danube carp *(Cyprinus carpio carpio)*. Prace Vurh Vodnany 18:27-33.

Lannan, J.E., G.A.E. Gall, J.E. Thorpe, C.E. Nash and B.E. Ballachey. 1989. Genetic resource management of fish. Genome 31:798-804.

Leary, R.F., F.W. Allendorf and K.L. Knudsen. 1984. Superior developmental stability of enzyme heterozygotes in salmonid fishes. Am. Nat. 124:540-551.

Lowe-McConnell, R. H. 1992. Tilapias in fish communities, p. 83-113. *In* R.S.V. Pullin and R.H. Lowe-McConnell (eds.) Biology and culture of tilapias. ICLARM Conf. Proc. 7, 360 p.

Lowe-McConnell, R.H. 1988. Ecology and distribution of tilapias in Africa that are important for aquaculture, p. 12-18. *In* R.S.V. Pullin (ed.) Tilapia genetic resources for aquaculture. ICLARM Conf. Proc. 16, 108 p.

Lusk, S. 1989. Gene pool of fishes in Czechoslovakia: present state and conservation efforts. Prace Vurh Vodnany 18:15-26.

Macaranas, J. and Y. Fujio. 1990. Strain differences in cultured fish - isozymes and performance traits as indicators. Aquaculture 85:69-82.

Macaranas, J.M., N. Taniguchi, M.J.R. Pante, J.B.

Capili and R.S.V. Pullin. 1986. Electrophoretic evidence for extensive hybrid gene introgression into commercial *Oreochromis niloticus* (L.) stocks in the Philippines. Aquacult. Fish. Manage. 17:249-258.

Meffe, G.K. 1986. Conservation genetics and the management of endangered species. Fisheries 11:14-23.

Nelson, K. and M.W. Soulé. 1987. Genetical conservation of exploited fishes, p. 345-368. *In* N. Ryman and F. Utter (eds.) Population genetics and fishery management. Washington Sea Grant Program, University of Washington, Seattle.

Nugent, C. 1988. The status of wild and cultured tilapia genetic resources in various countries: Côte d'Ivoire (freshwater culture), p. 22. *In* R.S.V. Pullin (ed.) Tilapia genetic resources for aquaculture. ICLARM Conf. Proc. 16, 108 p.

Nyman, L. and L. Norman. 1987. Genetic aspects on culture of Atlantic salmon and sea trout for stocking: guidelines for breeding methodology and management. Salmon Research Institute Report, Sweden. 20 p.

Prace Vurh Vodnany. 1989. Fish genetics and breeding. Papers of the Research Institute of Fishery and Hydrobiology, n. 18 (RIFH) Vodnany, Czechoslovakia. 60 p.

Pullin, R.S.V., Editor. 1988. Tilapia genetic resources for aquaculture. ICLARM Conf. Proc. 16, 108 p.

Pullin, R.S.V. 1990. Down-to-earth thoughts on conserving aquatic genetic diversity. Naga, ICLARM Q. 13(1):5-8.

Pullin, R.S.V., A.E. Eknath, T. Gjedrem, M.M. Tayamen, J.M. Macaranas and T.A. Abella. 1991. The genetic improvement of farmed tilapias (GIFT): the story so far. Naga, ICLARM Q. 14(2):3-6.

Rab, P. 1989. Genetic aspects of the conservation of fish gene resources. Prace Vurh Vodnany 18:9-14.

Regier, H.A. 1973. Sequence of exploitation of stocks in multispecies fisheries in the Laurentian Great Lakes. J. Fish. Res. Board Can. 30:1990-1992.

Reyment, R.A. 1991. Multidimensional palaeobiology. Pergamon Press, Oxford. 377 p.

Robinson, B.W. and R.W. Doyle. 1990. Phenotype correlations among behavior and growth variables in tilapia: implications for domestication selection. Aquaculture 85:177-186.

Ryman, N., Editor. 1981. Fish gene pools. Ecol. Bull., Stockholm. Vol. 34, 111 p.

Ryman, N. 1991. Conservation genetics considerations in fishery management. J. Fish Biol. 39 (suppl. A):211-224.

Ryman, N. and F. Utter, Editors. 1987. Population genetics and fishery management. Washington Sea Grant Program, University of Washington, Seattle. 418 p.

Saunders, R.L. 1981. Atlantic salmon *(Salmo salar)* stocks and management implications in the Canadian Atlantic provinces and New England, USA. Can. J. Fish. Aquat. Sci. 38:1612-1625.

Sbordoni, V., G. La Rosa, M. Mattoccia, M. Cobolli Sbordoni and R. De Mattheis. 1987. Genetic changes in seven generations of hatchery stocks of the Kuruma prawn, *Penaeus japonicus* (Crustacea, Decapoda), p. 143-156. *In* K. Tiews (ed.) Selection, hybridization and genetic engineering in aquaculture. Vol. 1. Springer-Verlag, Berlin.

Schuster, W.H. 1952. Fish culture in brackishwater ponds in Java. Indo-Pac. Fish. Counc. Spec. Publ. 1:1-12.

Seyoum, S. and I. Kornfield. 1992. Identification of the subspecies of *Oreochromis niloticus* (Pisces, Cichlidae) using restriction endonuclease analysis of mitochondrial DNA. Aquaculture 102:29-42.

Siler, R., I. Chudik and P. Majerciak. 1984. Towards the problem of gene resources conservation of farm and game animals in CSSR. Vestnik CSAZ 31 (9): 535-544.

Skaala, O., G. Dahle, K.E. Jorstad and G. Naevdal. 1990. Interactions between natural and farmed fish populations: information from genetic markers. J. Fish Biol. 36:449-458.

Smith, S.H. 1968. Species succession and fishery exploitation in the Great Lakes. J. Fish. Res. Board Can. 25:667-693.

Smitherman, R.O and D. Tave. 1987. Maintenance of genetic quality in cultured tilapias. Asian Fish. Sci. 1:75-82.

Stahl, G. 1983. Differences in the amount and distribution of genetic variation between natural populations and hatchery stocks of Atlantic salmon. Aquaculture 33:23-32.

Stahl, G. 1987. Genetic population structure of Atlantic salmon, p. 121-140. *In* N. Ryman and F. Utter (eds.) Population genetics and fishery management. Washington Sea Grant Program, University of Washington, Seattle.

STOCS. 1981. Stock Concept International Symposium. Can. J. Fish. Aquat. Sci. 38:1457-1921.

Taniguchi, N., J.M. Macaranas and R.S.V. Pullin. 1985. Introgressive hybridization in cultured tilapia stocks in the Philippines. Bull. Jap. Soc. Sci. Fish. 51:1219-1224.

Tave, D. 1986. Genetics for fish hatchery managers. AVI, Westport, Connecticut. 299 p.

Thomson, D.W. 1917. On growth and form. Cambridge University Press, Cambridge. 1116 p.

Thorpe, J.E. Atlantic salmon. *In* Managing aquatic genetic resources. National Research Council, Washington, DC. (In press).

Thorpe, J.E. and J.F. Koonce. 1981. Assessing and managing man's impact on fish genetic resources. Can. J. Fish. Aquat. Sci. 38:1899-1907.

Thys van den Audenaerde, D.F.E. 1988. Natural distribution of tilapias and its consequences for the possible protection of genetic resources, p.

1-12. *In* R.S.V. Pullin (ed.) Tilapia genetic resources for aquaculture. ICLARM Conf. Proc. 16, 108 p.

Turner, J. 1976. Changes in the size structure of cichlid populations of Lake Malawi resulting from bottom trawling. J. Fish. Res. Board. Can. 34:232-238.

Turner, G.E., Editor. 1987. Codes of practice and manual of procedures for consideration of introductions and transfers of marine and freshwater organisms. Prepared by the Working Group on Introduction and Transfer of Marine Organisms of ICES and by the Working Party on Introductions of EIFAC. ICES Doc. F:35A, 42 p.

Utter, F. 1991. Biochemical genetics and fishery management: a historical perspective. J. Fish Biol. 39 (suppl. A):1-20.

Vincke, M. 1971. Contribution à l'étude du tilapia "trois-quart" du lac Itasy (Madagascar), p. 34-53. *In* Centre Technique Forestier Tropical, Rapport Annuel 1970.

Vitturi, R., A. Mazzola, M. Macaluso and E. Catalano. 1986. Chromosomal polymorphism associated with Robertsonian fusion in *Seriola dumerili* (Risso, 1810) (Pisces: Carangidae). J. Fish Biol. 26:529-534.

Whitmore, D.H., Editor. 1991. Electrophoretic and isoelectric focusing techniques in fisheries management. CRC Press, Boca Raton, Florida. 350 p.

Wohlfarth, G.W. 1986. Decline in natural fisheries. A genetic analysis and suggestion for recovery. Can. J. Fish. Aquat. Sci. 43:1298-1306.

Wu, C. 1990. Restrospects and prospects of fish genetics and breeding research in China. Aquaculture 89:61-68.

Discussion

PULLIN: The lesson of history so far is that it is not possible to control fish transfers effectively whatever laws and regulations are made. The private sector especially, in developed and developing countries, usually finds a way to transfer or introduce fish irrespective of policy and regulations. I don't think this will change much. In fact, it may get worse. So, we have to live with this and consider how best to conserve fish genetic diversity. This paper clearly points out the *in situ* and *ex situ* approaches. For *in situ* conservation in natural habitats, however, we have to consider what is practically and politically possible. I recently wrote on this in Naga, the ICLARM Quarterly[1]. Conservation of large natural fish species assemblages in large natural waterbodies will become increasingly difficult as aquaculture grows. By analogy, conservation of natural plant succession is also difficult adjacent to intensive agriculture. What *can* work is giving more attention to conservation of all aquatic biota, including fish, in nature reserves and game parks and in small waterbodies that are sites of special scientific interest. Of course, some large waterbodies and their fish fauna are unique and very important and all possible efforts should be made to conserve and protect them from all potentially harmful disturbances, including harmful effects from aquaculture. This will probably require transnational agreements and their enforcement. Good examples are Lake Malaŵi and Lake Tanganyika.

ROSENTHAL: Although necessary, transnational control is not easy to achieve. In order to reduce risks it will be necessary to stimulate more awareness among farmers and traders on risks associated with transfers and introductions. In other words, the introduction of any regulatory measure will have to be accompanied by an adequate educational program to achieve the required acceptability of these measures by practitioners of aquaculture.

BILIO: We will have to learn how to cope with future fish transfers irrespective of regulations. We will have to do our best to anticipate their effects. Conservation of aquatic genetic diversity needs an overall long-term plan, recognizing where it is realistic or unrealistic to even try.

EDWARDS: I even have to watch some of the aquaculture faculty at AIT over plans for fish introductions - never mind the private sector!

PULLIN: Do you want that on record?

EDWARDS: Yes, I do, because it illustrates the complexity of the problem.

BILIO: Expert advice on introductions needs very close examination. In some culture-based fisheries projects, experts have recommended introduction of exotic species, like grass carp, to increase production. Often introductions are unsuccesful. For example, in Albania, introduction of exotic carps to supplement common carp production has been useless because the exotics are not wanted by consumers.

PULLIN: ICLARM has had one workshop on tilapia genetic resources[2] and is planning another on Asiatic carps in collaboration with the Asian Wetland Bureau and IUCN. We know from our Chinese colleagues that the wild genetic resources of Chinese carps are under threat.

---

[1]Pullin, R.S.V. 1990. Down-to-earth thoughts on conserving aquatic genetic diversity. Naga, ICLARM Q. 13(1):5-8.

[2]Pullin, R.S.V. Editor. 1988. Tilapia genetic resources for aquaculture. ICLARM Conf. Proc. 16, 124 p.

Aquaculture Development and Environmental Issues in the Developing Countries of Asia

IMRE CSAVAS
FAO Regional Office for Asia and the Pacific
39 Phra Atit Road
Bangkok 10200, Thailand

CSAVAS, I. 1993. Aquaculture development and environmental issues in the developing countries of Asia, p. 74-101. *In* R.S.V. Pullin, H. Rosenthal and J.L. Maclean (eds.) Environment and aquaculture in developing countries. ICLARM Conf. Proc. 31, 359 p.

Abstract

The status of aquaculture in the Asia-Pacific region is described with special emphasis on the developing countries of Asia. Characteristics of inland and coastal aquaculture are analyzed separately, and differences in the role and environmental compatibility of major aquaculture production systems are highlighted. The dominance of culturing aquatic organisms low in the food chain is documented in the developing countries of the region. Trends in culturing noncarnivorous and carnivorous finfishes, crustaceans, molluscs and seaweeds are studied individually; substantial differences in their impact on the environment are demonstrated. Major environmental issues raised by the rapid development of aquaculture are recounted and improved approaches are proposed to achieve sustainable growth in the future.

Introduction

Any attempt to analyze the status and trends of aquaculture production is hindered by a general lack of reliable data. FAO started only in 1984 to collect annual aquaculture statistical data and results of these enquiries were first published in 1989. Although the quality of data improved over the years, much confusion has arisen from the fact that the term "aquaculture" was not properly defined. To overcome this imprecision, in 1987 FAO adopted a definition of aquaculture production which follows closely the practical distinction between hunting/gathering and agriculture (FAO 1992a). Its most important impact was the exclusion of enhanced 'culture-based' fisheries (e.g., catch from stocked reservoirs, sea ranching) from aquaculture production. This led to the need to revise earlier data. Thus, figures quoted in this paper may not correspond fully with those published in the literature before 1990.

One continuing weakness of aquaculture data is the rather unreliable nature of their collection. A significant portion of aquaculture production never reaches a market where representative sampling may be possible. A general consequence of this is the underestimation of subsistence aquaculture, although this has an important impact on the nutrition of the poorest segments of the Asian population.

Nevertheless, the regional and global trends deduced from the data can be viewed with some confidence. This is primarily due to the efforts of those countries which are supplying the more reliable data (e.g., Indonesia, Japan, Republic of Korea, Malaysia, Taiwan, the Philippines and Thailand) or have recently improved their database (e.g., Bangladesh, China, Vietnam).

Another source of frustration is the timegap between the date of the attempted review and that of the latest available data. In this paper the latest data are for 1990.

Aquaculture Production

Volume and Value of Production

The share of Asia in global aquaculture increased from 80% in 1975 to 81% in 1980 and to 85% by 1990, despite significant breakthroughs on other continents (e.g., in salmon culture in Europe and in shrimp culture in Latin America). In 1990, out of the 15.3 million tonnes of the world total, 12.9 million tonnes were produced in the Asia-Pacific region (Table 1).

Developing countries of Asia produced 75% of the global aquaculture production in 1990. However, the value was only 63% of the world total (Table 2). The average value of aquaculture products in 1990 was US$1.44 per kg in the developing countries of the region, while it reached $2.42/kg in the rest of the world and as much as $2.89/kg in the developed countries of Asia and the Pacific (Japan, Australia and New Zealand). The reason behind these differences is that the developing countries of the region culture primarily aquatic organisms low in the food chain (e.g., seaweeds, molluscs and noncarnivorous finfishes). These commodities are usually significantly cheaper than aquatic animals higher in the food chain (e.g., crustaceans and carnivorous fish) preferred in developed countries.

Table 1. Global aquaculture production in 1990. (Source: FAO 1992a).

Continent/region	Production ($\times 10^3$ t)	Share (%)
Asia + Pacific	12,954.8	84.5
Europe + ex-USSR	1,628.2	10.6
North America	407.1	2.7
South America + Caribbean	216.6	1.4
Africa + Middle East	115.8	0.8
World total	15,322.5	100.0

Table 2. Volume and value of aquaculture production in 1990. (Source: FAO 1992a).

Countries/regions	Volume ($\times 10^3$ t)	Volume (%)	Value ($\times 10^6$ US$)	Value (%)
Developing Asia/Pacific	11,540.4	75.3	16,637.2	62.9
Developed Asia/Pacific*	1,414.4	9.2	4,083.9	15.4
Asia/Pacific total	12,954.8	84.5	20,721.1	78.3
Rest of the World	2,367.9	15.5	5,734.3	21.7
World total	15,322.7	100.0	26,455.4	100.0

*Note: Developed countries of Asia/Pacific are Australia, Japan and New Zealand.

Out of the 12.9 million t of aquatic organisms cultured in 1990 in Asia and the Pacific, 54% was finfish. Seaweeds made up 24% of the total, while molluscs represented 17% (Table 3). Crustacean aquaculture, which received most of the attention in the past decade, produced only 5% of the total volume in 1990. Despite some minor changes in the share of certain commodities, total aquaculture production in the Asia-Pacific region has shown a remarkable steady and balanced growth over the past decade (Fig. 1). Average annual growth of total production was 7% in the region between 1975 and 1990, with developing countries showing more rapid annual growth (8%) than the developed ones (3%).

It should be noted, however, that behind the generally bright picture there are significant differences among the individual countries of Asia and the Pacific. Eight out of 44 countries/territories of the region did not report aquaculture production at all in 1990 (Maldives and Mongolia in Asia; Cook Islands, Nauru, Western Samoa, Tonga, Tuvalu and Vanuatu in the Pacific) and ten others produced less than 1,000 t (Bhutan and Brunei in Asia; Fiji, French Polynesia, Guam, Kiribati, Federated States of Micronesia, New Caledonia, Papua New Guinea and the Solomon Islands in the Pacific). The top ten countries, on the other hand, all produced more than 100,000 t in 1990 (Table 4).

Table 3. Share of major commodity groups in the aquaculture of Asia and the Pacific. (Sources: FAO 1984, 1992a).

Commodity groups	1975 (x 10^3 t)	(%)	1990 (x 10^3)	(%)
Finfishes	1,842.5	44.1	6,995.7	54.0
Crustaceans	25.9	0.6	575.9	4.5
Molluscs	676.5	16.2	2,195.6	16.9
Seaweeds	1,630.8	39.0	3,144.0	24.3
Others	6.9	0.1	43.6	0.3
Total	4,182.6	100.0	12,954.8	100.0

Inland Aquaculture

The details of total aquaculture production by inland and coastal aquaculture in Asia and the Pacific are presented in Table 4. Inland

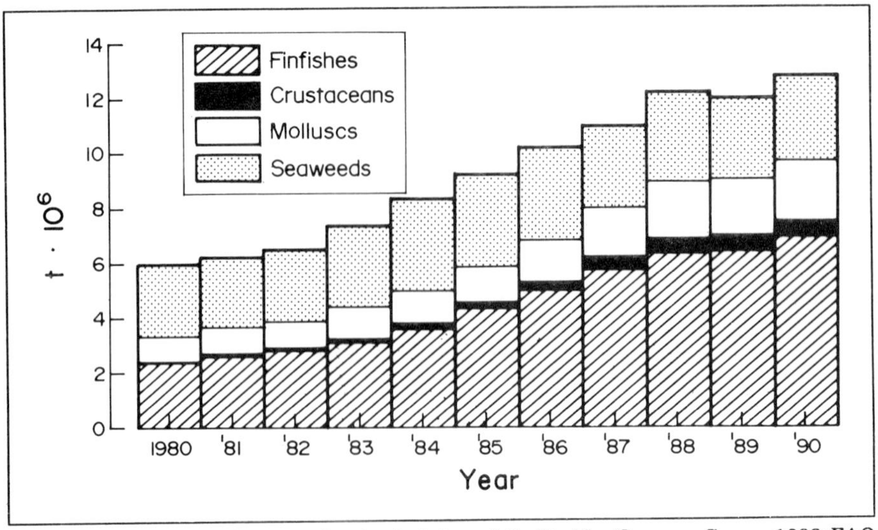

Fig. 1. Growth of aquaculture production in Asia and the Pacific. (Sources: Csavas 1988; FAO 1992a).

aquaculture produces almost half (49%) of the total. China is the leading inland aquaculture producer, followed by India and Indonesia.

The sheer volume of aquaculture production precludes finer analyses, therefore it is more appropriate to compare inland aquaculture production by land area (Table 5). Countries and territories with limited land area and high population densities (Hong Kong, Taiwan and Bangladesh) are outstanding in this respect (55.8, 41.1 and 11.6 kg·1,000 ha^{-1}, respectively), demonstrating the competitiveness of inland aquaculture in the use of scarce land resources.

Another indicator shown in Table 5 is the volume of inland aquaculture production per renewable freshwater resources in selected countries of Asia and the Pacific (unfortunately, data on renewable freshwater resources of some countries are not available). High values of China, Thailand and India (1,502, 841

Table 4. Inland and coastal aquaculture production in Asia and the Pacific in 1990. (Source: FAO 1992a).

Country/region	Total production (t)	Inland production (t)	Coastal production (t)	Share of coastal aq. (%)
Bangladesh	169,758	151,161	18,624	11.0
Bhutan	31	31	0	0
Brunei Darussalam	2	2	0	0
Cambodia	6,080	6,080	0	0
China	7,200,383	4,204,728	2,995,655	41.6
Hong Kong	10,256	5,525	4,731	46.1
India	1,011,136	982,136	29,000	2.9
Indonesia	558,795	242,625	316,170	56.6
Iran	45,134	45,134	0	0
Korea, D.P.R.	208,670	11,200	196,470	94.2
Korea, Rep. of	789,765	15,823	773,942	98.0
Laos	2,500	2,500	0	0
Malaysia	47,876	7,007	40,869	85.4
Myanmar	7,087	7,086	1	0
Nepal	9,258	9,258	0	0
Pacific Islands	1,442	172	1,270	88.1
Pakistan	40,057	40,016	41	0.1
Philippines	672,316	81,127	591,189	87.9
Singapore	1,857	0	1,857	100.0
Sri Lanka	5,700	5,000	700	12.3
Taiwan	343,954	148,795	195,159	56.7
Thailand	253,326	92,466	160,860	63.5
Vietnam	155,000	123,000	32,000	20.7
Developing countries	11,540,410	6,181,872	5,358,538	46.4
Australia	12,686	2,049	10,637	83.9
Japan	1,367,058	97,792	1,270,778	93.0
New Zealand	34,660	0	34,660	100.0
Developed countries	1,414,404	98,329	1,316,075	93.1
Regional total	12,954,814	6,280,201	6,674,613	51.5
Share (%)	100.0	48.5	51.5	-

Table 5. Indicators of inland aquaculture production in Asia and the Pacific in 1990.

Country/region	Inland aquaculture production (t)	Land area (x 10^3 ha)	Renewable inland aquaculture production per freshwater resources ($km^3 \cdot year^{-1}$)	land area ($kg \cdot 1{,}000\ ha^{-1}$)	water volume ($t \cdot km^3 \cdot year^{-1}$)
Bangladesh	151,161	13,017	1,357	11.6	111.4
Bhutan	31	4,650	.	0.0	.
Brunei Darussalam	2	527	.	0.0	.
Cambodia	6,080	17,652	88	0.3	69.1
China	4,204,500	929,100	2,800	4.5	1,501.6
Hong Kong	5,525	99	.	55.8	.
India	982,136	297,319	1,850	3.3	530.9
Indonesia	242,625	181,157	2,530	1.3	95.9
Iran	45,134	163,300	117	0.3	385.8
Korea, D.P.R.	12,200	12,041	.	1.0	.
Korea, Rep. of	15,823	9,873	63	1.6	251.2
Laos	2,500	23,080	270	0.1	9.3
Malaysia	7,007	32,855	456	0.2	15.4
Myanmar	7,086	65,754	1,082	0.1	6.6
Nepal	9,258	13,680	170	0.7	54.5
Pacific Islands	172	54,607	.	0.0	.
Pakistan	40,016	77,088	298	0.5	134.3
Philippines	81,127	29,817	323	2.7	251.2
Singapore	0	62	1	0.0	0.0
Sri Lanka	5,000	6,574	43	0.8	116.3
Taiwan	148,795	3,616	.	41.2	.
Thailand	92,466	51,089	110	1.8	840.6
Vietnam	123,000	32,536	.	3.8	.
Developing countries	6,181,872	2,100,299	(11,583)	2.9	(533.7)
Australia	2,049	761,793	343	0.0	6.0
Japan	96,280	37,652	547	2.6	176.0
New Zealand	0	26,867	397	0.0	0.0
Developed countries	98,329	826,312	1,287	0.1	76.4
Regional total	6,280,201	2,926,611	(12,870)	2.2	(488.0)

Note: Figures in parentheses are totals/averages of countries with available data, not those of the whole Asia/Pacific region.

and 531 $t \cdot km^{-3} \cdot year$, respectively) show that, although on a per hectare basis these countries do not belong to the top inland fish producers, they utilize very well their renewable freshwater resources with inland aquaculture. Putting it in a different way: some countries may have land constraints whereas others may have water constraints to developing further their inland aquaculture.

The advanced state of inland aquaculture could also be characterized by its share in the total inland fish production. Data in Table 6 demonstrate that in 1990, in the Asia-Pacific region, as much as 62% of the total inland fish production originated from aquaculture against 27% in the rest of the world. Aquaculture produced over 75% of the total inland fish supply in China, Hong

Table 6. Inland fisheries and aquaculture production in Asia and the Pacific in 1990. (Source: FAO 1992a).

Country/region	Total inland production (t)	Production from capture (t)	Production from culture (t)	%
Afghanistan	1,500	1,500	0	0.0
Bangladesh	594,377	443,216	151,161	25.4
Bhutan	1,000	969	31	3.1
Brunei Darussalam	130	128	2	1.5
Cambodia	65,100	59,020	6,080	9.3
China	5,237,621	1,032,893	4,204,728	80.3
Hong Kong	6,130	605	5,525	90.1
India	1,484,833	502,697	982,136	66.1
Indonesia	795,000	552,375	242,625	30.5
Iran	50,000	4,866	45,134	90.3
Korea, D.P.R.	110,000	97,800	12,200	11.1
Korea, Rep. of	33,000	17,177	15,823	48.0
Laos	20,000	17,500	2,500	12.5
Malaysia	14,664	7,657	7,007	47.8
Mongolia	131	131	0	0.0
Myanmar	144,586	137,500	7,086	4.9
Nepal	14,546	5,288	9,258	63.7
Pacific Islands	19,292	19,120	172	0.9
Pakistan	113,158	73,142	40,016	35.4
Philippines	585,766	504,639	81,127	13.9
Singapore	60	60	0	0.0
Sri Lanka	31,265	26,265	5,000	16.0
Taiwan	172,200	23,405	148,795	86.4
Thailand	200,000	107,534	92,466	46.2
Vietnam	240,000	117,000	123,000	51.3
Developing countries	9,934,359	3,752,487	6,181,872	51.6
Australia	4,306	2,257	2,049	47.6
Japan	208,120	111,840	96,280	46.3
New Zealand	103	103	0	0.0
Developed countries	212,529	114,200	98,329	46.3
Asia-Pacific total	10,146,888	3,866,687	6,280,201	61.9
Rest of World	4,298,508	3,123,098	1,175,410	27.3
World total	14,445,396	6,989,785	7,455,611	51.6

Kong, Iran and Taiwan. A significant shift from capture to culture fisheries and an increasing dominance of cultured fish in the inland fish catch are sure signs of diminishing natural freshwater habitats: the impoverishment of the original fish fauna due to overfishing and other human interventions.

Inland aquaculture is dominated by finfish production all over the world. Freshwater crustaceans, molluscs and aquatic plants contributed less than 1% to the total in the Asia-Pacific region and their share in the rest of the world is similarly very low (3%). Freshwater fish culture, especially in the developing countries of the region, produces primarily cheap food fish (e.g., cyprinids and tilapias) affordable for the poorer segments of the population. More expensive commodities (e.g., carnivorous fish species) are produced mainly in the developed or newly industrialized countries of the region.

Culture techniques in Asian inland aquaculture are characterized by simple, low input pond culture methods. Cage and pen culture in inland waters, although introduced in most countries of the region, seldom reaches commercially important proportions. However, in certain areas of Cambodia, Indonesia, Nepal, the Philippines, Thailand and Vietnam commercial freshwater cage or pen culture has demonstrated both economical and social feasibility. At the same time, these more intensive systems have the inherent threat of overloading the carrying capacity of the environment, as happened in the well-documented case of Laguna de Bay in the Philippines (Pullin 1981, this vol. p. 1). Raceway and tank culture are not really typical for Asia; few examples are found even in the developed and newly industrialized countries of the region.

Typical freshwater culture systems in the region are the extensive or semi-intensive polycultural systems with some fertilization and supplementary feeding. Suitable fish species (Chinese and Indian major carps and tilapias) for such systems were introduced in the 1960s and 1970s to practically every country of the region (De Silva 1989).

Freshwater aquaculture is generally a small-scale activity in Asia. Most of the ponds are operated by small farmers as part of their simple, rice-based farming systems. Thus, there are few problems with their environmental compatibility. Their competition for natural resources (land, water) or external inputs (manure, fertilizer, feed) remains also limited. Governments in many developing countries of the region, concerned about rice self-sufficiency, were afraid of converting too big a portion of ricefields to fishponds. Some of these countries even imposed bans on such conversions in the rapid growth phase of inland aquaculture. Later, however, these fears abated as it was proven that small farmers are unlikely to convert too high a portion of their ricefields to fishponds in order to maintain their simple but well balanced farming systems.

An interesting phenomenon was witnessed in several countries of the region where inland aquaculture was introduced only recently (e.g., in Bhutan, Laos, and Nepal). Although a significant portion of fishponds were constructed by converting ricefields, rice output of the affected communities increased rather than decreased. The explanation is that because of the introduction of aquaculture (which produces a high-value crop compared to rice) farmers had to improve on their traditional water and farm management practices. This, in turn, raised their rice yields.

Coastal Aquaculture

Coastal aquaculture in Asia received much less attention than freshwater fish culture in the past. Recently, however, the rapid growth of shrimp culture has focused attention on coastal areas and new perceptions have appeared which regard coastal aquaculture as a new, socially contradictory and environmentally incompatible phenomenon. A closer look at the history and status of Asian coastal aquaculture contradicts these views. Data given in Table 4 show that in 1990 as much as 51.5% of the total production of the region came from coastal aquaculture. This ratio has decreased rather than increased in the past 15 years; coastal aquaculture produced 63.3% of the regional total in 1975.

Coastal aquaculture dominates the scene in Australia, Japan, both Koreas,

Malaysia, New Zealand, the Philippines and Singapore by contributing more than 75% to the total output, while its share is between 50% and 75% in Indonesia, Taiwan and Thailand. If landlocked countries are disregarded, the least developed countries in this respect are Bangladesh, India, Cambodia, Myanmar, Pakistan, Sri Lanka and Vietnam with less than 25% of their total output produced by coastal aquaculture. Out of the 6.7 million t of coastal aquaculture products in the region, 81% comes from China, Japan, the two Koreas and Taiwan, showing a well defined geographical center of coastal aquaculture in Asia.

A more detailed analysis based on indicators showing coastal aquaculture production per land area and per length of coastline modulates but does not alter the above picture (Table 7). It is interesting to note that the two leading entities (China and Taiwan) produce more than 100 t of aquaculture products per every kilometer of their coastline; production exceeds 50 $t \cdot km^{-1}$ in Thailand and the Republic of Korea. The regional average is almost 25 $t \cdot km^{-1}$.

Coastal aquaculture in the major producing countries is dominated by seaweeds and molluscs. In 1990, the proportion of seaweeds was 47% in Asia, while molluscs contributed 33%. These ratios are typical for Asia. In the rest of the world, seaweed culture produces only 4% of the total coastal output whereas molluscs contribute as much as 65% to the total. The share of finfish is 11% in Asia and 23% in the rest of the world, while crustaceans contribute 9% of the total in Asia and other regions.

Despite significant recent advances in marine fish cage culture and intensive pond culture of shrimp in most of the developing countries of the region, traditional extensive or semi-intensive pond culture still dominate coastal fish and shrimp production. Extensive, trapping/growing ponds producing a mix of fish (mainly milkfish and mullets) and crustaceans are traditional in the tidal zones of many Asian countries. The general tendency over the past decade has been to intensify these and to shift from an uncontrolled "polyculture" to monoculture of shrimp by replacing trapping of wild juveniles with stocking. However, in the second half of the 1980s, construction of more intensive shrimp ponds started in several countries, both in the mangrove belt and behind it on higher grounds. This profit-driven development raised serious environmental and socioeconomic concerns, especially in Taiwan and Thailand (Csavas 1990; Phillips et al., this vol.).

Cage culture of carnivorous marine fish was practiced mainly in Japan and Hong Kong in the 1970s, but has proliferated in parts of Southeast Asia during the past decade. Simple methods of cage culture are accepted more easily by coastal fishing communities than pond culture of fish or shrimp, for which land and significant investments are needed. Expansion of cage culture was especially spectacular in Thailand in the early 1980s, but constraints in feed supply and marketing of the products soon slowed down further growth. Environmental problems related to marine cage culture are not too common in the developing countries of Asia due to the limited volume of production. Hazards of overloading the environment with marine cages, however, are well documented in Hong Kong and Japan (Anon. 1990; Davy 1991).

In mollusc and seaweed production, both on-bottom and off-bottom culture methods are common, although off-bottom methods which offer better control over the culture environment are gaining ground. Intensification of seaweed culture methods is also being pursued by the use of fertilizers in open coastal waters.

Whereas inland aquaculture in Asia is dominated by small-scale production, land

Table 7. Indicators of coastal aquaculture production in Asia and the Pacific in 1990.

Country/region	Coastal aquaculture production (t)	Land area (x 10^3 ha)	Coastline (km)	Coastal aquaculture production per area (kg·1,000 ha^{-1})	Coastal aquaculture production per coastline (t·km^{-1})
Bangladesh	18,624	13,017	700	1.43	26.6
Bhutan	0	4,650	-	-	-
Brunei Darussalam	0	527	130	0.00	0.0
Cambodia	0	17,652	435	0.00	0.0
China	2,995,655	929,100	18,000	3.22	166.4
Hong Kong	4,731	99	547	47.79	8.7
India	29,000	297,319	7,517	0.10	3.9
Indonesia	316,170	181,157	81,000	1.75	3.9
Iran	0	163,300	2,500	0.00	0.0
Korea, D.P.R.	196,470	12,041	18,313	16.32	10.7
Korea, Rep. of	773,942	9,873	13,200	78.39	58.6
Lao P.D.R.	0	23,080	-	-	-
Malaysia	40,869	32,855	3,432	1.24	11.9
Myanmar	1	65,754	2,278	0.00	0.0
Nepal	0	13,680	-	-	-
Pacific Islands	1,270	54,607	20,537	0.02	0.1
Pakistan	41	77,088	1,120	0.00	0.0
Philippines	591,189	29,817	34,600	19.83	17.1
Singapore	1,857	62	140	29.95	13.3
Sri Lanka	700	6,474	1,770	0.11	0.4
Taiwan	195,159	3,616	1,600	53.97	122.0
Thailand	160,860	51,089	2,624	3.15	61.3
Vietnam	32,000	32,536	3,260	0.98	9.8
Developing countries	5,358,538	2,100,299	213,703	2.55	25.1
Australia	10,637	761,793	20,000	0.01	0.5
Japan	1,270,778	37,652	29,751	33.75	42.7
New Zealand	34,660	26,867	5,400	1.29	6.4
Developed countries	1,316,075	826,312	55,151	1.59	23.9
Asia-Pacific total	6,674,616	2,805,842	268,854	2.49	24.8

tenure patterns in the coastal zone have produced a different picture. Tidal wetlands were traditionally common property resources with rather low perceived value, used primarily by small-scale fishers and other artisans. In most countries of the region, traditional, extensive trapping/growing ponds, the size of which far exceeded the dimensions of freshwater ponds, were constructed in the coastal zone. In Indonesia, for instance, almost 80% of freshwater fishponds are smaller than 0.1 ha, while more than 62% of the brackishwater ponds are bigger than 2 ha (Cholik 1988). In the recent rush for suitable shrimp pond sites, influential investors have had substantial advantages over small farmers (Hannig 1988). Thus, the ratio of small farms has further decreased in coastal areas.

Coastal aquaculture also differs from inland aquaculture in other ways. Whereas semi-intensive freshwater pond fish culture blends well with the rice-based rural economy of Asia, some negative environmental impacts of some coastal aquaculture systems are beyond doubt. Especially harmful are those pond systems which are constructed in mangrove areas and alter irreversibly the original rich ecosystem. Another significant difference lies in the nature of competition for the available resources. In inland areas, there may be some competition between aquaculture and crop production for land and water, usually within an integrated farm unit. In coastal aquaculture, the competition is usually between the traditional users of hitherto open-access resources and those who are encroaching on and expropriating these. This is because traditional users of coastal wetlands do not have property rights over these lands, which legally belong to the state in most countries of Asia. Moreover, small-scale fishers and other users of coastal resources (charcoal burners, gatherers, etc.) usually belong to the poorest segment of the population; they are easily outcompeted by "outsiders" in acquiring legal ownership over coastal lands and transforming them to shrimp/fish farms (Bailey 1988; Bailey and Skladany 1988).

It is also important to realize that the aspiration of coastal aquaculture is usually not to produce more food for local consumption. Coastal communities usually have a reasonable supply of cheap captured/collected seafood with which no cultured product can compete in price. In coastal areas, the dominant function of aquaculture is income generation, the production of cash crops sold in distant markets (often for export). This is as legitimate an ambition as cheap food fish production in inland areas. The real problem is that coastal communities are seldom direct beneficiaries of such aquaculture development. Benefits of shrimp culture development, for instance, trickle down to coastal fishing communities only by generating some additional employment and by enhancing the overall rural development of hitherto neglected, impoverished coastal areas. However, in countries with high population densities, coastal aquaculture can open up a new frontier without putting more pressure on the limited land resources. This is the reason why China has turned towards this type of aquaculture and has given its development high priority over the past decade. It is also important to realize that out of the broad range of coastal aquaculture systems only the pond culture of marine fish and shrimp is environmentally incompatible and socially controversial. These systems do not produce more than 13% of the total coastal aquaculture production volume (0.9 million t out of the 6.6 million t in 1990).

Finfish Culture

As already demonstrated, finfish species form the biggest commodity group in the aquaculture production of the Asia-Pacific region (54% of the total in 1990), and the most important one in the nutrition of the poor of the region's developing countries. Out of the 7 million t of cultured fish produced in 1990 as much as 6.1 million t or 87% were freshwater fish; the proportion of diadromous species was 9%, that of marine fish only 4%. These values are characteristic for the region. In the rest of the world, the shares of both freshwater and marine fish are lower (61% and 2%, respectively), at the same time diadromous fish (primarily salmonids) have a much higher share (38%).

Production of finfish shows a significant and steady growth in the region (Fig. 2). Between 1975 and 1990, the annual average growth rate was 8.7% per year, the developing countries of the region having a slightly higher rate of growth (9.0% per year). Information on the species composition of finfish culture was rather scanty before 1984 (when the regular data collection work of FAO started). However, the improvement of the database now makes possible a more detailed analysis.

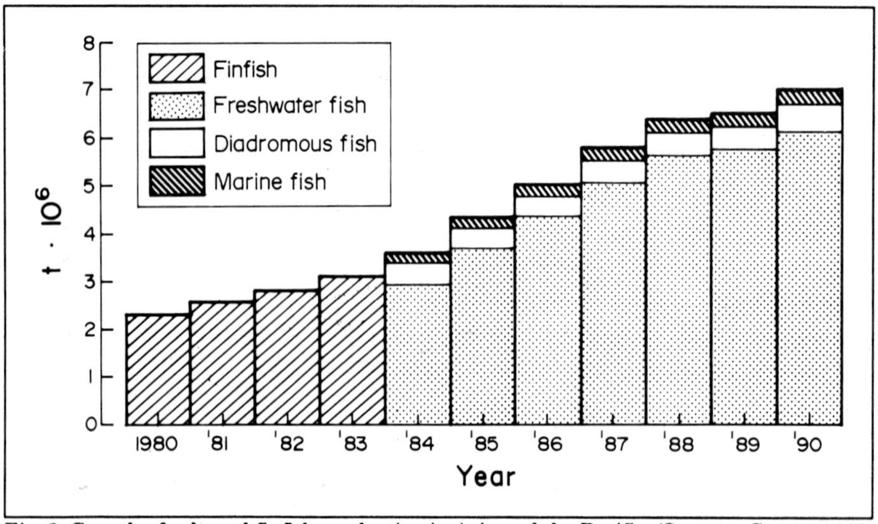

Fig. 2. Growth of cultured finfish production in Asia and the Pacific. (Sources: Csavas 1988; FAO 1992a).

The most important question is how big a portion of cultured finfish production belongs to the cheap category and what is the share of "luxury" species. Considering the diversity of cultured fish species, a distinction between carnivorous and noncarnivorous species seems to be the most suitable classification from this point of view. This reflects not only the difference in price of the product but also a difference in culture systems, because carnivores are in most cases cultured in intensive monoculture.

Table 8 shows that, out of the region's 7.0 million t of finfish production in 1990, 93% belonged to noncarnivorous species. Further, the vast majority (86%) of cultured noncarnivorous fish comprises cyprinid fish species (Table 9). Together with tilapias (5%), these relatively cheap food fishes add up to 91% of the noncarnivorous finfish and 84% of the total finfish volume cultured in the region. In the rest of the world, the combined share of cyprinids and tilapias is only 45% of the total cultured fish. Another feature of Asia and the Pacific is the 7% share of milkfish, a noncarnivorous species not cultured outside the region (Table 9).

Major producers of noncarnivorous fish are Bangladesh, China, India, Indonesia, the Philippines, Taiwan and Vietnam. Aquaculture was not traditionally practiced in Nepal, Laos and Bhutan (three landlocked countries of the region) but, during the past two decades, integrated fish culture was introduced to these countries, with international assistance, and was very well received. It is based on low-input polyculture of cyprinid fish species. In nutrient-rich lakes/reservoirs in Nepal and Laos, cage culture of filter-feeding Chinese carps was also introduced and has proven both economically and ecologically viable.

In some Asian countries (most characteristically in Bangladesh and India, but also in China), relatively small seasonal or perennial waterbodies, not originally meant for fish culture, are now widely utilized with low-input aquaculture. These "tanks" (as they are called in much of South Asia) were dug out as clay pits or constructed as small reservoirs for storing water for irrigation, domestic use and/or watering livestock. They are in most cases undrainable and have no additional water supply during the dry season. However,

Table 8. Cultured noncarnivorous and carnivorous fish in Asia and the Pacific in 1990. (Source: FAO 1992a).

Country/region	Total finfish production	Noncarnivorous (t)	(%)	Carnivorous (t)	(%)
Bangladesh	151,161	151,161	100.0	0	0.0
Bhutan	31	31	100.0	0	0.0
Brunei Darussalam	2	2	100.0	0	0.0
Cambodia	6,080	5,714	94.0	366	6.0
China	4,232,895	4,199,195	99.2	33,700	0.8
Hong Kong	9,451	5,700	60.3	3,751	39.7
India	981,938	942,327	96.0	39,611	4.0
Indonesia	381,425	374,055	97.9	7,370	1.9
Iran	45,134	44,241	98.0	893	2.0
Korea, D.P.R.	12,200	11,000	90.2	1,200	9.8
Korea, Rep. of	17,934	12,157	67.8	5,777	32.2
Lao P.D.R.	2,500	2,500	100.0	0	0.0
Malaysia	8,742	6,254	71.5	2,488	28.5
Myanmar	7,083	7,083	100.0	0	0.0
Nepal	9,258	9,258	100.0	0	0.0
Pacific Islands	204	192	94.1	12	5.9
Pakistan	40,016	40,000	99.9	16	0.1
Philippines	295,151	291,904	98.9	3,247	1.1
Singapore	560	7	1.2	553	98.8
Sri Lanka	5,000	5,000	100.0	0	0.0
Taiwan	236,540	172,241	72.8	64,299	27.2
Thailand	86,541	53,806	62.2	32,735	37.8
Vietnam	115,000	115,000	100.0	0	0.0
Developing countries	6,644,846	6,448,828	97.1	196,018	2.9
Australia	3,733	1	0.0	3,732	100.0
Japan	344,654	36,603	10.6	308,051	89.4
New Zealand	2,510	0	0.0	2,510	100.0
Developed countries	350,897	36,604	10.4	314,293	89.6
Asia-Pacific total	6,995,743	6,485,432	92.7	510,311	7.3
Rest of World	1,414,991	706,549	49.9	708,442	50.1
World total	8,410,734	7,191,981	85.5	1,218.753	14.5

Table 9. Cultured noncarnivorous fish by species groups in 1990. (Source: FAO 1992a).

Species groups	Asia-Pacific ($\times 10^3$ t)	(%)	Rest of World ($\times 10^3$ t)	(%)
Cyprinids	5,589.6	86.2	562.6	79.6
Tilapias	320.7	5.0	70.1	9.9
Milkfish	429.9	6.6	-	-
Mullets	9.2	0.1	14.3	2.0
Others	136.0	2.1	59.6	8.5
Total	6,485.4	100.0	706.6	100.0

with proper stocking, they can produce significant amounts of fish without endangering their other functions. Similarly, in Vietnam and Laos, aquaculture is successfully practiced in the millions of bomb craters which dot the war-torn landscape using the same (primarily photosynthesis-dependent) fish culture method. These methods do not compete for land, water, fertilizer and feed with crop or livestock production, blend well with their environment and produce the cheapest fish next to those from capture fisheries.

Culture of carnivorous fish, although it produces only 7% of the regional finfish production, cannot be neglected because it has a significant economic and environmental impact. Countries of the region can be classified into three groups in this regard: Australia, Japan, New Zealand and Singapore produce almost exclusively carnivorous species; Hong Kong, the Republic of Korea, Malaysia, Taiwan and Thailand produce significant amounts of carnivores (over 25% but below 50% of their total finfish production); while carnivorous fish production is insignificant in the rest of the region (Table 8). It is obvious that the production of the more expensive carnivorous finfish species is closely related to the prosperity of the country. Also, the total volume of cultured carnivores produced in a country is not necessarily consumed there. High-value export products play an important role in the foreign exchange earning of the Southeast Asian countries in which aquaculture is more developed.

While in other regions of the world, salmonid species and catfish dominate the production of carnivores, Asia and the Pacific present a different and more diverse picture (Table 10). Salmonids are only 11% of the total and catfish species represent 14%. The biggest contributor is the yellowtail *(Seriola quinqueradiata)*, with a 33% share of the total carnivorous fish production; eels amount to 19%; and various species of seabreams make up 10%. The "other" category comprises marine fish species too, of which the most important are the sea perch *(Lates calcarifer)*, called "seabass" in Asia and "barramundi" in Australia, and the bastard halibut or Japanese flounder *(Paralichthys olivaceus)*.

Despite the rapid growth in the production of some minor groups of cultured carnivores (e.g., salmon, groupers, flatfishes), development of carnivorous fish production from 1984 to 1990 has been slow (7% per year). The reason is the overwhelming dominance of Japan, where the required volume from one or another species is limited and market niches get saturated rather rapidly.

About 60% of the carnivorous fish in the region are cultured in marine cages (yellowtail, seabreams, salmons, seabass, groupers, snappers) and 40% in intensive ponds (eels, trouts, catfishes). Originally all these species were fed with so-called trashfish (by-catch of trawling) or, in Japan, with low-value marine fish (like sardines, anchovies, sand lances, mackerels). Producers in Japan, however, have shifted to formulated feeds, which

Table 10. Cultured carnivorous fish by species groups in 1990. (Source: FAO 1992a).

Species groups	Asia-Pacific (x 10^3 t)	(%)	Rest of World (x 10^3 t)	(%)
Yellowtails	169.1	33.1	0.3	0.0
Eels	96.1	18.8	8.0	1.1
Catfishes	70.4	13.8	169.7	24.0
Seabreams	53.0	10.4	5.1	0.7
Salmons	27.3	5.4	256.1	36.2
Trouts	27.2	5.3	266.3	37.6
Others	67.2	13.2	2.9	0.4
Total	510.3	100.0	708.4	100.0

are usually given in combination with fresh or frozen fish (some 10% of the total). In less developed countries, feeding of carnivores is still very much dependent on by-catch or low-value marine fish, which is increasingly in short supply. Fishing communities engaged in marine fish cage culture, therefore, often resort to capturing small-sized juveniles of economically valuable marine fish species in order to feed their stocks. Formulated feeds are also dependent on marine protein as they contain 40 to 70% fish meal. Thus, carnivorous fish culture is in direct competition with animal husbandry for the limited marine protein sources. Globally about 10% of the total fish meal production is used in aquaculture feeds (Pike 1989).

Due to the high stocking densities and heavy feeding, both cage culture and pond culture of carnivores are potential sources of pollution. Hazardous concentrations of cages/cage farms are seldom controlled by rules and regulations in the developing countries of Asia; even if such legislation exist, its enforcement is problematical. However, as feeding is based almost exclusively on fresh fish, the major problem is organic pollution, which can often be absorbed by the environment, provided the site selection is properly done.

Polluted effluents from high density eel or catfish ponds in inland areas are more problematical as they usually enter a common water supply network, from where they may be reused in neighboring ponds. Unfortunately, in developing countries of the region, separate drainage and irrigation networks rarely exist.

Another potential hazard is the unchecked (and in most cases unwarranted) use of various drugs, primarily antibiotics, in intensive pond culture of carnivores. Unscrupulous dealers persuade fish farmers to use these drugs as preventive measures (for instance in *Clarias* catfish ponds against the epizootic ulcerative syndrome) with very dubious benefits but with high risks of developing more virulent, antibiotic-resistant strains of pathogens (see Austin, this vol.).

Crustacean Culture

As noted earlier, until 1983 production of this commodity group did not play a significant role in Asian aquaculture, although its share increased from 0.6 to 1.6% of the total volume between 1975 and 1983. Then a spectacular growth started (Fig. 3). Between 1983 and 1988, the average annual growth of crustacean production was 41%, and by 1990 crustaceans reached 4.5% of the total volume of cultured aquatic organisms in the region.

Details of crustacean production in Asia and the Pacific in 1987 are presented in Table 11. Out of the 0.7 million t global output, the region produced 81%. Almost all came from the developing countries of Asia. Major producers were China, Indonesia, the Philippines, Taiwan, Thailand and Vietnam. Marine shrimp was the dominant species group in the region, mostly *Penaeus* and *Metapenaeus* spp. Freshwater prawns contributed 5% to the regional total and crabs, lobsters and other marine crustaceans provided the rest. Production of freshwater crayfish, which is an important species group in the rest of the world, is insignificant in the region, despite marked interest in the culture of *Cherax* species in Australia.

As crustacean aquaculture in Asia is dominated by brackishwater and marine species, it is primarily a coastal activity. When comparing the development of crustacean culture in various countries, the volume of production per kilometer of coastline is a suitable indicator (Table 12). In 1987, Taiwan (the leader in shrimp culture development) produced 56.7 t of crustaceans per kilometer of its coastline.

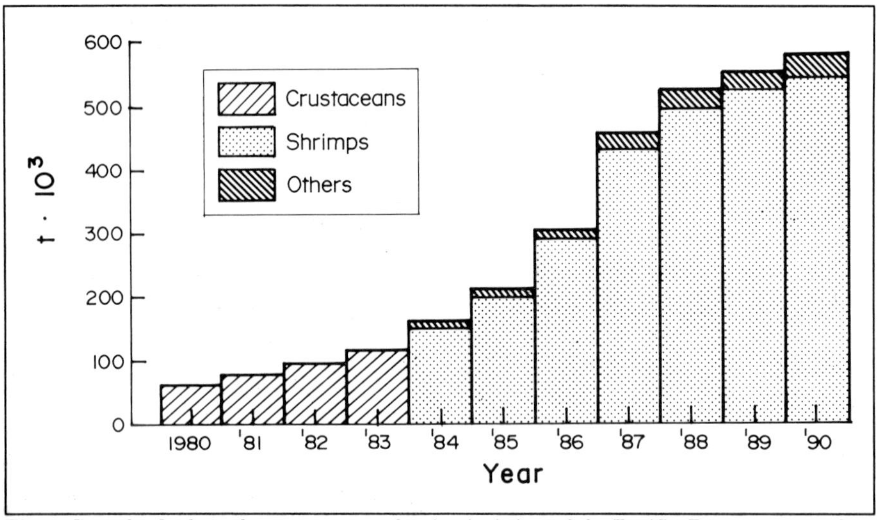

Fig. 3. Growth of cultured crustacean production in Asia and the Pacific. Data aggregated up to 1983. (Sources: Csavas 1988; FAO 1992a).

Table 11. Crustacean production in Asia and the Pacific in 1990. (Source: FAO 1992a).

Country/region	Freshwater crustaceans (t)	Shrimps (t)	Other marine crustaceans (t)	Total (t)
Bangladesh	-	18,624	-	18,624
China	-	186,181	7,833	194,014
India	198	26,000	-	26,198
Indonesia	-	94,960	2,410	97,370
Korea, D.P.R.	-	11,000	-	11,000
Korea, Rep. of	-	312	-	312
Malaysia	137	1,430	3	1,570
Myanmar	3	1	-	4
Pacific Islands	20	563	-	583
Pakistan	-	41	-	41
Philippines	-	53,989	1,000	54,989
Singapore	-	82	200	282
Sri Lanka	-	700	-	700
Taiwan	11,607	18,126	1,358	31,091
Thailand	8,000	89,300	30	97,330
Vietnam	8,000	30,000	-	38,000
Developing countries	27,965	531,309	12,834	572,108
Australia	109	594	-	703
Japan	50	3,000	30	3,080
Developed countries	159	3,594	30	3,783
Asia-Pacific total	28,124	534,903	12,864	575,891
Rest of World	37,717	101,904	74	139,695
World total	65,841	636,807	12,938	715,586

Table 12. Crustacean production per length of coastline in Asia. (Source: FAO 1992a).

Countries/regions	1987 production (t·km^{-1})	1990 production (t·km^{-1})
Taiwan	56.7	19.4
Bangladesh	21.1	26.6
Thailand	13.5	37.1
Vietnam	8.8	11.7
China	8.7	10.8
Singapore	3.2	2.0
India	2.0	3.5
Philippines	1.1	1.6
Indonesia	0.7	1.2
Malaysia	0.2	0.5
Developing Asia-Pacific	2.1	2.7
Developed Asia-Pacific	0.1	0.1
Regional average	1.7	2.1

By 1990, however, production fell to 19.4 t·km^{-1} in Taiwan, at the same time Thailand, Bangladesh, Vietnam and China increased their production to 37.1 t·km^{-1}, 26.6 t·km^{-1}, 11.7 t·km^{-1} and 10.8 t·km^{-1}, respectively. The rest of the major producers (including India, Indonesia and the Philippines) remained well below 10 t·km^{-1}. The high production level reached by Taiwan in 1987 has not proven sustainable, but the wide gap between the first five countries/territories and the rest shows that shrimp production potentials in the region are still far from being fully utilized.

Shrimp culture is presently regarded as the most obdurate destroyer of mangroves, which are now acknowledged as highly valuable coastal resources. Beyond doubt, pond culture not only eradicates the natural mangrove vegetation, but the construction of canals and dikes also alter irreversibly the hydrological characteristics of the areas. Much damage has been done to mangrove habitats in the major shrimp producing countries and the destruction continues despite major efforts of the governments involved (Mepham and Petr 1987). However, the development of shrimp culture is not as closely related to the availability of mangroves as often perceived. The relationship between the cultured shrimp production of a country and its mangrove resources is almost an inverse one (Fig. 4). One could argue that shrimp culture has already destroyed

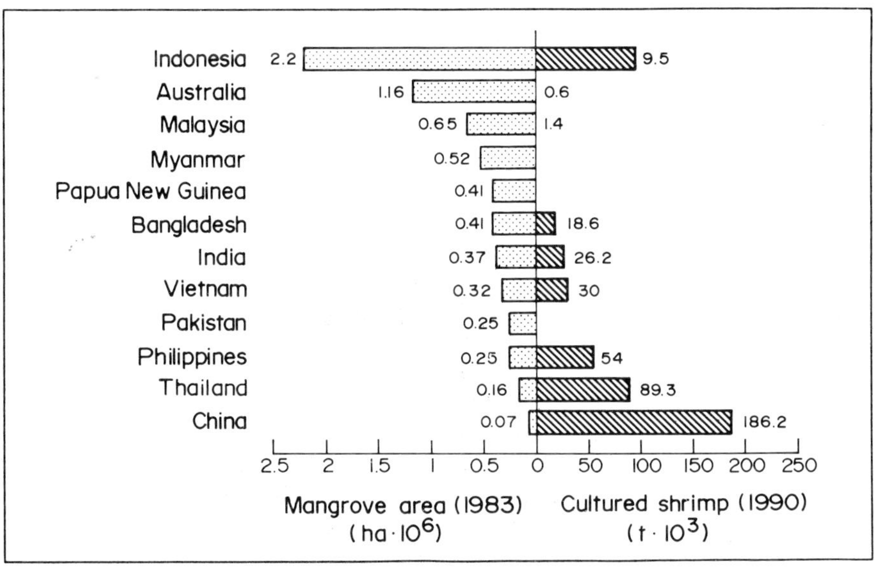

Fig. 4. Relationship between mangrove resources and cultured shrimp production. (Sources: Saenger et al. 1983; FAO 1992a).

the mangrove areas in the major producing countries, but data presented in the figure are from different years: mangrove areas are represented by figures published before the exponential growth of shrimp culture started (Saenger et al. 1983), whereas shrimp production data are from 1990.

Mangroves are in fact suboptimal sites for shrimp culture: construction is difficult to mechanize and expensive in the tidal zone and the soil is very often acidic and has to be improved at additional cost. There was only one advantage of selecting shrimp pond sites within mangrove areas, namely, the abundance of wild shrimp seed for trapping/growing operations. Once shrimp culture became independent of trapped wild seed, there was no good reason to place the ponds in mangrove areas; the more so because, in more productive semi-intensive systems, tidal water exchange is not enough and pumping has to be introduced anyway. Moreover, complete drainage of the ponds and drying of the pond bottom is also needed and this cannot be achieved by gravity alone in the tidal zone.

Disadvantages of converted mangroves as shrimp ponds - primarily the unfavorable experiences gained with acid sulfate soils - have turned the farmers' interests further inland. Pond construction costs on drylands are considerably lower than in swamplands and 30 to 40% savings have been reported from Thailand (FAO 1988). Marginal, salt-affected ricelands, uneconomic sugarcane fields and aged coconut plantations have been converted to shrimp ponds in many countries of the region (e.g., in Thailand, the Philippines, Indonesia and Malaysia). While this trend undoubtedly helped to conserve remaining mangroves, it involved another environmental hazard, namely the salinization of coastal subterranean freshwater aquifers and hitherto productive neighboring croplands (Phillips et al., this vol.).

Unfortunately, there are no reliable regional data on the siting of the new shrimp farms, but a recent survey implemented in the central part of Thailand provides some valuable information on this topic. Chaichavalit (1989) reported that only 32% of the new farms were located in converted mangroves (including nipa palm stands) and as much as 21% of them occupied former coconut plantations and other higher grounds. However, almost half of the new farms were constructed on the sites of former extensive trapping/growing ponds or salt farms, which were once also mangrove areas. The region had already in 1981 almost 0.5 million ha of trapping/growing ponds (called tambaks in Indonesia), while the total area of shrimp ponds in 1990 was estimated at about 0.9 million hectares (Rosenberry 1991).

In the Philippines, some 310,000 ha of mangroves were deforested since 1920 (Zamora 1989) but the total shrimp pond area was only about 200,000 ha in 1989 and a significant part of this occupies former sugarcane and coconut plantations, coastal ricefields. In Thailand, about 172,000 ha of mangroves were converted since 1961 (Chantadisai 1989). However, the total area under shrimp culture in 1989 was only 90,000 ha, a significant part of which is located outside the tidal zone. In Indonesia, about 20% of the mangroves or 800,000 ha are considered to be suitable for being converted to shrimp ponds (Adriawan and Jhamtani 1989), but the total shrimp pond area in 1989 was only 250,000 ha and Indonesia had already 155,000 ha of tambaks in 1980 (Soenodihardjo and Soerianegara 1989), most of them in use since the 15th century.

Shrimp farmers, after all, are usually converting mangrove areas already destroyed by the logging concessions or pulp-wood, wood-chip or charcoal industries which have expanded into the coastal areas since the 1970s. Much of the mangroves converted to shrimp ponds were already worthless shrub lands rather than high value forests, and would have needed systematic and expensive reforestation to become productive forests again. However, because of the entrenched interests of the logging industry in many countries of Asia, the least risky solution is to put the blame on the end-users of destroyed mangroves, that is, on shrimp farmers.

The environmental effects of shrimp culture are discussed in more detail in Phillips et al. (this vol.).

Mollusc Culture

Although several countries of East Asia have traditions in culturing molluscs, aquaculture production of this commodity group is not as widespread in the region as that of finfishes or crustaceans. The share of the Asia-Pacific region in the global production of cultured molluscs was 55% in 1975 and by 1990 it reached 74%. Several of the major producers of other aquaculture commodities (e.g., Bangladesh, Indonesia, Pakistan and Vietnam) do not report mollusc culture at all. This list of countries indicate cultural problems with the acceptance of molluscs. The growth of mollusc culture in the region was relatively slow until 1983 after which it accelerated to over 11% per year (Fig. 5). However, the share of molluscs in the total aquaculture output of the region has not changed since 1975; in 1990 it was 17% (Table 3).

Details of cultured mollusc production are given in Table 13. Major producers in 1988 in terms of total volume were China, Japan and the Republic of Korea, followed by Thailand and Taiwan. The comparison of countries is again more accurate on the basis of their production per length of coastline (Table 14). In 1990, China produced over 60 $t \cdot km^{-1}$ of cultured molluscs; the next three (Taiwan, Thailand, the Republic of Korea) between 25 and 40 $t \cdot km^{-1}$. These figures show the dominant role of cooler waters in the present production pattern and also the potentials in tropical countries.

As in the rest of the world, freshwater molluscs play only a minor role in the region, although interest in freshwater

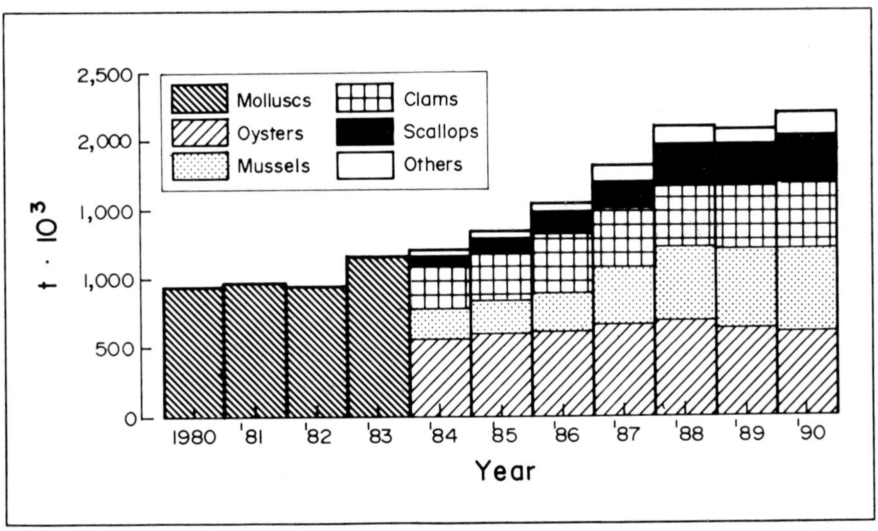

Fig. 5. Growth of cultured mollusc production in Asia and the Pacific. Data aggregated before 1984. (Sources: Saenger et al. 1983; FAO 1992a).

Table 13. Cultured mollusc production in Asia and the Pacific in 1990. (Source: FAO 1992a).

Country	Oysters (t)	Mussels (t)	Clams, cockles (t)	Other molluscs (t)	Total (t)
China	82,354	495,895	291,348	252,767	1,122,364
Hong Kong	805	0	0	0	805
India	-	-	-	3,000	3,000
Korea, D.P.R.	-	-	-	55,000	55,000
Korea, Rep. of	219,124	9,759	97,754	698	327,335
Malaysia	-	1,582	35,931	51	37,564
Pacific Islands	92	7	-	-	99
Philippines	13.485	17,515	-	-	31,000
Singapore	-	1,015	-	-	1,015
Taiwan	28,153	-	36,062	1,176	65,391
Thailand	1,400	56,050	12,000	-	69,450
Developing countries	345,413	581,823	473,095	312,692	1,713,023
Australia	7,171	929	-	-	8,100
Japan	248,793	-	1,500	192,042	442,335
New Zealand	2,100	30,000	-	50	32,150
Developed countries	258,064	30,929	1,500	192,092	482,585
Asia-Pacific total	603,477	612,752	474,595	504,784	2,195,608
Rest of World	273,152	469,022	25,093	2,165	769,432
World total	876,629	1,081,774	499,688	506,949	2,965,040

Table 14. Mollusc production per length of coastline in Asia in 1990. (Source: FAO 1992a).

Countries/regions	Mollusc production ($t{\cdot}km^{-1}$)
China	62.4
Taiwan	40.9
Thailand	26.5
Korea, Rep. of	24.8
Japan	14.9
Malaysia	11.0
Singapore	7.3
New Zealand	6.0
Korea, D.P.R.	3.0
Hong Kong	1.5
Developing Asia-Pacific	8.0
Developed Asia-Pacific	8.8
Regional average	8.2

pearl farming is increasing in several countries. Otherwise, the share of the major species groups in the region is distinctly different from the rest of the world. In other regions mussels dominate; their share in the total in 1990 was 61%, and almost all of the rest came from oyster species (36%). In the Asia-Pacific region, there are four almost equally important species groups: oysters (27.5%), mussels (27.9%), clams, cockles and arkshells (21.6%), and scallops (15.4%). Another special feature is the very high proportion of cultured molluscs in the total mollusc production. In 1990, 64% of mollusc landings (excluding squids, cuttlefishes and octopuses, the culture of which is not yet commercialized) came from aquaculture, while in the rest of the world this share was 36% (FAO 1992a, 1992b). In the case of some mollusc culture methods, the line between stock enhancement and aquaculture is somewhat blurred. However, the application of the ownership principle laid down in the FAO definition of aquaculture helps to make reasonably accurate distinctions. Therefore, the results of stock enhancement efforts, which have reached huge proportions in Japan, are not reported as aquaculture.

Mollusc culture is still almost entirely dependent on collected seed, although hatchery technologies are available for the most important cultured species. The high fecundity of commercially important bivalves makes mass collection of spat feasible in all those areas where breeding stocks of the required species are not yet depleted. Thanks to the hardy nature of juveniles of most species, long distance transport of seed from good breeding grounds to depleted ones is also feasible. However, in the case of some species (e.g., abalones, scallops), availability of natural seed is limited. Pioneering mollusc hatcheries are now in operation in some countries of the region but seed production from these hatcheries, although feasible technically, does not yet seem to be feasible economically. Governments are focusing on protecting good seed producing areas and on monitoring spatfall rather than on the establishment of hatcheries (FAO 1988).

Availability of suitable culture sites is becoming a serious constraint in those countries where mollusc culture is well developed. Most of the suitable sites in the two Koreas and Japan are already occupied; even in peninsular Malaysia and Thailand it is difficult to find new sites for cockle *(Anadara granosa)* culture. Overuse of suitable sites can lead to the deterioration of the culture environment. Growth rates of oysters in Japan have been decreasing, attributed to the self-pollution with the excreta of the cultured animals themselves. Similar problems are emerging in the Republic of Korea, where the government plans to limit further access to culture sites (Park 1988). Mollusc culture in both of these countries has been declining since 1987/88.

Industrial pollution resulting in heavy metal accumulation in cultured mussels, is not yet a widespread problem in the developing countries of the region. On the other hand, bacterial contamination of cultured molluscs originating from pollution with domestic sewage is a widespread public health concern in the region. Such pollution of marine areas is much less documented than that of the freshwaters in the region, although the situation in Southeast Asia is now under increasing scrutiny (Ruddle 1981; Alabaster 1986). Untreated sewage from the burgeoning coastal population centers is well known to contaminate otherwise ideal mollusc culture sites. Moreover, with the rapid development of tourism, hitherto pristine sites are rapidly becoming polluted with the untreated wastes of hotels and beach resorts (Chua and Garces 1992). Bacterial contamination is often reported to occur also as a result of poor postharvest handling of molluscs (FAO 1988). Increasing occurrences of red tides are further affecting culture sites in the region (Maclean, this vol.).

Molluscs produced in the Asia-Pacific region are in general low-priced. Supply and demand are balanced and in recent years production has decreased slightly in Hong Kong, Japan, the Republic of Korea, Malaysia, the Philippines, Singapore and Thailand. The demand, however, would increase if hygienic standards and the general image of the product in this respect were improved. Depuration, followed by certification (already mandatory for some export markets) is increasingly recognized as an important means of boosting consumption. Protecting production sites from pollution and improving the hygiene of postharvest handling seem to be more viable options for the time being. Export markets, however, cannot be developed without the introduction of efficient depuration and inspection/certification methods. In Malaysia, it has been estimated that depuration would add about 10% to the production costs and at the same time shorten somewhat the shelf life of cockles (FAO 1988).

In order to establish mollusc culture or to diversify the available commodities, several introductions of exotic mollusc species have been made into the region over the past two decades, especially in the Pacific Islands. This has raised international concern because of the danger of introducing new diseases and parasites. In order to avoid such problems in the future, introductions are now controlled much more carefully. Future proposed transfers will, hopefully, be subject to international scrutiny under codes of practice that are gaining wider acceptance (Turner 1988; Coates 1992). Thorough appraisal and safe methods for transfers of molluscs are especially important for the Pacific Islands, where the track record of fish and shrimp aquaculture development is not very impressive (Uwate 1988). In most of the island countries, seafood prices are not high enough to make aquaculture competitive with capture fisheries. However, some commodities produced by cultured molluscs (e.g., pearl and pearl shell, trochus shell, green snail shell) are nonperishable, relatively high priced products with good market potentials. There is also a considerable interest in the region in the culture of giant clams, the technology of which was developed recently with international inputs (see Munro, this vol.).

Seaweed Culture

Several species of marine macroalgae (called somewhat misleadingly seaweeds) have been consumed by coastal populations since ancient times in Asia, particularly in Japan, Korea and China. Cultivation of the purple laver *(Porphyra tenera)* started in Japan as early as the 17th century. As seaweed consumption and cultivation remain Asian specialities, aquaculturists outside the region tend to underestimate the role and importance of cultured seaweed production. Aquaculture statistics are often quoted with the exclusion of seaweeds, although this commodity group represented 25% of the total aquaculture production in the Asia-Pacific region in 1990 (Table 3).

Culture of seaweeds is restricted geographically even within the Asia-Pacific region, although during the past decade several countries outside the traditional culture areas have introduced it: Indonesia, the Philippines and Vietnam in Asia; Fiji, Kiribati and the Federated States of Micronesia in the Pacific. Despite these achievements, only ten out of the 34 countries/territories of the region reported commercial seaweed culture in 1990. The growth of cultured seaweed production (Fig. 6) shows surges and recessions, which indicate marketing rather than technological constraints. The annual average growth rate of seaweed culture between 1975 and 1990 was only 4.2%, the slowest among the major aquaculture commodities.

Seaweed production data in Asia-Pacific region for 1990 are presented in Table 15. The major producers were China, Japan, the two Koreas and the Philippines. The indicator showing the volume of production per length of coastline (Table 16) demonstrates the intensity of seaweed culture in relation to the available coastal resources. China, with 91.7 $t \cdot km^{-1}$, shows an outstanding achievement; the production of the Republic of Korea and Japan is also impressive (31.2 and 19.9 $t \cdot km^{-1}$, respectively). The rest of the countries are below the regional average of 11.7 $t \cdot km^{-1}$.

Half the regional cultured seaweed production in 1990 was kelp *(Laminaria*

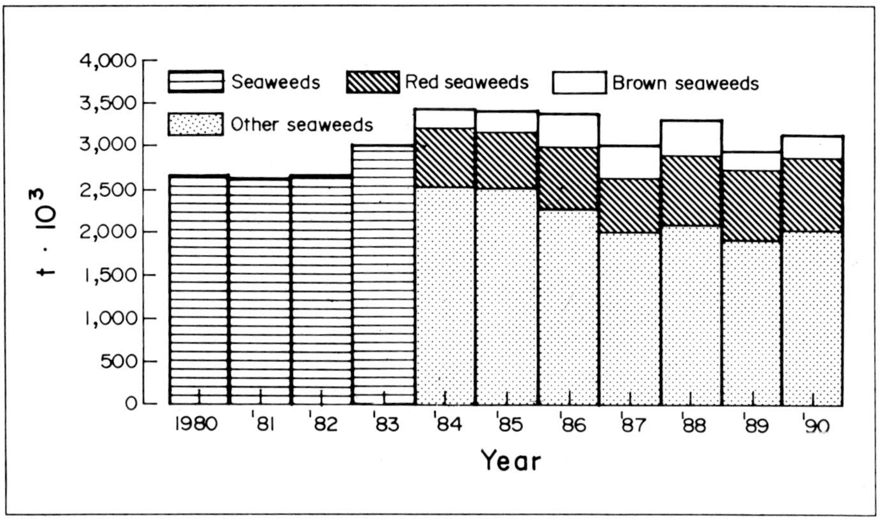

Fig. 6. Growth of cultured seaweed production in Asia and the Pacific. Data aggregated up to 1983. (Sources: Csavas 1988; FAO 1992a).

Table 15. Cultured seaweed production in Asia and the Pacific in 1990. (Source: FAO 1992a).

Country/region	Brown seaweeds (t)	Red seaweeds (t)	Other seaweeds (t)	Total (t)
China	1,465,836	52,938	132,336	1,651,110
Indonesia	-	-	80,000	80,000
Korea, D.P.R.	117,500	2,970	10,000	130,470
Korea, Rep. of	277,417	97,637	36,828	411,882
Pacific Islands	-	550	-	550
Philippines	-	291,176	-	291,176
Taiwan	-	10,614	1	10,615
Vietnam	-	2,000	-	2,000
Developing countries	1,860,753	457,885	259,165	2,577,803
Japan	172,974	387,245	6,000	566,219
Developed countries	172,974	387,245	6,000	566,219
Asia-Pacific total	2,033,727	845,130	265,165	3,144,022
Rest of World	5,388	38,021	260	43,669
World total	2,039,115	883,151	265,425	3,187,691

spp.). *Laminaria* is traditionally cultured in Japan and both Koreas, but it is in China where kelp culture has developed into a huge operation of around 1.5 million t (wet weight) annually about a third of which is not for direct human consumption but for extraction of alginates and iodine. *Undaria* and *Porphyra* spp. form the second and third most voluminous groups. Both commodities are produced

Table 16. Seaweed production per length of coastline in Asia and the Pacific in 1990. (Source: FAO 1992a).

Countries/regions	Seaweed Production ($t \cdot km^{-1}$)
China	91.7
Korea, Rep. of	31.2
Japan	19.0
Philippines	8.4
Korea, D.P.R.	7.1
Taiwan	6.6
Indonesia	1.0
Vietnam	0.6
Developing Asia-Pacific	12.1
Developed Asia-Pacific	10.3
Regional average	11.7

traditionally for human consumption (primarily in Japan and the two Koreas) but, since the middle of the 1980s, their production has declined with increasing competition from other users of resources. The next group consists of *Eucheuma* species, cultivated only since the 1970s, and almost exclusively for extraction of carrageenans. The well developed seaweed industry in the Philippines and the fledgling operations in Indonesia and some Pacific Island countries are based on these species. *Gracilaria* species are produced primarily for agar extraction; the major producer is Taiwan. Other species (mostly green algae, like *Monostroma, Caulerpa, Enteromorpha, Ulva*) are produced in smaller quantities for human consumption.

Seaweed supply in the Asia-Pacific region is now entirely dependent on aquaculture. Over the past five years, more than 90% of the total regional landings came from culture operations, compared to less than 5% in the rest of the world. However, even in Asia, only two decades ago, most of the supply still came from the exploitation of natural stocks which were rapidly and, in many cases irreversibly, depleted. Seaweed culture started with simple methods to replace natural stocks but it is now being rapidly transformed into a "high-tech" industry with indoor "hatcheries", genetic manipulation of stocks and fertilization of growout areas (Trono 1987; Chen et al. 1990; Suo and Wang 1992). Intensively farmed coastal areas in China, Japan or the Koreas are impressive examples of how humans can open up such a new frontier. The ecological implications of such changes, however, have not yet been addressed.

International demand for phycocolloids derived from various seaweed species (agar, alginates and carrageenans) has grown rapidly and has led to the rapid depletion of natural seaweed stocks. Culturing seaweed was promoted by the processing industry itself. One of the most sought after species was *Laminaria*, which is the raw material for alginate extraction. This seaweed was already cultivated for direct human consumption, and its production was easily expanded to supply the industrial demand. Presently there seems to be an oversupply of *Laminaria*, with production decreasing slightly since 1985, especially in China, the major producer in the region. China has developed its own alginate extracting industry, but about 70% of the global alginate production is in the hands of two companies, one in the UK and one in the USA. Norway, Japan and France are also well known producers (Santos et al. 1988).

Carrageenans were previously manufactured from *Chondrus* and *Gigartina* species. However, as natural stocks of these seaweeds became rapidly depleted, the processing industry turned to alternative species. It is now estimated that nearly half of the world supply of these phycocolloids comes from *Eucheuma* species, cultured primarily in the Philippines. The carrageenan industry, which is closely linked to the food and pet food processing industry, is dominated by six major manufacturers in the USA, Denmark and France, together with some smaller ones in Japan, Portugal, Spain and the Republic of Korea. The Philippines also commenced producing semi-refined carrageenan (Santos et al. 1988). Production potentials of *Eucheuma* species are immense in the tropical areas of Asia-Pacific region. Moreover, culturing this seaweed has proven socially and culturally acceptable to both the Southeast Asian fishing communities and to the inhabitants of the Pacific island countries. This was reflected in the rapid establishment of *Eucheuma* culture in the Philippines, Indonesia, Fiji and Kiribati. Unfortunately, the market demand for carrageenan has a moderate growth rate (not more than about 5% per year), and this has resulted in an oversupplied, buyers' market for the raw material. Spectacular increases in the culture of *Eucheuma* species are therefore unlikely.

Furthest from saturation is the global market for agar-agar, although this gel has been used in Japan since the 17th century (McHugh 1987). Originally agar was manufactured from *Gelidium* species, but after the depletion of natural stocks in Japan, Korea and China, the attention turned towards *Gracilaria* species found also in tropical seas of southern China, the Philippines, Indonesia, India, Sri Lanka, Thailand and Vietnam. While significant amounts of cultured *Gelidium* have yet to be produced, a successful pond culture method of *Gracilaria* was developed in Taiwan (Trono 1987). Vietnam and Thailand have launched promising programs in order to develop *Gracilaria* culture, together with the establishment of domestic agar extraction industries. As processing technology of agar-bearing seaweeds is not yet monopolized by a few companies, joint development of *Gracilaria* culture and processing seems to be a seaweed-related development with potentially wide social benefits for producers.

Seaweeds, being autotrophic, are able to synthetize high-energy organic compounds from low-energy inorganic ones. This makes them less problematical from an ecological/environmental point of view. Seaweeds are capable of removing significant amounts of nutrients from overloaded waterbodies and may be used very efficiently in complex integrated systems. *Gracilaria* is cultured in southwestern Taiwan in semi-intensive ponds stocked with shrimp and/or mangrove crabs (Trono 1987). Experiments in shrimp ponds are underway in Thailand. Also in Thailand, agar-bearing seaweeds are harvested from marine fish cages where they grow exceptionally well because of the nutrient-rich environment (Chandrkrachang and Chinadit 1988). In Korea, a multilevel polyculture is being developed to produce kelp, abalone and flounders (B.H. Park, pers. comm.).

Future Perspectives

Giving serious consideration to environmental issues related to aquaculture is a rather new phenomenon in the general development process in the Asia-Pacific region. The current awareness of aquaculturists in this respect, however, exceeds that of their counterparts working in terrestrial food production sectors, due to the strong dependence of the cultured aquatic species on a healthy aquatic environment. Re-reading the report of the 1976 FAO Technical Conference on Aquaculture held in Kyoto, Japan (FAO 1976a) is an instructive exercise, because participants of this historic meeting had already raised all the major environmental concerns raised by aquaculture development.

An aquaculture planning workshop organized by the Aquaculture

Development Coordination Programme of FAO in Bangkok in 1975 (FAO 1976b) set a target of 3 million t of aquaculture production by 1985 for ten Asian countries, excluding China. However, actual production in 1985 was only 53% of the targeted amount in the selected countries. Only two countries out of the ten (Indonesia and Malaysia) reached their target within the deadline, two more (Bangladesh and Singapore) by 1987. By 1990, two more countries (Nepal and Thailand) reached the targeted outputs, but Hong Kong, India, the Philippines and Sri Lanka are not likely to realize their ambitious plans in the near future.

The reasons why the overly optimistic expectations of the 1970s did not come true were not simply technical or economic or environmental. The projected average annual growth rate was 11%, not unattainable in the growth phase of production, as demonstrated earlier. The main problem was that some production techniques and/or choice of species were not socially or culturally acceptable or feasible in all of the countries of the region. The marketability of certain produce proved to be also much more limited than expected, because the local markets (which are the destination of most of the aquaculture products in Asia) are rather conservative. Nutritional traditions are deeply embedded, especially in the rural communities of the Asia-Pacific region. A nontraditional commodity, however cheap and valuable nutritionally, may not be marketable in significant amounts.

On the other hand, the demonstrated trends in aquaculture development clearly indicate that production is going to increase steadily (although probably with a decelerating pace) all over the region in the forthcoming decade. In the developed countries of the region, further development will most likely continue to be driven by the expansion of culturing high-value species. Developing countries are not expected to increase significantly their share of export-oriented aquaculture production of crustaceans and high-value fish species, although their volume may continue to increase. The proportion of species lower in the food chain for domestic consumption is much more likely to increase. The steady increase in intensities witnessed over the past decade is also expected to continue.

When analyzing the potentials of aquaculture development at the World Conference on Aquaculture held in 1986 in Venice, Italy, Kinne remarked: "The production of food for some 8 billion people is a nightmare for an ecologist" (Kinne 1986). This global figure would mean an average population density of 5.4 persons per hectare of arable land. Asia, however, had to feed 6.1 persons from every hectare of its arable land already in 1987 and, according to present predictions, this indicator will reach 7.6 persons per hectare by the year 2000. Asians have had to learn to cope with the nightmare of feeding 55% of the world's population on 30% of the arable land of the globe.

Considering the relatively low growth rates of food production through the cultivation of terrestrial crops and also the fact that inland and marine capture fisheries are not expected to grow significantly during the forthcoming decade, continuing growth of both inland and coastal aquaculture production will be instrumental in keeping the nutritional standards of the Asian population improving. Luckily, present trends indicate that total aquaculture production in the region may reach as much as 20 million t by they year 2000, even if growth rates decelerate to some 3% per year by the end of the century.

Entrepreneurs of the private sector, producing cash-crops in profit-oriented aquaculture systems, will have a marginal role in Asia in the enormous task of producing enough cheap food for the common people, although their role in the technical development of certain production techniques, in the establishment of marketing structures, processing facilities, etc., will remain essential. However, combating hunger cannot be the primary responsibility of the entrepreneur, as Kinne (1986) pointed out. This daunting task cannot be resolved by the public sector either, although government support for research, extension, demonstrations and training is indispensable in this respect. Only the hundreds of millions of small farmers/fishers in Asia can cope with the mass production of cheap fish and seafood required by the poorer segments of the population.

Luckily, what may not be profitable enough for the entrepreneur, may turn out an attractive cash crop for the small farmer/fisher. Producing cheap planktonivorous/herbivorous fish, molluscs or seaweeds in a commercial enterprise is quite different from producing the same commodities within the framework of a well balanced, complex farming system with hitherto underutilized family labor. This type of smallholder aquaculture should not be mistaken for subsistence farming. Those familiar with Asian aquaculture know well that only the most entrepreneurial small farmers embark on aquaculture and the successful ones sell a substantial part of their product instead of consuming it. In fact, these farmers must know very well the absorption capacity of their local markets and adapt to it both with the produced volume and the cultured species.

Increasing substantially the amount of aquatic products low in the food chain is getting more and more difficult by expanding the area of production, because of the increasing population pressure and the stiffening inter- and intra-sectoral competition for land and water. Options for increasing the yields of photosynthesis-dependent systems without additional external inputs are also limited, even in the tropics. However, integrated systems, based on recycling wastes of other food production branches (which may include intensive fish or shrimp culture) and/or domestic and communal wastes, offer efficient and environmentally acceptable solutions for boosting productivity of smallholder aquaculture. Such systems are traditional in China and other countries of the region (e.g., India, Indonesia, Thailand and Vietnam), but there is a need to improve their efficiency and safety. This cannot be achieved without the support of the public sector for research, development and extension of these systems.

One of the major problems of the expanding aquaculture production, which caused most of the negative environmental impacts in the developing countries of the region, is the basically unplanned and unregulated nature of the development process. In view of the increasing competition for suitable sites and water resources both in inland and coastal areas, future development of Asian aquaculture will depend on its inclusion in watershed management, irrigation and coastal zone management plans. Another constraint is that while in Japan, Australia and New Zealand both the establishment and the operation of aquaculture ventures are strictly regulated and the regulations enforced, usually this is not the case in the developing countries, at least not until environmental problems emerge. Methodologies of aquaculture planning, guidelines for site selection and models for rules and regulations are much more needed in the present phase of development in most of these countries than further technology transfer projects of the traditional type. These would help to keep the negative environmental impacts of aquaculture development at a reasonable level.

References

Alabaster, J.S. 1986. Review of the state of water pollution affecting inland fisheries in Southeast Asia. FAO Fish. Tech. Pap. 260, 25 p.

Adriawan, E. and J. Jhamtani. 1989. Mangrove forest destruction in Indonesia and its impact on community system, p. 215-223. *In* I. Soerianegara et al. (eds.) Symposium on Mangrove Management: Its Ecological and Economic Considerations, 9-11 August 1988, Bogor, Indonesia. BIOTROP Spec. Publ. 37. SEAMEO-BIOTROP, Bogor.

Anon. 1990. Report highlights pollution problem in Hong Kong. Fish Farmer November/December 1990: 23.

Bailey, C. 1988. The social consequences of tropical shrimp mariculture development. Ocean Shoreline Manage. 11(1): 31-44.

Bailey, C. and M. Skladany. 1988. The political economy of aquaculture development in the third world. Paper presented at the Seventh World Congress for Rural Sociology, 26 June 1988, Bologna, Italy. 13 p.

Chaichavalit, S. 1989. An economic analysis of intensive giant tiger prawn culture: a case study of Samut Songkram. Thammasart University, Bangkok. 92 p. M.E. thesis.

Chandrkrachang, S. and U. Chinadit. 1988. Seaweed production and processing - a new approach. INFOFISH Intl. 4/88: 22-25.

Chantadisai, T. 1989. Environmental impact of aquaculture on mangrove ecosystem, p. 215-223. *In* I. Soerianegara et al. (eds.) Symposium on Mangrove Management: Its Ecological and Economic considerations, 9-11 August 1988, Bogor, Indonesia. BIOTROP Spec. Publ. 37. SEAMEO-BIOTROP, Bogor.

Chen, C.M., I. McCracken and M.A. Ragan. 1990. Biotechnology potential for hoshi-nori production. World Aquaculture 21(3): 106-107.

Cholik, F. 1988. Review of the status of aquaculture in Indonesia. Paper presented at the Seventh Session of the IPFC Working Party of Experts on Aquaculture, 1-6 August 1988, Bangkok. 36 p.

Chua, T.E. and L.R. Garces, Editors. 1992. Waste management in the coastal areas of the ASEAN region. ICLARM Conf. Proc. 33, 218 p.

Coates, D. 1992. Implementation of the EIFAC/ICES codes of practice: experiences with the evaluation of international transfers into the Sepik River basin, Papua New Guinea. Paper presented at the World Fisheries Congress, 3-8 May, Athens. 26 p.

Csavas, I. 1988. Problems of inland fisheries and aquaculture, p. 137-180. *In* Fishing industry in Asia and the Pacific. Asian Productivity Organization, Tokyo.

Csavas, I. 1990. Shrimp aquaculture developments in Asia, p. 207-222. *In* M.B. New, H. de Saram and T. Singh (eds.) Technical and economic aspects of shrimp farming. Proceedings of the AQUATECH '90 Conference, 11-14 June 1990, Kuala Lumpur, Malaysia.

Davy, F.B. 1991. Mariculture research and development in Japan - an evolutionary review. IDRC-MR298e, 97 p. International Development Research Centre, Ottawa.

De Silva, S.S., Editor. Exotic aquatic organisms in Asia. Asian Fish. Soc. Spec. Publ. 3, 154 p. Asian Fisheries Society, Manila.

FAO. 1976a. Report of the FAO Technical Conference on Aquaculture, 26 May-2 June 1976, Kyoto. FAO Fish. Rep. 188, 47 p.

FAO. 1976b. Aquaculture planning in Asia - Report of the Regional Workshop on Aquaculture Planning in Asia, 1-17 October 1975, Bangkok. Rep. Aquacult. Dev. Coord. Programme 2, 154 p.

FAO. 1984. A study of methodologies for forecasting aquaculture development. FAO Fish. Tech. Pap. 248, 47 p.

FAO. 1988. Report of the Seventh Session of the Indo Pacific Fishery Commission (IPFC) Working Party of Experts on Aquaculture, 1-6 August 1988, Bangkok. FAO Fish. Rep. 411, 37 p.

FAO. 1992a. Aquaculture production (1984-1990). FAO Fish. Circ. 815. Rev. 4, 206 p.

FAO. 1992b. FAO yearbook - fishery statistics - catches and landings, 1990. Vol. 70, 646 p.

Hannig, W. 1988. Innovation and tenant survival: brackishwater pond culture in Java. Naga, ICLARM Q. 11(2): 5-6.

Kinne, O. 1986. Realism in aquaculture - the view of an ecologist, p. 11-24. *In* M. Bilio, H. Rosenthal and C.J. Sindermann (eds.) Realism in aquaculture: achievements, constraints, perspectives. European Aquaculture Society, 21-25 September 1981, Italy.

McHugh, D.J. 1987. Production and utilization of products from commercial seaweeds. FAO Fish. Tech. Pap. 288, 189 p.

Mepham, R.H. and T. Petr, Editors. 1987. Papers contributed to the Workshop on Strategies for the Management of Fisheries and Aquaculture in Mangrove Ecosystems and Country Status reports on inland fisheries presented at the Third Session of the Indo-Pacific Fishery Commission Working Party of Experts on Inland Fisheries, 23-25 June 1986, Bangkok. FAO Fish. Rep. 370 (Suppl.), 248 p.

Park, B.H. 1988. Review of the status of aquaculture in the Republic of Korea. Working Party of Experts on Aquaculture, 1-6 August 1988, Bangkok. 28 p.

Pike, I.H. 1989. Global use of fish meal and developments leading to changes in future use, p. 10-22. *In* Proceedings of the 29th Annual Conference of the International Association of Fish Meal Manufacturers, 21-24 November 1989, Hong Kong.

Pullin, R.S.V. 1981. Fish pens of Laguna de Bay, Philippines. ICLARM Newsl. 4(4): 11-13.

Rosenberry, B., Editor. 1991. World shrimp farming 1991. Aquaculture Digest, San Diego, California. 45 p.

Ruddle, K. 1981. Pollution in the marine coastal environment of ASEAN countries, p. 136-176. *In* L.S. Chia and C. MacAndrews (eds.) Southeast Asian seas - frontiers for development. McGraw Hill, Singapore.

Saenger, P., E.J. Hegerl and J.D.S. Davie, Editors. 1983. Global status of mangrove ecosystems. IUCN Comm. on Ecol. Pap. 3, 88 p.

Santos, C.A.L.D., G.L. Roessink, P.N.R. Rajadurai, C.T. Taylor and I. Kano. 1988. Seaweed processing and marketing in Asia/Pacific. Paper presented at the Seventh Session of the IPFC Working Party of Experts on Aquaculture, 1-6 August 1988, Bangkok. 82 p.

Soenodihardjo, S. and I. Soerianegara. 1989. The status of mangrove forests in Indonesia, p. 215-223. *In* I. Soerianegara et al. (eds.) Symposium on Mangrove Management: Its Ecological and Economic considerations, 9-11 August 1988, Bogor. BIOTROP Spec. Publ. No. 37. SEAMEO-BIOTROP, Bogor.

Suo, R. and Q. Wang. 1992. *Laminaria* culture in China. INFOFISH Intl. 1/92: 40-43.

Trono, G.C. Jr. 1987. Seaweed culture in the Asia-Pacific region. RAPA Publ. 1987/8, 41 p. Regional Office for Asia and the Pacific, Bangkok.

Turner, G.E., Editor. 1988. Codes of practice and manual of procedures for consideration of introductions and transfers of marine and freshwater organisms. FAO, EIFAC/CECPI Occas. Pap. 23, 44 p.

Uwate, K.R. 1988. Viable aquaculture technologies in the Pacific islands region. IPFC WPA/WP 18, 28 p. Working Paper of the Seventh Session of the IPFC Working Party of Experts on Aquaculture, 1-6 August 1988, Bangkok.

Zamora, P.M. 1989. Philippine mangroves: their depletion, conversion and decreasing productivity. Wallaceana 58: 1-5.

Discussion

BILIO: We should discuss further the interpretation of these statistics and the question of definitions, especially for coastal aquaculture, when time permits.

The Environmental Consequences of Intensive Coastal Aquaculture in Developed Countries: What Lessons Can Be Learnt

RICHARD J. GOWEN[a]
Natural Environment Research Council
Dunstaffnage Marine Laboratory
P.O. Box 3
Oban, Argyll, Scotland, UK

HARALD ROSENTHAL
Fischereibiologische Abteilung
Institut für Meereskunde
an der Universität Kiel
Düstenbrooker Weg 20
D-2300 Kiel 1, Federal Republic of Germany

GOWEN, R.J. and H. ROSENTHAL. 1993. The environmental consequence of intensive coastal aquaculture in developed countries: what lessons can be learnt, p. 102-115. *In* R.S.V. Pullin, H. Rosenthal and J.L. Maclean (eds.) Environment and aquaculture in developing countries. ICLARM Conf. Proc. 31, 359 p.

Abstract

The initial development of salmonid farming in Europe and North America took place in sheltered embayments with little regard for environmental consequences. These trends in development, copied in other countries, have raised similar concerns regarding environmental impact. Ecological effects such as enrichment of the seabed ecosystem, nutrient enrichment of the water column (and potentially eutrophication) and the effects of chemicals have implications for long-term development. Other potential effects such as genetic interaction and disturbance of wildlife communities may conflict with nature conservation. Social and economic impacts generally relate to competition for space with other activities such as recreation, tourism and traditional fishing. The complex nature of the potential environmental impact which requires integrated planning, combining considerations of ecological, social and economic factors, has been recognized and employed to resolve conflicts. It is suggested that proactive coastal zone planning with the careful selection of appropriate locations may assist in the sustainable development of coastal aquaculture in developing countries and avoid many of the problems experienced in Europe and North America.

Introduction

In northern Europe and North America, the expansion of coastal marine aquaculture has been predominantly through intensive floating cage production of salmonids. Data on the levels of production attained in a number of countries are presented in Table 1. In some countries the increase in production has been dramatic. In Canada (Atlantic

[a]Present address: Aquatic Sciences Research Division, Department of Agriculture for Northern Ireland, Newforge Lane, Belfast, BT9 5PX.

Table 1. Marine cage salmonid production (in tonnes) in Canada, Norway, Scotland and Ireland (Source of material, country reports contained in reports of the ICES Working Group on the Environmental Impact of Mariculture)

	Country				
	Canada		Norway	Scotland	Ireland
Year	Pacific (Salmon/trout)	Atlantic (Salmon)	(Salmon/trout)	(Salmon/trout)	(Salmon/trout)
1982	273		12,906		
1983	128		14,954		
1984	107		22,287		
1985	120		25,936	7,000	
1986	397		33,833		1,215
1987	(900)		49,740	15,928	2,552
1988	5,825	3,542	89,875	18,660	5,200
1989	12,083	4,750	119,278	29,039	6,400
1990			161,675	32,350	6,271
1991	16,500	9,414	160,655	40,593	8,690

and Pacific coasts) for example, production increased by 263% and 283%, respectively, during the period 1988 to 1991. In Norway salmonid production has increased from an annual production of 12,906 to 160,655 tonnes (representing an average increase of 138% per year) during the period 1982 to 1991. This expansion, often without any apparent control, has brought aquaculture into conflict with other users of aquatic resources, particularly with government and nongovernmental organizations promoting nature conservation. It is known that substantial amounts of waste are produced from intensive fish farming. For example, from Gowen and Bradbury (1987) it can be estimated that approximately 32 kg of soluble nitrogen and 300 kg of particulate carbon waste are generated per tonne of fish produced. The potential for large amounts of waste to be generated together with increased human activity associated with fish farming has often been cited as the reason why aquaculture development is not in accord with the requirements for coastal nature conservation.

It is clear that in some cases the ways in which the industry began and expanded, have compounded environmental change associated with intensive fish farming. In some locations, ecological change has had a negative effect on farm production and forced the fish farmer to move to an alternative site. In addition, the rapid development of cage farming in sheltered coastal waters has brought the industry into conflict with other activities such as tourism and recreation. In this paper some of the environmental issues associated with intensive fish farm development are reviewed. Solutions are suggested which may be of value in managing the environmental impact associated with the development of coastal aquaculture in developing countries.

The Development and Expansion of Intensive Cage Culture

During the initial development phase of intensive fish farming in Norway and Scotland, individual production units were generally small, with an annual production in the order of 50 to 100 tonnes. Furthermore, at this time the primary requirements for a site were: sufficient water depth to accommodate cage netting; shelter from potential storm damage, because cages were generally small (5 m^2) and of wood construction; and proximity to supporting infrastructure such as roads and distribution networks. In many cases therefore, farms were located in sheltered embayments, often at the head of fjords and sea-lochs.

Improvements in technology, especially cage design and construction, for example robust 15 m^2 aluminium cages with 8 m deep nets allowed expansion into more open sites. The early success of the industry was built upon by developing new sites although as the availability of new sites declined existing sites were expanded. Thus, in many cases the original production of 50 to 100 tonnes per year was increased four or five fold with some sites producing three to four hundred tonnes per year.

The development of farming in coastal waters of southern Europe and North America has generally followed development in northern Europe. Rather than an initial start-up period with small production units, the tendency has been for the establishment of large production units, but following the European example of often being located in sheltered coastal waters. This has created environmental issues similar to those experienced by the cage farming industry in northern Europe.

The Physical Environment of Embayments Used for Cage Culture

In a number of European countries the initial development of fish farming took place in coastal embayments. While these embayments provide suitable locations for cage farming in terms of the proximity of deep water to the shore and shelter, they have a number of distinct physical features which can compound the ecological effects of the waste released from cage farms.

A common feature of many of the embayments used for cage farming is a restricted entrance. Such restrictions, which are narrow or shallow or both, restrict the exchange of water between the embayment and more open coastal waters. In addition, shallow entrance sills prevent the penetration of seawater at depth; thus, deep water within the embayment may become isolated for a period of time which in some fjords can be in the order of years (Lazier 1963; Gade and Edwards 1980).

In general, water currents are weaker inshore and since currents are due in part to wind, may be considerably reduced in sheltered inshore locations. For example, Gowen et al. (1988) found that at a number of fish farm sites in Scottish sea-lochs there was little evidence of a tidal component to water movement and that at such sites the maximum current speed was in the order of 0.16 $m \cdot s^{-1}$. Further offshore and outside these embayments, tidal currents might be expected to be higher, and for the Scottish west coast, are in the order of 1 $m \cdot s^{-1}$. In addition, it has been shown that for some coastal regions there is often a residual flow of water. On the west coast of Scotland and Norway for example, there is a northerly flow of coastal water. Such flow of water might be expected to aid the transport and dilution of fish farm waste.

Many coastal embayments have rivers which discharge into them. Within such embayments there is generally an estuarine circulation with a net seaward flow of surface brackishwater and a compensating landward flow at depth. Estuarine circulation may aid dispersal of waste from the immediate vicinity of the farm, but there is also the potential for recirculation of embayment water. This is particularly the case when tidal energy causes mixing at entrance shallows or at narrows. Gowen et al. (1983) estimated that approximately 50% of the water leaving Loch Ardbhair (a small sea-loch on the west coast of Scotland) during the ebb returned during the flood tide. Such recirculation can increase the residence time of water within the basin and this has implications for the dilution of soluble waste from fish farms and the response of the embayment ecosystem to the waste.

An understanding of the physical characteristics of the coastal region in which development is likely to take place is critical for gauging the likelihood of the accumulation or dispersal of waste from intensive cage farming. In recognition of this some workers (Weston 1986; Hakanson et al. 1988; Lumb 1989) have formulated simple principles, based on coastal topography and bathymetric features, for characterizing coastal areas in terms of the potential for waste to accumulate.

Ecological Change Associated with Waste from Intensive Cage Culture

In recent years a considerable amount of research has been conducted into the effects of fish farm waste on the coastal marine ecosystem (see reviews by Rosenthal et al. 1988 and Gowen et al. 1990). In this paper, the intention is to present a brief overview of some of the main findings of these studies.

Fish farm waste can bring about enrichment of the coastal marine ecosystem through the release of soluble dissolved nutrients and particulate organic waste. There have been a number of attempts to quantify the output of waste from marine salmonid farms (Ervik et al. 1985; Gowen et al. 1988; Hall et al. 1990; Ackefors and Enell 1990; Holby and Hall 1991; Mäkinen 1991). From these studies a reasonable estimate of the type (particulate or soluble) and quantities of waste released can be derived and used to estimate the loading (amount per unit volume) to the recipient waterbody.

It is important to realize that data from the studies quoted above are derived from salmonid cage culture in north temperate coastal waters and as such might not be appropriate for cage culture of warmwater fish. For example, the amount of food wasted has an important bearing on the severity of sedimentary enrichment beneath the farm. Given the improvements in feed composition and digestibility and husbandry practice which are reflected in improved food conversion ratios (in excess of 2.0:1 some six to eight years ago to current ratios of about 1.8:1) the early estimates of dry feed wastage of 20% (Beveridge 1984) are probably too high and a value of 12% might be considered more realistic. With respect to some warmwater fish, however, food conversion ratios in the order of 3.0:1 or higher have been reported. Clearly, adopting a value of 12% for food wastage in such circumstances would be inappropriate.

Enrichment of the Seabed Ecosystem

Enrichment of the seabed ecosystem resulting from the deposition of particulate organic waste released from fish farms has been studied by workers in a number of countries: Brown et al. (1987) in Scotland; Ritz et al. (1989) in Australia (Tasmania); Weston (1990) in the USA (Washington State). The results show that the changes which take place are similar to and consistent with other forms of organic enrichment such as woodpulp and domestic sewage sludge (Pearson and Rosenberg 1978).

The succession from undisturbed to enriched sediment and the final equilibrium conditions that result are dependent on the quantity of material deposited. The latter is, in turn, dependent on the size of the farm, husbandry practice and the topographic and hydrographic characteristics of the site. The accumulation of particulate waste causes an increase in the oxygen demand by the sediment ecosystem, probably due to increased chemical oxygen demand and microbial activity. Increased oxygen demand has been measured in the vicinity of freshwater cage farms and in earthen marine ponds. With respect to the former, Enell and Löf (1983) measured a rate of between 34 and 41 mmol.$0_2 \cdot m^{-2} \cdot day^{-1}$ and Blackburn et al. (1988) measured an oxygen consumption rate of between 42 and 60 mmol$\cdot m^{-2} \cdot day^{-1}$ in earthen ponds. Similar rates have been measured in sediments beneath salmon cage farms in Norway (A. Ervik, pers. comm.).

Enhanced consumption of oxygen by the sediment can result in depletion of oxygen in the water overlying the sediment (Tsutsumi and Kikuchi 1983; Gowen et al. 1988). In both studies, however, the periods of oxygen depletion were short, only lasting for a few months during the summer. In general, intensive cage culture of fish is unlikely to cause widespread deoxygenation of bottom water in coastal waters of northern Europe and North America. The exceptions to this are some low energy coastal marine environments, particularly in those coastal embayments in which bottom water remains trapped for a period in excess of several months and in which natural deoxygenation is likely to take place (see, for example, Lazier 1963).

At those locations where particulate waste accumulates on the seabed, the amount of oxygen within the sediment declines (and may fall to zero) due to an imbalance between the supply and consumption of oxygen. As a result, the balance between oxidation and reduction processes changes and the latter become the dominant pathway for the turnover of organic material. Of these, sulfate reduction is likely to be the most important, at least initially, but since gas bubbles released from enriched sediments beneath fish farms contain methane in addition to hydrogen sulfide (Samuelsen et al. 1988), it seems probable that methanogenic bacteria also play an important role in the decomposition of particulate waste.

Bacterial activity within enriched sediments beneath fish farms can be sufficiently high to cause outgassing from the sediment (Braaten et al. 1983). The many anecdotal observations of the smell of hydrogen sulfide in sediment samples collected from beneath fish farms and the presence of hydrogen sulfide in gas bubbles leaving the sediment surface has been confirmed (Samuelsen et al. 1988). Since hydrogen sulfide is toxic to fish there has been considerable debate regarding the potential for 'self pollution' and 'souring of sites' to limit the production potential. It is clear that problems have arisen in some locations and fish farmers have been forced to abandon their sites.

Braaten et al. (1983) attributed damage to the gills of farmed fish to hydrogen sulfide released from the sediment and a similar effect was suggested by Rosenthal and Rangeley (1989). In addition, it is known that continual use of some sites for four or five years or more has caused a deterioration in fish health and a decline in productivity (Gowen, unpubl. data). At those shallow water sites where there is a substantial accumulation of waste, with increased particulate loading in the water, short-term reduction in dissolved oxygen in bottom water and vigorous outgassing, it would be easy to link such changes with a deterioration in fish health and reduced production. However, the relationship between the enrichment of sediments (particularly hydrogen sulfide release) and fish health remains unclear. Furthermore, experience in British Columbia has shown that the production potential has declined at some locations where the water depth is 300 m (E.A. Black, pers. comm.), indicating that perhaps benthic enrichment is only one of a number of potentially negative interactions between intensive cage culture and the coastal marine ecosystem.

The accumulation of particulate waste together with changes in the physical structure of the sediment, reduced levels of oxygen and the presence of hydrogen sulfide, result in significant changes in the ecology and community structure of the benthic macrofauna. In extreme cases the macrofauna disappear altogether and in many situations there is a reduction in the biomass, abundance and species composition, with only a few 'opportunistic' species persisting. Most studies have shown that the effects of enrichment are limited to the immediate vicinity of the farm. In Scotland, for example, Gowen et al. (1988) found that the effects of enrichment could not be detected beyond a distance of 30 to 40 m from farms. The work of Weston (1990), however, has shown that in some locations more subtle effects of enrichment can be detected at distances of up to 100 m from the cages.

Enrichment of the Water Column

The introduction of anthropogenic nutrients into coastal waters can cause hypernutrification (a substantial and measurable increase in the concentration of a nutrient). In turn, this could result in eutrophication, that is, an increase in the biological productivity of a water body (see Jaworski 1979 and references cited therein). The most likely first step in the eutrophication process is an increase in phytoplankton production and biomass (Barlow et al. 1963; Caperon et al. 1971; Eppley et al. 1972). In addition to this direct effect on phytoplankton growth, more subtle changes may occur in the succession of phytoplankton species. For example, where the growth of diatoms is limited by the availability of silicate, other species such as dinoflagellates which do not require silicate may dominate the phytoplankton (Officer and Ryther 1980).

The increase in phytoplankton production can have undesirable consequences. In the Baltic, enhanced phytoplankton growth has contributed to deoxygenation of deep Baltic water; enrichment of the benthos (Baden et al. 1990); and changes in natural fisheries (Hansson and Rudstam 1990). Hypernutrification has been linked to an increase in the frequency of algal blooms in some coastal waters (for example, Lam and Ho 1989; Maclean, this vol.) and also an increase in the occurrence of blooms of species which are toxic to other marine organisms (and indirectly humans and other animals which may feed on such organisms) Lam et al. (1989).

The recognition that intensive cage culture of salmonids generates substantial quantities of dissolved nutrients together with the rapid expansion of this industry in oligotrophic coastal waters of countries like Scotland and Norway has been viewed with concern by some government and nongovernmental organizations (Anon. 1988; NCC 1989). Simple assessments of the potential for large-scale hypernutrification can be made (Gowen and Ezzi 1992) and suggest that this is unlikely at the current level of farming in coastal waters of most countries. Nevertheless, localized increases in the concentration of ammonia in the immediate vicinity of cage farms have been observed by many researchers, and nutrient enrichment of individual embayments could occur. Gowen and Ezzi (1992) found clear evidence of a fish farm having increased the nutrient status of a Scottish sea-loch basin and found that changes in the concentration of ammonia were related to the activity of the fish farm. During the operation of the farm, ammonia concentrations were significantly higher (statistically) than concentrations in neighboring west coast sea-lochs but after the farm had ceased operation the differences were no longer apparent.

There is anecdotal evidence from Norway (K. Tangen, pers. comm.) of an increase in phytoplankton production as a result of fish farming activity. Persson (1991) suggests that there is clear evidence of eutrophication resulting from fish farming in coastal waters of Finland. In the latter case, however, the measured increase in phytoplankton biomass may have been the result of accumulation due to the ponding of near surface inshore water, rather than an increase in growth stimulated by fish farm waste (T. Mäkinen, pers. comm.). Gowen and Ezzi (1992) were unable to measure any changes in phytoplankton production or biomass in relation to the nutrient enrichment they observed and this raises an important point: anthropogenic nutrients, including fish farm waste, will only stimulate phytoplankton growth if growth is limited by the availability of nutrients. In those coastal regions where phytoplankton growth is limited by light or the accumulation of biomass is restricted by dilution, eutrophication is unlikely.

Use of Chemicals in Intensive Cage Culture

Various therapeutants are used in intensive cage culture to control disease and other chemicals are used to control external parasites such as sea lice. Concerns over the use of these chemicals relates to their ecotoxicology, the potential for bioaccumulation and in the case of antimicrobial compounds, the development of disease resistance in target and nontarget bacteria.

The problems experienced with the antifouling compound tributyl tin, (used on cage nets) serves as a good example of why full evaluation and strict control is necessary before a compound is licensed for use in fish farming. Tributyl tin (now banned for use in aquaculture in most European countries) has been shown to be toxic to nontarget organisms (Stepaenson et al. 1986); accumulate in the flesh of the farmed fish and cause mortality of farmed fish (Short and Thrower 1986); and accumulate through the food chain (Laughlin 1986).

Current concerns center on the use of dichlorvos (Nuvan or Aquaguard) an organophosphorus compound use against parasitic copepods (sea lice). This compound is a general pesticide and therefore is nonspecific, being toxic to a range of crustacean larvae. Extensive use of this compound in intensive cage culture appears to have resulted in the development of resistance in sea lice populations and where repeated treatment of fish has been carried out, increased the sensitivity of fish to the compound. These two factors have reduced the efficacy of the treatment.

Interaction with Wildlife

In addition to the ecological effects of the waste released from fish farms, the physical presence of fish farms, the presence of high biomasses of fish and human activity can interact with wildlife in a number of ways.

There have been a number of accusations that fish farm development has had a negative impact on wildlife, in particular predatory birds and mammals (Anon. 1988; Whilde 1990). While there is clear evidence that many predatory birds (such as cormorants, herons and shags) and mammals (for example, seals) are killed at fish farms (either as a result of becoming trapped in netting or deliberately killed), no objective studies into the effects of these mortalities on breeding populations have been carried out.

A number of studies have shown that wild fish often congregate in the vicinity of marine cage farms. In addition, the presence of fish farms may influence the population structure and act as areas for recruitment because of the additional supply of food (Henriksson 1991). The significance of these findings is not known.

The potential for the transfer of genetic material from escaped farmed fish to wild fish has recently caused considerable debate (NASCO 1989). It has been suggested that breeding programs in aquaculture have resulted in significant differences between farmed and wild fish (Cross 1991). A consequence of interbreeding between significant numbers of escaped farmed fish and wild fish could result in the latter losing important traits and becoming less well adapted. It is clear that large numbers of farmed fish do escape and enter rivers. Gudjonsson (1991) found that 30% of the fish caught in the River Ellidaar in Iceland were of farm origin. Furthermore, it has been shown that genetic material can pass from farmed fish into wild populations (Crozier, in press). At the present time, the ecological significance of such transfers is not known, but in recognition of the potential problem, several countries have established regulations excluding development from important river estuaries.

Social and Economic Effects of the Development of Intensive Cage Culture

The potential social and economic benefits of aquaculture development are clear. In Scotland, it was estimated that in 1990 approximately 1,600 people were in full-time employment in aquaculture (mainly finfish farming). Despite this, expansion of fish farming in Scotland has brought this industry into conflict with other users of aquatic resources. This trend of increasing conflict with other forms of coastal development has also been observed in southern Europe and North America.

Some of the more obvious potential conflicts are listed in Table 2. Only a few attempts have been made to assess the effects of aquaculture development on tourism (Anon. 1988; Sargeant 1990) and recreation (McNab et al. 1987). In general however, such studies are subjective and it is difficult to obtain a clear understanding of the effects of aquaculture development on other users of the coastal environment. In recognition of some of the problems which have arisen, fish farming associations have produced codes of practice. The Scottish Salmon Growers' Association has produced a voluntary code for avoiding visual impact of developments on the landscape. Such codes of practice

Table 2. A summary of some of the main social issues associated with aquaculture development.

Competition for space

- Traditional fishing
- Navigation
- Anchorages and marinas for recreational boating
- Different forms of aquaculture and between aquaculture and other industries (e.g., wood pulp).

Amenity, recreation and tourism

- Visual impact and loss of wilderness aspects of the countryside
- Restriction on access to land, foreshore and inshore areas which may affect outdoor activities (water sports) and harvesting of shellfish for noncommercial purposes
- Reduction in amenity value of freshwater for recreational fishing
- Reduction in the value of property

have been criticized (Sergeant 1990) and some organizations have gone further to suggest that such codes should be integrated into planning procedures and that if this were the case such issues would be properly evaluated (Anon. 1988).

It is only recently that some countries have attempted to consider coastal zone management as an appropriate tool for ensuring the equitable and sustainable use of coastal marine resources. The Norwegian Government has implemented a scheme referred to as LENKA (see Pedersen et al. 1988) which includes an evaluation of the ecological, social and economic implications of all potential activities in the coastal region. The Provincial Government in British Columbia (Canada) has also developed a coastal inventory scheme for minimizing conflict between different activities by identifying potential locations for aquaculture and evaluating existing demands (industrial development; natural fisheries; tourism and recreation and nature conservation) on the coastal marine environment (Black 1991).

Any coastal zone management scheme should be designed to ensure that there is equitable use of coastal resources and therefore include an environmental impact assessment of all potential developments. In relation to aquaculture, it is clear that localized ecological change brought about by the farm itself, can limit long-term production. For this reason a detailed assessment of the potential ecological effects of development is desirable. The necessary steps in undertaking such an evaluation have been discussed by ICES (1989) and Gowen et al. (1990) and adapted. An important feature of such an assessment is that it is proactive, the aim being to anticipate or predict the degree of ecological change and stop or modify the type and scale of production prior to development.

Such an approach requires a full understanding of the interaction between aquaculture and the coastal marine ecosystem, an ability to model and hence make quantitative predictions about the scale of these interactions, and finally the establishment of ecologically based acceptable levels of change. With respect to the effects of the waste from intensive cage culture of fish, the interactions are known and models have been developed to predict the scale of these effects (see Gowen et al. 1990 and references cited therein). At the present time, there are few standards for acceptable levels of ecological change, although there have been some attempts to identify appropriate variables (Jaworski and Orterio 1979) and develop trophic indices based on nutrient concentration scales for coastal waters (Ignatiades et al. 1992). In most European countries there are strict controls governing the licensing of chemicals for use in aquaculture but in relation to the interaction between aquaculture and wildlife (including possible genetic interaction) there are few, if any, objective criteria for controlling aquaculture development.

One of the benefits of appropriately formulated coastal zone management schemes should be that the social, economic and ecological implications of each development are considered in parallel. Furthermore, each development must be regarded as part of the total rather than as a discrete development which has no effect on existing or future activities. Properly formulated therefore, coastal zone management schemes should allow the equitable and sustainable use of coastal marine resources, based on a broad range of activities.

Implications for Developing Countries

Because of the potential for the accumulation of waste in sheltered sites within coastal embayments, such sites have a limited potential for large-scale sustainable fish farm development compared to more open coastal waters. The development of simple guidelines together with technological advances in cage design and construction should ensure that in developing countries selection of appropriate sites is achieved and that 'overloading' of sheltered inshore sites with aquaculture waste does not occur.

Extensive use of chemicals has caused a number of problems to the cage farming industry in northern Europe, including: adverse publicity from the perceived ecological threat of this industry; direct loss of revenue because of high mortalities; and indirect loss of revenue due to the image of high chemical usage reflecting on the marketability of the product. Reduction of the quantity of chemicals used for intensive cage culture of fish has taken place in some countries as a result of improvements in husbandry (such as the separation of year classes), and a recognition that high density farming in sheltered locations may increase disease problems through stress to the fish and increased susceptibility to disease. To avoid the problems noted above, methods of reducing chemical usage should be recognized and fully exploited in coastal fish farming in developing countries.

Many of the conflicts between fish farm development and other users of coastal marine resources have arisen because of the failure of planning procedures to ensure that there is equitable use of the resources. In developing countries therefore, full use should be made of integrated coastal zone management procedures, such as the Norwegian LENKA program and the Canadian coastal inventory scheme.

References

Ackefors, H. and M. Enell. 1990. Discharge of nutrients from Swedish fish farming into adjacent sea areas. Ambio 19: 28-35.

Anon. 1988. Marine fish farming in Scotland: a discussion paper. Scottish Wildlife and Countryside Link. Perth, Scotland. 70 p.

Baden, S.P., L. Loo, L. Pihl and R. Rosenberg. 1990. Effects of eutrophication on benthic communities including fish: Swedish west coast. Ambio 19: 113-122.

Barlow, J.P., C.J. Lorenzen and R.T. Myren. 1963. Eutrophication of a tidal estuary. Limnol. Oceanogr. 8: 251-262.

Beveridge, M.C.M. 1984. Cage and pen fish farming. Carrying capacity models and environmental impact. FAO Fish. Tech. Pap. 255, 131 p.

Black, E.A. 1991. Coastal resource mapping - a Canadian approach to resource allocation for salmon farming, p. 441-459. *In* N. De Pauw and J. Joyce (eds.) Aquaculture and the environment. European Aquaculture Society. Spec. Publ. 16. Gent, Belgium.

Blackburn, T.H., B.A. Lund and M.D. Krom. 1988. C- and N- mineralization in the sediments of earthen marine fish ponds. Mar. Ecol. (Prog. Ser.) 44: 221-227.

Braaten, B., J. Aure, A. Ervik and E. Boge. 1983. Pollution problems in Norwegian fish farming. ICES CM 1983/F:26, 11 p.

Brown, J.R., R.J. Gowen and D.S. McLusky. 1987. The effects of salmon farming on the benthos of a Scottish sea-loch. J. Exp. Mar. Biol. Ecol. 109: 39-51.

Caperon, J., S.A. Cattell and G. Krasnick. 1971. Phytoplankton kinetics in a sub-tropical estuary: eutrophication. Limnol. Oceanogr. 16: 599-607.

Cross, T. 1991. Potential genetic interactions betwee reared and wild fish in Europe with particular emphasis on Atlantic salmon, p. 299-308. *In* N. De Pauw and J. Joyce (eds.) Aquaculture and the environment. European Aquaculture Society, Spec. Publ. 16. Gent, Belgium.

Crozier, W.W. Electrophoretic evidence of genetic interaction between escaped farm salmon and wild Atlantic salmon (*Salmo salar* L.). Aquaculture (in press).

Enell, M. and J. Löf. 1983. Environmental impact of aquaculture sedimentation and nutrient loadings from fish cage culture farming. Vatten/Water 39 (4):364-375.

Eppley R.W., F. Carlucci, O. Holm-Hansen, D. Kiefer, J.J. McCarthy and P.M. Williams. 1972. Evidence for eutrophication in the sea near southern California coastal sewage outfalls - July 1970. Calif. Mar. Res. Comm. CalCOFI Rep. 16: 74-83.

Ervik, A., P. Johannessen and J. Aure. 1985. Environmental effects of marine Norwegian fish farms. ICES C.M. 1985/F:37, 13 p.

Gade, H.G. and A. Edwards. 1980. Deep water renewal in fjords, p. 453-489. *In* H.J. Freeland, D.M. Farmer and C.D. Levings (eds.) Fjordic oceanography. Plenum Publishing Corp., New York.

Gowen, R.J. and N.B. Bradbury. 1987. The ecological impact of salmon farming in coastal waters: a review. Oceanogr. Mar. Biol. Ann. Rev. 25: 563-575.

Gowen, R.J. and I.A. Ezzi. 1992. Assessment and prediction of the potential for hypernutrification and eutrophication associated with cage culture of salmonids in Scottish coastal waters. Dunstaffnage Marine Laboratory, Oban, Scotland. 136 p.

Gowen, R.J., P. Tett and K.J. Jones. 1983. The hydrography and phytoplankton ecology of Loch Ardbhair: a small sea-loch on the west coast of Scotland. J. Exp. Mar. Biol. Ecol. 71: 1-16.

Gowen, R.J., J.R. Brown, N.B. Bradbury and D.S. McLusky. 1988. Investigations into benthic enrichment, hypernutrification and eutrophication associated with mariculture in Scottish coastal waters (1984-1988). Dept. Biol. Sci., Univ. Stirling, Scotland. 289 p.

Gowen, R.J., H. Rosenthal, T. Makinen and I. Ezzi. 1990. Environmental impact of aquaculture activities, p. 257-283. *In* N. De Pauw and R. Billard (eds.) Aquaculture Europe '89 - Business joins science. European Aquaculture Society. Spec. Publ. 12. Bredene, Belgium.

Gudjonsson, S. 1991. Farmed salmon in natural salmon rivers in Iceland, p. 183-196. *In* E. Hoffmann, R. Persson, E. Gaard and G.S. Jonsson (eds.) Havbrug og Miljo. Nordisk Ministerrad, Kobenhavn K. Denmark. Nord 1991:10, 291 p.

Hakanson, L., A. Ervik, T. Makinen and B. Moller. 1988. Basic concepts concerning assessments of environmental effects of marine fish farms. Nordic Council of Ministers, Copenhagen, Denmark. 103 p.

Hall, P.O.J., L.G. Anderson, O. Holby, S. Kollberg and M.A. Samuelsson. 1990. Chemical fluxes and mass balance in a marine cage farm. I. Carbon. Mar. Ecol. (Prog. Ser.) 61: 61-73.

Hansson, S. and L.G. Rudstam. 1990. Eutrophication and Baltic fish communities. Ambio 19: 123-125.

Henriksson, S. 1991. Effects of fish farming on natural Baltic fish communities, p. 85-104. *In* T. Makinen (ed.) Marine aquaculture and environment. Nordic Council of Ministers, Copenhagen, Denmark. 126 p.

Holby, O. and P.O.J. Hall. 1991. Chemical fluxes and mass balances in a marine fish cage farm. II. Phosphorus. Mar. Ecol. (Prog. Ser.) 70: 263-272.

ICES (International Council for the Exploration of the Sea). 1989. Report of the Working Group on Environmental Impact of Mariculture. ICES C.M. 1989/F:11.

Ignatiades, L., M. Karydis and P. Vounatsou. 1992. A possible method for evaluating oligotrophy and eutrophication based on nutrient concentration scales. Mar. Pollut. Bull. 24: 238-243.

Jaworski N. 1979. Sources of nutrients and the scale of eutrophication problems in estuaries, p. 83-110. *In* B.J. Neilson and L.E. Cronin (eds.) Estuaries and nutrients. Humana Press, Clifton, New Jersey.

Jaworski, N.A. and V. Orterio. 1979. A suggested approach for developing estuarine water quality criteria for management of eutrophication, p. 499-516. *In* B.J. Neilson and L.E. Cronin (eds.) Estuaries and nutrients. Humana Press, Clifton, New Jersey.

Lam, C.W.Y. and K.C. Ho. 1989. Red tides in Tolo Harbour, Hong Kong, p. 49-52. *In* T. Okaichi, D.M. Anderson and T. Nemoto (eds.) Red tides. Biology, environmental science and toxicology. Elsevier, New York.

Lam, C.W.Y., M. Kodama, D.K.O. Chan, T. Ogata, S. Sato and K.C. Ho. 1989. Paralytic shellfish toxicity in shellfish in Hong Kong, p. 455-458. *In* T. Okaichi, D.M. Anderson and T. Nemoto (eds.) Red tides. Biology, environmental science and toxicology. Elsevier, New York.

Laughlin, R.G. 1986. Bioaccumulation of tributyl tin: the link between environment and organism, p. 1206-1209. *In* Proceedings of Oceans '86, Vol. 4, Organotin symposium. Marine Technology Society, Washington, DC.

Lazier, J.R.N. 1963. Some aspects of the oceanographic structure of the Jervis Inlet system. Department of Physics, University of British Columbia, Vancouver, British Columbia, Canada. 54 p. M.Sc. thesis.

Lumb, C.M. 1989. Self pollution by Scottish salmon farms? Mar. Pollut. Bull. 20: 375-279.

Mäkinen, T. 1991. Nutrient load from marine aquaculture, p. 1-8. *In* T. Makinen (ed.) Marine aquaculture and environment. Nordic Council of Ministers, Copenhagen, Denmark. 126 p.

McNab, A., J. Martin and P. Landless. 1987. An environmental assessment of fish farms. (Cobham Resource Consultants with Fisheries Development Ltd.) Final report to The Countryside Commission for Scotland, Crown Estate Commissioners, Highlands and Islands Development Board and Scottish Salmon Growers' Association. 100 p.

NASCO (North Atlantic Salmon Conservation Organisation) 1989. The genetic threat to wild salmon posed by salmon aquaculture. Report of a meeting held on 23rd May 1989, Dublin, Ireland. NASCO, Edinburgh. 18 p.

NCC (Nature Conservancy Council). 1989. Fish farming and the safeguard of the natural marine environment. Nature Conservancy Council Scottish headquarters, Edinburgh. 136 p.

Officer, C.B. and J.H. Ryther. 1980. The possible importance of silicon in marine eutrophication. Mar. Ecol. (Prog. Ser.) 3: 83-91.

Pearson, T.H. and R. Rosenberg. 1978. Macrobenthic succession in relation to organic enrichment and pollution of the marine enviroment. Oceanogr. Mar. Biol. Ann. Rev. 16: 229-311.

Pedersen, T., J. Aure, B. Berthelsen, S. Elvestad, A. Ervik and H. Kryvi. 1988. LENKA - a nationwide analyisis of the suitability of the Norwegian coast and watercourses for aquaculture. A coastal zone management programme. ICES C.M.1988/ F:11, 18 p.

Persson, G. 1991. Eutrophication resulting from salmonid fish culture in fresh and salt waters: Scandinavian experiences, p. 163-185. *In* C.B. Cowey and C.Y. Cho (eds.) Nutritional strategies and aquaculture waste. Proceedings of the First International Symposium on Nutritional Strategies in Management of Aquaculture Waste. University of Guelph, Guelph, Ontario.

Ritz, D.A., M.E. Lewis and M. Shen. 1989. Response to organic enrichment of infaunal macrobenthic communities under salmonid sea cages. Mar. Biol. 103: 211-214.

Rosenthal, H., D.P. Weston, R.J. Gowen and E.A. Black. 1988. Report of the ad hoc study group on environmental impact of mariculture. ICES (International Council for the Exploration of the Sea) Coop. Res. Rep. 154, 81 p.

Rosenthal, H. and R.W. Rangeley. 1989. The effect of salmon cage culture on the benthic community in a largely enclosed bay (Dark Harbor, Grand Manan Island, NB, Canada), p. 207-223. *In* H. Rosenthal and K. Lillelund (eds.) Fish health protection strategies. Bundesministerium für Forschung und Technologie, Hamburg/Bonn, Germany.

Samuelsen, O.B., A. Ervik and E. Solheim. 1988. A qualitative and quantitative analysis of the sediment gas and diethylether extract of the sediment from salmon farms. Aquaculture 74: 277-285.

Sargeant, K. 1990. Tourism and aquaculture, p. 60-63. *In* P. Oliver and E. Colleran (eds.) Interactions between aquaculture and the environment. An Taisce: The National Trust for Ireland, Dublin. 75 p.

Short, J.W. and F.D. Thrower. 1986. Tri-n-butyltin caused mortality of chinook salmon, *Onchorhynchus tshawytscha*, on transfer to a TBT-treated marine net pen, p. 1202-1212. *In* Proceedings of Oceans '86, Vol.4, Organotin symposium. Marine Technology Society, Washington, DC.

Stepaenson, M.D., D.R. Smith, J. Goetzl, G. Ichikawa and M. Martin. 1986. Growth abnormalities in mussels and oysters from areas with high levels of tributyl tin in San Diego Bay, p. 1246-1251. *In* Proceedings of Oceans '86, Vol. 4, Organotin symposium. Marine Technology Society, Washington, DC.

Tsutsumi, H. and T. Kikuchi. 1983. Benthic ecology of a small cove with seasonal oxygen depletion caused by organic pollution. Publ. Amakusa Mar. Biol. Lab. Kyushu Univ. 7: 17-40.

Weston, D.P. 1986. The environmental effects of floating cage mariculture in Puget Sound. School of Oceanography, College of Oceans and Fishery Science, University of Washington, Seattle. 148 p.

Weston, D.P. 1990. Quantitative examination of macrobenthic community changes along an organic enrichment gradient. Mar. Ecol. (Prog. Ser.) 61: 233-244.

Whilde, T. 1990. Sharing the environment with the original inhabitants - some thoughts on aquaculture and the environment, p. 56-59. *In* P. Oliver and E. Colleran (eds.) Interactions between aquaculture and the environment. An Taisce: The National Trust for Ireland, Dublin. 75 p.

Discussion

The following discussion on a decisionmaking schema regarding aquaculture development and environmental impact devised by ICES (1989) led to a revised schema (see p. 336, this vol.)

BILIO: This kind of decisionmaking procedure is very similar to that used by GTZ for planning technical cooperation projects. It is good that such schemes exist and could perhaps be modified for use in developing countries. Certainly donor agencies must follow such approaches and the developed countries must show how they avoid mistakes and then not take courses of action that would result in mistakes in the developing countries that they assist.

MACLEAN: Donor-funded projects may well require planning with regard to environmental impact but the projects themselves are not usually for commercial purposes and the developments often stop as soon as the donor's money runs out. As Dr. Ruddle has said, it is the entrepreneurs who race ahead with new development. The reality, from my experience in developing countries, is that governments rarely, if ever, *begin* these projects with consideration of processes that will safeguard the environment. They merely try to regulate opportunistic entrepreneurs 'after the fact'. This is very hard to do in developing countries. The entrepreneurs always argue that they are producing for the benefit of the people. The environment takes second place. There are some rare exceptions. For example, in the Philippines, a seaweed farming company recently took over a very important area, previously designated by the government as a marine park: the Tubbataha reefs. The company proposed a 30,000 ha seaweed farm on a national heritage site. The government said no. The company said that there are lots of people to feed. The company subsequently occupied the site but was evicted by a conservationist crew and the farm buildings were burnt down. It remains a rare exception to the general rule. I cannot see that processes and procedures, such as are being discussed here, will have much of a near-term future in developing countries, given the pressure on governments from entrepreneurs.

GOWEN: But isn't it a question of education of both the governments and the entrepreneurs? The entrepreneurs run the risk of negative feedback if they cause damage.

MACLEAN: Yes. But we are talking about the near future - perhaps the next 20 years.

PULLIN: Jay Maclean's point is important, but I also like the idea of devising the schemes. We will probably need several schemes - say, one for *policy* evaluation, another for project evaluation.

Governments may be ineffectual in stopping or controlling individual projects, but they can develop policies; for example, whether to encourage rice-fish culture, or indeed aquaculture at all. Governments need guidelines, decision trees, etc., for policy formulation. The scheme presented here is excellent but needs modifications for application in developing countries because of the urgency with which they must change. The option of doing little or nothing is not there.

The question of acceptable environmental impact is therefore very difficult. The scheme probably needs feedback loops and decision points so that, if environmental impact looks unacceptable, steps to reduce it can be considered and the process gone through again on an iterative basis. The environmental costs need to weighed against benefits. Where aquaculture can be well integrated with other sectors, the benefits can be great. In a developing - country situation, the decision whether or not to encourage an aquaculture development will likely face a very different environmental cost-benefit scenario than say a decision whether or not to put salmon cages in a sea loch or fjord. My main point, however, is that the scheme should not just have a single pass - a yes/proceed versus no/foreclose outcome. The option of doing nothing, say in rural Bangladesh, is just not there. The resources systems that are there *must* generate more food and more cash and this needs urgent change. Nevertheless, this scheme is an excellent start.

BILIO: I agree with the pessimism and the reservations expressed and that the scheme presented makes a good start. GTZ seeks a political dialogue with a government before proceeding with a project. If there is no agreement on how to proceed, nothing proceeds. The procedure used for planning is now very strict and involves the target groups and all proper authorities concerned. We then often have an orientation (or pilot) phase where planning is continued. Changes are made according to results achieved. If a major environmental impact is foreseen, a special environmental compatibility study has to be carried out. Its results are decisive for the continuation of the project.

KING: I agree very much with Dr. Pullin. In African countries with urgent needs it is practically impossible to get environmental issues to the top of the agenda when considering development. I remember assisting a panel considering the development of a tannery. I advocated stopping the credit for this development because of its likely environmental impact, but the view prevailed that the development should proceed despite its environmental impact. We need to raise the level of such debates to get a proper balance taking acceptable impacts and benefits into account.

BILIO: I agree with you in principle, but in Africa most development, even if successful, is so slow that the process *is* delayed. Also, there are many examples of how wrong it can go. Therefore, in an increasing number of countries, there is a consciousness that the ways things were done before cannot be repeated. We have, of course, to be realistic - as in the case of introductions of exotic species. On the one hand we have to try to formulate and enforce Codes of Practice; on the other hand we have to be prepared to cope with what really happens. We still have the responsibility to try to do something serious about these problems.

RUDDLE: The weak link in this interesting scheme is evaluation, which is hardly ever conducted effectively, if at all, for development projects - why they went wrong, etc., especially the sociocultural aspects. Evaluation must be built into such systems. Also, I don't really like the yes/proceed vs. no/foreclose option. It would be better to say foreclose this/these (aquaculture) option but consider others such as raising chickens or other forms of protein.

BILIO: Yes. The GTZ planning process is not dichotomous. It has multiple options.

RUDDLE: The real world is too complex for a simple dichotomous system.

ROSENTHAL: One clarifying comment - the foreclosure option indicated in the ICES flow chart on impact management simply indicates that such an option must exist in principle. There should be other options as well and these should not only be concerned with the aquatic environment. It may well be that the principle scheme may have to be modified from region to region. The chart presented by ICES represents the approach taken in Europe not only for aquaculture but also for other industrial developments. Stringent environmental regulatory conditions were often considered to be a constraint to industrial development by many industries. It was often claimed that jobs would be lost if environmental safeguards would have to be met, but industries could seldom prove that this is true and attitudes have changed. Often industries once forced to comply with environmental quality standards become more efficient and competitive as a result of these enforced changes.

BILIO: In this context, what monitoring parameters are used for cage culture? There remains confusion on the use of socioeconomic as well as biotechnical parameters. Some parameters may be useless. How to get an overall assessment of such a complex situation? That is the question.

ROSENTHAL: I agree with you entirely.

CATAUDELLA: Examples from developed countries are important, but their political stability has enabled there to be a continuum from investment in industry to more investment in environmental protection. They can invest in both directions. In developing countries there is a lack of such continuity. The scheme proposed and the procedures used by donors cannot escape the fact that environmental protection requires *investment* for that purpose. Projects that have a limited budget (say US$2 million for small-scale aquaculture) are less likely to be able to generate investment for environmental protection than larger projects (say US$25 million for some hyperintensive ponds) It will be difficult to educate the decisionmakers.

PULLIN: The word 'acceptable' has been used here. This begs the question 'acceptable to whom?' ICLARM and its collaborators have learned some hard lessons over the years. One is that we should talk to *farmers*, not last, but first. They often have very clear and accurate perspectives on what impacts changes would bring to their families and communities and to the natural environment. This sounds obvious, but many agencies, institutions, scientists and consultants simply do not do it effectively. Farmers should have the major voice in what is or is not acceptable. This is particularly true in the rural areas or developing countries where small-scale farmers and fishers are not like some of the larger, corporate concerns that *will* try to cut corners with no regard for the environment or sustainability. Farmers' perspectives and indigenous knowledge are of paramount importance in considering environmental issues.

BILIO: I wholeheartedly agree. GTZ has examples of this in small-scale fisheries projects. Some of the best suggestions came from the target groups.

BURAS: If we are really concerned about the environment, then education should be the first priority, starting with the farmers, the people. This education should be with simple methods and concepts, not sophisticated schemes. We need to work on such simple schemes in our discussion period.

BILIO: I do not quite agree. Training and education are very important, but *first* the confidence of the target group must be secured. Second, we have to learn what they already know. They know much better than us what their needs and possibilities are and the environmental constraints. This is certainly so for rural developments. For industrial operations, which we should also consider, a different approach may be necessary.

KING: We seem to be focusing on development financed by projects and forgetting that entrepreneurial ativities work differently. There is often a very well-developed indigenous entrepreneurial capacity.

Aquaculture Development and Environmental Issues in Africa

Hannah R. King
7, Duchy Wood, Daisy Hill
Bradford, West Yorkshire
BD9 6D2, United Kingdom

KING, H.R. 1993. Aquaculture development and environmental issues in Africa, p. 116-124. *In* R.S.V. Pullin, H. Rosenthal and J.L. Maclean (eds.) Environment and aquaculture in developing countries. ICLARM Conf. Proc. 31, 359 p.

Abstract

Aquaculture development in Africa and related environmental issues are reviewed in the context of different types of culture systems: extensive, semi-intensive and, to a lesser extent, intensive aquaculture. Environmental impact is discussed in terms of the intensity of production of each system and interactions with surrounding ecosystems.

The general paucity of empirical data, partly due to low levels of investment in aquaculture and the weaknesses of scientific research in the region, necessitates approaching discussions in terms of perceived current and future development trends. The most important issues of environmental impact are identified and discussed: health hazards to humans, water quality, ecosystem degradation, fish disease and effects on genetic resources. Measures to control adverse environmental impacts are discussed.

Introduction

In considering aquaculture development and associated environmental issues in Africa, it is necessary to specify perceived current and future trends because of the dearth of information. On the one hand, the present low level of investment suggests that most aquaculture in Africa is extensive, uses low inputs and, apparently, has limited impact on the environment; on the other hand, the shortcomings of related scientific research in the region render it difficult to substantiate interactive relationships between aquaculture and the environment.

Specifying problems requires answering the following questions:

- Do problems exist now and, if so, to what extent?
- Are there future potential problems and, if so, what are their possible sources?
- Can present and likely future scenarios be studied together (for example, by case studies) so that the seeds of future problems are recognized and can be avoided or solved?

Such questions are all valid and relevant to aquaculture development in Africa.

This paper attempts to identify the status of aquaculture in relation to environmental issues in Africa, focusing on the impact of one on the other and suggesting strategies for the control of negative impacts. The proper management of aquaculture and the environment also raises fundamental development issues and these are also discussed here.

Aquaculture is defined here as any human intervention in the production of aquatic food organisms, through stocking, feeding and/or disease control. The term environment includes all living and nonliving resources, their relationships and interdependencies.

Fundamental Problems

The literature on aquaculture development in relation to environmental issues in Africa is sparse: general, often incomplete, with few empirical data. This suggests deep structural causes, remedies for which appear to be and will probably remain beyond the reach of most institutions, unless necessary incentives are created. It also means that identifying and tackling problems are severely constrained.

It is clear, however, that development trends in many areas could pose potential problems. For example, coastal aquaculture of crustaceans (and perhaps molluscs, under some circumstances) could pose a threat to fragile ecosystems. Likewise, the uncontrolled use of agricultural and vector control pesticides which wash into river systems and reservoirs may pose serious problems for the quality of water sources for inland aquaculture. The poorly developed treatment systems for domestic sewage and industrial effluents may cause similar water quality problems. Authorities in many African countries appreciate the need for aquatic pollution control, especially for provision of safe drinking water and wastewater disposal, but have yet to act effectively. Alabaster (1981), Calamari (1985) and CIFA (1987) identify some of the underlying structural constraints. These include the poor distribution of expertise, poor research facilities, and the need for new and improved legislation along with the means for its enforcement.

It is impossible not to perceive causal links between the prevailing, depressed macroeconomic climate in many countries in Africa and the difficulties experienced with what would otherwise be straightforward administrative procedures. Attracting qualified staff is hampered, not necessarily by poor financial rewards but more significantly by poor working conditions. This is reflected, for example, in the paucity or low quality of data available to assess adequately the status of aquatic pollution of fisheries data in general.

This scenario pervades any in-depth discussion on any means of increasing food production in Africa and related environmental issues. Such deep structural weaknesses are evident from the worsening capabilities in many African fisheries research institutions and extension services over the last decade, in spite of a multitude of studies and recommendations and substantial financial and technical assistance. Small as it is, aquaculture production in Africa actually declined by about 10 per cent between 1974 and 1985, even though some US$150 million had been spent on aquaculture development over the same period (FAO/UNDP 1987).

Aquaculture in Africa

The degree to which problems of environmental impact might exist depends on the type of aquaculture systems used and particularly on the level of inputs (i.e., intensity). The systems in use vary within countries and among countries reflecting the existence of some aquaculture traditions and more recent trends in aquaculture practices from other regions.

Traditional Systems

Balarin (1984a, 1984b, 1985, 1986a, 1986b, 1988) and ICLARM-GTZ (1991) have described some of the traditional systems of fish farming in Africa. These include:

THE BARACHOIS SYSTEM

This is unique to Mauritius and is practiced in coastal lagoons. Fringing coral reefs from sheltered lagoons, the inlets of which are cut-off by stone walls to create enclosures, known as barachois. These are stocked at variable rates with fingerlings of a variety of lagoon fish. Stocking generally approaches 1,000 fish per hectare and is performed annually, as is the harvesting. The system relies entirely on natural productivity, since intertidal exchanges preclude the use of fertilizers.

THE ACADJA OR BRUSH-PARK SYSTEM

This is indigenous to Benin and Ghana and has up to eight variations, depending upon the size and configuration of the system and the materials used. Acadjas are made by placing wood (hard woods and tree branches) in marked-off areas of some 150-300 m^2 in shallow lagoons and lakes. The wood forms an aggregation device which attracts the fish, mainly tilapias (e.g., *Sarotherodon melanotheron*) and catfishes (e.g., *Chrysichthys* spp.), due to increased growth of epiphytes on the wood and provision of refuges. Fish are harvested about twice each year in the larger brush parks and as much as ten times per year in smaller varieties. The main limitation of this system is the high requirement for wood, with the obvious potential consequence of deforestation.

THE WHEDOS OR 'FISH HOLE' SYSTEM

This is practiced mainly in Benin and is also found in Cameroon and Togo. Whedos are artificial ponds or river channels. Their production is dependent on size: annual yields of 1.5-2.1 $t \cdot ha^{-1}$ have been recorded. Species composition also varies with size, since oxygen depletion in smaller fish holes with excessive plant growth makes them better suited to airbreathing species like *Clarias gariepinus*.

THE HOWASH SYSTEM

This is indigenous to the Nile delta of Egypt and comprises three different types, based on their location: coastal, lakeshore and lakewater. It is a system of shallow ponds constructed with earthen dikes. The coastal howash are found along the Mediterranean, constructed on land between the deltaic lakes and sea and filled and flushed by tidal movements. The lakeshore howash are located around two lakes, Manzala and Burullous, and depend on the discharge from irrigation systems. The lakewater howash are built within the lake to a depth of some 2 m. Annual production from howash ranges from 0.5 to 5.0 $t \cdot ha^{-1}$ based on initial stocking densities of carps, mullets and tilapias, i.e., between 1,000 and 5,000 fish per hectare.

OTHERS

Apart from these familiar and often cited traditional systems, others also exist. These tend to be seasonal activities largely performed by women, at a subsistence level, supplementary to other work. For example, in Ghana, women of the lower Volta used to collect juvenile freshwater clams (*Egeria radiata*) by diving, for subsequent restocking in more fertile and less saline areas. The creation of the Akosombo Dam has curtailed this activity. Also, in Ghana, catfish (e.g., *Clarias gariepinus*) used to be reared in earthen pots. In Gabon, women catch fingerlings for stocking in ponds. In Liberia, juvenile catfish and turtles have been stocked in barrage ponds.

Introduced Systems

The introduced aquaculture systems to Africa can be traced back at least seventy years: for example, the culture of tilapias in Kenya in 1924 (Balarin 1985) and the introduction of rainbow and brown trout *(Oncorhynchus mykiss* and *Salmo trutta)* in Kenya between 1910 and 1921 (Welcomme 1981) and in Swaziland, 1914-15 (Chondoma 1988). More generally, the genesis of aquaculture in most countries in Africa dates from the late 1940s and 1950s (Table 1).

Introduced aquaculture systems are practiced in different aquatic environments, ranging from freshwater to marine, and with the intensity of production ranging from extensive to intensive.

Extensive and semi-intensive freshwater aquaculture systems are the most widely practiced, and which account for about 97% of Africa's total aquaculture production [about 10,500 t for 1985 (Satia 1989)]. Predominantly rural, such aquaculture is normally practiced in earthen ponds which range in size from 100 to 1,000 m^2 surface area. Inputs are limited to domestic wastes and plant wastes or by-products, such as compost. Annual yields vary with quality and quantity of inputs, ranging from 0.5 to 5.0 $t \cdot ha^{-1}$, the lower limit being more common.

Semi-intensive integrated farming systems are becoming more prevalent, finding acceptance in countries including Benin, the Central African Republic (CAR), Côte d'Ivoire, Egypt, Kenya, Malaŵi, Nigeria, Rwanda and Zambia. In addition to natural productivity, the systems receive applications of manure, inorganic fertilizers and, where possible, artificial feeds or wastes from farm produce such as bagasse from sugar mills. Combined systems of integrated poultry-fish or livestock-fish farming, with duck, chicken or pig excreta fertilizing the ponds is also practiced as in the CAR, Madagascar, Nigeria and Zambia - whereas combined rice and fish production has had success in Madagascar, where tilapias and carp *(Cyprinus carpio)* are raised in ricefields.

The most commonly cultured species are indigenous tilapias, whose popularity stems from their hardiness and prolific breeding. Predatory species such as *Clarias gariepinus* and *Hemichromis fasciatus* are sometimes stocked to control excessive tilapia reproduction, but with mixed success. In the absence of adequate management, pond culture of tilapias has

Table 1. Dates of the initiation of introduced aquaculture systems in Africa.

Country	System(s)	Year	Country	System(s)	Year
Benin		1955	Mozambique		1950
Burkina Faso		1952	Nigeria		1944
Cameroon		1947	Rwanda		1950s
Central African Republic		1952	Sierra Leone		1970s
Côte d'Ivoire		1955	Sudan		1950
Egypt		1934	Swaziland		1972
The Gambia		1979	Tanzania		1972
Ghana		1953	Togo		1954
Kenya		1924	Uganda		1957
Madagascar		1951	Zaïre		1943
Malaŵi		1954	Zimbabwe		1950s

usually proved disastrous to the chagrin of fish farmers and extension agents alike, and has led to the abandonment of tilapia farming by many rural fish farmers in most African countries. It is estimated that some 300,000 fishponds were in use in the 1950s - a figure which had dropped to about 50,000 by the mid-1980s.

Apart from pond culture, the stocking of impoundments, irrigation reservoirs, lakes and rivers has been tried out in most subSaharan African countries, often following species transfers within and between countries. Thys van den Audenaerde (1988) observed that from 1945, many transfers of tilapia were made in Africa giving rise to confusion about their natural patterns of distribution and posing serious consequences for genetic resources conservation.

Intensive systems are the least common systems in Africa. Where they exist, they are largely experimental. Satia (1989) reported that commercial initiatives in Benin, Burkina Faso, Côte d'Ivoire, Congo and Nigeria have proved largely nonviable. However, there are some commercial intensive pond systems in Benin, Côte d'Ivoire, Kenya and Nigeria. Cage culture and pen culture have been tried in Benin, Egypt, Côte d'Ivoire, Kenya, Niger and Nigeria. Benin has had success with pen culture as a replacement for the wood-dependent acadja system.

Intensive aquaculture in tanks and raceways in Africa is very limited, Kenya and Zimbabwe being the pioneers. In Kenya, tilapia *(Oreochromis niloticus* and *O. spirulus)* are produced commercially at a private farm.

Hatcheries are rare and mostly owned and operated by private farms. The majority of government-owned establishments, on which rural farmers depend, are run down and unproductive. In some countries, this has motivated some farmers tc produce their own fingerlings, mainly tilapias and carp. No doubt these limitations further constrain the advancement of aquaculture.

Impact on the Environment

The relationships between aquaculture and adjacent ecosystems are unique, compared to other food production systems. Pullin (1989) observed that 'the aquatic medium is in direct and intimate contact with the metabolic processes of fish'. Where relevant, soil structure and composition, the impact of introduced substances, and natural environmental factors (e.g., temperature, salinity, oxygen concentration) all influence the quality of water which in turn determines the health of fish and their performance as cultured organisms.

Similarly, the environmental consequences of adoption of aquaculture systems must be weighed against their relative costs and benefits and pressing food production and livelihood needs.

Extensive and Semi-Intensive Systems[1]

Given their low input requirements and productivity, these systems have relatively few direct negative impacts on the environment. Nevertheless, they do involve the loss of habitats, particularly coastal zone wetlands, that support diverse fauna and flora. The unwitting destruction of mangroves for shrimp and bivalve culture is an example. The culture of penaeid shrimps is being actively pursued in Côte d'Ivoire, the Gambia, Guinea Bissau, Madagascar, Mozambique and Sénégal. In Mozambique, Pauly et al. (1989) reported that the main emphasis on marine aquaculture research was shrimp culture. They suggested, however, that the 680,000 ha of mangrove (exclusive of 170,000 ha of tidal swamplands) should

[1] See p. 2, 141-142 for definitions - Eds.

remain as natural nursery grounds for various species, including shrimps.

Health hazards from extensive and semi-intensive aquaculture can derive from two sources: the aquatic medium itself and the consumption of unsafe produce. Risks from the aquatic environment are of particular importance in Africa, where aquatic vectors and intermediate hosts of human diseases are widespread; for example, the snail intermediate hosts of schistosomiasis, black flies for onchocerciasis and mosquitos for malaria, filariasis and viral infections.

Schistosomiasis is endemic to specific areas and is considered an 'occupational disease' amongst fishers of the Volta and Niger deltas and amongst farmers in Sudan and Egypt. Aquaculture could create new habitats for snail intermediate hosts and new foci of infection. Fishponds, their water supply and drainage systems may be colonized by snails prevalent in adjacent natural habitats or by species whose natural habitats have been reclaimed and subsequently occupied by fish farmers.

Although the very successful practice in Southeast Asia of using human excreta for pond fertilization is very limited or nonexistent in Africa, it is worth pointing out some of its associated risks with regard to schistosomiasis. The survival of schistosome eggs in human excreta applied to ponds facilitates infestation of the snail host. The cercariae, which are shed by the snails into the aquatic environment, bore into the skin of humans in contact with the pond water. Edwards (1985) suggested that avoiding the use of fresh excreta, in which the eggs can survive for up to a week, is a useful means of avoiding infection.

Onchocerciasis is prevalent in communities of West Africa settled along fast flowing parts of rivers; for example, in the Niger, Sénégal and Volta systems. The vector is the black fly *(Simulium)*. It does not occur in slow-flowing or stagnant waters such as fishponds, but may be found in associated raceways and water channels. Interactions between the vector and aquaculture concern mainly the effects of the insecticides used in vector control programs which may be toxic to fish, although the use of temephos (Abate) to control *Simulium* has had no discernible long-term effects on fish populations (CIFA 1987).

Habitats for mosquito larvae vary with species; for example, *Anopheles funestus* prefers vegetated swamps and river margins whereas *A. gambiae* prefers open pools. Recently built ponds are a good habitat for the latter.

Risks from the consumption of contaminated fish produce result from the invasion of fish by spoilage organisms due to poor handling and processing methods, the accumulation of enteric bacterial and viral organisms, and the bioaccumulation of pesticides. The accumulation in fish of enteric bacteria, such as *Salmonella* and *Shigella* spp., and viral hepatitis is possible where the fish have been cultured in waters polluted by human excreta. Furthermore, the widespread use of pesticides and insecticides in Africa for disease vector control and agriculture increases the likelihood of the accumulation of these substances in fish cultured in contaminated sources, although supporting data are sparse. Calamari (1985) reported that several chlorinated hydrocarbons, including DDT, have been banned in some West African countries and replaced by less toxic organophosphorus compounds, carbamates and natural and artificial pyrethroids.

Intensive Systems

Some of the concerns discussed above for extensive and semi-intensive systems are also applicable to intensive culture. One area of concern is the prevalence of antibiotics and other antimicrobial drugs for prophylaxis and disease management in fish hatcheries. This may have far-reaching consequences in the evolution and spread of antibiotic-resistant strains of potentially harmful pathogens. Moreover, intensive aquaculture and associated fish stress may create situations in which otherwise benign organisms become pathogenic.

The use of antibiotics to control fish mortalities, by reducing the levels of pathogens in the water, may impact not only on the fish, but also on humans. They may ingest pathogenic bacteria associated with the culture system and these may be drug-resistant. There are also risks associated with consumption of fish with drug residues and contamination from direct exposure to the drugs used. Of particular interest is chloramphenicol, whose attractiveness as a wide-spectrum antibiotic has found "overwhelming preference among hatchery managers" (Brown 1989). Alderman (1989) reported that such general use of this drug was strongly discouraged in the United Kingdom because of the drug's level of toxicity and its value in human disease therapy, particularly against typhoid. Direct chloramphenicol contamination can cause potentially irreversible aplastic anemia: fatal in 70% of cases, with the increased incidence of leukemia in survivors (Brown 1989). This is a sobering thought, given the lax attitude towards protective clothing in many establishments in Africa.

Apart from their effects on human health, the indiscriminate use of antibiotics and other antimicrobial drugs also constitutes a source of pollution. For example, only 20-30% of the oxytetracycline used as medication in food pellets is taken up by fish. The remainder reaches the environment and is detectable 3-6 months after its administration (Samuelsen 1989).

In addition, fecal material and the excess, uneaten feeds find their way into the environment through effluent discharges, causing eutrophication and oxygen depletion of adjacent waters and reducing their quality and scope for aquaculture or other uses.

Impact of the Aquatic Ecosystem on Aquaculture

The quality of the aquatic medium largely determines success in aquaculture. Earlier discussions already referred to some of the potential impact of pesticide use. For Africa, Alabaster (1981) and Calamari (1985) have identified pesticides and organic pollutants with high biological oxygen demand (BOD), as the major potential sources of aquatic pollution of inland waters. Saleh et al. (1988) in a study of manmade reservoirs used for the drainage of agricultural wastewater in Egypt, reported the potential dangers from inorganic pollutants (pesticides and fertilizers). They concluded that although concentrations of heavy metals were low at that time, evaporation could raise them to dangerous levels which could affect fish production.

Strategy for Environmental Impact Control

CIFA (1987) identified the need for regional cooperation in effecting pollution control measures for the maintenance of water quality to protect the aquatic environment in Africa. There were calls for action in three areas of concern: toxic substances (such as pesticides and heavy metals) from agricultural use and vector control programs; discharges of organic matter with a high BOD; and discharges of suspended solids and nutrients. The coordination and implementation of such measures depend on the availability of results from field studies and the political will to legislate and enforce guidelines.

Established criteria for aquatic pollution control are rigid (as they should be) and often difficult to maintain in industrialized countries, necessitating the adoption of intermediate measures. This would be even more of a problem in Africa. CIFA (1987) proposed a general scheme and a strategy for the management of effluent discharge which would offer the possibility for starting pollution monitoring and control at levels with less demanding technical and financial inputs.

The absence in many African countries of the necessary legal framework to ensure adherence to pollution control measures generally, but particularly within fisheries legislation, is indicative of the urgent need for such action.

References

Alabaster, J.S. 1981. Review of the state of aquatic pollution of East African inland waters. CIFA Occas. Pap. 9, 36 p. Committee for Inland Fisheries of Africa, FAO, Rome.

Alderman, D.J. 1989. Fisheries chemotheraphy: a review, p. 1-61. *In* J.F. Muir and R.J. Roberts (eds.) Recent advances in aquaculture. Croom Helm, London.

Balarin, J.D. 1984a. Etudes nationales pour le développement de l'aquaculture en Afrique. 4. Togo. FAO Circ. Pêches 770.4, 66 p.

Balarin, J.D. 1984b. Etudes nationales pour le développement de l'aquaculture en Afrique. 5. Benin. FAO Circ. Pêches 770.5, 52 p.

Balarin, J.D. 1985. National reviews for aquaculture development in Africa. 6. Cameroon. FAO Fish. Circ. 770.6, 88 p.

Balarin, J.D. 1986a. National reviews for aquaculture development in Africa. 7. Kenya. FAO Fish. Circ. 770.7, 96 p.

Balarin, J.D. 1986b. National reviews for aquaculture development in Africa. 8. Egypt. FAO Fish. Circ. 770.8, 128 p.

Balarin, J.D. 1988. National reviews for aquaculture development in Africa. 18. Ghana. FAO Fish. Circ. 770.18, 121 p.

Brown, J. 1989. Antibiotics: their use and abuse in aquaculture. World Aquacult. 20: 34-35, 38-39, 42-43.

Calamari, D. 1985. Review of the state of aquatic pollution of West and Central African inland waters. CIFA Occas. Pap. 12, 26 p. Committee for Inland Fisheries of Africa, FAO, Rome.

Chondoma, E.C. 1988. National reviews for aquaculture development in Africa. 20. Swaziland. FAO Fish. Circ. 770.20, 104 p.

CIFA. 1987. Report of the First Session of the Working Party on Pollution and Fisheries, 16-20 June 1986, Accra, Ghana. Committee on Inland Fisheries for Africa. FIRI/R369(En), FAO Fish. Rep. 369, 23 p.

Edwards, P. 1985. Aquaculture: a component of low cost sanitation technology. World Bank Tech. Pap. 36. The World Bank, Washington, DC.

FAO/UNDP. 1987. Thematic evaluation of aquaculture. A joint study by UNDP, the Norwegian Ministry of Development Cooperation and FAO. FAO, Rome. 44 p.

ICLARM-GTZ. 1991. The context of small-scale integrated agriculture-aquaculture systems in Africa: a case study of Malaŵi. ICLARM Stud. Rev. 18, 302 p.

Pauly, D., T.O. Acere and M. Vincke. 1989. On research and aquaculture in southeastern Africa. Report of a mission to assess the fishery research needs of Kenya, Malaŵi, Mozambique and Zimbabwe. Study on International Fisheries Research. The World Bank, Washington, DC. (mimeo).

Pullin, R.S.V. 1989. Third-world aquaculture and the environment. Naga, ICLARM Q. 12(1): 10-13.

Saleh, A.S., M.A. Saleh, M.M. Fouda, M.A. Saleh, M.S. Abdel Lattif and B.L. Wilson. 1988. Inorganic pollution of the man-made lakes of Wadi El-Raiyan and its impact on aquaculture and wildlife of the surrounding Egyptian desert. Arch. Environ. Contam. Toxicol. 17: 391-403.

Samuelsen, O.B. 1989. Degradation of oxytetracycline in seawater of two different temperatures and light intensities, and the persistence of oxytetracycline in the sediment from a fish farm. Aquaculture 83: 7-16.

Satia, B. 1989. A regional survey of the aquaculture sector in Africa south of the Sahara. Aquaculture Development and Coordination Programme. ADCP/REP/89/36. FAO, Rome.

Thys van den Audenaerde, D.F.E. 1988. Natural distribution of tilapias and its consequences for the possible protection of genetic resources, p. 1-12. *In* R.S.V. Pullin (ed.) Tilapia genetic resources for aquaculture. ICLARM Conf. Proc. 16, 108 p.

Welcomme, R.L. 1981. Register of international transfers of inland fish species. FAO Fish. Tech. Pap. 213, 120 p.

Discussion

CATAUDELLA: We should compare the traditional use of acadjas with the modern situation. The technique requires the use of tree branches. In some locations, increasing salinity is also killing trees. Both acadjas and howash are good examples of traditional systems that can be modified for future use provided that conflicts between farmers and fishers are solved.

BILIO: The requirements for wood need careful appraisal. Afforestation is very important; for example, mangrove replanting. In a GTZ project in Benin this has been started on the initiative of the local community target group.

EDWARDS: Integrated livestock-fish culture, which was mentioned, nearly always implies integration of intensive livestock raising with semi-intensive fish culture. Are such systems really viable for small-scale farmers in Africa? My impression is that such ideas don't survive the development projects that introduce them.

KING: You are right. The problem is that such projects have made demonstrations using a well-chosen farmer and supporting the activities with aid money. Once this finishes, the available resources cannot sustain the system. There is, however, an interesting system on the campus of the International Institute for Tropical Agriculture (IITA), Ibadan, Nigeria, which uses in-house resources for simple, extensive fish culture. There is also a livestock-fish integrated system. The Institute is situated in a very dynamic farming area, with substantial input from private and public sectors, so some of the ideas might spread.

PULLIN: It is encouraging that the CGIAR centers, like IITA and the West African Rice Development Agency (WARDA), Bouaké, Côte d'Ivoire, and other agricultural and forestry institutions in Africa are beginning more and more to consider how aquaculture can be integrated with their operations. In many rural areas in Africa incomes are very low, as low as US$20/month, and sections of the rural community may be outside the cash economy for part of the year. We should bear these socioeconomic constraints in mind when considering how aquaculture can be integrated with some African farming systems. The constraints are not just technical.

Aquaculture Development and Environmental Issues in the Tropical Pacific*

JOHN L. MUNRO
International Center for Living Aquatic Resources Management
Coastal Aquaculture Centre
P.O. Box 438
Honiara, Solomon Islands

MUNRO, J.L. 1993. Aquaculture development and environmental issues in the tropical Pacific, p. 125-138. *In* R.S.V. Pullin, H. Rosenthal and J.L. Maclean (eds.) Environment and aquaculture in developing countries. ICLARM Conf. Proc. 31, 359 p.

Abstract

Commercial aquaculture in Oceania is currently restricted to the red alga *Eucheuma* sp., the blacklip pearl oyster, *Pinctada margaritifera* and penaeid shrimp, principally *Penaeus monodon*. There have been numerous attempts to cultivate exotic species of bivalves, crustaceans and fish, many of which have been unsuccessful in terms of aquaculture but which have resulted in wild stocks of the exotic species becoming established. Owing to a paucity of knowledge of the natural composition and dynamics of the aquatic communities, the effects of the exotic species on these communities are largely unknown. Few physical or chemical changes in the environment of Oceania are attributable to aquaculture but it is clear that the coral reefs and lagoons which make up the principal nearshore biotopes are sensitive to environmental perturbations and care will need to be taken to ensure that future developments in giant clam cultivation, reef and lagoon ranching systems and cage cultivation of finfish or crustaceans do not lead to environmental degradation. Almost all introductions of exotic species or transfers of indigenous species have been of unquarantined stocks with concomitant dangers of the introduction or spread of pathogens, parasites and predators.

Introduction

This paper concentrates on the situation in the developing island countries of Oceania, bounded by French Polynesia in the southeast, Palau, Papua New Guinea and Australia in the west, the Northern Marianas, Federated States of Micronesia and Marshall Islands in the north and Tonga and New Caledonia to the south (Fig. 1).

Aquaculture is in its infancy in Oceania. There have been numerous attempts at developing various forms of fish and invertebrate culture in past years, but few have come to fruition (Uwate et al. 1984). Most efforts have focused on the marine sector, as lentic and lotic environments are very limited in the islands except for the larger landmasses comprised of Papua New Guinea, Solomon Islands and Fiji. Most past efforts have been based on introductions of familiar species such as oysters, shrimp, mussels, tilapia, carp but there have been a few

*ICLARM Contribution No. 752.

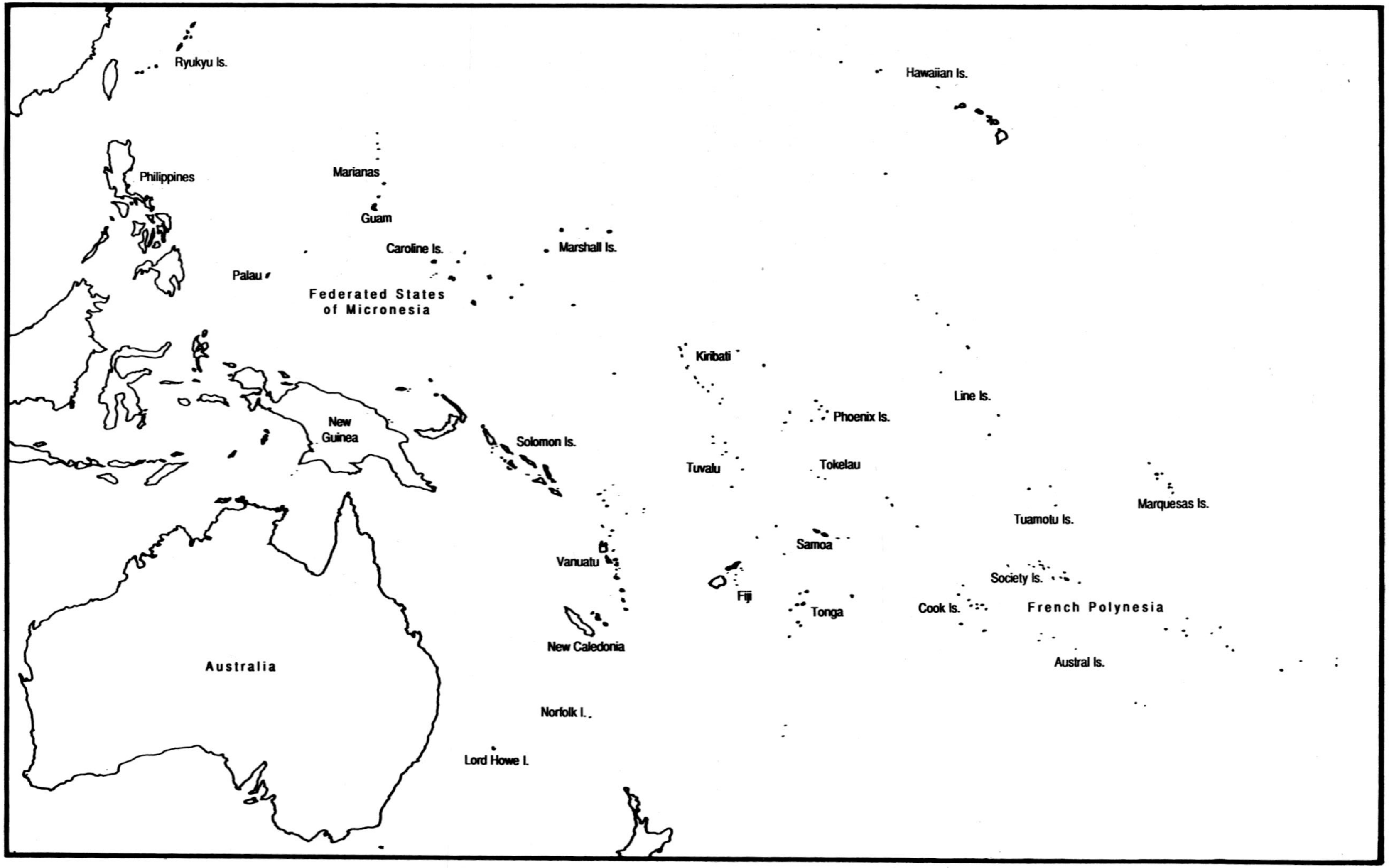

Fig. 1. Countries and territories of the South Pacific, showing place names mentioned in the text.

commercial successes. Remnants of introduced stocks survive and are exploited in many areas.

Successful commercial aquaculture in Oceania is currently confined to the cultivation of the red algae *(Eucheuma* sp.), pearl oysters *(Pinctada margaritifera)* and penaeid shrimp. These activities can still be described as marginal in some areas, depending very much on scale, access of markets and level of technology and only the pearl oysters can be said to represent a substantial industry at the present time; this being directed primarily to the cultivation of cultured pearls, with the pearl shell being an important secondary product.

Previous reviews of the status and prospects for aquaculture have been done by Uwate and Kunatuba (1984), Uwate et al. (1984) and Glude (1984). For the most part these have been rather negative, suggesting rather dismal prospects for most species groups.

Current Status of Commercial Aquaculture

Species of the red alga *Eucheuma* are of principal interest at present and farming trials have been conducted in Fiji, Kiribati and Solomon Islands, and there is much interest elsewhere in the Region (Why 1985). However, Fiji is the only country producing *Eucheuma* on a consistent basis. There are development trials in progress in Kiribati and Solomon Islands. Other species of algae which have been tried include *Gracilaria* sp. (FitzGerald 1982) and *Spirulina* sp. (Uwate et al. 1984).

Pearl oysters are the primary bivalve product in the South Pacific Region, being harvested principally for their shells. However, in French Polynesia, the Cook Islands and Fiji there are also substantial pearl culture industries based on the blacklip pearl oyster *Pinctada margaritifera*. The goldlip pearl oyster, *P. maxima*, is used for pearl culture in Australia and *P. fucata* is the dominant commercial species and the foundation of the Asian cultured pearl industry.

The only other bivalves of significance are tridacnid clams (family Tridacnidae) comprised of the genera *Tridacna* and *Hippopus*. Substantial progress has been made in recent years in the development of cultivation systems (Copland and Lucas 1988) and there are a number of organizations operating on a commercial basis, although none have yet claimed to operate at a profit.

Penaeid shrimp cultivation is established on a small scale in some areas and, where the local conditions are favorable, farms appear to be making substantial profits on local markets. Hatcheries operate on a commercial scale in New Caledonia, French Polynesia and Guam (FitzGerald 1982; AQUACOP 1984).

Milkfish *(Chanos chanos)* are reared on a quasi-commercial scale in a government-operated fish farm in Tarawa, Kiribati. This was developed primarily to supply live bait for the skipjack tuna fishery (Uwate et al. 1984).

Apart from the above-mentioned species, cultivation of all other species is or has been on an experimental or trial basis; mostly on a small scale. This has meant that direct environmental impacts have been minimal. Few habitats have been significantly modified and none have been physically or chemically degraded by aquaculture activities.

Introductions and Transfer of Organisms for Aquaculture

While the physical and chemical effects of aquaculture on the aquatic environment have been minimal, the tropical Pacific Ocean and the freshwaters of the islands of Oceania have been subjected to the widest possible range of transplantations and introductions of exotic species in pursuit of aquaculture development or, to a lesser degree, fisheries enhancement. Many of these species have not survived but there is no record of the effects of concurrent introductions of species of parasites, pathogens or predators.

A summary of introductions for aquaculture purposes up to 1984 is given by Uwate et al. (1984). Since 1984 the major recorded transfers have been of *Eucheuma* spp. and of *Tridacna derasa* and *T. gigas*. Table 1 summarizes the situation.

A very active role in these introductions was played by the South Pacific Commission and by the South Pacific Islands Fisheries Development Agency (SPIFDA) in 1970-73. With very few exceptions these transfers have violated all accepted protocols (Turner 1987; Munro et al. 1985; ICES, n.d.) for introductions and transfers of aquatic organisms, despite the known risks (Humphrey 1988). It should be added that there is currently no evidence that these transfers have had any negative effects on the ecosystems of Oceania, but so little is understood about these systems that it is unlikely that any but the grossest effect would be recognizable.

Geographical distributions of Pacific marine organisms have been substantially altered in a number of cases. In the case of *Tridacna derasa* the distribution of stocks on the basis of early scientific records was from southern Japanese waters, the Philippines, Palau and Indonesia, southwards to Australia and south-eastwards from Palau to Tonga, thus extending to 175°W longitude in the South Pacific, but only to 135°E longitude in the northern hemisphere (Munro 1989). It is suggested that the eastward extension to Tonga might be anthropogenic in origin, as a result of early seafarers carrying live *T. derasa* in the bilges of canoes and jettisoning their cargoes on arrival in new territories. Whatever the case, this curious distribution has now been dramatically altered by the transfer of cultivated stocks from the Micronesian Mariculture Demonstration Center (MMDC) in Palau throughout Micronesia and to American Samoa, Hawaii and the Cook Islands (Heslinga et al. 1988; CTSA 1990). Some MMDC stocks have recently been transferred to the Caribbean (North 1990; Grashof and Hensen 1991). Unfortunately, no quarantine procedures were used at the MMDC before 1988 and the current procedures (Heslinga et al. 1988) do not conform with the protocol recommended by the South Pacific Commission (Munro et al. 1985) and are therefore a potential source of introduced parasites and pathogens (Pernetta 1987; Humphrey 1988).

Pests and parasites affecting giant clam cultivation are as yet poorly known, but a number of species of the gastropod *Cymatium* and various members of the Pyramidellidae are emerging as significant pests. Current evidence is that the important species are by no means uniformly distributed around the Pacific and inadvertent introductions are possible if ocean to ocean transfers are effected. For example, juvenile *Cymatium muricinum* are reported to embed themselves in the tissues around the byssal orifice of *Tridacna derasa* (Perron et al. 1985) and could therefore easily be inadvertently introduced to new areas along with juvenile clams which have been reared in the ocean. Other species of *Cymatium*, possibly more voracious than *C. muricinum*, have recently been identified (H. Govan, unpubl. data) and might not be as widespread as *C. muricinum*.

Pyramidellid snails, ectoparasitic opisthobranchs which feed on the body fluids of their hosts, have been recorded from almost all giant clam hatcheries and nurseries in the Indo-Pacific but the genera and species implicated in infections appear to be different at almost every location

Table 1. Species introduced for the purpose of aquaculture to territories and nations of Oceania.

Taxon Genus	Country	Species	Source	Date	Outcome	Reference
Rhodophyta						
Eucheuma						
	Fiji	?	?	?	Significant industry	Uwate et al. (1984)
	F.S.M. (Pohnpei)	?	?	1982	No industry yet	Uwate et al. (1984)
	French Polynesia	?	?	?	No industry yet	Why (1985)
	Kiribati	*striatum*	Hawaii	1977	No industry yet	Why (1985)
	Kiribati	*spinosum*	?	?	No industry yet	Unpublished
Pteriidae						
Pinctada						
	Cook Islands	*margaritifera*	?	1905	Stock established	Uwate et al. (1984)
	Kiribati	*fucata*	?	1902	Unsuccessful	Uwate et al. (1984)
	Tonga	*maxima*	?	1902	Unsuccessful	Tanaka (1990)
	Tonga	*margaritifera*	?	1902	Unsuccessful	Tanaka (1990)
	Tonga	*fucata*	?	1902	Unsuccessful	Tanaka (1990)
Pteria						
	Tonga	? *penguin*	?	1975-79	Possible establishment of wild stocks	Tanaka (1990)
Ostreidae						
Crassostrea						
	Fiji	*iredalei*	Philippines	1977	Unsuccessful	Angell (1986)
	Fiji	?	?	1895	Inconclusive	Glude (1984)
	Fiji	*gigas*	?	1965	Cultivation failed but wild stocks found	Angell (1986)
	French Polynesia	*gigas*	California		Successful trials	Coeroli et al. (1984)
	New Caledonia	*gigas*		1967	Some commercial development	Coeroli et al. (1984)
	Palau	*gigas*		mid-70s	Failed	Angell (1986)
	Papua New Guinea	?			Failed	Glude (1984)
	Tonga	*gigas*			Failed	Glude (1984)
	Vanuatu	*gigas*	?		Failed	Glude (1984)

continued

Table 1 continued

Taxon Genus	Country	Species	Source	Date	Outcome	Reference
Saccostrea						
	French Polynesia	*echinata*	New Caledonia	?	Declined after 1975	Coeroli et al. (1984)
	Guam	*cucullata*	Solomon Islands	?	Failed	Braley (1984)
	Guam	*echinata*	Solomon Islands	?	Established	Braley (1984)
Mytilidae						
Perna						
	Fiji	*viridis*	Philippines	1975	?	Uwate et al. (1984)
	French Polynesia	*viridis*		1978	Succesful trials	Coeroli et al. (1984)
	New Caledonia	*viridis*		1972	Successful but stocks lost in cyclone	Uwate et al. (1984)
	Western Samoa	*viridis*		1981	?	Uwate et al. (1984)
	Western Samoa	*viridis*	Tahiti	1982-83	Successful trials	Uwate et al. (1984)
Tridacnidae						
Tridacna						
	American Samoa	*derasa*	Palau	1984	Continuing	CTSA (1990)
	Cook Islands	*derasa*	Palau	1984	Continuing	Sims and Howard (1988)
	Federated States of Micronesia	*derasa*	Palau	1984	Continuing	Price and Fagolimul (1988)
	Fiji	gigas	Australia	1986	Successful reintroduction	Ledua and Adams (1988)
	Marshall Islands	*derasa*	Palau	1984	Continuing	CTSA (1990)
	Tuvalu	*derasa*	Palau	1984	Continuing	CTSA (1990)
	Western Samoa	*derasa*	Palau	1984	Continuing	CTSA (1990)
Penaeidae						
Penaeus						
	American Samoa	?	?	1979	Inconclusive	Uwate et al. (1984)
	French Polynesia	*aztecus*	Texas, USA	1972	Inconclusive	AQUACOP (1975)
	French Polynesia	*indicus*	?	?	Successful trials	AQUACOP (1984)
	French Polynesia	*japonicus*	Hawaii, USA	1972	Successful trials	AQUACOP (1975)
	French Polynesia	*japonicus*	New Caledonia	1972	Successful trials	AQUACOP (1975)
	French Polynesia	*merguiensis*	New Caledonia	1972	Successful trials	AQUACOP (1975)
	French Polynesia	*monodon*	New Caledonia	1972	Successful trials	AQUACOP (1975)

continued

Table 1 continued

Taxon Genus	Country	Species	Source	Date	Outcome	Reference
	French Polynesia	*monodon*	Fiji	1975	Successful trials	AQUACOP (1977b)
	French Polynesia	*stylirostris*	?	?	Successful trials	AQUACOP (1984)
	French Polynesia	*vannamei*	?	?	Successful trials	AQUACOP (1984)
	Guam	*monodon*	?	1978	Initial hatchery operations failed (but recently reopened)	Uwate et al. (1984)
	Kiribati	?	?	?	?	Uwate et al. (1984)
	New Caledonia	?	?	?	Successful	
	Palau	?	?	<1975	?	Uwate et al. (1984)
	Solomon Islands	*monodon*	Australia	1987	Successful	Unpublished
Metapenaeus						
	French Polynesia	*ensis*	New Caledonia	1972	Inconclusive	AQUACOP (1975a)
Palaemonidae						
Macrobrachium						
	French Polynesia	*rosenbergii*	Hawaii	1973	Successful trials	AQUACOP (1977a)
	Guam	*rosenbergii*	Hawaii	1974	Commercial hatchery failed (but recently reopened)	Uwate et al. (1984)
	Kiribati	*rosenbergii*	?	?	?	Uwate et al. (1984)
	New Caledonia	*rosenbergii*	?	1972	Technically feasible	Uwate et al. (1984)
	Palau	*rosenbergii*	?	?	Limited freshwater supplies	Uwate et al. (1984)
	Solomon Islands	*rosenbergii*	Tahiti	1985	Abandoned in favor of penaeids	Unpublished
	Vanuatu	*rosenbergii*	?	1955	?	Uwate et al. (1984)
	Western Samoa	*rosenbergii*	?	1982	Some success	Uwate et al. (1984)
Anguillidae						
Anguilla						
	Guam	*rostrata*	South Carolina	1977	Failed	FitzGerald (1982)
	Guam	*japonica*	Taiwan	1973	Discontinued	FitzGerald (1982)
	Guam	*japonica*	Hong Kong	>1973	Discontinued	FitzGerald (1982)
	Guam	*japonica*	China	>1973	Discontinued	FitzGerald (1982)
Poeciliidae						
Poecilia						
	American Samoa	?	?	1972	Uneconomic baitfish culture	Uwate et al. (1984)
	Fiji	?	?	1975	Unfeasible	Uwate et al. (1984)

continued

Table 1 continued

Taxon Genus	Country	Species	Source	Date	Outcome	Reference
	Palau	?	?	mid 70's	Discontinued	Uwate et al. (1984)
	Tonga	?	?	1974	Inconclusive	Uwate et al. (1984)
	Western Samoa	?	?	1978	Commercially infeasible	Uwate et al. (1984)
Cichlidae						
Oreochromis						
	American Samoa	*mossambicus*	?	1950-60	Wild stocks established	Uwate et al. (1984)
	Cook Islands	*mossambicus*	?	1955	Wild stocks established	Uwate et al. (1984)
	Fiji	*mossambicus*	?	1949	Small fishery developed	Uwate et al. (1984)
	Fiji	*niloticus*	Philippines	1980+	Broodstock liberated by floods	Uwate et al. (1984)
	French Polynesia	*mossambicus*	?	1975	?	Uwate et al. (1984)
	Guam	*mossambicus* x *niloticus*	Taiwan	1954	Unsuccessful	FitzGerald (1982)
	Guam	red hybrid	Taiwan	1974	Commercial culture	FitzGerald (1982)
	Kiribati	*mossambicus*	?	1963	Pests in milkfish ponds	Uwate et al. (1984)
	Nauru	*mossambicus*	?	1961	Not acceptable as food	Uwate et al. (1984)
	New Caledonia	*mossambicus*	?	1955		Uwate et al. (1984)
	Papua New Guinea	*mossambicus*	Malaya	1954	Wild stocks established	West and Glucksman (1976)
	Solomon Islands	*mossambicus*	?	1957	Wild stocks established	Uwate et al. (1984)
	Tonga	*mossambicus*	?	?	Now widespread	Uwate et al. (1984)
	Tuvalu	*mossambicus*	?	?	Widespread	Uwate et al. (1984)
	Western Samoa	*mossambicus*	?	1961	Stocked in natural waters	Uwate et al. (1984)
Belontiidae						
Trichogaster						
	Papua New Guinea	*pectoralis*	?	?	Established in lowlands	West and Glucksman (1976)
Osphronemidae						
Osphronemus						
	Papua New Guinea	*goramy*	?	?	Failed in highlands	West and Glucksman (1976)
Pangasiidae						
Pangasius						
	Guam	*sutchi*	?	1973	Unfavourable	FitzGerald (1982)

continued

Table 1 continued

Taxon Genus	Country	Species	Source	Date	Outcome	Reference
Cyprinidae						
Cyprinus						
	Fiji	*carpio*	?	1986	Inconclusive	Uwate et al. (1984)
	Guam	*carpio*	Taiwan	?	?	
	Papua New Guinea	*carpio*	?	>1958	Widespread in natural waters	West and Glucksman (1976)
Ctenopharyngodon						
	Fiji	*idella*	?	?	Broodstock liberated by floods	Uwate et al. (1984)
	Guam	*idella*	Taiwan	1974	?	FitzGerald (1982)
Aristichthys						
	Fiji	*nobilis*	?			Uwate et al. (1984)
	Guam	*nobilis*	Taiwan	1974	?	
Hypophthalmichthys						
	Fiji	*molitrix*	?		?	Uwate et al. (1984)
	Guam	*molitrix*	Taiwan	1974	?	
Puntius						
	Fiji	*gonionotus*	?	?	Released into natural waters	Uwate et al. (1984)
Salmonidae						
Oncorhynchus						
	Papua New Guinea	*mykiss*	NZ and Australia		Some commercial success	West and Glucksman (1976)
Salmo						
	Papua New Guinea	*trutta*	Australia		Stock established in highlands	West and Glucksman (1976)
Testudines						
Trionyx						
	Guam	*sinensis*	Taiwan	1977	Stock escaped	FitzGerald (1982)
	Guam	*sinensis*	Taiwan	>1977	Marketed in Hawaii	FitzGerald (1982)

(Cumming 1988). The pyramidellids retreat to the lower surfaces of the clam or into gravel substrata by day and adults or egg masses could easily be inadvertently transferred with juvenile clams which have been exposed to untreated seawater. In the past year, shipments of giant clam spat to the Philippines, Guam and Fiji are reported to have been contaminated in this way.

The pearl oyster cultivation industry at Takapoto Atoll in French Polynesia was devastated in 1985 by an unidentified disease (Hauti et al. 1987), as a result of which transfers of live pearl oysters from the atoll were prohibited. A decade earlier a pearl oyster farm in Papua New Guinea, using regular shipments of *Pinctada maxima* from Western Australia, was abandoned because of high mortality rates. These are all indications that transfers between ocean areas should not be lightly undertaken.

Physical Environmental Changes

Physical environmental changes attributable to aquaculture have been minimal in Oceania up to the present. Destruction of coastal mangroves and shoreline vegetation for pond excavation has been limited. To some extent this is attributable to the communal ownership of land and marine estates and the sheer difficulty of obtaining agreement about development projects. However, in other cases the opportunities for aquaculture development have only recently been perceived and it appears that some environmental lessons have been well learned and that some of the mistakes made in other parts of the world will not be repeated in Oceania.

Nevertheless, where changes in the physical environment have been wrought in the interest of development the consequences have been serious. This is best observed in the US affiliated territories where environmental modifications in the interests of harbor, airfield and road construction, principally for military purposes, have destroyed huge tracts of reef, seagrass and mangrove habitats. In other areas, the construction of causeways for linking the small islands in atoll systems have resulted in blockages of vital passes whereby fresh oceanic water is flushed into the atoll lagoon on each high tide, and has led to severe changes in hydrographic regimes, caused disruption of spawning runs of important fish species and, possibly, reduced the numbers of recruits of various fish species (Johannes 1975). Logging, mining and poor land management have contributed to siltation of reefs and lagoons in some areas but this is not a widespread problem in Oceania. Such changes are not attributable to aquaculture development but do highlight the point that the island environment is a sensitive instrument.

Future Developments

It is likely that aquaculture will develop at a moderate pace in Oceania, depending mostly on technical breakthroughs, technology transfer and development of marketing opportunities, particularly through new air and shipping routes.

Established marine aquaculture technologies are for cultivation of *Eucheuma*, pearl oyster and penaeid shrimps. Giant clam cultivation is in a rapidly developing phase and research is currently underway or planned on cultivation of various species of algae, sponges, sea cucumbers, sea urchins, crabs, spiny lobsters, topshell, green snails, various species of reef fishes and a limited variety of freshwater, brackishwater and oceanic fish species, including anguillid eels.

Many of the species under consideration are coral reef or coral lagoon species which, like giant clams and pearl oysters, simply will not survive in highly eutrophic environments. Although there is some environmental degradation in many areas, particularly near to urban centers in Oceania, there is a positive attitude towards environmental conservation in many areas. In the Solomon Islands, for example, coastal villagers readily perceive the link between an undegraded reef environment and giant clam cultivation and it is very likely that *Eucheuma*, giant clam or pearl oyster cultivation can be used as positive incentives towards reef and lagoon conservation in many areas.

For many of the species under consideration such as sea urchins, sea cucumbers, gastropods and species of reef-bound reef fishes, effort is likely to be directed towards the reef or lagoon ranching concept (Yamaguchi 1977; Munro and Williams 1985) where selected desirable species of reef organisms will be propagated and stocked into reef and lagoon habitats, isolated from other reef areas by tracts of deepwater or soft bottom habitats. This is analogous to stocking a lake or reservoir. The fishes stocked have nowhere else to go and provided the trophic resources are adequate will survive to be harvested in due course. An important feature of coral reef systems is that it appears that, for the most part, coral reef ecosystems are recruitment limited and can support far greater numbers of organisms than are commonly encountered on a given reef (Munro et al. 1973; Williams 1980; Doherty 1982). The only difference between this concept and the much older concept of restocking marine systems, as practiced in Japan, is that where level bottom marine habitats are stocked the habitat is often virtually limitless and as in the Japanese experience, beneficial effects are difficult to detect.

Heavy stocking of a reef or lagoon system with a particular species might be expected to have substantial effects on the dynamics of the biotic community. Some of the anticipated effects can probably be modeled using a program such as ECOPATH (Christensen and Pauly 1991) but others will undoubtedly be discovered empirically. The transfer of pathogens from hatchery-reared stocks to wild stocks will represent a significant hazard. On the other hand, the relative isolation of island systems does offer the opportunity of restricting the spread of pathogens by quarantining islands where pathogens have appeared, as was successfully done in the case of the pearl oyster disease in French Polynesia.

More important threats to the marine environment come from the prospects of cage or pen cultivation of desirable species in coral reef lagoons. The protected waters of coral reef lagoons offer near-ideal hydrographic conditions for cage cultivation (AQUACOP 1975b), particularly if the lagoon is not entirely enclosed. The technology for cultivation of high-valued predatory species such as snappers, groupers, bream, bass, seabass and oceanic dolphinfish (*Coryphaena hippurus*) is currently available and constraints lie in hatchery production techniques, nutrition and marketing. Countries of Oceania with canneries or processing plants for skipjack and yellowfin tuna are now producing relatively low-cost fish meal from tuna wastes and hence the production of low-cost pelletized feeds is currently feasible in Fiji and Solomon Islands. As is the case in salmon cultivation, poor siting of cages or high density cage cultivation can lead to heavy sedimentation of the subadjacent benthic habitat, local anoxic conditions, algal blooms and environmental degradation (Beveridge 1987).

A final point is that there are virtually no legal provisions in countries of Oceania concerning aquaculture, the installation of moored or seabed structures, disposal of organic wastes or modification of the marine environment. Without such safeguards, lagoon and reef areas are vulnerable to uninformed, shortsighted or even unscrupulous modifications of the environment.

Conclusion

The atolls and high island lagoon systems of Oceania will, without doubt, come to be extensively used for coastal aquaculture because of their favorable hydrographic features. Constraints on aquaculture development are imposed by economic features and by slow technology transfer but those will be eroded in time. Freshwater aquaculture will be constrained in most areas by the limited availability of lentic and lotic environments.

Owing to their relative isolation, the countries of Oceania have unparalleled opportunities for the development of pest- and disease-free stocks of cultivated species. Careless transfers between island systems or unthinking introductions of exotic species pose a major threat to this opportunity.

In Oceania, there has been no significant environmental degradation which is attributable to aquaculture and it is likely that the errors of habitat destruction which have taken place elsewhere will not be repeated.

However, widespread introductions of exotic species and transfers of indigenous species for aquaculture occurred in the past and continue and pose a major threat to the natural ecosystems and to the prospects of developing disease- and pest-free stocks.

References

Angell, C.L. 1986. The biology and culture of tropical oysters. ICLARM Stud. Rev. 13, 42 p.

AQUACOP. 1975a. Maturation and spawning in captivity of penaeid shrimps *Penaeus merguiensis* de Man, *Penaeus japonicus* Bate, *Penaeus aztecus* Ives, *Metapenaeus ensis* de Haan and *Penaeus semisulcatus* de Haan. Proc. 6th Annu. Meet. World Maricult. Soc. 6: 123-132.

AQUACOP. 1975b. First experiments on the rearing of tropical Carangidae in floating cages. Proc. 6th Annu. Meet. World Maricult. Soc. 6: 271-284.

AQUACOP. 1977a. *Macrobrachium rosenbergii* (de Man) culture in Polynesia: progress in developing a mass intensive larval rearing technique in clear water. Proc. 8th Annu. Meet. World Maricult. Soc. 8: 311-326.

AQUACOP. 1977b. Reproduction in captivity and growth of *Penaeus monodon*, Fabricius, in Polynesia. Proc. 8th Annu. Meet. World Maricult. Soc. 8: 927-945.

AQUACOP. 1984. Ten years of experimental rearing of penaeid shrimp in Tahiti and New Caledonia (South Pacific). Report to the 15th Meeting, World Mariculture Society. 32 p.

Beveridge, M. 1987. Cage Culture. Fishing News Books, Farnham, UK. 352 p.

Braley, R.D. 1984. Mariculture potential of introduced oysters *Saccostrea cucullata tuberculata* and *Crassostrea echinata* and a histological study of reproduction of *C. echinata*. Aust. J. Mar. Freshwat. Res. 35:129-141.

Christensen, V. and D. Pauly. 1991. A guide to the ECOPATH II program (version 2.0). ICLARM Software 6, 71 p.

Coeroli, M.D., Gaillande, J.P. Landret and A.D. Coatanea. 1984. Recent innovations in cultivation of molluscs in French Polynesia. Aquaculture 39:45-67.

Copland, J.W. and J.S. Lucas, Editors. 1988. Giant clams in Asia and the Pacific. ACIAR Monogr. 9, 274 p.

CTSA. 1990. CTSA continues work in giant clam aquaculture. Center for Tropical and Subtropical Aquaculture, Hawaii. Regional Notes 1(4), 5 p.

Cumming, R.L. 1988. Pyramidellid parasites in giant clam mariculture systems, p. 231-236. *In* J.W. Copland and J.S. Lucas (eds.) Giant clams in Asia and the Pacific. ACIAR Monogr. 9, 274 p.

Doherty, P.J. 1982. Coral reef fishes: recruitment-limited assemblages?. Proc. 4th Int. Coral Reef Symp. 2:465-470.

FitzGerald, W.J. 1982. Aquaculture development plan for the territory of Guam. Economic Development Administrations Section Guam. 182 p.

Glude, J.B. 1984. The applicability of recent innovations to mollusc culture in the western Pacific islands. Aquaculture 39:29-43.

Grashof, M. and R. Hensen. 1991. Cultivation of endemic and non-endemic bivalves on Bonaire, Netherlands Antilles. (Abstract). World Aquaculture Conference and Exposition. 22. San Juan, Puerto Rico.

Hauti, A.G., A. Tehoiri, M. Tchoung and G. Matitai. 1987. L'aquaculture nacrière et perlière à

Takapoto en 1986. Ministère de la Mer, Service de la Mer et de l'Aquaculture. 64 p.

Heslinga, G.A., T.C. Watson and T. Isamu. 1988. Giant clam research and development in Palau, p. 49-50. *In* Giant clams in Asia and the Pacific. ACIAR Monogr. 9, 274 p.

Humphrey, D.J. 1988. Disease risks associated wiıh translocation of shellfish, with special reference to the giant clam *Tridacna gigas*, p. 241-244. *In* J.W. Copland and J.S. Lucas (eds.) Giant clams in Asia and the Pacific. ACIAR Monogr. 9, 274 p.

ICES. n.d. Guidelines for implementing the ICES code of practice concerning introductions and transfers of marine organisms. 8 p. (mimeo report).

Johannes, R.E. 1975. Exploitation and degration of shallow marine food resources in Oceania, p. 47-71. *In* R.W. Force and B. Bishop (eds.) The impact of urban centers in the Pacific. Pacific Science Association, Honolulu.

Ledua, E. and T.J.H. Adams. 1988. Quarantine aspects of the reintroduction of *Tridacna gigas* to Fiji, p. 237-240. *In* J.W. Copland and J.S. Lucas (eds.) Giant clams in Asia and the Pacific. ACIAR Monogr. 9, 274 p.

Munro, J.L. 1989. Fisheries for giant clams (Tridacnidae: Bivalvia) and prospects for stock enhancement, p. 541-558. *In* J.F. Caddy (ed.) Marine invertebrate fisheries: their assessment and management. John Wiley, New York.

Munro, J.L., V.C. Gaut, R. Thompson and P.H. Reeson. 1973. The spawning seasons of Caribbean reef fishes. J. Fish Biol. 23: 184-194.

Munro, J.L., J. Lucas, A. Alcala, E.D. Gomez, A.D. Lewis and J.C. Pernetta. 1985. Considerations regarding the introduction or transfer of tridacnid clams. Proc. 17th South Pacific Commission Regional Technical Meeting on Fisheries, Noumea New Caledonia. Working Paper 24, 2 p.

Munro, J.L. and D.M. Williams. 1985. Assessment and management of coral reef fisheries: biological, environmental and socio-economic aspects. Proc. 5th International Coral Reef Congress 4: 545-581.

North, D. 1990. The return of the giant clam. Pacific Island Monthly 60(5): 32-34.

Pernetta, J.C. 1987. Giant clams: a new source of food or biological contamination? The Siren (United Nations Environment Programme) 34: 18-25.

Perron, F.E., G.A. Heslinga and J.O. Fagolimul. 1985. The gastropod *Cymatium muricinum*, a predator on juvenile tridacnid clams. Aquaculture 48:211-221.

Price, C.M. and J.O. Fagolimul. 1988. Reintroduction of giant clams to Yap State, Federated States of Micronesia, p. 41-43. *In* J.W. Copland and J.S. Lucas (eds.) Giant clams in Asia and the Pacific. ACIAR Monogr. 9, 274 p.

Sims, N.A. and N.T. Howard. 1988. Indigenous tridacnid clam populations and the introduction of *Tridacna derasa* in the Cook Islands, p. 34-40. *In* J.W. Copland and J.S. Lucas (eds.) Giant clams in Asia and the Pacific. ACIAR Monogr. 9, 274 p.

Tanaka, H. 1990. Winged pearl shell newly found in Tonga. SPC Pearl Oyster Info. Bull. 1:15.

Turner, G.E., Editor. 1987. Codes of practice and manual of procedures for consideration of introductions and transfers of marine and freshwater organisms. ICES Doc. F35-A, 42 p.

Uwate, K.R. and P. Kunatuba. 1984. Aquaculture development in the Pacific islands region. Pacific Islands Development Program Report to to Pacific Congress on Marine Technology, Honolulu, Hawaii. 21 p.

Uwate, K.R., P. Kunatuba, B. Raobati and C. Tenakanai. 1984. A review of aquaculture activities in the Pacific islands region: general overview by political entity. Pacific Island Development Program, East-West Center, Honolulu, Hawaii. 23 p. (mimeo report).

West, G.J. and J. Glucksman. 1976. Introduction and distribution of exotic fish in Papua New Guinea. PNG Agric. J. 27: 19-48.

Why, S. 1985. *Eucheuma* seaweed farming in Kiribati, Central Pacific. UK Overseas Development Administration, Research Project R3775. 91 p.

Williams, D.M. 1980. The dynamics of the pomacentrid community on small patch reefs in One Tree Lagoon (Great Barrier Reef). Bull. Mar. Sci. 30: 159-170.

Yamaguchi, M. 1977. Conservation and cultivation of giant clams in the tropical Pacific. Biol. Conserv. 11: 13-20.

Discussion

RUDDLE: There is evidence from Kiribati, where aquaculture of milkfish *(Chanos chanos)* is an old-established tradition, that the introduction of tilapia has damaged this severely and that tilapia is regarded there as a pest and a trash fish[1].

MUNRO: That is true. Kiribati is in central Oceania and has enormous stocks of juvenile milkfish which have traditionally been gathered. They might have been 'husbanded' to a small extent but I know of no extensive pond systems other than those at Christmas Island which was uninhabited until recently. Kiribati people will not eat tilapia and these have infested modern milkfish culture ponds and taro swamps in Tarawa. The species is *Oreochromis mossambicus* which adapts readily to saltwater. It is regarded as a pest.

D. PHILLIPS: Can you comment on the choice of tridacnid clam species for aquaculture and also on the economics of giant clam farming? Does it depend upon sales of the adductor muscle?

MUNRO: The three species of interest are *Tridacna gigas, Tridacna derasa* and *Hippopus hippopus* - in descending order of size. There are two separate camps: One camp believes that *T. derasa* is the preferred species because of its hardiness. The other, which includes ICLARM, favors *T. gigas* because of its more rapid growth rate and larger asymptotic size. *T. gigas* can grow to a length of about 50 cm (30 kg weight) in six years. It is undoubtedly a bit more fragile than *T. derasa* but, given the right techniques, this problem can be overcome. *Hippopus hippopus* is a very robust species but it is quite small. The question of choice remains open but the weight of evidence is for *T. gigas*.

The whole of the animal flesh is edible and has been eaten in Oceania for thousands of years. The adductor muscle is about 10% of the saleable parts - which means 500 g of sashimi-grade adductor muscle from a 50-cm clam. The adductor muscle is prized in Taiwan and whole small tridacnids are used for sashimi in Okinawa. We are not discouraged by the market prospects relative to production - which is very low at present.

---

[1]Iuta, T. 1989. Formal address and personal letter from the Minister of Natural Resources to participants of the Kiribati Applied Atoll Research Consultation. *In* R. Thaman (ed.) Applied atoll research for development. Proceedings of the Kiribati Applied Atoll Research for Development Consultation, 27 February-2 March 1989, Tarawa. Ministry of Natural Resources Development, Tarawa, Kiribati.

Environmental Issues in Integrated Agriculture-Aquaculture and Wastewater-Fed Fish Culture Systems

Peter Edwards
Division of Agricultural and Food Engineering
Asian Institute of Technology
P.O. Box 2754
Bangkok 10501, Thailand

EDWARDS, P. 1993. Environmental issues in integrated agriculture-aquaculture and wastewater-fed fish culture systems, p. 139-170. *In* R.S.V. Pullin, H. Rosenthal and J.L. Maclean (eds.) Environment and aquaculture in developing countries. ICLARM Conf. Proc. 31, 359 p.

Abstract

There are few data on the environmental impact of semi-intensive fish culture systems. Comparisons are made between a theoretical intensive aquaculture system using a complete diet (food conversion ratios, FCR 1.0-2.5) which has complete water exchange or flowthrough, and experimental semi-intensive fertilized fishpond systems in which water is released to the environment only on draining the ponds at the end of the culture cycle. Data are presented from two experimental semi-intensive systems, involving bagged chicken manure supplemented with inorganic fertilizers and clarified sewage effluent: only 15 and 11% of the fertilizer nitrogen (N) and 8 and 6% of the fertilizer phosphorus (P), respectively, were removed in harvested fish; only 2 and 6% of the N and 6 and 10% of the P, respectively, were released to the environment on draining the ponds at the end of the experiment; the remaining 83% of the N and 86 and 84% of the P, respectively, "disappeared" within the culture system, and probably most accumulated in the sediments. While 21-53% of the N and 11-28% of the P in feed are removed in harvested fish in the intensive system (with a higher feed nutrient conversion efficiency to fish than the semi-intensive system), the remainder of the nutrients in the feed are voided to the environment continuously throughout the culture cycle. In terms of weight of N and P released to the environment per kg of fish produced, the intensive system is 7-31 and 3-11 times more polluting than the semi-intensive system, respectively, depending on the FCR of complete feed.

The feeds in intensive systems are consumed directly by fish, whereas fertilizer nutrients are utilized indirectly through natural feed production in the semi-intensive pond system. However, the nutrient conversion efficiency of the semi-intensive system (with nitrogen supplied as fertilizer) is only slightly less than that of the intensive system (with nitrogen supplied in the form of feed protein): less than one-tenth the difference that might be expected from the extra step in the food chain in the fertilized system. This finding has major implications for reducing the cost of fish production because N and P are several times more expensive in the form of pelleted feed than as fertilizer. Interdisciplinary research to promote integrated agriculture-aquaculture farming systems is required. Promotion of wastewater-fed fish culture to recycle nutrients from urban areas would alleviate eutrophication and produce fish. Many developing countries lack a tradition of excreta reuse. The challenge is to implement excreta reuse systems which do not pose an unacceptable risk to public health in those countries where excreta reuse is not traditional but where the need for increased efficiency of use of nutrient resources is greatest.

Introduction

A consideration of environmental issues of integrated agriculture-aquaculture and wastewater (or human excreta)-fed fish culture systems together might seem inappropriate at first glance. However, it is logical to consider them together because both involve fertilization of ponds to produce natural food for fish. Furthermore, the prototype integrated farming system with fish culture, which probably originated in China, involved the reuse of human excreta as well as livestock manure and crop residues. It is only relatively recently, and in particular in developed countries, that the cycling of nutrients between agriculture and other human activities has been broken, leading to eutrophication.

A holistic or systems approach is taken in this paper in which the environment is defined in a broad sense with the ecosystem/agroecosystem comprising all nonliving and living resources, including humans who are central to wastewater-fed fish culture. The scope is restricted to a consideration of nutrients in general, and to fertilized pond systems in particular.

Certain terms and concepts (integrated farming, intensity of fish culture) germane to fertilized systems are widely used but with various meanings and interpretations. Rather than avoid their usage, which is impractical because they are in vogue, they are first defined as used in this paper with a brief discussion of alternative meanings. To aid in the identification of low cost, sustainable fish culture systems, the derivation of fish culture from the natural aquatic ecosystem is presented, followed by a discussion of fertilized aquaculture systems. Nutrient cycling is placed in an ecological perspective to contrast natural ecosystems which have more or less closed nutrient cycles with agroecosystems characterized by more open nutrient flows.

Integrated Farming

Integrated farming is defined here in the sense of diversification of agriculture with the development of aquaculture as a subsystem on a farm with existing crops, or crops and livestock, subsystems: "an output from one subsystem in an integrated farming system which otherwise may have been wasted becomes an input to another subsystem resulting in a greater efficiency of output of desired products from the land/water area under a farmer's control" (Edwards et al. 1988).

The main biological feature of integrated farming from a nutrient flow point of view is by-product recycling with diminished reliance on inter-farm or agroindustrial inputs. However, integrated farming has been defined in other ways. With the concern in developed countries to reduce the adverse environmental impact of agriculture, integrated farming systems have been defined as those that combine lower inputs of "conventional agriculture" such as low inputs of commercial fertilizers and pesticides with more traditional methods of mixed agriculture such as crop rotations and increased use of organic residues (Edwards 1989). Similarly, integrated farming aims to satisfy a complex of environmental and social aims rather than only being concerned with maximizing profit through predominantly agricultural objectives as in "conventional farming" (Vereijken 1986).

Muir (1986) used the term integrated fish farming in a much broader sense to include, in addition to integration with agriculture, the association of aquaculture with industry to provide nutritional inputs from agroindustrial by-products or heated water from power stations, and with sanitation in terms of sewage inputs to ponds. While such a definition indicates the wide spectrum of human activities that can be linked with aquaculture, it is too broad for the purpose of this paper because it includes intensive aquaculture systems.

Yet another type of integration is vertical integration, characteristic of intensive agriculture, in which the various inputs and stages of farming of an organism, including processing, are all controlled by a single company, in contrast to within-farm integration as defined here.

Intensity of Fish Farming Systems

Intensification of fish farming can be classified into three categories as previously recognized by Ling (1967), who did not name them, and presented more fully by Edwards et al. (1988):

a. Extensive systems rely on natural feed produced without intentional inputs. By definition they are excluded from integrated farming systems except for integrated rice-fish farming in which fish may derive benefits from inputs added solely for rice.
b. Semi-intensive systems depend on fertilization to produce natural feed *in situ* and/or on feed given to the fish, supplementary feed, to complement the natural feed which develops. A significant amount of the fish nutrition is derived from natural feed. Integrated crop-livestock-fish farms and wastewater-fed fishponds have semi-intensive pond systems.
c. Intensive systems depend on nutritionally complete feeds, either in moist formulations or in dried pelleted form, with fish deriving little to no nutrition from natural feed produced *in situ*.

The degree of intensification is defined according to feeding practice but intensification may be accompanied by increasing amounts of capital, labor and mechanization. The classification is particularly useful for this paper because fertilized systems are synonymous with semi-intensive systems in which a significant part of the fish diet is supplied by natural feed.

The terms extensive, semi-intensive and intensive have been widely used in the literature but rarely defined precisely. There appears to be a general consensus of opinion that at one end of the spectrum, extensive culture, fish rely on natural feed without inputs and at the other end of the spectrum, intensive culture, fish derive nutrition entirely from complete feed, with an intermediate area of semi-intensive systems in which natural feed is important. However, in contrast to the definition used in this paper, both extensive (Hickling 1971) and intensive systems (Schaeperclaus 1933; Hora and Pillay 1962) have been defined as involving fertilization.

Extrapolated fish yields that can be expected in well-managed aquaculture systems with different degrees of intensification are presented in Fig. 1. There is a distinct boundary in terms of nutrition between extensive and semi-intensive systems as defined here. However, if a semi-intensive system is given feed as well as fertilizer as the biomass of individual fish and the total weight of fish in the pond increase, the proportion of fish nutrition derived from natural food in a semi-intensive system declines relative to that of the feed (supplementary and/or complete feed). Aeration may also be required later in the culture cycle to maintain adequate dissolved oxygen in the pond.

The avoidance of aeration by less intensive culture in which nutritional inputs are in balance with fish growth may be a useful boundary to define the practical upper limit of semi-intensive systems for small-scale farmers, without

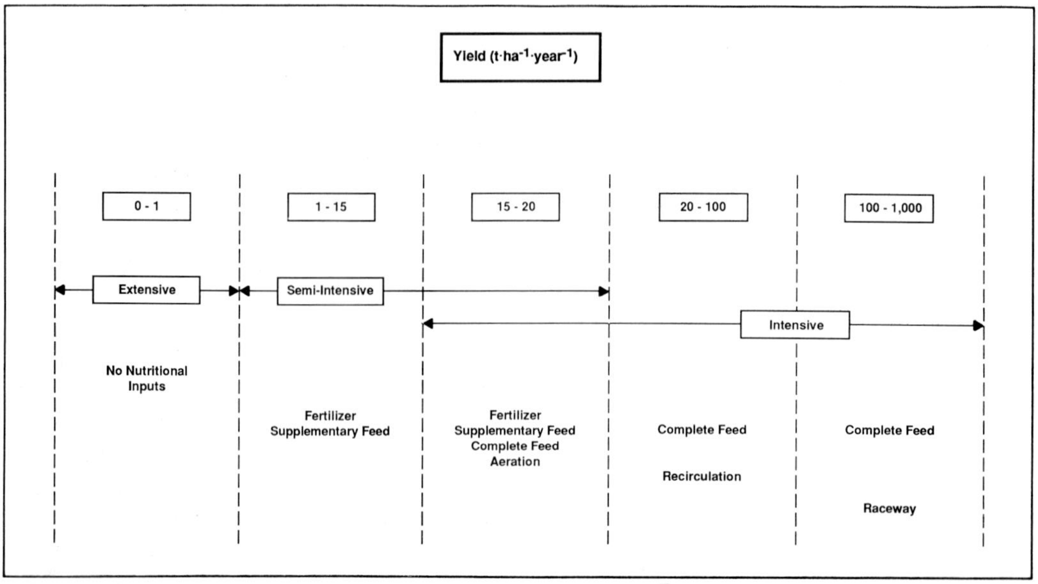

Fig. 1. Intensification of aquaculture systems. Modified from Edwards et al. (1988).

further intensification involving mechanical aeration and complete diets.

These definitions are discussed here only with respect to growout systems but they can be applied to hatchery and nursery systems.

Fish Nutrition in Freshwater Aquaculture

In the quest for low-cost sustainable aquaculture systems, systems which depend on natural food chains in aquatic ecosystems should be considered before those which rely on fish feed imported into the system as the former have less subsidies than the latter. The former should be more cost effective in terms of nutritional inputs and have less adverse environmental impact than the latter. A useful analogy which was fully appreciated more than 50 years ago is the comparison of a fertilized fishpond with cattle or dairy production on grass as animals feed at the base of the food chain in both systems (Schaeperclaus 1933). Schroeder (1977, 1980) stressed the importance of using the pond not only as a medium in which to grow the fish but also as an "external rumen" in which nutrients bound in a relatively indigestible form could be released by microbial activity and provide substrates for both heterotrophic and autotrophic production that could subsequently serve as high protein natural food for the fish.

Derivation of Aquaculture Systems from Natural Aquatic Ecosystems

There are three basic types of fish culture systems that have been derived from natural aquatic ecosystems (Fig. 2) (Edwards et al. 1988):

System 1 - vegetation-fed systems. Aquatic macrophytes in natural aquatic ecosystems are grazed by macrophagous feeding fish but in a well-managed fishpond with adequate inputs to support good fish growth they are rarely present as they are shaded out by water turbidity (plankton, detritus, and silt suspended in the water column by fish disturbing the sediments). Terrestrial plants and/

or aquatic macrophytes from outside the fishpond are fed to herbivorous fish and large amounts of fish excreta produced by the inefficient digestion of vegetation produce natural feed for plankton/detritus filtering fish and omnivorous/detritus benthic feeding fish. The central species in the traditional Chinese carp polyculture is the grass carp (*Ctenopharyngodon idella*). The culture of other macrophagous feeding fish such as giant gourami (*Osphronemus goramy*) and silver barb or tawes (*Puntius gonionotus*) is also a traditional practice in Southeast Asia.

System 2 - excreta or manure-fed systems for plankton/detritus filtering fish such as the Chinese carps, [silver carp (*Hypophthalmichthys molitrix*) and bighead carp (*Aristichthys nobilis*)], the Indian major carps [catla (*Catla catla*) and rohu (*Labeo rohita*)], and tilapias such as Nile tilapia (*Oreochromis niloticus*) and blue tilapia (*O. aureus*), and omnivorous/detritivorous benthic feeding fish such as common carp (*Cyprinus carpio*), the Chinese mud carp (*Cirrhinus molitorella*) and the Indian major carp mrigal (*Cirrhinus mrigala*).

System 3 - trash-fish fed systems for carnivorous fish such as snakehead (*Channa striata*) and walking catfish (*Clarias* spp.).

Whereas carnivorous fish are usually cultured in monoculture, fish are normally cultured in polyculture in systems 1 and 2 with a considerable overlap of feeding niches as indicated in Fig. 2. Fish feeding in a minimum of three niches has been recommended for type 1 systems (Edwards et al. 1988). Inefficient digestion of vegetation by macrophyte-feeding fish results in copious feces which stimulate food production for both water column feeders (plankton/detritus filtering fish)

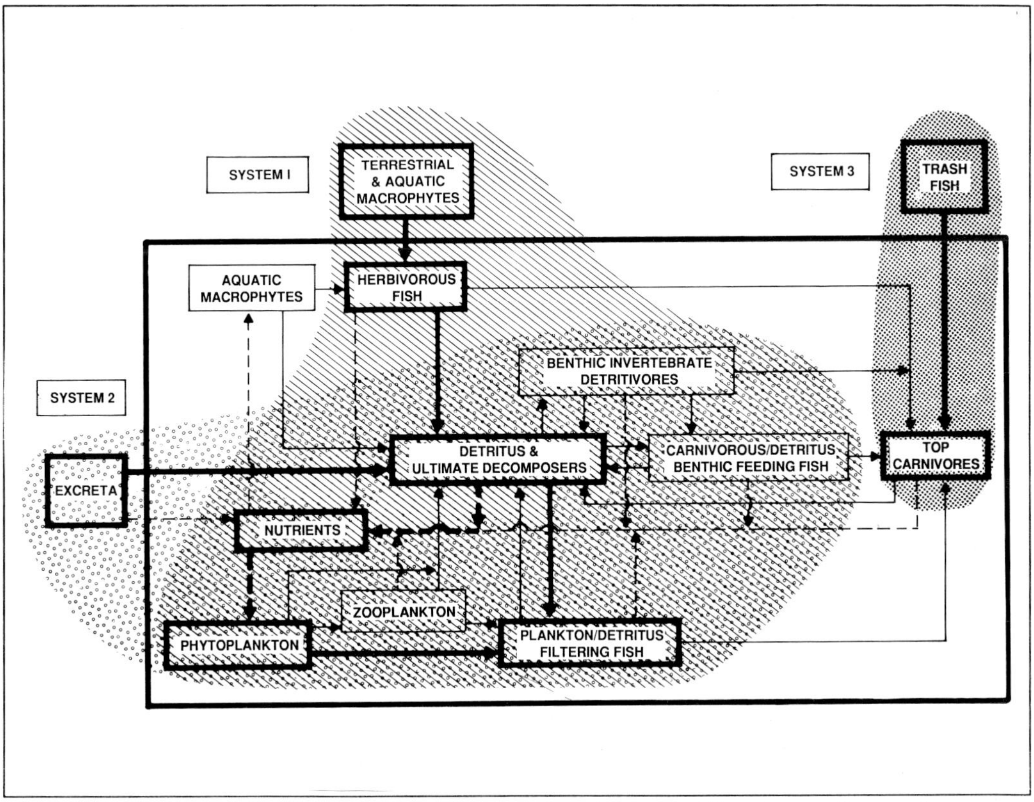

Fig. 2. The relationship between a natural pond ecosystem and three traditional aquaculture systems.

and benthic feeding fish. Fish feeding in two niches would be appropriate for type 2 systems that receive only manure. In the traditional pond culture of common carp in Europe and Israel, an omnivorous/ detritus benthic feeding fish has been made more productive by stocking plankton/detritus filtering fish (Reich 1975; Opuszynski 1978).

Pond Fertilization

The fertilization of fishponds with organic matter, particularly human excreta and livestock manure, is widely practiced in Asia (Ling 1967; Prowse 1967). According to Huet (1972), fish "must rely to the greatest possible extent on natural feed which contains all the constituents of a complete diet - amino acids, vitamins, minerals, etc". Natural food contains about 50-60% protein on a dry weight basis (Hepher 1979). It has also been appreciated for a long time that the culture of common carp in Europe could not be conducted economically with total dependence on artificial feed but only with at least 50% of nutrition from natural feed; the pond cannot be only a "stall" or feedlot, it is always "stall and pasture" in one and "the natural nutrition in the carp fishery is the basis for profitable feeding" (Schaeperclaus 1933). However, the promotion of organic fertilization is constrained by sociological and public health considerations for human excreta reuse and by the need to confine livestock in feedlots to utilize their manure in fishponds in integrated farming systems.

Commercial inorganic fertilizers are used infrequently in aquaculture despite their widespread use in agriculture in the developed world, and to an increasing degree in many developing countries. Part of the reason for the limited use of commercial fertilizers in fish culture is there are few scientific data on how to promote and maintain a rich growth of natural food or on the optimal amount required for growth of different fish species at different stages of their life history, at different stocking densities and in various environmental conditions (Ling 1967). Nutrient dynamics in fertilized fishponds are poorly understood and strategies to improve yields in such systems have not been developed. However, proper fertilization of ponds for the optimum and manageable growth of algae could bring about a substantial increase in fish production (Barica 1976). While we now have sufficient understanding of fishpond nutrient dynamics to pose relevant questions, their solutions through research and practical strategies to implement them at farm level are still wanting.

Supplementary Feed

Supplementary feed, by definition, is meant to augment natural food. When feeding on natural feed only, part of the natural feed proteins are used for energy which is wasteful. Furthermore, when natural feed becomes scarce, energy becomes limiting first and not protein. Relatively cheap cereal grains which are rich in carbohydrates as energy sources should be used as the main supplementary feeds (Hepher 1975, 1978; Opuszynski 1987). A corollary is that plant foods such as cereals (and legumes), which are the main supplementary feeds for common carp would be nutritionally unbalanced if fed alone; they must be given with natural food, to be utilized effectively as mentioned above (Opuszynski 1987). Rice bran is widely used as a supplementary feed in fishponds but many ponds have low fertility with limited natural feed, so much is utilized inefficiently.

Complete Feed

Intensive systems depend either on dry pelleted feed or on fish from capture fisheries. Dry feed formulations have been used increasingly in intensive aquaculture since about 1960 (Hickling 1971; Huet

1972) and have gradually replaced fresh fish as feed except for the culture of certain marine and freshwater carnivores (Csavas 1989). As dry feed formulations are usually dispensed from a bag, it is more difficult to appreciate readily the relatively more precarious supply of some of the ingredients, particularly fishmeal, than in former times. Dried marine fish (*Chatoessus nasus*) was imported from India before World War II to feed trout in Germany (Schaeperclaus 1933). Before the use of pelleted feed formulations permitted the widespread use of marine fish in feed, inland salmonid farms used fresh meat: cattle and horse meat, abattoir waste, slaughtered condemmed animals and also healthy animals purchased in markets (Huet 1972). The total fresh meat requirement of trout culture was often considerable and a farm producing 10,359 pounds (4,550 kg) of trout annually required a whole cattle or horse daily in the summer (Schaeperclaus 1933).

Pelleted feed formulations incorporate fishmeal derived from marine capture fisheries as the animal protein source. As capture fisheries are limited, there is increasing competition with livestock diets, the major consumer (Wijkström and New 1989; Csavas 1989). Reservations about the general relevance of intensive systems in developing countries were expressed about 30 years ago by Hora and Pillay (1962) concerning intensive eel culture in Japan: "if these methods are followed in other countries, the price of fish may increase beyond the reach of the common man".

Fertilized Aquaculture Systems

Integrated Farming Systems

Traditional small-scale integrated agroecosystems involving crops, livestock and fish, as well as humans, probably evolved in China. The major possible interactions between the various subsystems in such systems, excluding off-farm inputs and produce, are presented in Fig. 3. Livestock and human excreta or manure may be used to fertilize crops or the fishpond. Crops may be fed to livestock or fish (or humans). Fishpond water may be used to water crops or provide drinking water for livestock. Mud removed from the fishpond may be used to fertilize crops.

Integrated agriculture-aquaculture farming systems are most developed in China which also has the greatest variety of systems. Perhaps the best known system is the mulberry dike-fishpond system in which mulberry is raised on the pond dike and the leaves are fed to silkworms. Silkworm excreta and pupae are fishpond inputs and pond mud is used to fertilize mulberry (Ruddle and Zhong 1988; Zhong 1989). Livestock are often integrated with aquaculture. Some of their feed is provided from crops grown on the farm and manure is used to fertilize both crops and fishponds. Crops are grown on fishpond dikes either as vegetables for humans with the waste leaves fed to fish or as crops grown specifically for fish feed, e.g., grasses.

The most common integrated farming system outside China is feedlot livestock-fish (Fig. 4). Integration of crops with fish is less common outside China (Pullin and Shehadeh 1980). Nor is it general for pond mud to be removed to fertilize crops. Fish culture in ricefields is an example of integration of crops with fish but, in contrast to popular supposition, relatively few farmers are involved to date.

Small-scale farmer interest in tropical Asia in the culture of fish is increasing as wild fish decline in abundance but well-established strategies for semi-intensive fish culture do not exist. The majority of small-scale farms in tropical Asia are crop-dominated with most of the produce of the major crop, rice, being consumed domestically (Fig. 5). Crop by-products are fed to relatively small

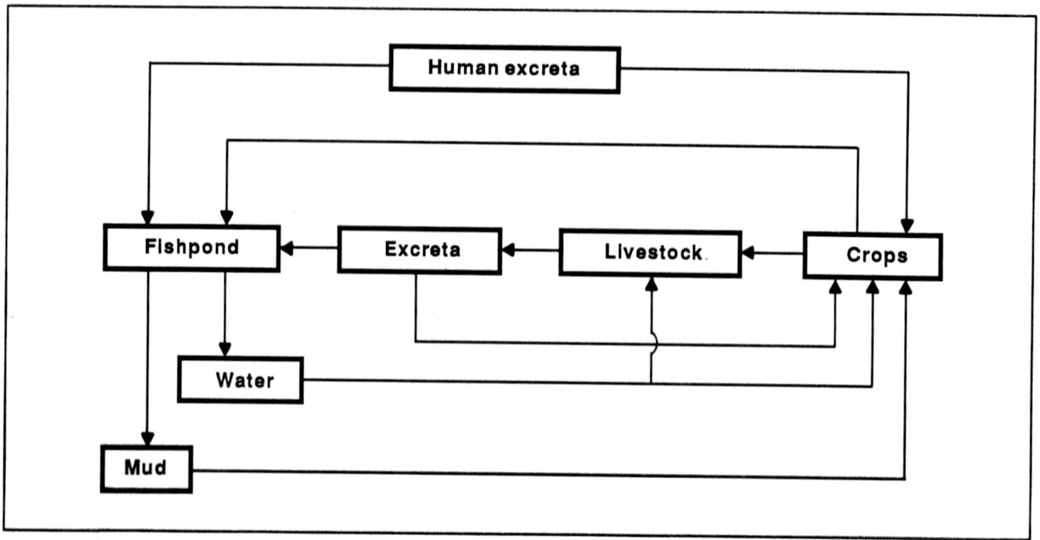

Fig. 3. **Schema of the major interactions between the various subsystems in a crop/livestock/fish integrated farming system.** Modified from Edwards et al. (1988)

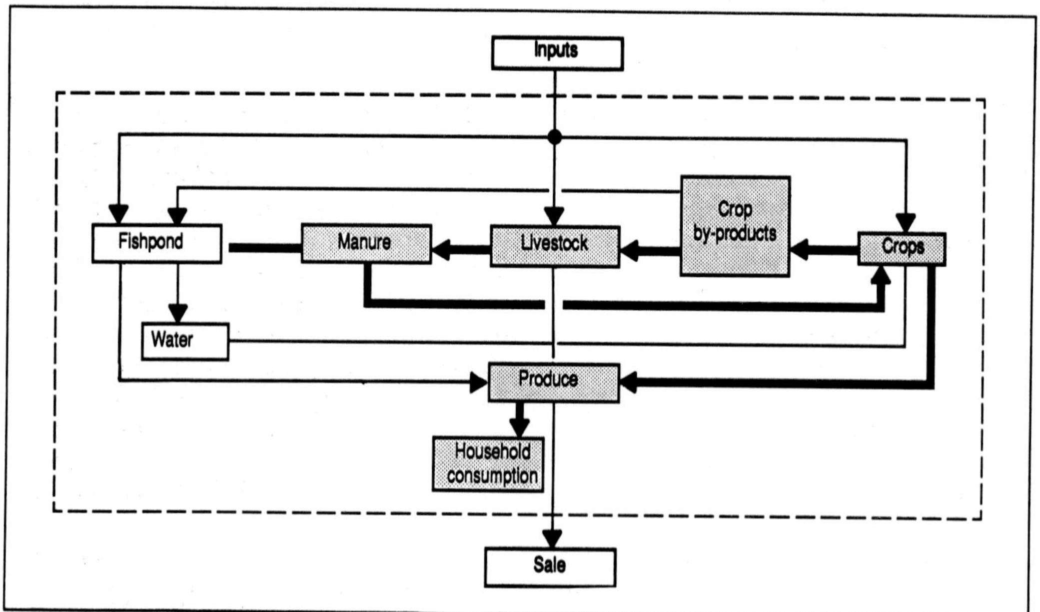

Fig. 4. **Major (often the only) interactions between the various subsystems (thicker lines and shaded boxes) in a feedlot-fish integrated farm.**

numbers of livestock, the manure of which fertilizes the crops. Most small-scale farmers are unable to raise feedlot livestock integrated with fish because they cannot afford this method of livestock raising. Moreover, farmers may dig ponds and stock them with fish (research has successfully addressed the problem of seed production of most species cultured in semi-intensive systems) but cost-effective fertilization and feeding strategies remain to be developed. As a consequence, farmers' ponds usually produce very low yields of fish. Fishponds commonly become

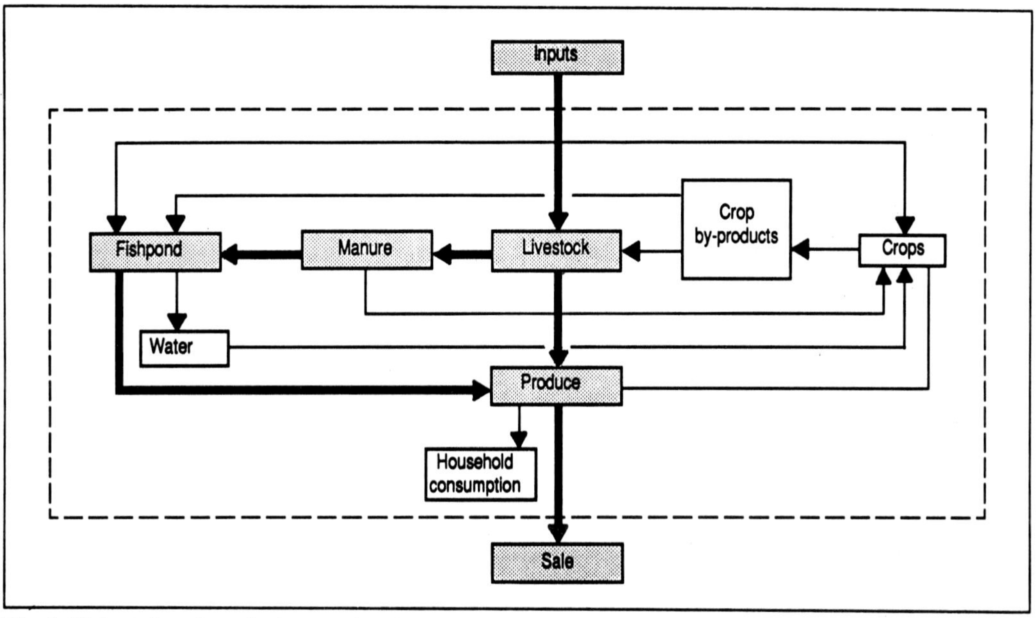

Fig. 5. Major (often the only) interactions between the various subsystems (thicker lines and shaded boxes) in a crop-dominated small-scale farm.

infested with emergent aquatic macrophytes stimulated by penetration of light to the pond bottom or contain silty water, both of which reflect the low nutrient status of the pond water.

Extant Wastewater-fed Aquaculture Systems

Wastewater is used here in the broad sense to include the various forms of human excreta: fresh or recently collected excreta or nightsoil, fecally polluted surface water, septage from septic tanks, and sewage. Wastewater reuse in fishponds occurs mainly in Asia but there is a well-known sewage-fed fishpond system in Munich, Germany, which is now used only for the tertiary treatment of activated sludge effluent (Prein 1990).

Most systems use fresh or only partially treated wastewater with potential public health problems. Cartage, the removal of nightsoil by bucket, cart or vacuum truck, is prevalent in China although the nightsoil is now stored for four weeks prior to use in fishponds. The overhung latrine, a superstructure built on piles over a fishpond, is widespread in China, Indonesia (West Java) and southern Vietnam. Fecally polluted surface water and conventional sewage for fertilizing fishponds are used in China where there are a total of 8,500 ha (Zhang 1990); India with 4,000 ha (Edwards 1992); and Vietnam with 1,400 ha in Hanoi alone (Tuan and Trac 1990).

Wastewater reuse in aquaculture is receiving increasing scientific attention (Edwards and Pullin 1990; Edwards 1992) as its potential to recycle nutrients in productive, socially acceptable systems has been realized.

Nutrient Cycling in Agroecosystems

An agroecosystem is simply an ecosystem which is used for agricultural purposes. The term agroecosystem is somewhat contradictory as an ecosystem is usually defined as a closed biological system, i.e., separated from its surroundings by boundaries through which no transport of material occurs (Fig. 6, upper). However, the main characteristic of an agroecosystem is that it is a system which produces food or fiber which is passed through its boundaries; thus it is an open system as far as nutrient transportation across its boundaries is concerned (Frissel 1977; Tivy 1987). Agroecosystems are characterized by a larger and more rapid turnover of nutrients than unmanaged ecosystems (Fig. 6, lower). Nutrients removed in farm produce need to be compensated by nutrient inputs, the magnitude of which depends on the nature and intensity of the farming system (Tivy 1987).

A significant event in our understanding of nutrients in agriculture was a 1976 symposium on the cycling of mineral nutrients in agrecosystems (Frissel 1977). Sixty-five agroecosystems were described in the proceedings although none included aquaculture. Agroecosystem boundaries were those of individual farms and contained plants, soils and animals but excluded humans. Nutrients crossing the farm boundary as inputs in the form of fertilizer and feed and outputs in the form of crop or animal produce were determined as well as other gains and losses. The models for nutrient cycling in agroecosystems consisted of three main compartments or pools - plant, livestock and soil. Nutrient transfers among pools within the system were indicated. Thirty-one fluxes were used to describe inputs, outputs and transfers with the system.

Nutrient cycles in agroecosystems involving fish in addition to crops and livestock involve even more nutrient fluxes but these remain to be studied. However, a preliminary attempt has been made to identify nutrient cycles for an agroecosystem involving crops, livestock and fish (Fig. 7) (ICLARM 1988). Furthermore, there is a need for a model for wastewater reuse in aquaculture which includes humans.

The two most important criteria selected to classify the systems from an agricultural point of view were type of farming and yield, the latter expressed as nitrogen (N) output (Frissel 1977). Extensive agroecosystems depend to a large extent on the natural or little modified soil nutrient reservoir, e.g., livestock grazing where physical constraints limit arable crop production and shifting agriculture. Yields are low with an output of consumable N of less than 20 kg $\cdot ha^{-1}$. Such an output in the form of fish corresponds to an annual production of 781 $kg \cdot ha^{-1}$ of fish (assuming fish are 16% protein, fresh weight basis). This figure is close to the upper productivity of extensively managed common carp ponds in Europe in which natural productivity without fertilization ranges from 20 to 600 $kg \cdot ha^{-1}$ of fish, although maximal values are rarely produced because most ponds are not situated on high quality soils (Opuszynski 1987). In general, yields in extensive culture of common carp in Europe range from 120 to 240 kg $\cdot ha^{-1} \cdot year^{-1}$ (O'Grady and Spillet 1985). Average fishpond yields of 690 kg $\cdot ha^{-1} \cdot year^{-1}$ in 1973 (compared to 2,070 $kg \cdot ha^{-1} \cdot year^{-1}$ in 1988) in China also indicate that the ponds received few inputs (Guan and Chen 1989).

At the opposite end of the spectrum, intensive agroecosystems have high productivity but are maintained by a correspondingly large input of nutrients, e.g., continuous and wholly arable systems and intensive livestock systems. The latter can be based on grass production with high inputs of fertilizers or can be feedlots with all feed supplied from off-farm (Frissel 1977). The highest output of consumable N reported by Frissel (1977) was 400 $kg \cdot ha^{-1}$ (for French intensive grass production, without animals), an output which corresponds to a fish production

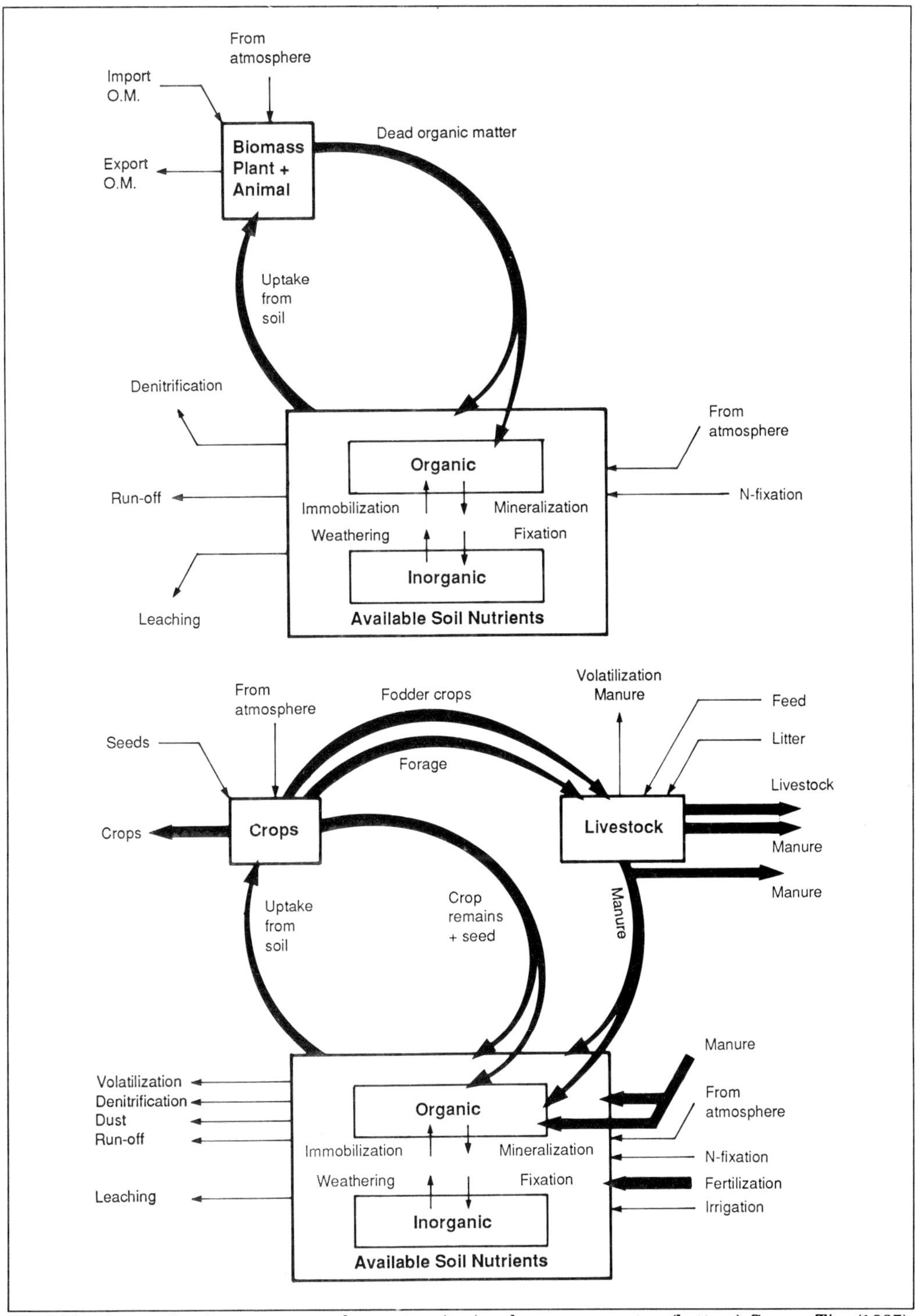

Fig. 6. Nutrient cycling in an unmanaged ecosystem (top) and an agroecosystem (bottom). Source: Tivy (1987). (This figure was first published in *Appl. Geogr.* 7(2):93-113, and is reproduced here with the permission of the author and Butterworth-Heinemann, Oxford, UK.)

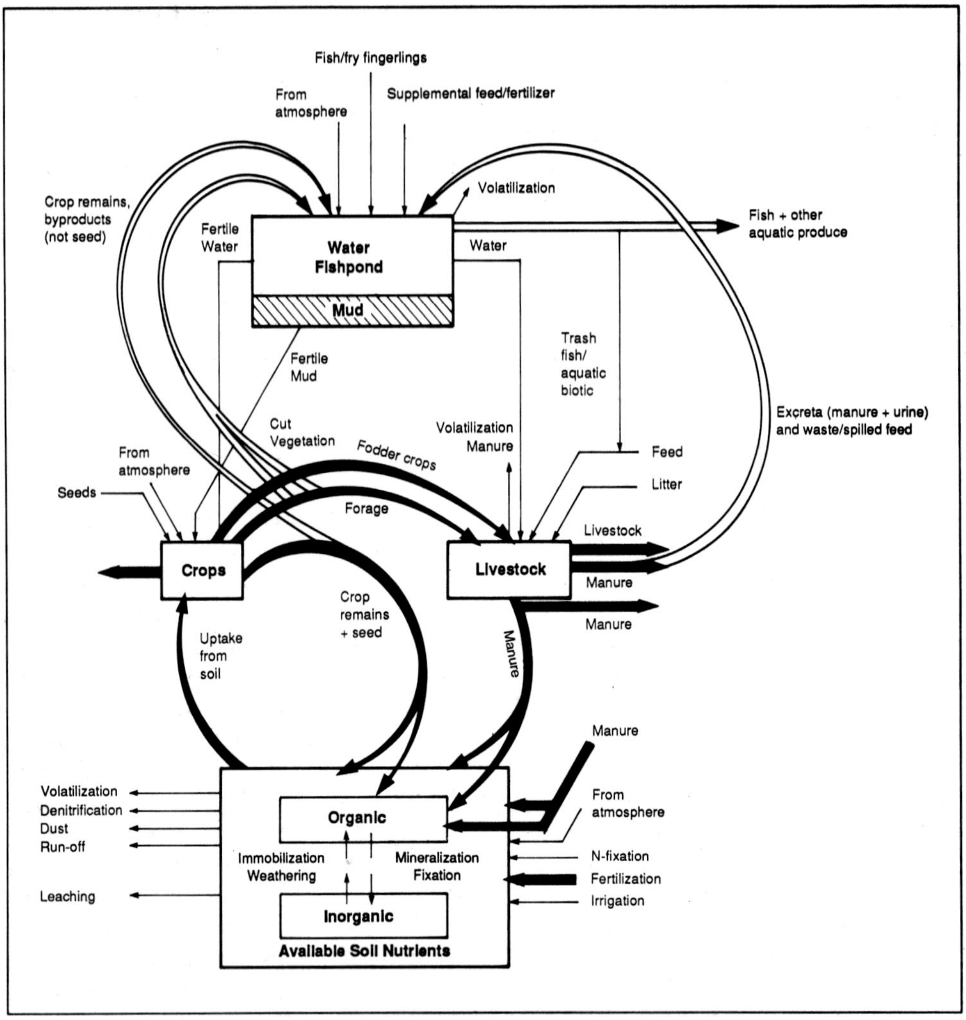

Fig. 7. Nutrient cycles for an agroecosysytem involving crops, fish and livestock. The crops, livestock and soil pathways, for which the major linkages are broad black arrows, are reproduced unmodified from Tivy (1987). (This figure was first published in *Appl. Geogr.* 7(2):93-113, and is reproduced here with the permission of the author and Butterworth-Heinemann, Oxford, UK.) The fishpond, its major linkages with crops and livestock (broad white arrows), other inputs and outputs have been added here. Nutrient recycling within and between the fishpond water and mud are omitted for clarity. In some traditional Asian systems, human (households) excreta is an important additional input to crops and fishponds and kitchen waste to livestock and fishponds.

of 15.6 $t \cdot ha^{-1}$, although yields in intensive feedlot systems raising either fish or livestock can greatly exceed this value.

Traditional, and formerly more widespread, mixed farms characterized by livestock production on the basis of arable crops and improved grassland, are intermediate between the extensive and intensive systems defined by Frissel (1977) although they were classified as moderately intensive agroecosystems by Tivy (1987). They are essentially integrated farms which receive low to moderate inputs of nutrients from outside the farm. Losses of nutrients are small, particularly on farms in which only livestock are produced, and there is a close approximation to an almost closed, self-sustaining, nutrient-conserving cycle (Tivy 1987).

Perhaps the only detailed study of an integrated agriculture-aquaculture system is that by Ruddle and Zhong (1988) of the dike-pond system of the Zhujiang Delta in southern China. The authors claimed that the system, when studied, was a relatively closed ecological cycle "based almost entirely on the tightly managed recycling of materials" and "the bulk of the inputs have always been generated from within the system itself". However, this is difficult to reconcile with the major objective of the system, the production and export of commercial crops (fish, silk and sugar). The mean yield of fish, about 7 $t \cdot ha^{-1}$, equates to the export of about 180 kg consumable $N \cdot ha \cdot year^{-1}$. The maximum efficiency of N conversion to fish in the form of high protein pelleted feed is about 25% (see Table 1) which means that an absolute minimum of 700 $kg \cdot N \cdot ha \cdot year^{-1}$ is required to support the fish yield. Most of the N could not have been regenerated within the system. In communes where the dike-system was the dominant land use, 76% of the agricultural land was devoted to the system (43% fishponds, 15% mulberry, 18% sugar cane) with 12% to rice and 11% to mixed and miscellaneous crops. Rates of N fixation are low in fertilized fishponds (Colman and Edwards 1987) and most of the crops in this southern China example were not N-fixing plants. Data presented from a detailed study of four households indicate that 66-88% of the total inputs to the fishpond were pig and human manure, with the former far exceeding the latter in three out of four households. With an extremely high rural human population density of 1,700 persons/km^2 and only 12% of the agricultural land under rice, a considerable amount of human food must have been imported into the system. Furthermore, it is unlikely that the feed requirements of the pigs were met by "a diet of greens, particularly water hyacinth, sugar cane tops and vegetable waste". While greens are traditionally fed to pigs in China, pigs are monogastrics and require more digestible sources of nutrients for adequate growth. The large flow of nutrients through the system was most likely provided mainly by the import of human and pig food which was not considered in the study, i.e., the system was, therefore, not a closed agroecological system.

Problems

The developing world is confronted with two interrelated problems: how to stimulate agricultural productivity and in particular to increase the productivity and welfare of small-scale farmers, the single most populous group in the world; and how at the same time to safeguard the environment from pollution. Pollution is best defined broadly as an undesirable change in the physical, chemical or biological characteristics of air, water, or land that may or will be harmful to humans and other living organisms (Odum 1975).

The problems facing the developing and the developed world are to a certain extent different. Most agricultural systems in developing countries, including pond culture of fish, are characterized by low productivity whereas developed world agriculture is highly productive. However, developed-country agriculture (and aquaculture) depends to a large degree on high inputs from industry. Maximizing agricultural yields without regard to other consequences is producing a serious environmental backlash (Loehr 1970; Alexander 1974; Killey-Worthington 1980). A ten-fold increase in agricultural chemicals (fertilizers, pesticides) and energy is required to double agricultural crop yield (Odum 1971). Thus, developed-country agriculture (e.g., Japan) produces

Table 1. A comparison of nitrogen and phosphorus efficiences (in terms of nutrient inputs in feed or fertilizer relative to nutrient content of fish at harvest) in a theoretical intensive system and experimental semi-intensive aquaculture systems. AIT, Asian Institute of Technology; DM, dry matter; FCR, feed conversion ratio; TSP, triple superphosphate; CRSP, Collaborative Research Support Program.

			Nitrogen		Phosphorus			
Culture system	Treatment	Extrapolated net fish yield ($kg \cdot ha^{-1} \cdot year^{-1}$)	g to produce 1 kg fish	Conversion efficiency of N input to fish N (%)	g to produce 1 kg fish	Conversion efficiency of P input to fish P (%)	N:P ratio by weight	Reference
Intensive systems								
Complete diet	FCR 1.0	-	48	53	12	28	4 : 0	theoretical[1]
Comlete diet	FCR 1.5	-	72	36	18	19	4 : 0	theoretical[1]
Complete diet	FCR 2.0	-	96	27	24	14	4 : 0	theoretical[1]
Complete diet	FCR 2.5	-	120	21	30	11	4 : 0	theoretical[1]
Semi-intensive systems								
Integrated pig/fish	100 animals/ha	7,121	103	25	-	-	-	Hopkins and Cruz (1982)[2]
Integrated chicken/fish	5,000 animals/ha	10,475	124	21	-	-	-	Hopkins and Cruz (1982)[2]
Integrated duck/fish	1,500 animals/ha	10,000	133	19	32	11	4 : 1	AIT (1986)[2]
Bagged chicken manure plus urea and TSP	8.5 kg manure DM (total 4.0 N and 1.0 P/ha/day)	8,601	170	15	42	8	4 : 1	CRSP/AIT (unpubl. data)[2]
Buffalo manure	300 kg manure DM/ha/day	3,120	491	5	70	5	7.2 : 1	AIT (1986)[2]
Buffalo manure plus urea and TSP	100 kg manure DM/ha/day + total 4.4 kg N and 0.4 kg P/ha/day	3,487	460	6	46	7	10 : 0	AIT (unpubl. data)[2]

[1]Assumptions: pelleted complete feed, crude protein content 30% on dry weight basis, 1.2% phosphorus (Le 1989); fish 16% crude protein (Hastings 1979) and 0.34% P (Beveridge 1987) on fresh weight basis; protein contains 16% nitrogen.
[2]Assumptions: fish same nitrogen and protein contents as above.

4 times the areal yield of developing-country agriculture as in India but is 100 times more demanding of resources and energy with much greater consequential adverse environmental impacts. Agriculturalists in general have "ignored the large ecological framework in which farming is conducted and, as a result, agricultural production has often (perhaps unconsciously) exploited the natural environment" (Dent and Anderson 1971).

Developing-country agriculture also has an adverse effect on the environment but one that is rather different. Population pressure leads to the expansion of agriculture into marginal forested areas with deforestation leading to soil erosion and more unstable water regimes which can threaten the very existence of agriculture. The problem for developing countries is essentially how to stimulate agricultural (and aquaculture) productivity and profitability without further environmental degradation, in contrast to the need to reduce the level of intensification to a sustainable level in the developed world.

Until relatively recently, nutrient flows were more closed than they are in much of the world today (Fig. 8). They have become more open over the past 150 years

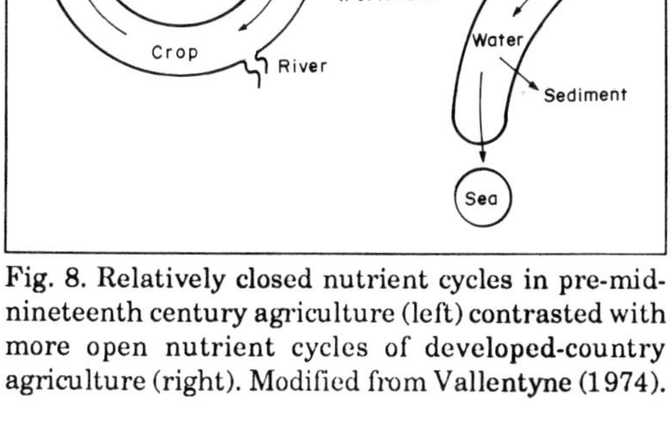

Fig. 8. Relatively closed nutrient cycles in pre-mid-nineteenth century agriculture (left) contrasted with more open nutrient cycles of developed-country agriculture (right). Modified from Vallentyne (1974).

because of increased urbanization and because food production, processing and distribution systems, particularly in the developed world, have become increasingly dependent on industrial products. Borgstrom (1973) has referred to the plethora of ecological problems associated with the interruption of the flow of nutrients between man and the soil as "the breach in the flow of mineral nutrients". All are part of the unsuccessful attempt to cope with the increasing rate of urbanization over the past century and the resulting large urban ecosystems. Human food chains and nutrient cycles have become exceedingly complex and unstable because only half of the world's population now lives on farms where agricultural by-products and wastes and human excreta are returned to the land (Duckham et al. 1976). Urbanization has a higher rate of growth than that of the human population (Odum 1971). Only 17% of the population of the less developed countries was urban in 1950 but over 50% is expected to be urban by 2020 (Keyfitz 1989).

The ultimate cause of the short-circuiting of nutrients from humans to the soil was the development in the mid-nineteenth century of a waterborne transport system or sewerage for toilet waste, which led to a unidirectional flow of nutrients to receiving water bodies (Vallentyne 1974). The increasing rate of population growth and urbanization led to the development of "artificial" or mineral fertilizers to stimulate crop production. The guano trade of the nineteenth century was the first step in the development of a "geographic imbalance" in the distribution of nutrients (Borgstrom 1973), later followed by industrial manufacture of fertilizers. Livestock were originally fed with on-farm feed but in the development of feedlot livestock (and intensive aquaculture) there has been a dissociation of plant and animal production. A good example is intensive pig rearing in the Netherlands with "a mountain of manure" to dispose of (Armstrong 1988); feed comes from other continents: energy in cassava from Thailand and fishmeal protein from Peru.

Nutrient enrichment of water from intensive agriculture, agroindustry and human sewage lead to eutrophication. Natural eutrophication occurs without human intervention in the natural succession of a waterbody over geological time due to weathering of rocks but human intervention, referred to as artificial or cultural eutrophication, can accelerate the process alarmingly. Eutrophication leads to the development of blooms of phytoplankton which may be considered undesirable because of increased treatment costs to provide domestic water, decomposition of algae to produce unpleasant smells and decreased recreational value for swimming and boating, and occasional mass mortality of fish when blooms collapse.

However, eutrophication needs to be placed in a proper context because it also provides the nutritional basis for fertilized fishponds (Colman and Edwards 1987). Ryther (1971) proposed the term "controlled eutrophication" for the use of phytoplankton in artificially fertilized systems, which include fertilized fishponds, the basis of much of the world's production of farmed fish. In the development of controlled eutrophication systems, there has to be both organism and chemical balance in which rates of natural food production and oxygen regeneration lead to sustainable fish production (Colman and Edwards 1988). Similar concepts have been expressed in the pollution literature. Barica (1981) described the ultimate stage of eutrophication, which must be avoided in fish culture, as hypertrophy when the system becomes unstable due to a large biomass of phytoplankton which is light- rather than nutrient-limited. Mee (1988) defined the term "critical eutrophication" to indicate the point at which deleterious effects of nutrient enrichment begin, when the rate of production of new organic matter by algae exceeds the natural net rate of oxygen to oxidize it.

Environmental Impact of Aquaculture Systems

Nutrient Efficiency of Aquaculture Systems

A comparison is made here of the conversion efficiency to fish of N and phosphorus (P) inputs in intensive systems (in the form of protein and P in nutritionally balanced or complete pelleted feed) and semi-intensive systems (in the form of livestock manure with or without commercial fertilizer supplementation). Theoretical data are used for the intensive system and experimental data are used for the semi-intensive systems (Table 1). Intensive aquaculture has the highest nutrient conversion efficiencies because feed is ideally formulated according to the nutritional needs of the target species. Nutrient conversion efficiencies in terms of nutrient inputs and nutrients in fish range of 21-53% for N and 11-28% for P with food conversion ratios (FCR) of 1.0-2.5, respectively. Nitrogen and P conversion efficiences of semi-intensive systems are 5-25% and 5-11%, respectively.

Nutrient conversion efficiencies for semi-intensive systems are lower because there is at least one extra step involved in the conversion of nutrients to fish through natural feed production in the pond. Animals in semi-intensive systems also expend more energy in feeding than those in intensive systems, which lowers the efficiency further. However, the nutrient conversion efficiences of semi-intensive systems using pig and poultry manure are much less than one order of magnitude lower than those for intensive systems as might be expected with at least one additional step in the food chain involving natural food production as feed for fish.

Efficiencies at the upper end of the semi-intensive range for integrated/ livestock fish systems are remarkably close to those for fish fed complete diets, particularly for nitrogen (Table 1). While spilled livestock feed undoubtedly contributes to fish growth in an integrated system, N conversion efficiencies only a little lower than those in pelleted feed may be due in large part to nutrient cycling in the static water system and to the high nutritional quality of natural food in ponds manured with high nutrient quality livestock manure from feedlot livestock.

Similar yields to those obtained in integrated feedlot livestock/fish culture systems have been obtained recently in experimental systems fertilized with aged (bagged) layer chicken manure supplemented with urea and TSP (triple superphosphate) (Table 1), with N and P conversion efficiencies about 50% lower than those of the intensive system fed with a pelleted feed at a FCR of 2 (CRSP/ AIT, unpubl. data).

As the economics of raising confined livestock are usually inappropriate for small-scale farmers (Edwards 1983), research is being carried out with buffalo manure available on small-scale farms. However, much lower yields were obtained with buffalo manure (AIT 1986), about 3 t compared to about 10 $t{\cdot}ha^{-1}{\cdot}year^{-1}$ with integration with feedlot livestock (Table 1). This was due to the relatively high dry matter (DM) loading rates of buffalo manure required to provide reasonable N and P fertilization rates because of the low quality of the manure, which caused poor water quality. Reduction of the buffalo DM loading rate and supplementation with commercial fertilizers increased the yield a little but research is required to elucidate the rate limiting factor(s) in such a system.

Comparison of Nutrient Pollution of Semi-intensive and Intensive Systems

In contrast to most intensive systems which are "open" in the sense that they have water exchange and therefore contribute nutrients to the adjacent environment, semi-intensive systems are usually "closed" or static water systems with little to no exchange of water with the surrounding area except when the pond is drained. Without a static water system it would be impractical to fertilize the pond to produce natural food for fish as a flowthrough system would flush out nutrients before they had time to stimulate natural food production. Nutrient rich water in fertilized ponds may be used to water crops grown on pond dikes in integrated farming systems but this should have little to no adverse environmental impact.

There is a dearth of information in the literature concerning the environmental impact of semi-intensive fish culture systems. A tentative nutrient balance for an intensively fed but static water fishpond in Israel had a yield of 7.3-12.3 $t{\cdot}ha^{-1}{\cdot}year^{-1}$ over a 4-year period (Avnimelech and Lacher 1979), similar to those reported from semi-intensive systems fertilized with the manure of feedlot livestock. Common carp were raised in the 0.3-ha, 1-m deep pond and were fed with sorghum and pelleted feed. The water N and P contents did not show increasing concentrations with time but fluctuated around mean values of 1.5 and 0.5 $mg{\cdot}l^{-1}$, respectively. About 75% of the N and about 80% of the P supplied in feed were not utilized by the fish, about 1.0 $t{\cdot}ha^{-1}$ of N and 0.17 $t{\cdot}ha^{-1}$ of P, and most accumulated in the sediments.

A similar absence of accumulation of nutrients occurred in the water of a series of fishponds stocked with Nile tilapia fertilized with urea and TSP supplemented chicken manure (Table 2). The 280-m^2 x 0.95-m deep ponds were stocked with Nile tilapia and fertilized with 46.0 g N and 8.9 g P per square meter for 147 days which gave a net fish yield of 97.0 kg, equivalent to 8.6 t·ha^{-1}·year^{-1}. The fish removed only 15.1 and 8.1% of the N and P added to the ponds, respectively. The total N and total P (TP) concentrations of pond water at the end of the experiment were 1.12 and 0.91 mg·l^{-1}, respectively, from which it can be calculated that only 1.8 and 5.9 % of the N and P remained in the water, respectively. The difference represents the N (83.1%) and the P (86.0%) which disappeared within the system. Data were collected on sewage treatment in wastewater-fed aquaculture over a 5-year period from a 10.7-ha complex of experimental ponds with a mean depth of 1.0-1.3 m in Hungary (Kovacs and Olah 1982, 1984). Raw sewage was subjected to physical treatment involving primary filtering by a wire net, a sand trap to remove finer particles and a sedimentation tank. The mechanically treated clarified effluent was then pumped daily under high pressure into the ponds through a series of sprinklers. The system was batch fed without an effluent except when the ponds were drained at the end of the fish growing period. The optional fish stocking structure was 50-65% silver carp, 22-30% bighead carp, 8-10% grass carp, and 5-10% common carp. The initial stocked weight of fish was 0.8-1.1 t·ha^{-1} with a net yield of 1.8-2.0 t·ha^{-1}·120-day growing season, or an extrapolated net yield of 5.5-6.1 t·ha^{-1}·year^{-1}. The optimal sewage dosage of settled sewage with a biological oxygen demand (BOD_5) of 110-120 mg·l^{-1} was 100 m^3·ha^{-1}·day^{-1}, equivalent to a domestic sewage production of 800-1,200 persons, which gave an organic loading of 11-12 kg·BOD_5·ha^{-1}·day^{-1}. Over the 120-day period, 400-500 kg of N and 80-120 kg of P were sprayed into the system. Treated water on draining the ponds at the end of the fish growing season had 2-3 mg·l^{-1} total Kjeldahl N and 0.7-1.0 mg·l^{-1} TP. Based on mean data, 11% of the N and 6% of the P added to the system were removed in fish, 6% of the N and 10% of the P were drained from the pond in water, and 83% of the N and 84% of the P were removed by the system.

A comparison of the nutrients released from semi-intensive and intensive systems is presented using the data for semi-intensive systems from Table 2 (chicken manure + urea + TSP) and the hypothetical data for intensive systems from Table 1 (complete diet, FCR 2). To facilitate

Table 2. Preliminary nutrient budget for a semi-intensive pond fertilized with bagged chicken manure supplemented with urea and triple superphosphate and stocked with Nile tilapia (*Oreochromis niloticus*). The 280-m^2 ponds gave a net fish yield of 96.3 kg fish over the 146-day experiment. See text for more details: Section on Comparison of Nutrient Pollution of Semi-intensive and Intensive Systems. Figures in parentheses are percentages of nutrients (CRSP/AIT, unpubl. data). N, total nitrogen; P, total phosphorus.

Nutrient	Added in fertilizer (kg)	Utilized by fish (kg)[1]	Remaining in water (kg)	Disappeared within the system (kg)
N	16.352 (100)	2.466 (15.1)	0.298 (1.8)	13.588 (83.1)
P	4.088 (100)	0.328 (8.1)	0.242 (5.9)	3.518 (86.0)

[1]Assumptions: fish 16% crude protein and protein 16% N; fish 0.34% P, both on fresh weight basis.

comparison, extremes are assumed with water released to the surrounding environment only on draining the system at the end of the culture cycle for the static water, semi-intensive pond system but with complete water exchange or flowthrough for the intensive system as in a cage or raceway systems (Fig. 9). Relatively small amounts of nutrients are removed in harvested fish, 15% N and 8% P in the semi-intensive, and 27% N and 14% P in the intensive system of the total nutrients added to the systems. Amounts removed in fish differ in the two systems despite the same hypothesized yield because nutrient conversion efficiencies of N and P are higher in the intensive than in the semi-intensive system. The nutrients that are removed by the system are assumed to be the difference between those added and accounted for in water immediately prior to fish harvest and in the harvested fish for the semi-intensive system. Eighty-three per cent of N and 86% of P are removed by the system and do not pollute the environment. Most were probably immobilized in the sediments. In contrast, no nutrients in the intensive system are sequestered by the culture system and pass through the system in the water, 73% N and 86% P, to pollute the external environment. Relatively few nutrients occur in the drainage water of the semi-intensive system, only 2% N and 6% P, although in practice some nutrients

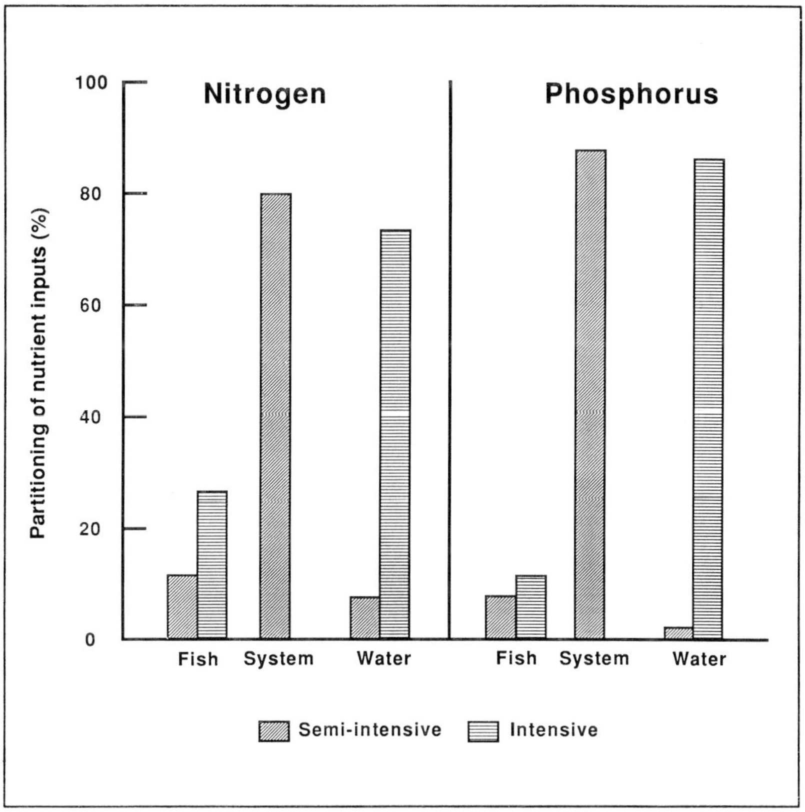

Fig. 9. A comparison of the environmental impact of semi-intensive aquaculture systems with the same fish yield. Partitioning of added nitrogen and phosphorus nutrients are expressed as a per cent of those added in fertilizer and feed for semi-intensive and intensive systems, respectively. Nutrients are removed in fish at harvest, disappear within the system, or are removed in water when the system is drained in the semi-intensive system or continually in the intensive system.

retained in sediments would be flushed out on draining the pond. While the initial N:P ratios by weight were 4.0 for both systems, the final N:P ratios in the water of 0.3 and 3.4 for the semi-intensive and intensive systems, respectively, indicate that both systems are much more efficient at removing N than P, but the semi-intensive more so than the intensive system.

A comparison is also made of the polluting capacity of the above semi-intensive system (chicken manure plus urea and TSP) and the intensive system at FCRs ranging from 1.0 to 2.5 (Table 3). The intensive system is 26-44 times more polluting than the semi-intensive system in terms of N, and 12-15 times in terms of P, expressed as a percentage of the nutrients required to produce 1 kg of fish. However, the pollution differential is reduced based on the weights of nutrients released to the environment because the intensive system has higher N and P conversion efficiencies than the semi-intensive system (Table 1). In terms of weight of N and P released to the environment in water, the intensive system is 7-31 and 3-11 times more polluting than the semi-intensive system, respectively (Table 3).

It is stressed that the comparison above is based on data from a single experiment. However, further study is unlikely to disprove the conclusion that semi-intensive systems are much less polluting than intensive systems. Furthermore, as intensive systems usually have a far higher per unit area fish production capacity than semi-intensive systems, the latter clearly have considerably less adverse environmental impact than the former.

Guidelines for Research and Development of Semi-intensive Aquaculture Systems

Misconceptions Concerning Developing-Country Aquaculture

There is considerable misunderstanding concerning the distribution, the amount of production, and the types and degree of development of various systems of Asian aquaculture which constrains the provision of support for research and development.

Aquaculture is far less widespread in Asia than is widely supposed; although Asia accounts for about 85% of the world's total production, more than other

Table 3. A comparison of the pollution capacity (per kg of fish) of intensive and semi-intensive aquaculture systems. Nitrogen (N) and phosphorus (P) leave the system continuously in a flowthrough intensive system but only when the pond is drained at the end of the culture cycle in the static water semi-intensive system. Pollution potential data are derived from Tables 1 and 2.

Food conversion ratio of intensive system	Intensive system more polluting than semi-intensive system (number of times)			
	Percentage of nutrients added		Weight of nutrients added (g)	
	N	P	N	P
1.0	26	12	7	3
1.5	36	14	15	5
2.0	41	15	23	8
2.5	44	15	31	11

continents, aquaculture is dwarfed by the contributions of agronomy and animal husbandry to food production. Perhaps less than 1% of farmers are involved in aquaculture in the region as a whole which indicates considerable scope for development. Huet (1972) in his seminal "Textbook of fish culture" was certainly mistaken when he wrote "In the Far East where all farmers are fish farmers and vice versa"... and ... "In general, all fish culturists are farmers and all farmers cultivate fish". By the Far East, Huet meant East, Southeast and probably South Asia. More recently, the "Thematic Evaluation of Aquaculture", a joint study by the United Nations Development Programme, the Norwegian Ministry of Development Cooperation, and the Food and Agriculture Organization of the United Nations overstated the status of aquaculture in Asia as follows: "the region's aquaculture sector is highly developed" and "aquaculturally advanced countries of Southeast Asia" (Anon. 1987). Such misunderstanding has probably arisen because of the short time spent by the authors in developing countries and by their hosts taking them to see the relatively few areas in the various countries in the region where aquaculture operations are common. Extrapolation from limited observations to the region as a whole is grossly misleading.

Overseas visitors are invariably interested in, and taken to see, the limited number of more glamorous intensive systems. In fact, carnivorous finfish production in the Asia-Pacific Region was only 6.4% of the total finfish production in 1986. Furthermore, for all countries in the region except three (Japan 80%, Taiwan 11% and Thailand 5% of the regional output), carnivorous finfish production was less than 1% of the total fish production (Csavas 1989). When only inland aquaculture with the greatest potential for integrated farming is considered, herbivorous/omnivorous fish comprised about 99% of the total finfish production in the Asia-Pacific Region.

Need for a Systems Approach

A holistic or systems approach is required to promote aquaculture on small-scale farms in developing countries. Future fish farmers are already engaged in the production of crops and livestock and nutrient inputs to the fishpond need to be produced on-farm as much as possible to reduce cost of production and adverse environmental impact. From this it follows that aquaculturists need to study existing agricultural systems.

A framework has been proposed for an interdisciplinary approach to research and education in integrated agriculture-aquaculture farming systems (Edwards et al. 1988). This is no mean task as it involves studying not only the fishpond but also the integrated crop and livestock subsystems. To study these chains and cycles it is necessary to draw on at least 20 academic disciplines to identify the various links, to quantify them, and to strengthen or bypass weak links to improve the efficiency of nutrient cycling (Duckham et al. 1976).

A schema of the possible evolutionary development of integrated farming systems is presented in Fig. 10. Settled agriculture, which has aquaculture development potential (in contrast to pastoral nomadism and shifting cultivation, two other major types of agriculture with limited aquaculture development potential), has been divided into three phases to illustrate the potential role of integrated farming in the development of aquaculture for small-scale farmers in developing countries. In settled agriculture phase 1 (crop dominated), characteristic of densely populated pre-industrial societies, most land is under food crops and livestock are kept mainly for draught but there may be scavenging pigs and poultry. There is

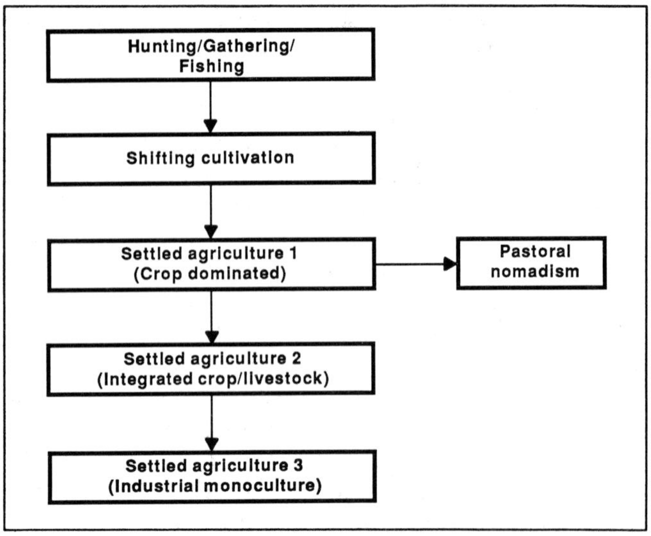

Fig. 10. A schema of possible evolutionary phases in farming systems development to highlight the potential role of integrated farming systems in developing-country agricultural development. Source: Edwards et al. (1988).

little or no integration between crops and animals. Large ruminants depend on rough grazing and although they also feed on stubble in the fields after crop harvest, are fed straw, and some of their manure fertilizes the field, the farming system is mainly crop based because of the limited number of livestock. It was characteristic of much of western Europe until about 1850; it applies to most small-scale farmers in developing countries today except those who "leap-frogged" to settled agriculture phase 3 through the green revolution.

In settled agriculture phase 2 (integrated crop/livestock), crops are integrated with livestock. It was a major feature of much of western Europe and the eastern USA from about 1850 to 1945 and some still exists today. Such farms are usually referred to as mixed farms and are characterized by livestock production on the basis of arable crops and improved grassland. Livestock production is clearly integrated with crop production because the former feed on the latter and animal manure helps to maintain soil fertility together with nitrogen fixing legumes. They are a "close approximation to an almost closed, local or farm-based, self-sustaining nutrient-conserving cycle" (Tivy 1987).

The trend towards settled agriculture phase 3 (industrial monoculture), in which farming has become increasingly dependent on industrial inputs derived from science and engineering, began about 1850; but it is only since World War II that traditional mixed farming in western Europe has been largely replaced. The major components of industrial monoculture or "conventional farming" as it has come to be called are: improved varieties, chemical fertilizers, herbicides, pesticides, pharmaceutical chemicals, feed concentrates, pelleted feed and mechanization. Most farms now raise only a single product because of increasing technical complexity and economies of scale. Much aquaculture in Japan and the West is industrial monoculture. In fact, the development of feedlot systems for fish (salmonids in Europe and eels in Japan) occurred about 100 years ago (Hora and Pillay 1962; Huet 1972), prior to the development of feedlots for animal husbandry (pigs, poultry) which took place only about 50 years ago (Shaw 1936, 1938), possibly because the only efficient way to raise carnivorous fish is in feedlots. Dependence on industrial inputs has made the need for integrated farming redundant (Duckham 1959, 1966) but the drawbacks to industrial monoculture, including a much greater adverse environment impact, are now apparent.

A major premise of the Edwards et al. (1988) study is that major increases in developing-country farm productivity and profitability can be made by moving farmers up from crop dominated settled agriculture phase 1 to settled agriculture phase 2 characterized by integrated crop-livestock-fish farming systems, without the need for wholesale adoption of environmentally damaging industrial monoculture (settled agriculture phase 3).

The development of a fishpond on a small-scale farm is a *de facto* diversification of farming activities. For the promotion of integrated farming systems involving fish, ponds need to be built on agricultural land to facilitate integration. It has been stated that fish culture can use poor soils or land of marginal agricultural potential such as sandy or sour soils, and undrainable wetlands (Hickling 1971; Pantulu 1980). Fishponds in continental Europe were usually built in the past on such poor soils or by damming a small stream or a spring in an open valley occupied by wetland (Korinek et al. 1987; Matena and Berka 1987; Opuszynski 1987). Such a practice militates against integration. While fish culture might compete with agriculture in terms of land area, the relatively small area occupied by a fishpond compared to the larger area of land required for meaningful integration should be more than compensated for by the overall increased production of the farm through synergism associated with integrated farming. The use of good agricultural land for fishponds should not be conceived as misuse but rather as an integral part of strategies to increase food production.

To optimize output from the system, the fishpond subsystem should be integrated as much as possible with existing farming activities. However, small-scale farmers in resource-poor regions in the tropics and with limited capital require new solutions to the problem of increasing the output of their farms (Francis et al. 1986), even before integration of aquaculture is considered. The key to increased agricultural production on such small-scale farms is a thorough understanding of the ecology of traditional agroecosystems upon which more productive systems can be based (Gliessman et al. 1981).

Altieri and Anderson (1986) proposed a strategy to develop small-scale farms in developing countries based on a grass roots or bottom-up approach. They suggested that the development and diffusion of improved agricultural practices for small-scale farmers must meet four criteria: 1) be based on small-scale farmers' needs as they perceive them; 2) use local technologies; 3) involve active participation of the farmers; and 4) emphasize local and indigenous resources. They recommended detailed study of the highly heterogenous and localized nature of small-scale farms in developing countries because of variations in more technical aspects such as local climate, crops and soils; social factors such as demography and social organization; and economic factors such as availability of capital and credit, farm gate prices and marketing. A holistic approach to account for this complexity is essential as many different causal factors interact.

Crop diversification is required to stabilize the environment of which the fishpond is an integral part as well as to provide a source of nutrients for fish. According to Harrison (1987) in his book "The Greening of Africa", agroforestry, which he defines simply as "tree planting on and around farms" is arguably "the single most important discipline for the future of sustainable development in Africa". Trees counteract soil erosion, particularly on slopes, draw up nutrients from deeper soil layers and deposit them on the surface in leaf litter, and increase the organic content, fertility and water holding capacity of the soil. The promotion of intercropping is recommended, the combining of a range of species in the same field, a characteristic of many traditional agricultural systems. Nitrogen-fixing legumes can be combined with plants having a high nitrogen requirement. The rooting systems of different crops can tap nutrients at different levels in the soil profile leading to greater total nutrient uptake (Charlton 1987).

Crops are likely to play an increasingly important role in the development of small-scale integrated farming systems involving fish culture because of major constraints in the development of livestock on small-scale farms, particularly limited feed (Devendra 1983). Aquaculture cannot wait until livestock promotion has successfully provided a source of manure as a potential fishpond input. Nor are livestock going to play a major role in the development of small-scale aquaculture in much of Africa where many farming systems are crop based.

Extensive, Semi-intensive or Intensive Systems?

Extensive pond aquaculture was widespread in the past but, with the exception of stocking fish in ricefields, it should not be promoted today except for culture-based fisheries in small waterbodies such as farm dams and reservoirs which are not primarily fishponds. Extensive and semi-intensive fishponds are usually between 10 and 100 ha in Europe although they can be several hundred ha (Korinek et al. 1987). The largest drainable fishpond in Czechoslovakia is 742 ha (Matena and Berka 1987). However, fishponds in developing countries where farm holdings may be less than 1 ha are usually much smaller, and both the size of the pond as well as the intensity of fish culture need to be balanced with the resources at the farmer's disposal. For semi-intensive fish culture on a small-scale farm, this equates to a pond size measured in ares (100s m^2) rather than hectares (10,000s m^2).

With decreasing land area/caput due to increasing population growth, agricultural production (including aquaculture) must be intensified. Furthermore, because of the relatively high cost of pond construction, the use of nutrient inputs is essential to justify its use. The presence of an extensively managed fishpond with minimal fish production, readily recognized by containing either silty water or clearer water with a dense growth of emergent and/or floating aquatic macrophytes, is an indication of improper management due usually to lack of knowledge concerning fertilization and feeding.

Semi-intensive systems can be recommended for promotion on the basis of minimal environmental impact (see Section on Comparison of Nutrient Pollution of Semi-intensive and Intensive Systems), cost of production of fish, and protein production efficiency. While not all fish species can be raised in fertilized systems, the similarity of the nutrient conversion efficiencies between fertilized systems and those fed complete diets has important implications in producing fish at least cost. Using Thai data, it can be calculated that both N and P cost 24 times more in the form of pelleted feed than in urea and TSP, respectively (assumptions: pelleted feed 30% protein, 1.2% P, Baht 9.5 (US$0.38)/kg; urea Baht 4.5 ($0.18)/kg; TSP Baht 8 ($0.32)/kg). Nitrogen and P are provided together in pelleted feed rather than separately as in urea and TSP and the N conversion efficiency of pelleted feed is slightly higher than that of a fertilized system but the cost of fish production per unit weight using pelleted feed is still about 7 times higher than that of using fertilizer (assumptions: FCR of 2 for pelleted feed with corresponding nutrient conversion efficiencies; decreased but conservative nutrient conversions for the fertilized system with 150 g N and 40 g P per kilogram of fish produced; and all N and

P supplied by urea and TSP, respectively, Table 1).

Furthermore, Hepher (1978) showed that there is a net loss of protein in intensive fish culture with high protein pelleted feed: a 1-kg trout contains 150 g protein but 600 g of protein are required to produce the fish protein assuming an FCR of 1.5 of a 40% protein pellet. The system may be profitable to the individual farmer because of the cost differential between feed and the high market value fish cultured but intensive systems do not generate a net gain in protein.

Semi-intensive methods of fish culture are thus more appropriate for the current socioeconomic conditions of developing countries in Asia than are intensive systems because they utilize natural resources better, generate more employment, are less dependent on off-farm inputs and are net protein producers. Furthermore, they are more flexible in adjusting to the marketing constraint of production outstripping demand, a characteristic of the culture of high market value fish (Csavas 1989). In contrast, intensive fish culture systems are analogous to "feedlot" methods of livestock rearing (Wohlfarth and Schroeder 1979). It has been established for some time that such methods are energy intensive (Purdom and Preston 1977; Weatherly and Cogger 1977) and protein intensive (Hepher 1978; Schroeder 1980) and there is increasing concern about their adverse environmental impact.

The Need for Research in Semi-intensive Aquaculture

Tropical aquaculture has a weak research base in comparison to the other two major food producing activities, agronomy and animal husbandry (Pullin and Neal 1984). Research in tropical aquaculture has been constrained by most aquaculture scientists, being concerned with only the fish and their requirements and not the complete resource system upon which most developing-country aquaculture depends. This is a result of most aquaculture scientists being educated as zoologists or fisheries biologists with the consequence that the emphasis in research on nutrition of fish has been on complete diets which are more appropriate for intensive aquaculture. There has been much less emphasis on pond fertilization to produce natural feed and its augmentation by supplementary feed which are essential features of semi-intensive aquaculture.

It was pointed out more than 50 years ago that the development of integrated agriculture-aquaculture systems in China, which were developed empirically by farmers over thousands of years, were not supported by scientific study (Chen 1934; Hoffmann 1934). While research is proceeding on such systems, they are still little understood scientifically (Li 1987). It has been suggested that the principles of Chinese integrated aquaculture be more widely applied in developing countries and quantified under tropical conditions (Edwards 1987).

Research Needs on Nutrient Dynamics in Semi-intensive Aquaculture Systems

A major research effort on the production of natural feed in fertilized ponds and its utilization by fish was called for at the Bellagio Conference on "Detritus and Microbial Ecology in Aquaculture" (Moriarty and Pullin 1987). To the list of research topics recommended by the conferees should be added those pertaining to nutrient dynamics and the environmental impact of semi-intensive systems.

Analyses of nutrient budgets in fertilized fishponds have revealed that 83% of the N and 84-86% of the P added to fishponds in the form of fertilizer are not accounted for in either harvested fish or in water drained from the pond but are removed by the system (see Section on Comparison of Nutrient Pollution of Semi-intensive and Intensive Systems). Studies in fishponds (Shilo and Rimon 1982) and in waste stabilization ponds (Ferrara and Avci 1982) both indicate that most of the N and P accumulate in the sediments through sedimentation of organic matter. A major question is how can the large amounts of nutrients tied up in the sediments be brought back into productive use? A traditional method in fish culture is mineralizing the sediments through exposure to the air by draining the ponds (Schaeperclaus 1933). In China, pond muds are excavated to repair the dikes and fertilize crops but the practice is being abandoned as it is very labor intensive (Li 1987). The relationship between the sediments and the water is extremely complex with various biological, chemical, mechanical and physical processes involved. Stirring the bottom sediments during the fish culture cycle has been proposed to improve the efficiency of nutrient use (Pullin 1987; Costa-Pierce and Pullin 1989) but experimentation is required before this can be recommended in view of the complexity of factors involved.

Research is still required on the relative rates of N, P and carbon fertilization to promote primary productivity to sustain adequate fish production. The nutrient content and nutrient availability of diverse potential on-farm pond inputs and commercial fertilizers and their effectiveness in ponds with sediments and water of differing physicochemical characteristics require elucidation.

The Case for Use of Commercial Fertilizers in Aquaculture

The adverse environmental impact of developed-country agriculture with its high levels of industrial inputs has led to the study of systems which use lower external inputs such as the old European integrated system of mixed crop and livestock farming (Frissel 1977). Other systems have been studied but these either relied on organic matter built up in the soil during previous years when fertilizers were used or imported manure from elsewhere (Frissel 1977). It has even been argued that the use of inorganic fertilizers should be suspended (Killey-Worthington 1981). However, low-input sustainable agriculture implies lower use of external industrial inputs rather than their total elimination, with minimal adverse environmental impact (Ikerd 1990).

Most tropical soils are low in fertility. The productivity of tropical virgin soils declines rapidly in the first few years after they have been brought into cultivation and thereafter liberal and judicious application of fertilizers is essential for the maintenance of soil fertility (Chang 1977). Subsistence agriculture with very low levels of production may be self sustaining from an environmental point of view but it is not from a human aspect with rapid increase in population growth (Evans 1976). Low nutrient inputs into developing-country agroecosystems contrast markedly with the intensively farmed systems of the developed world (Tivy 1987) but small-scale farms must intensify as they are currently not sufficiently productive to support a decent standard of living. It can be argued that, for the small-scale developing country farmer, the main concern is where the nutrients will come from rather than excessive fertilizer use.

A recent FAO study on the potential capacities of developing countries to support their burgeoning populations indicated that the largest annual increase in world population in history will take place from 1980 to 2020 and the fastest rates will occur in areas where the land resources are least able to meet the demand for increased food production (FAO 1984). The study pointed out that increased levels of inputs are essential. While it was recognized that improved agricultural methods must be appropriate to national social conditions and safeguard the environment, the study recommended increased use of commercial fertilizers although complemented by reuse of organic residues and biological N fixation. A balance is required between the slower but sustainable traditional methods of "organic farming" and more rapid "conventional agriculture" involving use of inorganic fertilizers to increase the flow of nutrients in developing-country agroecosystems.

Until recently it was believed that the greatest potential for integrating fish into small-scale farming systems lay with integration with feedlot livestock as the manure from such livestock supports high yields of fish (Edwards 1983, 1986). However, small-scale farmers face constraints to raising monogastrics in large enough numbers to integrate with fish, due in part to underdeveloped on-farm supplies of livestock feed (Devendra 1983). In contrast, the manure of buffalo and cattle used for draught on small-scale farms is nutrient poor as they feed mainly on volunteer (wild, uncultivated) vegetation, rice stubble, and rice straw. It was necessary to use a high dry matter loading rate of buffalo manure to provide reasonable N and P inputs to the pond. However, this led to a poor dissolved oxygen regime and low levels of plankton which resulted in a low fish yield (AIT 1986).

The supplementation of nutrient-poor, on-farm pond inputs with commercial fertilizers is recommended for fishpond subsystems of small-scale integrated farms. However, broader studies concerning nutrient cycling through crop and livestock as well as fish subsystems of small-scale farms are required. Research should determine whether to fertilize the fishpond or crops, whether the crops should be consumed by fish, livestock or humans, and whether livestock manure should fertilize the fishpond or crops.

Promotion of Wastewater-fed Fish Culture Systems

The recommended system for wastewater treatment in developing countries, provided that land is available at reasonable cost, is the stabilization pond system (Mara and Pearson 1986; Pescod and Mara 1988). This comprises a series of shallow artificial lagoons with the earlier ponds (anaerobic and facultative ponds) in the series loaded with wastewater at too high a rate for fish to survive. However, fish culture is feasible in later ponds in the series (maturation ponds) which are aerobic and function principally in pathogen removal.

Fish have been assessed for their ability to improve wastewater treatment in stabilization ponds. The limited evidence to date indicates that they do not reduce phytoplankton biomass significantly, a desirable objective as maturation pond effluents contain too high a concentration of suspended solids to conform to effluent standards for secondary treatment (Edwards 1992). As fish cause a more rapid circulation of nutrients in the pond by mixing the sediments and by accelerating the mineralization of seston (living and nonliving floating matter), they may increase the phytoplankton biomass in the pond, which may be increased further by grazing on zooplankton which also filter phytoplankton. This has been called ichthyoeutrophication by Opuszynski (1987).

It has been proposed that fish be raised in the final ponds, the maturation ponds, in stabilization pond series (Mara and Cairncross 1989). However, this may not lead to much fish production with the available wastewater because most of the nutrients will have been removed from the water during treatment in the earlier ponds in the series and the concentration reaching the maturation ponds in the flow may be insufficient for adequate natural feed production for fish. Assuming that public health can be safeguarded (for a review of the public health aspects of wastewater-fed fish culture systems see Edwards 1992), it may be more sociologically as well as economically desirable, depending on the cost and availability of land, to optimize a system for maximum fish production. This would essentially entail loading wastewater which has received a lower degree of pretreatment (but contains more nutrients) into fishponds at a much lower organic loading rate although this would require an increase in the total area of land needed.

With the rapid growth and urbanization of the population in developing countries, wastewater reuse in aquaculture may have a role to play in closing the nutrient cycle again and alleviating eutrophication. It may be useful to go to a higher order of systems level than a single farm and thus broaden the definition of an agroecosystem (Frissel 1977). This would include all agricultural produce (and only those) consumed by the local population with all wastes (plant, animal and human) returned to the agricultural area. Such a higher order system would be a closed system regarding nutrients. Borgstrom (1978) predicted that the wastewater treatment plants of rapidly growing cities will become huge centers for food production. It has even been suggested that the reuse of organic wastes from cities might lead eventually to the development of alternative agricultural systems (Newcombe and Bowman 1978).

As discussed earlier, wastewater reuse and food production systems are not separated today in some countries, nor were they separated in Europe and Japan in the past. Nightsoil was carted from cities in Europe for agricultural reuse until the introduction of sewage in the mid nineteenth century. Attempts were made to continue the tradition of excreta reuse but sewage farming declined at the turn of the twentieth century due in part to flooding the soil which became anaerobic (Hamlin 1980; Dean and Lund 1981). Great increases in urban population also made it difficult to expand the farms as land values rose. Excreta reuse continued longer in today's more developed countries of the Far East (Japan, Korea, Taiwan) but was superceded for economic reasons as commercial fertilizers became available and nightsoil became scarcer as septic tanks, which require less frequent desludging, replaced storage in watertight storage pits (conservancy vaults).

There appears to be a correlation between high population density in recent historical time with the occurrence of excreta reuse which might be called the nutritional imperative: northern Europe, India in South Asia, China, Japan and Korea in the Far East, and Java in Southeast Asia all reuse excreta in agriculture and/or aquaculture either today or in the recent past (Edwards 1992). It is now the less developed countries of Africa, Asia and Latin America that were not densely populated until relatively recently, and which are now expanding the most rapidly, that mostly do not have a tradition of excreta reuse. Excreta reuse may also be related to whether excreta were collected in receptacles prior to the invention of sewage or water borne disposal of wastes, and this was more likely to happen in densely populated societies. There would have been less tendency to collect and perhaps reuse excreta in sparcely populated societies where indiscriminate defecation in nature would have been more likely. The challenge today is to implement excreta reuse in countries in which it is not traditional but which need it the most to augment food production and safeguard the rapidly declining quality of their surface waters.

Wastewater-fed fish culture may be more sociologically acceptable if excreta were used to produce high-protein animal feed rather than for culture of fish for direct human consumption (Edwards 1990). This may be feasible in brackishwater as well as freshwater ponds. While there are conflicting uses of coastal pond areas, particularly for shrimp culture which has expanded rapidly recently, it has been suggested that reuse of wastewater in brackishwater fishponds in Southeast Asia could lead to a marked increase in fish production (Goldman and Ryther 1976). Filter-feeding tilapias which can be cultured in brackishwater ponds have been suggested as a likely species for culture in fertilized coastal aquaculture systems (Edwards 1992). While the tilapia could be raised for either human food or for high-protein animal feed, the latter may be a more feasible option for sociological reasons with a portion of the vast brackishwater milkfish pond systems of Indonesia and the Philippines used to culture tilapia as high-protein animal feed to partially replace fishmeal in carnivorous finfish and shrimp diets.

Acknowledgements

The author thanks Dr. C.F. Knud-Hansen for providing unpublished nutrient data from the CRSP/AIT project. Drs. J.A. Gartner, C.F. Knud-Hansen and D.C. Little reviewed the paper and provided invaluable comments and constructive criticism. The author is seconded to the Asian Institute of Technology by the Overseas Development Administration (ODA), London, England.

References

AIT. 1986. Buffalo/fish and duck/fish integrated systems for small-scale farmers at the family level. AIT Res. Rep. No. 198. Asian Institute of Technology, Bangkok.

Alexander, M. 1974. Environmental consequences of rapidly rising food output. Agro-Ecosystems 1:249-264.

Altieri, M.A. and M.K. Anderson. 1986. An ecological basis for the development of alternative agricultural systems for small farmers in the third world. Am. J. Alternative Agric. 1:30-38.

Anon. 1987. Thematic evaluation of aquaculture. Joint study by the United Nations Development Program, Norwegian Ministry of Development Cooperation and the Food and Agriculture Organization of the United Nations, Rome. 44 p.

Armstrong, S. 1988. Marooned in a mountain of manure. New Scientist, 26 November 1988:51-55.

Avnimelech, Y. and M. Lacher. 1979. A tentative nutrient balance for intensive fish ponds. Bamidgeh 31:3-8.

Barica, J. 1976. Nutrient dynamics in eutrophic inland waters used for aquaculture. SCS/76/WP/24, 29 p. South China Sea Fisheries Development and Coordinating Programme, Manila.

Barica, J. 1981. Hypereutrophy - the ultimate stage of eutrophication. Water Qual. Bull. 6:95-97, 155-156.

Beveridge, M. 1987. Cage aquaculture. Fishing News Books Ltd., Farnham, Surrey.

Borgstrom, G. 1973. The breach in the flow of mineral nutrients. Ambio 2:129-135.

Borgstrom, G. 1978. The contribution of freshwater fish to human food, p. 469-491. *In* S.D. Gerking (ed.) Ecology of freshwater fish production. Blackwell Scientific Publications, Oxford.

Chang, J.H. 1977. Tropical agriculture: crop diversity and crop yields. Econ. Geogr. 53:241-254.

Charlton, A. 1987. Problems and prospects for sustainable agricultural systems in the humid tropics. Appl. Geogr. 7 : 153-174.

Chen, T.P. 1934. A preliminary study on association of species in Kwangtung fish ponds. Lingnan Sci. J. 13:275-283.

Colman, J.A. and P. Edwards. 1987. Feeding pathways and environmental constraints in waste-fed aquaculture: balance and optimization, p. 240-281. *In* D.J.W. Moriarty and R.S.V. Pullin (eds.) Detritus and microbial ecology in aquaculture. ICLARM Conf. Proc. 14, 420 p.

Costa-Pierce, B.A. and R.S.V. Pullin. 1989. Stirring ponds as a possible means of increasing aquaculture production. Aquabyte 2(3):5-7.

Csavas, I. 1989. Development of semi-intensive and intensive aquaculture in Asia and the Pacific. Paper presented at the 29th annual conference of the International Association of Fish Meal Manufacturers, 21-24 November 1989, Hong Kong.

Dean, R.B. and E. Lund. 1981. Water reuse: problems and solutions. Academic Press, London.
Dent, J.B. and J.R. Anderson. 1971. Systems, management and agriculture, p. 3-14. *In* J.B. Dent and J.R. Anderson (eds.) Systems analysis in agricultural management. John Wiley & Sons Australasia, Sydney.
Devendra, C. 1983. Small farm systems combining animals with mixed cropping, p. 173-191. *In* Proc. Fifth World Conference on Animal Production. Vol. 1. Japanese Society of Zootechnical Science, Tokyo.
Duckham, A.N. 1959. The current agricultural revolution. Geography 44:71-78.
Duckham, A.N. 1966. Forty years on: agriculture in retrospect and prospect. J. Royal Agric. Soc. 127L:7-16.
Duckham, A.N., J.G.W. Jones and E.H. Roberts. 1976. Introduction: the function and efficiency of human food chains, p. 3-11. *In* A.N. Duckham, J.G.W. Jones and E.H. Roberts (eds.) Food production and consumption: the efficiency of human food chains and nutrient cycles. North-Holland Publishing Company, Amsterdam.
Edwards, C.A. 1989. The importance of integration in sustainable agricultural systems. Agric. Ecosystems Environ. 27:25-35.
Edwards, P. 1983. The future potential of integrated farming systems in Asia, p. 273-286. *In* Proc. Fifth World Conference on Animal Production. Vol. 1. Japanese Society of Zootechnical Science, Tokyo.
Edwards, P. 1986. Duck/fish integrated farming systems, p. 267-291. *In* D.J. Farrell and P. Stapleton (eds.) Duck production science and world practice. University of New England, Armidale.
Edwards, P. 1987. Use of terrestrial vegetation and aquatic macrophytes in aquaculture, p. 311-335. *In* D.J.W. Moriarty and R.S.V. Pullin (eds.) Detritus and microbial ecology in aquaculture. ICLARM Conf. Proc. 14, 420 p.
Edwards, P. 1990. An alternative excreta-reuse strategy for aquaculture: the production of high-protein animal feed, p. 209-221. *In* P. Edwards and R.S.V. Pullin (eds.) Wastewater-fed aquaculture. Environmental Sanitation Information Center, Asian Institute of Technology, Bangkok.
Edwards, P. 1992. Reuse of human waste in aquaculture, a technical review. Water and Sanitation Report No. 2. UNDP-World Bank Water and Sanitation Program. The World Bank, Washington, DC.
Edwards, P. and R.S.V. Pullin, Editors. 1990. Wastewater-fed aquaculture. Environmental Sanitation Information Center, Asian Institute of Technology, Bangkok.
Edwards, P., R.S.V. Pullin and J.A. Gartner. 1988. Research and education for the development of integrated crop-livestock-fish farming systems in the tropics. ICLARM Stud. Rev. 16, 53 p.
Evans, L.T. 1976. The two agricultures: renewable or resourceful. J. Aust. Inst. Agric. Sci. 1976:222-230.
FAO. 1984. Land, food and people. FAO, Rome. 79 p.
Ferrara, R.A. and C.B. Avci. 1982. Nitrogen dynamics in waste stabilization ponds. J. Water Pollut. Control Fed. 54:361-369.
Francis, C.A., R.R., Harwood and J.F. Parr. 1986. The potential for regenerative agriculture in the developing world. Am. J. Alternative Agric. 1:65-74.
Frissel, M.J. 1977. Cycling of mineral nutrients in agricultural ecosystems. Agro-Ecosystems 4:1-315.
Gliessman, S.R., R.E. Garcia and M.A. Amador. 1981. The ecological basis for the application of traditional agricultural technology in the management of tropical agro-ecosystems. Agro-Ecosystems 7:173-185.
Goldman, J.C. and J.H. Ryther. 1976. Waste reclamation in an integrated food chain system, p. 197-214. *In* J. Tourbier and R.W.J. Pierson (eds.) Biological control of water pollution. University of Pennsylvania Press, Philadelphia.
Guan, R.J. and Y.D. Chen. 1989. Reform and development of China's fisheries. FAO Fish. Circ. No. 822, 33 p.
Hamlin, C. 1980. Sewage: waste or resource? Environment 22:16-20, 38-42.
Harrison, P. 1987. The greening of Africa, breaking through in the battle for land and food. Paladin Grafton Books, London.
Hastings, W.H. 1979. Fish nutrition and fish feed manufacture, p. 568-574. *In* T.V.R. Pillay and W.A. Dill (eds.) Advances in aquaculture. Fishing News Books Ltd., Farnham, Surrey.
Hepher, B. 1975. Supplementary feeding in fish culture, p. 183-198. *In* Proc. 9th Int. Congr. Nutrition 3. S. Karger, Basel.
Hepher, B. 1978. Ecological aspects of warm-water fishpond management, p. 447-468. *In* S.D. Gerking (ed.) Ecology of freshwater fish production. Blackwell Scientific Publications, Oxford.
Hepher, B. 1979. Supplementary diets and related problems in fish culture, p. 344-347. *In* J.E. Halver and K. Tiews (eds.) Finfish nutrition and fishfeed technology. Heenemann Verlagsgesellschaft, Berlin.
Hickling, C.F. 1971. Fish culture. Faber and Faber, London.
Hoffman, W.E. 1934. Preliminary notes on the fresh-water fish industry of South China, especially Kwangtung Province. Lingnan Univ. Sci. Bull. No. 5, 70 p. Lingnan University, Canton, China.
Hopkins, K.D. and E.M. Cruz. 1982. The ICLARM-CLSU integrated animal-fish farming project: final report. ICLARM Tech. Rep. 5, 96 p.

Hora, S.L. and T.V.R. Pillay. 1962. Handbook on fish culture in the Indo-Pacific region. FAO Fish. Biol. Tech. Pap. 14, 204 p.

Huet, M. 1972. Textbook of fish culture, breeding and cultivation of fish. Fishing News Books Ltd., Farnham, Surrey.

ICLARM. 1988. Research for the development of tropical aquaculture: a proposed new venture for the Consultative Group on International Agricultural Research (CGIAR). International Center for Living Aquatic Resources Management, Manila. 28 p.

Ikerd, J.E. 1990. Agriculture's search for sustainability and profitability. J. Soil Water Conserv. 45:18-23.

Keyfitz, N. 1989. The growing human population. Sci. Am. 261(3):71-77A.

Killey-Worthington, M. 1980. Problems of modern agriculture. Food Policy 5:208-215.

Killey-Worthington, M. 1981. Ecological agriculture. What it is and how it works. Agric. Environ. 6:349-381.

Korinek, J., J. Fott, J. Fuksa, J. Lellak and M. Prazakova. 1987. Carp ponds of Central Europe, p. 29-62. *In* R.G. Michael (ed.) Managed aquatic ecosystems. Ecosystems of the world 29. Elsevier, Amsterdam.

Kovacs, G. and J. Olah. 1982. Silver carp (*Hypophthalmichthys molitrix* Val.) dominated domestic sewage oxidation fish pond technology, p. 32-38. *In* Proc. 2nd Int. Symp. on Herbivorous Fish, Faculty of Agriculture, Novi Sad, Yugoslavia.

Kovacs, G. and J. Olah. 1984. Silver carp (*Hypophthalmichthys molitrix* Val.) dominated domestic sewage oxidation fish pond technology. Aquacultura Hungarica (Szarvas) 4:149-155.

Le, Thanh Hung. 1989. Evaluation of soybean meal as a supplementary feed for red tilapia (*O. niloticus* and *O. mossambicus*). Thesis No. AE-89-66. Asian Institute of Technology, Bangkok, Thailand.

Li, S.F. 1987. Energy structure and efficiency of a typical Chinese integrated fish farm. Aquaculture 65:105-118.

Ling, S.W. 1967. Feeds and feeding of warm water fishes in ponds in Asia and the Far East. FAO Fish. Rep. 44:291-309.

Loehr, R.C. 1970. Changing practices in agriculture and their effect on the environment. CRC Crit. Rev. Environ. Control 1:69-99.

Mara, D. and S. Cairncross. 1989. Guidelines for the safe use of wastewater and excreta in agriculture and aquaculture. World Health Organization, Geneva.

Mara, D.D. and H.W. Pearson. 1986. Artificial freshwater environment: waste stabilization ponds, p. 177-206. *In* H.J. Rehm and G. Reed (eds.) Biotechnology 8. VCH Verlagsgesellschaft, Weinheimk.

Matena, J. and R. Berka. 1987. Freshwater fish-pond management in the world, p. 3-27. *In* R.G. Michael (ed.) Managed aquatic ecosystems. Ecosystems of the world 29. Elsevier, Amsterdam.

Mee, L.D. 1988. A definition of "critical eutrophication" in the marine environment. Rev. Biol. Trop. 36:159-161.

Moriarty, D.J.W. and R.S.V. Pullin, Editors. 1987. Detritus and microbial ecology in aquaculture. ICLARM Conf. Proc. 14, 420 p.

Muir, J.F. 1986. Integrated carp farming in Western Europe, p. 391-399. *In* R. Billard and J. Marcel (eds.) Aquaculture of cyprinids. Institut National de la Recherce Agronomique, Service des Publications, Versailles.

Newcombe, K. and K. Bowman. 1978. Intensive animal production at the crossroads. Search 9:351-359.

Odum, E.P. 1971. Fundamentals of ecology. 3rd edition. Saunders, Philadelphia.

Odum, E.P. 1975. Ecology: the link between the natural and the social sciences. 2nd edition. Holt Rinehart and Winston, London.

O'Grady, K.T. and P.B. Spillett. 1985. A comparison of the pond culture of carp, *Cyprinus carpio* L., in Britain and mainland Europe. J. Fish Biol. 26:701-714.

Opuszynski, K. 1978. The influence of the silver carp (*Hypophthalmichthys molitrix* Val.) on eutrophication of the environment of carp ponds. Part VII. Recapitulation. Roczniki Nauk Rolniczych H 94(2):127-151.

Opuszynski, K. 1987. Freshwater pond ecosystems managed under a moderate European climate, p. 63-91. *In* R.G. Michael (ed.) Managed aquatic ecosystems. Ecosystems of the world 29. Elsevier, Amsterdam.

Pantulu, V.R. 1980. Aquaculture in irrigation systems, p. 35-44. *In* R.S.V. Pullin and Z.H. Shehadeh (eds.) Integrated agriculture-aquaculture farming systems. ICLARM Conf. Proc. 4, 258 p.

Pescod, M.B. and D.D. Mara. 1988. Design, operation and maintenance of wastewater stabilization ponds, p. 93-115. *In* M.B. Pescod and D.D. Mara (eds.) Treatment and use of sewage effluent for irrigation. Butterworths, London.

Prein, M. 1990. Wastewater-fed fish culture in Germany, p. 13-47. *In* P. Edwards and R.S.V. Pullin (eds.) Wastewater-fed aquaculture. Environmental Sanitation Information Center, Asian Institute of Technology, Bangkok.

Prowse, G.A. 1967. A review of the methods of fertilizing warmwater fish ponds in Asia and the Far East. FAO Fish. Rep. 44:7-12.

Pullin, R.S.V. 1987. General discussion on detritus and microbial ecology in aquaculture, p. 368-381. *In* D.J.W. Moriarty and R.S.V. Pullin (eds.) Detritus and microbial ecology in aquaculture. ICLARM Conf. Proc. 14, 420 p.

Pullin, R.S.V. and R.A. Neal. 1984. Tropical aquaculture. Need for a strong research base. Mar. Policy 8:217-228.

Pullin, R.S.V. and Z.H. Shehadeh, Editors. 1980. Integrated agriculture-aquaculture farming systems. ICLARM Conf. Proc. 4, 258 p.

Purdom, C.E. and A. Preston. 1977. A fishy business. Nature (London) 266:396-397.

Reich, K. 1975. Multispecies fish culture (polyculture) in Israel. Bamidgeh 27:85-99.

Ruddle, K. and G.F. Zhong. 1988. Integrated agriculture-aquaculture in South China, the dike-pond system of the Zhujiang delta. Cambridge University Press, Cambridge.

Ryther, J.H. 1971. Recycling human waste to enhance food production from the sea. Environ. Lett. 1:79-87.

Schaeperclaus, W. 1933. Textbook of pond culture. Book Publishing House, Paul Parey, Berlin.

Schroeder, G.L. 1977. Agricultural wastes in fish farming. A commercial application of the culture of single-celled organisms for protein production. Water Res. 11:419-420.

Schroeder, G.L. 1980. Fish farming in manure loaded ponds, p. 73-86. *In* R.S.V. Pullin and Z.H. Shehadeh (eds.) Integrated agriculture-aquaculture farming systems. ICLARM Conf. Proc. 4, 258 p.

Shaw, E.B. 1936. Swine production in the corn belt of the United States. Econ. Geogr. 12:359-372.

Shaw, E.B. 1938. Swine industry of Denmark. Econ. Geogr. 14:23-37.

Shilo, M. and A. Rimon. 1982. Factors which affect the intensification of fish breeding in Israel. Bamidgeh 34:101-114.

Tivy, J. 1987. Nutrient cycling in agro-ecosystems. Appl. Geogr. 7:93-113.

Tuan, P.A. and V. Trac. 1990. Reuse of wastewater for fish culture in Hanoi, Vietnam, p. 69-71. *In* P. Edwards and R.S.V. Pullin (eds.) Wastewater-fed aquaculture. Environmental Sanitation Information Center, Asian Institute of Technology, Bangkok.

Vallentyne, J.R. 1974. The algal bowl, lakes and man. Department of the Environment, Fisheries and Marine Service, Ottawa.

Vereijken, P. 1986. From conventional to integrated agriculture. Neth. J. Agric. Sci. 34:387-393.

Weatherley, A.H. and B.M.G. Cogger. 1977. Fish culture: problems and prospects. Science 197:427-430.

Wohlfarth, G.W. and G.L. Schroeder. 1979. Use of manure in fish farming - a review. Agric. Wastes 1:279-299.

Wijkström, U.N. and M.B. New. 1989. Fish for feed: a help or a hindrance to aquaculture in 2000. Infofish 89:48-52.

Zhang, Z.S. 1990. Wastewater-fed fish culture in China, p. 3-12. *In* P. Edwards and R.S.V. Pullin (eds.) Wastewater-fed aquaculture. Environmental Sanitation Information Center, Asian Institute of Technology, Bangkok.

Zhong, G. 1989. The structural characteristics and effects of the dyke-pond system in China. Outlook on Agriculture 18:119-123.

Shrimp Culture and the Environment: Lessons from the World's Most Rapidly Expanding Warmwater Aquaculture Sector

M.J. PHILLIPS*
Institute of Aquaculture
University of Stirling
Stirling FK9 4LA
Scotland, UK

C. KWEI LIN
Division of Agricultural and Food Engineering
Asian Institute of Technology
P.O. Box 2754, Bangkok 10501
Thailand

M.C.M. BEVERIDGE
Institute of Aquaculture
University of Stirling
Stirling FK9 4LA
Scotland, UK

PHILLIPS M.J., C. KWEI LIN and M.C.M. BEVERIDGE. 1993. Shrimp culture and the environment: lessons from the world's most rapidly expanding warmwater aquaculture sector, p. 171-197. *In* R.S.V. Pullin, H. Rosenthal and J.L. Maclean (eds.) Environment and aquaculture in developing countries. ICLARM Conf. Proc. 31, 359 p.

Abstract

Shrimp culture is the world's most rapidly expanding warmwater aquaculture sector. This paper reviews the environmental impacts of shrimp culture and discusses some of the issues raised by the interactions between shrimp culture and the coastal environment. The main impacts are destruction of coastal mangroves, loss of land resources and deterioration of water quality. There is evidence that some of these impacts have adversely affected the long-term sustainability of shrimp culture itself. For example, loss of mangroves has been linked to decline in availability of shrimp postlarvae in Ecuador, which has resulted in shortage of seed for stocking culture ponds. Farms constructed on mangroves in Asia and Latin America have also suffered from reduced production and diseases linked to acid-sulfate soils, in some cases resulting in ponds being abandoned. Loading of ponds and the surrounding environment with shrimp pond effluent has also been linked to outbreaks of disease and poor pond productivity. These problems could be reduced by application of a number of techniques, including appropriate site selection, improved pond management, effluent treatment and more effective planning and monitoring. Effective and balanced planning, based on a clearer understanding of the interactions between shrimp culture and the environment, is the key to use of coastal environments for shrimp culture, without which there is a distinct possibility that shrimp culture will not be sustainable.

*Present address and address for correspondence: Division of Agricultural and Food Engineering, Asian Institute of Technology, P.O. Box 2754, Bangkok 10501, Thailand.

Introduction

The culture of shrimps, which for the purposes of this review are defined as brackishwater and marine penaeid shrimps, has a long history in Asia and some parts of Latin America. Traditional forms of shrimp culture, such as the 'tambaks' of Java (Schuster 1952), 'bheris' of Bengal (Naamin 1987) and tidal ponds in Ecuador, have been in operation for many years, but these traditional systems, characterized by extensive culture methods, minimal management and low yields, are now being superceded by the modern shrimp culture industry seen in Asia and Latin America, and to a lesser extent in some other tropical and subtropical regions, characterized by semi-intensive or intensive culture methods with higher yields.

Global production of cultured shrimp has grown rapidly over the past few years, from around 90,000 t in 1980 to an estimated 565,000 t in 1989, representing 26% of the global supply of marine shrimp (Anon. 1990a) (Fig. 1). Between 1985 and 1988, production from shrimp culture increased by 146%, compared to 41% for finfish and 38% for marine molluscs (FAO 1989), making shrimp culture the fastest growing warmwater aquaculture sector. This spectacular growth is the result of several factors, including a strong demand for shrimp, slow growth in wild shrimp catches, development of technology and desire for foreign revenue in tropical and subtropical developing countries (Hirasawa 1988). The bulk of cultured shrimp production (85%) comes from Asia (Fig. 2), but there are growing contributions from Latin America (dominated by Ecuador) and other subtropical and tropical regions. The main cultured species are giant tiger shrimp *Penaeus monodon*, which made up 33% of total output in 1988, followed by Chinese white shrimp *P. chinensis* (22%), western white shrimp *P. vannamei* (18%) and several other species (Fig. 3).

There is evidence from production trends that the exponential growth period for shrimp culture is drawing to a close (see Fig. 1). There was a significant downturn in prices of shrimp in 1989 due to overproduction and market saturation, particularly in the major Japanese market (Ferdouse 1990). The future of the industry is difficult to predict, but the present expansion looks set to continue at a slower rate. The Aquatech '90 conference in May 1990 predicted that farm raised production in Asia may rise to between 800,000 and 900,000 t by the year 2000, with a total production of 1.5 million t, or 50% of the global shrimp production (Anon. 1990b).

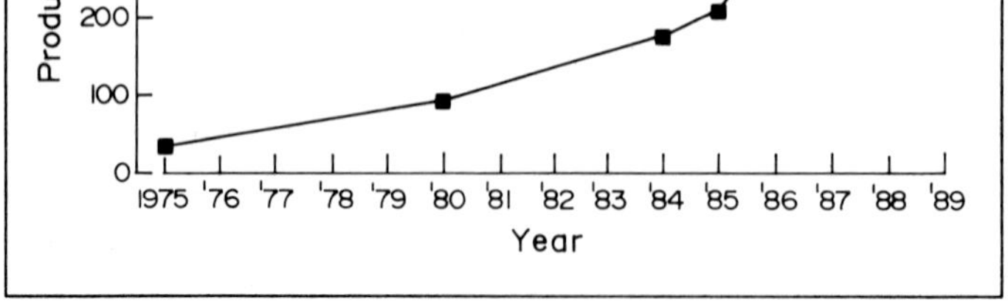

Fig. 1. World production from shrimp culture (from FAO 1989; Anon. 1990a; Ferdouse 1990).

This dramatic expansion has raised a number of concerns over the environmental impact of shrimp culture (e.g., Primavera 1989; Chua et al. 1989a). The aim of this paper is to review the interactions between shrimp culture and the environment and to highlight some of the lessons which may be learned from the explosive development of the industry.

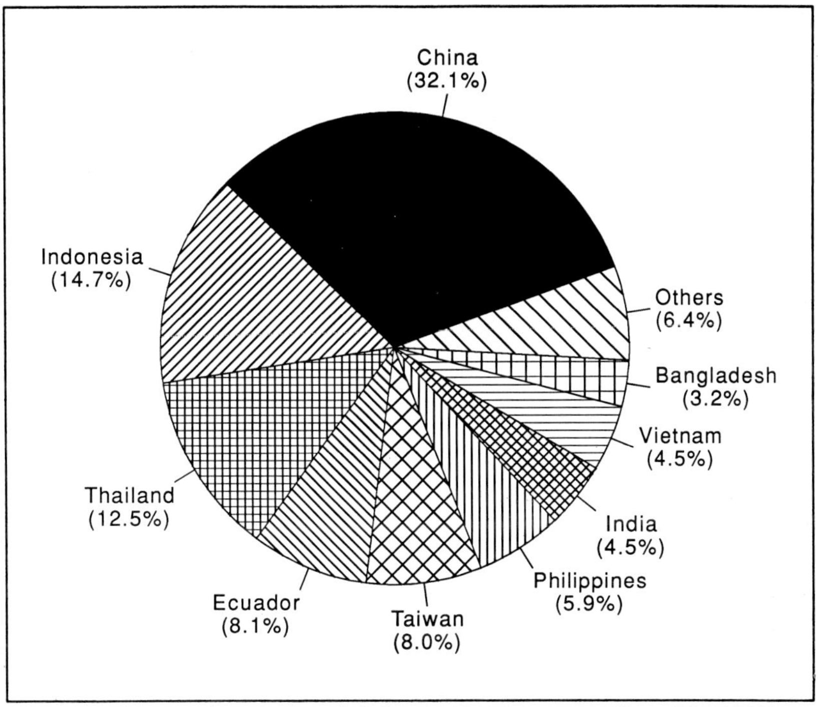

Fig. 2. World production of cultured shrimp in 1988, by country (from FAO 1989).

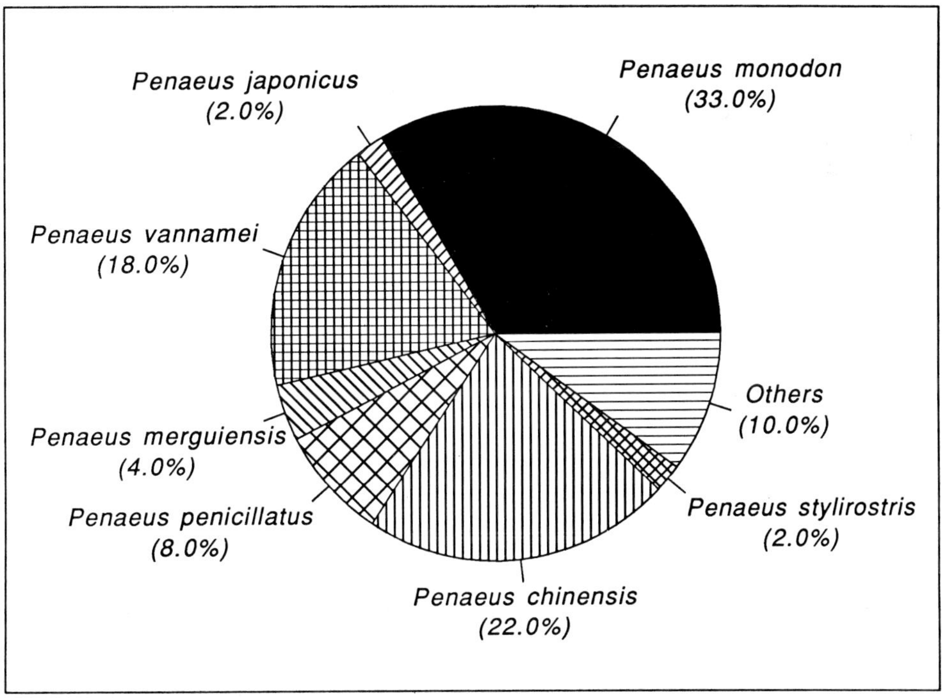

Fig. 3. World production of cultured shrimp in 1988, by species (from FAO 1989).

Shrimp Culture Methods

There are many different techniques used to culture shrimps, ranging from very extensive operations to super-intensive systems. The reader is referred to Tseng (1987), Apud et al. (1989) and Main and Fulks (1990) for details, but some basic understanding of shrimp culture systems is necessary to understand environmental impacts.

Extensive Culture Methods

Extensive shrimp culture has a long history in Asia. The systems rely on natural food within the pond and tidal fluctuations for water exchange (Table 1). Fertilizer and stocking may be used to increase pond productivity (Apud et al. 1989), but overall extensive systems are characterized by low inputs and low yields (Table 1). The traditional Indian 'bheris' are ponds constructed in naturally inundated land in the Sundarbans after clearing of mangroves, some of which may extend up to several hundred hectares (Silas 1987). Stocking is traditionally by natural inputs of seed through tidal movement. The 'tambaks' of Indonesia date back to the 16th century (Schuster 1952) and are similar in that stocking is by the natural inflow of seed and pond yields are derived from natural productivity plus some supplementary fertilization (Naamin 1987). Supplementary stocking is now more usual in these traditional culture systems, which are still very common in Asia, although productivity is low, at less than 500 $kg \cdot ha^{-1} \cdot year^{-1}$ (Csavas 1988).

Semi-Intensive and Intensive Methods

Extensive culture methods are being superceded by semi-intensive or intensive pond culture methods in many shrimp producing countries, involving monoculture, greater inputs of feed or fertilizer, supplementary stocking and improved water management. This trend started around the early 1970s in Taiwan, but gradually spread to other parts of Asia in the mid-1970s and early 1980s as hatchery and other technology developed (Liao 1990). The result has been a gradual increase in annual pond yields, from less than 0.5 $t \cdot ha^{-1}$ to greater than 15 $t \cdot ha^{-1}$ in Taiwan (Wickins 1986) This move towards more intensive aquaculture has not been uniform, and in most countries in Asia and Latin America, production still comes from a mixture of traditional extensive systems and semi-intensive and intensive culture systems, although the overall trend is towards higher average productivity (Fig. 4)

Table 1. Outline of basic shrimp culture methods (from Apud et al. 1989).

	Extensive	Semi-intensive	Intensive
Feed	Natural	Natural + supplement	Formulated diets
Water management	Tidal	Tidal + pump	Pump + aeration
Pond size (ha)	2-20 ha	1-5 ha	0.1-1 ha
Stocking density ($PL \cdot m^{-2}$)	0.1-1	1-5	5->25
Production ($kg \cdot ha^{-1} \cdot year^{-1}$)	100-500	500-4,000	4,000->15,000

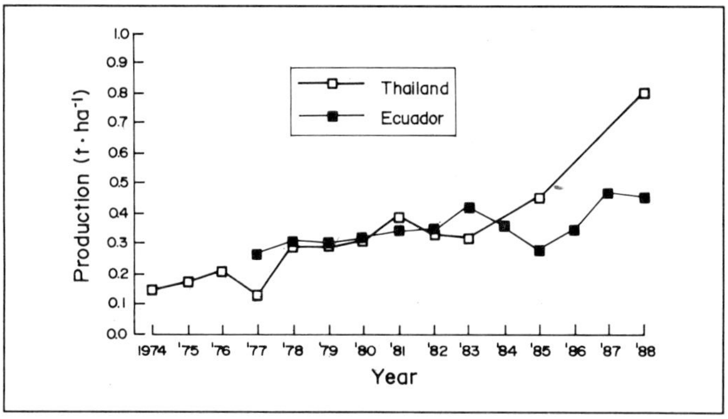

Fig. 4. Productivity of shrimp culture in Thailand and Ecuador.

Table 2. Average productivity (kg $\cdot ha^{-1}$) from shrimp culture in major shrimp producing countries (1988 figures from Anon. 1989).

Country	Productivity ($kg \cdot ha^{-1}$)
China	1,000
Ecuador	700
Taiwan	1,500 (1988)
	4,500 (1989)
Indonesia	250
Thailand	800
Philippines	430
India	600
Vietnam	250

(Csavas 1988). The overall productivity levels for different countries are given in Table 2.

There is an indication that the trend towards intensification in some countries has moderated, as the culture industry has become concerned about the impact of disease in high density culture ponds as well as their economic sustainability. For example, in the Philippines, shrimp farmers have reduced stocking densities and levels of feeding to reduce stress (Anon. 1990c) and in other countries (e.g., India), government policy is to promote the development of semi-intensive methods rather than capital-intensive methods. Technological inputs have also limited the move towards intensification in many countries (Csavas 1988).

Land Resource Requirements

Most shrimp culture is carried out in ponds, although *Penaeus monodon* and *Penaeus indicus* have been cultured in pens and cages in some countries (Beveridge 1984; Walford and Lam 1987). The size of the ponds varies depending on the culture system. Extensive culture is usually carried out in larger ponds, commonly between 2 and 20 ha in surface area (Table 1), but up to 200 ha in some traditional culture systems (Silas 1987). Intensive and semi-intensive culture normally uses smaller ponds (Table 1) to facilitate management and harvesting. Limited land availability and the high cost of such land in some countries, such as Taiwan (Liao 1990), has stimulated development of more intensive shrimp culture. In others, such as Indonesia (Cholik and Poernomo 1986), low competition for land has allowed space for more extensive and semi-intensive methods to predominate.

Total land area covered by shrimp ponds has been increasing in parallel with the developing shrimp industry and estimates suggest that 765,500 ha worldwide were used in 1988 with 639,000 ha in Asia and 126,500 ha in Latin America (Table 2; Fig. 5).

The siting of shrimp ponds is governed by many factors, including climate, elevation, water quality, soil type, vegetation, supply of postlarvae, support facilities and legislative aspects, as described in detail in various manuals including Tseng (1987) and Apud et al. (1989). The result is that ponds for shrimp culture have been constructed in a variety of different habitats, including salt pans, rice paddies, sugar fields, other agricultural land, abandoned coastal land and mangrove forests. As an example, the land-use pattern for shrimp

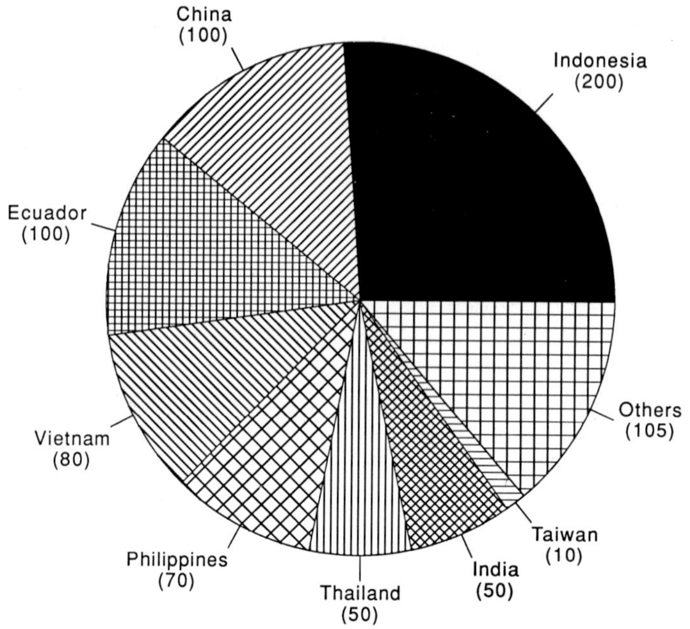

Fig. 5. Surface area of shrimp culture ponds in major shrimp producing countries in 1988 (units: 1,000 ha) (from Anon. 1989).

Table 3. Distribution of shrimp farms on the west coast of Sri Lanka in 1988, categorized by land use categories (from Jayasinghe and De Silva 1990).

Type of land	Area (ha)	(%)	No. of farms	(%)
Bare land	294.5	(29.3)	13	(29.5)
Agricultural				
Homesteads	30.4	(3.0)	3	(6.8)
Coconut	97.2	(9.7)	5	(11.4)
Cropland	53.8	(5.3)	4	(9.1)
Others	16.1	(1.0)	4	(9.1)
Wetland				
Marshlands } Mangroves	340.0	(34.7)	9	(20.5)
Unclassified	159.2	(16.4)	6	(13.6)
Total	1,006.2		44	

farms on the west coast of Sri Lanka is shown in Table 3. The patterns of land uşe may also change depending on availability and conflicts with other users. For example, when the industry was developing in Ecuador, the favorite sites for construction of ponds were the salt pans. Once these areas were used, the industry moved to less favorable areas, initially substandard agricultural land and then mangroves (Doumenge 1989; Aiken 1990). Although precise figures are difficult to obtain, a substantial area of shrimp ponds in Asia has been constructed on previous mangrove forests (Kapetsky 1985, 1986).

Water Resource Requirements

The quality and quantity of available water are major factors controlling shrimp production. Water quality is critical during the early life stages in the hatchery and during growout in ponds. In general, the water for shrimp culture should be free from agricultural and industrial pollution and be within the necessary salinity and temperature range for the shrimp species being cultured (Apud et al. 1989). The water quality requirements for shrimp culture have been broadly defined in several publications (Wickins 1986; Boyd 1989). There is evidence that shrimp culture itself is threatened by growing water pollution in Asia and Latin America (Chua et al. 1989b; Aiken 1990).

The amount of water required for shrimp culture depends on the nature of the culture system. Shrimp hatcheries normally require only modest amounts of water. The 'backyard' shrimp hatcheries in Thailand can produce 1 million postlarvae per year with a tank capacity of 20 m^3 (Kungvankij et al. 1986). The modest water requirements are such that hatcheries can be established many miles from the sea, with hypersaline water supplied by tanker from salt pans a considerable distance away (Kungvankij et al. 1986). Pond culture requires considerably more water than does hatchery operation.

Extensive pond culture relies largely on tidal flow and sites are usually selected such that there is adequate inflowing water and the ponds are drainable during periods of low tide (Csavas 1988). The requirement for water increases with intensification and most intensive shrimp culture ponds will pump 30-40% of pond volume per day to remove waste metabolites. Water requirements for pond culture, not surprisingly, are very variable depending on management practices and available water. Studies of shrimp culture in Taiwan show a water requirement of 11,000-21,430 $m^3 \cdot t^{-1}$ production for semi-intensive culture and 29,000-43,000 $m^3 \cdot t^{-1}$ for intensive culture (Table 4). The demand for water is higher in intensive culture because of the need to remove substantially more waste metabolites and supply oxygen at high stocking and feeding rates (Phillips et al. 1991).

Seed Requirements

Shrimp pond culture requires a source of postlarvae for stocking of ponds. In traditional extensive systems, such as the

Table 4. Water requirements for semi-intensive and intensive shrimp culture in comparison with other forms of aquaculture (from Phillips et al. 1991).

System and species	Production ($t \cdot ha^{-1}$)	Water use ($m^3 \cdot t^{-1}$)	Source
Penaeid shrimps semi-intensive ponds (Taiwan)	4.1-11.0	11,000-21,430	Wickins (1986); Chien et al. (1986)
Penaeid shrimps intensive ponds (Taiwan)	12.6-27.4	29,000-43,000	Wickins (1986); Chien et al. (1986)
Salmonids (UK) tank culture	-	252,000	Solbe (1987)
Clarias batrachus intensive ponds	100-200	50-200	Muir (1981)
Common carp raceway (Japan)	1,443	740,000	Hepher (1985)

Indian bheris or Indonesian tambaks, farmers originally relied on natural stocking of ponds through inflowing wild postlarvae (Schuster 1952). This practice is somewhat haphazard and also introduces potential predators and competitors into the pond. In recent years, shrimp culturists have relied more on deliberate stocking of extensive, semi-intensive and intensive ponds. The postlarvae come from two sources, wild-caught and hatchery-reared stock. Growing demand for postlarvae, combined with declining or unpredictably fluctuating natural catches, has led to the development of hatcheries in many countries with semi-intensive and intensive culture. Indeed, the development of hatchery technology for *P. monodon* in the late 1960s is seen as a key factor in the subsequent growth of shrimp culture in Asia (Liao 1990).

Stocking rates per square meter for ponds are dependent on culture methods, from 0.1 to 1.0 in extensive systems to greater than 25 in intensive systems, going up to 70 or more in some very intensive systems (Table 1). Postlarvae are usually stocked from 5 to 25 days after metamorphosis (PL 5 to PL 25) depending on species and management practices (Csavas 1988; Apud et al. 1989; Main and Fulks 1990).

Wild-caught postlarvae are thought to be better suited for pond stocking than hatchery-reared stock in many countries. In Ecuador, some pond farmers leave ponds empty rather than stock with hatchery-reared stock, due to the poor quality of hatchery stock and high probability of poor survival and disease (Hirono and van Eys 1990). The poor quality of hatchery-reared postlarvae is partly blamed for the severe decline in the Taiwanese shrimp industry (Lin 1989; Chen 1990), related to overuse of chemicals, poor hatchery management and lack of stock selection (Lin 1989).

There has been a trend towards declining numbers of wild-caught postlarvae in many countries, due to several factors including overfishing, coastal pollution (some from the shrimp industry itself) and removal of mangroves. A declining catch of wild *P. monodon* postlarvae has been reported in India (Silas 1987) and Bangladesh (Mahmood 1986) where there is concern over overfishing of postlarvae. The practice of fishing wild postlarvae may have impacts on other finfish or shrimp resources, as a large number of finfish and lesser-valued shrimps accidentally harvested along with the valuable *P. monodon* postlarvae may be subsequently destroyed. Silas (1987) estimates that 10 kg of finfish and shrimp larvae or fry are killed for each 1 kg of *P. monodon* postlarvae harvested in the Sundarbans, India. The natural stocks of shrimp postlarvae may also be affected by development of shrimp culture itself. In Ecuador, the industry has been limited by supply of postlarvae, believed by some to be due to removal of mangroves and shrimp nursery areas by shrimp ponds, overfishing of postlarvae and coastal pollution (Kapetsky 1986; Hirono and van Eys 1990). Shrimp farmers in the Chakaria Sundarbans in Bangladesh have found that natural stocking of extensive ponds is decreasing, one of the reasons for which is thought to be the destruction of coastal mangrove ecosystems for shrimp pond construction (FAO 1984, cited in Kapetsky 1986).

Feed and Fertilizer Requirements

Extensive culture systems rely largely on natural productivity but semi-intensive and intensive systems require inputs of fertilizer and feed (Table 1). The use of formulated diets is common in *P. monodon* culture in Southeast Asia and becoming more so in other parts of Asia and Latin America. The use of trash fish and locally available 'fresh' diets is more common in some countries. For example, pond cultured shrimp are fed living marine or brackishwater invertebrates such as molluscs, small crustaceans and polychaetes in China (Liu 1990).

The trend towards intensive shrimp culture means that reliance on formulated diets will increase. New and Wijkström (1990) estimate that shrimp culture in Asia will require 1.1 million tonnes of feed in the year 2000. Shrimp farming alone consumed 180,000 t of the 650,000 t of fishmeal used in intensive and semi-intensive aquaculture in 1988, which accounts for 10% of global fishmeal supply. There is concern that growing requirements of aquaculture, including shrimp, for fishmeal could exacerbate shortages of global fishmeal supplies, although the environmental impact of such a demand is uncertain.

Impacts on Coastal Mangrove Resources

Mangrove Clearance For Shrimp Culture

Mangroves occur throughout the tropics and in some subtropical areas on sheltered shores bearing soft intertidal sediments (Macintosh 1982). The total global mangrove resource has been estimated to cover an area of 160,000 km^2 (Saenger et al. 1983) with associated estuaries and coastal lagoons covering another 83,000 km^2 (Kapetsky 1985). If it is assumed that all the 765,500 ha of shrimp ponds in 1988 were converted from mangroves, then shrimp ponds would account for less than 5% of the global resource. In fact, in many countries (e.g., China), shrimp ponds have been developed in nonmangrove areas and many of the traditional extensive systems have been in operation for many years. Therefore, conversion of mangroves for shrimp culture represents considerably less than 5% of the original resource. Even so, various authors point out that a substantial proportion of ponds has been converted from original mangrove forest in some countries and the issue has become a major concern in both Asia and Latin America (Ong 1982; Kapetsky 1986; Csavas 1988; Chua et al. 1989a, 1989b).

The area of mangrove converted to shrimp ponds is very difficult to ascertain, but in the Philippines, Thailand, Malaysia, Indonesia and Ecuador, shrimp culture has certainly contributed significantly to the loss of mangrove resources, although much of the change may be attributed to earlier conversion of mangroves to traditional extensive culture ponds. In the Philippines, a combination of shrimp and milkfish culture is responsible for reducing the mangrove area from 448,000 ha in 1968 to 110,000 ha in 1988 (Primavera 1989). In Thailand, 38.3% of mangrove removed between 1979 and 1986 was removed for aquaculture, equivalent to an area of 38,000 ha or 13% of the 287,300 ha of the mangrove resource in 1979 (Arbhabhirama et al. 1988). In Malaysia, shrimp farms have encroached onto mangrove reserves (Chua et al. 1989a) and in Indonesia, a significant part of the estimated 200,000 ha of shrimp ponds has been developed from former mangrove forests. Total pond area is still less than 5% of the massive Indonesian mangrove resource of 4 million ha, but in Java, Sulawesi and Sumatra, construction of ponds for shrimp and milkfish culture has contributed to significant local denudation. In Singapore, large tracts of mangrove were transformed into ponds for production of *P. indicus*, although the shrimp ponds themselves are now being lost to urban and industrial development (Fortes 1988). In Ecuador, 24% of the 113,530 ha of shrimp ponds in 1987 were on former mangrove areas, representing 13-14% of the original mangrove area. In some provinces of Ecuador, it is reported that up to 36% of mangroves have been removed for shrimp ponds and, in some cases, ponds have subsequently been abandoned or taken out of production because of the suboptimal culture conditions prevailing in the mangrove areas (Terchunian et al. 1986; Doumenge 1989; Aiken 1990). There is little quantitative information from other countries, but anecdotal evidence suggests that mangrove clearance is ongoing in other locations with developing shrimp culture industries.

Impact of Mangrove Clearance

The impact of mangrove clearance is difficult to quantify, but there is no doubt that removal of mangroves can have a significant ecological, economic and social impact. Mangroves are important in nutrient cycling, as a source of nutrients for adjacent coastal ecosystems and a breeding ground, nursery area and ongrowing environment for many commercially important finfish, crustaceans and molluscs.

Studies have demonstrated a net export of organic matter from mangrove swamps into the nearshore area, where they may form an important source of food for invertebrates and fish (Christensen 1978), even as an important source of food for cultured molluscs. There is also a wide range of plants and animals associated with mangroves (Macintosh 1982), many of which are known to be of significant economic importance (Matthes and Kapetsky 1988), and may be lost following mangrove removal.

There is evidence that removal of mangroves leads to a decline in fisheries production and loss of potential for development of integrated aquaculture and fisheries within and adjacent to the mangroves themselves (Kapetsky 1986). Several studies have demonstrated that coastal mangroves are important habitats for fry and larval stages of commercially important finfishes, molluscs and crustaceans (Martosubroto and Naamin 1977; Turner 1977) and productivity of some commercially important shrimp fisheries appears to be positively correlated with abundance of mangrove forests adjacent to the fishing grounds (Martosubroto and Naamin 1977), indicating the potential for adverse impacts on commercial fisheries production following large-scale mangrove clearance.

The loss of shrimp nursery areas from mangrove areas may also adversely affect the shrimp culture industry itself through reduced availability of broodstock and wild postlarvae. *P. monodon* culture still relies on wild-caught broodstock because techniques for maturation and spawning in captivity have yet to be perfected. A loss of shrimp resources represents a loss of broodstock and potential genetic material for future selective breeding programs. In Ecuador, removal of mangroves for shrimp aquaculture has been linked to serious shortages of wild *P. vannamei* postlarvae for stocking the ponds, such that 60% of ponds were forced out of production in 1986 (Snedaker et al. 1986), although other factors such as water temperature play an important role in determining availability of wild shrimp larvae (Aiken 1990).

Mangroves are also important in coastal protection and removal may cause coastal erosion (Carter 1959), changes in patterns of sedimentation and shoreline configuration (Snedaker and Getter 1985). Mangrove forests are also used for other products, including lumber, thatching material, firewood and a variety of foodstuffs. Thus, their removal can have far reaching economic and social impacts (Bailey 1988).

These impacts, applied generally to mangrove removal, have yet to be quantified in relation to shrimp culture, but clearly indicate the problems for both natural fisheries and shrimp culture, and multi-user conflicts that may occur through indiscriminate removal of mangroves. There is a need for more detailed ecological (and economic) studies to assess impacts of shrimp culture on mangroves, but even without such studies, the value of mangroves is such that a 'cautionary' approach seems warranted to ensure that the risk of such problems developing is reduced.

Impacts on Land Resources

Shrimp pond culture may affect other land resources, which may cause adverse ecological impacts or result in multi-user conflicts. Coastal pond construction inevitably results in some loss of terrestrial habitat and the development of extensive, semi-intensive and extensive shrimp culture has resulted in losses of agricultural land, such as grazing land, sugar land or rice paddies, in several countries (Mahmood 1986). The construction of shrimp ponds on marginal land has been welcomed as a more economical use of coastal land resources in some countries but in others, such as India, there is concern that conversion of rice paddies to shrimp ponds may adversely affect local rice production.

Land resources may be affected during farm operation beyond the boundary of the shrimp farm. In the Philippines, abstraction of freshwater from underground aquifers has resulted in subsidence of coastal land (Primavera 1989; SEAFDEC 1989). In Taiwan, subsidence around the Pingtuug coastal area during 1970-83 was 0.3-2.0 m, mainly due to shrimp and eel ponds (Chiang and Lee 1986). These problems have resulted in very significant conflicts between coastal inhabitants.

Saltwater intrusion into some of these freshwater aquifers may also be accompanied by salinization of soils (Jayasinghe and De Silva 1990), resulting in further devaluation of often already marginal agricultural land.

Impacts on Ground- and Surfacewater Resources

The abstraction of groundwater for intensive shrimp farming in coastal areas of Taiwan, the Philippines and Thailand has resulted in saltwater intrusion and salinization of freshwater aquifers (Primavera 1989). This salinization has resulted in a degradation of potable and agricultural water supplies which, combined with the land subsidence mentioned above, has caused significant conflicts with local farmers and residents. The economic impact of these environmental changes has not been assessed, but seems likely to offset some of the economic benefits of shrimp farm development (SEAFDEC 1989). The problems of groundwater salinization have led to the use of groundwater for shrimp culture being banned in some countries, e.g., Taiwan and parts of Thailand.

The channeling or diversion of water supplies for shrimp culture can lead to changes in local hydrology, with resultant effects on salinity, sedimentation and other water quality parameters. Salinization of surfacewaters due to shrimp culture has been reported in several countries, including Thailand, Bangladesh (Mahmood 1986), Indonesia (Cholik and Poernomo 1986), India and the Philippines (SEAFDEC 1989). Shrimp culture may increase salinity through facilitating the flow of saline water inland, discharge of saline effluent (which may be more saline through evaporation in ponds) or diversion of freshwater flow. Salinization can cause problems for other agricultural users and salinity changes may also influence stenephaline species, such as some mangroves (Cholik and Poernomo 1986), but in most instances these impacts have not been studied.

Waste Production

The output of waste from shrimp culture ponds and hatcheries has become an important issue in many countries in Asia and Latin America as concern over the impact of shrimp pond effluent on coastal water quality has grown (Chua et al. 1989a; Liao 1990). The main components of this waste are dissolved and particulate nutrients and organic matter, plus smaller quantities of chemicals, microorganisms and other detritus. The bulk of the waste is nutrient and organic matter derived largely from the input of feed and fertilizer.

Extensive Culture

Traditional extensive culture systems throughout Asia are characterized by low stocking densities and little or no supplementary feeding or fertilizer, and as such are unlikely to provide any significant loading of nutrients or organic matter to the coastal ecosystem. The reliance on natural feed is such that extensive systems may be net removers of nutrients from the coastal zone. There are examples where extensive finfish and mollusc aquaculture has significantly reduced aquatic ecosystems of nutrients (Beveridge 1984), but no such problems have been reported for extensive shrimp culture. The fact that extensive ponds in some areas have been used for culturing shrimps and finfish for many years testifies to the potentially sustainable nature of such aquaculture systems.

The only effluent problems reported with extensive systems are with very acidic discharges from newly constructed ponds sited on potential acid sulfate soils (Cholik and Poernomo 1986).

Semi-Intensive and Intensive Culture

The progression from extensive to semi-intensive and intensive shrimp culture is marked by increasing inputs, of fertilizers (as in many semi-intensive systems) or supplementary feed (as in intensive systems) and consequently the potential for increasing nutrient and organic matter loads per unit area or per unit weight of shrimp production. The potential nutrient and organic loads from intensive culture tend to be higher than semi-intensive culture, because of the more efficient use of these materials in semi-intensive aquaculture systems (Beveridge and Phillips, this vol.; Edwards, this vol.).

Nutrient and Organic Wastes from Intensive Shrimp Culture

Nutrient and organic wastes produced by intensive shrimp pond culture consist of solid matter (mainly uneaten feed, feces and phytoplankton) and dissolved metabolites (mainly ammonia, urea and carbon dioxide). The dissolved and solid fraction will also be supplemented by fertilizers when these are also applied (Apud et al. 1989).

There have been very few studies of shrimp culture waste. Some figures from laboratory studies of ammonia output are given in Table 5, but such figures are of limited use in building a budget for intensive shrimp culture. One approach is to estimate total loadings of both nitrogen and phosphorus from the concentration of nitrogen and phosphorus in feeds and shrimp carcasses as shown in Table 6. This approach suggests total nitrogen loads of 57.3-118.1 kg and total phosphorus loads of 13.0-24.4 kg per tonne of *P. monodon* shrimp production, with feed conversion ratios of 1.2:1-2.0:1, respectively. This analysis suggests that 63-78% of nitrogen and 76-86% of phosphorus fed to shrimps in intensive pond culture is probably lost to the environment, either within the pond ecosystem, or at some stage discharged into the external environment.

These figures should be treated with caution because they are based on limited data, and there is a need to generate good quantitative information on loadings of nutrients and organic matter under realistic farm conditions, as available for

Table 5. Excretion of ammoniacal nitrogen by penaeid shrimps.

Species	Excretion rate	Source
P. monodon (1.6-27 g) aquarium, dry pellets	0.93 -0.30 mg NH_3-N·g^{-1}·day^{-1}	Wickins (1985)
P. esculentus (18 g) aquarium, fresh diet	1.0 -1.43 mg NH_3-N·g^{-1}·day^{-1}	Dall and Smith (1986)
P. monodon (0.2-70 g) aquarium	0.77 -0.33 mg NH_3-N·g^{-1}·day^{-1}	Mohanty et al. (1989)
P. semisulcatus (0.6-43.8 g) aquarium, pelleted feed	2.17 -0.50 mg NH_3-N·g^{-1}·day^{-1}	Wajsbrot et al. (1989)

*NH_3 refers to the sum of NH_3 + NH_4.
**Excretion refers to the wet weight of shrimp.

Table 6. Nitrogen and phosphorus budgets for an intensive *Penaeus monodon* pond [data collected from Wickins (1985), Civera and Guillaume (1989) and Sedgwick (1979)].

Production related loadings (kg·t^{-1} produced)				
	Food conversion ratio			
	1.2:1		2.0:1	
	N	P	N	P
Feed input*	91.2	17.0	152.0	28.4
Shrimp harvest**	33.9	4.0	33.9	4.0
Waste load	57.3	13.0	118.1	24.4

*Diet 76.0 gN·kg^{-1} wet weight (5% moisture) (Sedgewick 1979); 14.2 gP·kg^{-1} wet weight (5% moisture) (Civera and Guillaume 1989)

**Shrimp 33.9 gN·kg^{-1} wet weight (73% moisture) (Wickins 1985); 4.0 gP·kg^{-1} wet weight (73% moisture) (Civera and Guillaume 1989)

temperate finfish culture (NCC 1990). Even so, such data are helpful in assessing the contribution of nutrient loadings from intensive shrimp culture in comparison with other loadings.

The calculations also show the importance of food conversion ratio (FCR) in waste output. Several other factors will also be important, as judged from experiences with other intensive culture systems.

Feed Quality

The quality of feed plays an important role in waste output in finfish culture (NCC 1990) and is likely an important factor in shrimp culture. Loadings of nutrients and organic matter are higher for finfish fed with trash fish and fresh diets than with pelleted moist or dry diets (Warrer-Hansen 1982). Similar principles will apply in shrimp culture. Fresh diets, infrequent feeding and high stocking densities increase nitrogen loads from shrimp held in recirculating tank systems (Wickins 1985) and regular feeding with pelleted diets is known to maximize growth of prawns (Sedgwick 1979). The use of trash fish and invertebrates, advocated as an economic feed in some countries (Liu 1990), could therefore bring about a higher waste load, with the risk of deteriorating pond conditions and increased loadings to the environment.

Feed Wastage

The amount of feed wasted plays an important role in the total waste loadings. Because feed settles directly onto the pond bottom, feed wastage can have a significant effect on sediment quality and ultimately the health of bottom-living shrimp (Boyd 1989). There is little information on wastage of feed in shrimp culture. Laboratory studies by Wickins (1985) show that an average of 11% of a mixture of wet and dry pelleted feed remained uneaten by *P. monodon*. Shrimps are known to be less efficient feeders on pelleted diets because they masticate food externally (Wickins 1985). The widely varying FCRs reported from intensive *P. monodon* culture ponds suggest that feed wastage may at times be higher than 11% and very variable. A common FCR in Thai *P. monodon* culture is estimated at 2.0:1, but with careful monitoring of feed input it is possible to reduce it to 1.2:1 (C. Kwei Lin, pers. obs.). The higher FCR represents an additional load of 0.8 t of uneaten or poorly converted food, with obvious potential for increased pond water and sediment quality problems and higher loadings to receiving waters, as well as economic implications.

Effluent from Shrimp Culture

The nutrient and organic wastes influence the quality and quantity of effluent discharged from shrimp farms, and the subsequent impact on the external environment. The dissolved nutrients and organic solids stimulate the rapid growth of microorganisms, including bacteria, phytoplankton and zooplankton, as well as the benthos. The bacterial production in intensive shrimp rearing ponds takes place mainly in the water column with a doubling time of only a few hours, increasing linearly with increasing feed inputs (Moriarty 1986). Despite the action of mechanical aerators in intensive ponds, a considerable amount of detrital material and dead planktonic organisms often accumulates on the pond bottom, especially in areas where water circulation is slow. This accumulation is associated with decreasing redox potentials, when release of harmful gases such as hydrogen sulfide and methane can be a significant cause of stress to bottom feeding shrimp (Boyd 1989). Sediment accumulations may also interfere with harvesting (Williamson 1989).

The effluent discharged from intensive ponds will reflect these internal processes. Krom and Neori (1989) reported that effluent from an intensive finfish pond with a water retention time of two days carried the major loss of particulate nutrients in the form of heterotrophic and autotrophic microplankton, and that bottom deposition only accounted for 10-15% of the particulate nutrients. Phytoplankton dynamics also played a major role in the quality of effluent, with significantly higher dissolved nutrient loads following a 'crash' in phytoplankton blooms. Similar internal processes will probably affect effluent quality in intensive and semi-intensive shrimp ponds.

The quality of effluent during normal operation will be similar to the quality of water in the pond, which if managed effectively will tend to be well mixed, with water quality within acceptable ranges for shrimp. The data from one 5-month study are shown in Table 7. The wide range in certain values, such as nutrients and chlorophyll *a*, represent an increasing trend over the culture period rather than sporadic fluctuations. This quality can be compared with the quality of effluent during the final stages of harvesting and cleaning of ponds (Table 8) which shows extremely high concentrations of both nutrients and organic matter. The loadings at this time are substantially more than loadings during the culture period, because of discharge of material previously bound to sediment particulate matter. Impacts on receiving waters are likely to be most significant at this time. Similar findings have been reported for finfish culture by Boyd (1978), who showed high levels of suspended solids during harvesting of channel catfish ponds. Bergheim et al. (1984) have also demonstrated high 'shock' loads of solids, BOD, nitrogen and phosphorus during tank cleaning, several times the load at other times. A similar situation is anticipated in semi-intensive and intensive shrimp culture during harvesting, pond draining and cleaning.

The effluent released from ponds during construction may also be very poor in acid sulfate soil areas. In Sri Lanka, waters draining ponds at this time have been shown to be extremely acidic and high in iron, aluminium and manganese (Table 9). These effluents are likely to be highly toxic to aquatic life in receiving waters, as all parameters except manganese are well beyond 'acceptable' criteria for aquatic life (Alabaster and Lloyd 1982).

Table 7. The ranges of effluent water quality recorded at an intensive shrimp farm in Thailand during a 5-month growout period (data from C. Kwei Lin).

Pond size (ha)	0.48 - 0.56
Pond depth (m)	1.5 - 1.8
Salinity (ppt)	10 - 35
Temperature (°C)	22 - 31
pH	7.5 - 8.9
Total phosphorus ($mg \cdot l^{-1}$)	0.05 - 0.4
Total nitrogen ($mg \cdot l^{-1}$)	0.50 - 3.4
Total ammonia ($mg \cdot l^{-1}$)	0.05 - 0.65
Dissolved oxygen ($mg \cdot l^{-1}$)	4.0 - 7.5
Chlorophyll *a* ($mg \cdot l^{-1}$)	20 - 250
Total suspended solids ($mg \cdot l^{-1}$)	30 - 190
Water exchange frequency ($\% \cdot day^{-1}$)	5 - 40

Table 8. Effluent quality during cleaning of four intensive shrimp ponds in Thailand (data from C. Kwei Lin).

Total nitrogen ($mg \cdot l^{-1}$)	2,600	1,900	2,400	2,600
Total phosphorus ($mg \cdot l^{-1}$)	110	60	40	70
Organic carbon (%)	13.6	7.3	10.4	13.7

Table 9. Concentration of metal ions and pH in water draining shrimp ponds during pond preparation on acid sulfate soil areas in Sri Lanka (data courtesy of J.M.P.K. Jayasinghe).

Farm site	pH	Iron ($mg \cdot l^{-1}$)	Aluminium ($mg \cdot l^{-1}$)	Manganese ($mg \cdot l^{-1}$)
A	3.5	19.3	2.5	<0.16
B	3.9	12.7	1.8	1.12
C	2.7	17.6	2.35	3.18

Impacts of Shrimp and Effluent

The discharge of effluent rich in nutrients and organic matter can have several consequences for receiving waters, although in most instances impacts have been very poorly quantified for shrimp culture. Studies with intensive finfish farming (reviewed in NCC 1990) suggest that nutrients and organic matter in shrimp pond effluent have potential for the following impacts:

i) reduced dissolved oxygen in receiving waters, due to discharge of effluent low in dissolved oxygen and breakdown of dissolved and particulate organic matter and other waste materials (BOD and COD);

ii) hypernutrification and eutrophication of receiving waters, resulting in increased primary productivity (with potential risks of phytoplankton blooms), alteration of community structure and secondary productivity; and
iii) increased sedimentation due to organic matter, leading to changes in productivity and benthic community structure, plus possible siltation.

Particulate material settles out of the water column once the settling velocity is greater than the velocity in the receiving canal or stream. This material can accumulate below the point of effluent discharge and there is anecdotal evidence from Thailand, Sri Lanka and other countries of irrigation canals becoming silted as a result of solid matter discharged from shrimp farms (C. Kwei Lin, pers. obs.; J.M.P.K. Jayasinghe, pers. comm.). There is concern also that coastal environments are being subject to hypernutrification and eutrophication as a result of shrimp culture (SEAFDEC 1989), but so far such impacts have not been quantified (Chua et al. 1989a).

In practice, impacts depend on the mass flow of effluent and the capacity of the receiving environment to assimilate waste materials. It is essential to match loads with the capacity of the environment to accept waste materials. Models for integrating intensive finfish culture into coastal and freshwater environments are now being developed (NCC 1990; Rosenthal et. al., this vol.) and a similar understanding of the capacity of coastal environments for shrimp culture is also necessary.

Chemotherapeutants and Other Chemicals

Chemicals may be used in shrimp culture as medicaments or therapeutants for prevention or treatment of disease, as feed additives, disinfectants, piscicides and for soil or water treatment. It is very difficult to obtain accurate data on the type and amounts used because of some reluctance by farmers to disclose information, together with the lack of monitoring and regulations in many countries involved in shrimp culture. A list of chemicals commonly used in shrimp culture is given in Table 10, although the types and amounts used are known to vary considerably from country to country and this list should not be regarded as complete. Such chemicals may find their way into natural waters in pond or hatchery effluent as no methods are yet in use to treat or remove such materials from effluent water.

Lime and Fertilizers

Lime is widely used to neutralize acidity and sterilize pond sediments between crops. The amounts used depend on sediment type (Boyd 1989) but several tonnes per ha may be required on acid sulfate soils. The environmental impacts of liming are likely minimal, although there are reports that long-term liming can harden pond sediments and make them less suitable for shrimp culture (Poernomo and Singh 1982).

Fertilizers are widely used in semi-intensive culture systems to promote the growth of shrimp food organisms, particularly for the early postlarval stages (Apud et al. 1989; Boyd 1989). Inorganic and organic fertilization of ponds may contribute to the nutrient load in receiving waters although such effects have not been quantified.

Piscicides and Molluscicides

Piscicides are widely used to remove potential predators and competitors from semi-intensive and intensive shrimp ponds. Rotenone, saponin, nicotine and calcium hypochlorite have been widely used (Apud et al. 1989) (Table 10). Chlorinated hydrocarbons such as DDT, endrin and aldrin have also been used and highly toxic organotins have been used as molluscicides in some Southeast Asian countries (Baticados et al. 1986). The biodegradable organic plant extracts rotenone, saponin and nicotine degrade fairly easily in water and are probably less harmful than chlorinated hydrocarbons and organotins, both of which are highly toxic, persistent compounds. The latter compounds pose a threat to shrimp health, product quality, human health and the wider environment and their use should be discouraged (Apud et al. 1989). Baticados et al. (1986) found that use of organotin molluscides by Philippine shrimp farmers caused soft-shell syndrome in cultured shrimps, emphasizing the potential of some of these compounds to affect adversely shrimp culture itself, as well as the external environment.

Disinfectants and Water Treatments

Sodium hypochlorite and benzalkonium chloride are widely used in hatcheries to sterilize water and equipment between stocking (Table 10). Calcium carbide and calcium hydroxide have also been used to sterilize pond sediments between harvests. The impact of these on the pond ecosystem and the wider environment remains to be quantified.

Sodium-ethylenedinitrotetraacetic acid (Na-EDTA) is quite widely used in shrimp hatcheries to improve hatching and survival of larvae (Licop 1988), because of its potential to complex and reduce the toxicity of heavy metals known to contaminate water supplies in many parts of Asia and Latin America. Zeolites and other materials have also been used in attempts to remove ammonia and improve water quality in growout ponds, particularly in Taiwan and other countries practicing very intensive culture methods (Tseng 1987). Algicides such as copper sulfate and various proprietary brands are also used to control algal growth in growout ponds. The impact of most of these water treatment compounds on the wider environment is probably minimal, although would merit more detailed investigation.

Chemotherapeutants

Many chemotherapeutants are used by shrimp culturists for the prevention and treatment of disease (Table 10). Antibiotics and antimicrobials are in widespread use in hatcheries and growout ponds in Asia and Latin America. Apart from antibiotics, the most commonly used chemotherapeutants in hatcheries are formalin (for treatment of parasitic diseases) and malachite green (fungal diseases). Formalin is also widely used in ponds in some parts of Southeast Asia. The impact of these compounds on tropical environments has been little studied. The lethal concentration of formalin is 0.7-1.2 $mg \cdot l^{-1}$ for algae, 5 $mg \cdot l^{-1}$ for zooplankton and 35 $mg \cdot l^{-1}$ for certain bacteria. Malachite green is also toxic to algae at concentrations greater than 0.3 $mg \cdot l^{-1}$ (NCC 1990), below normal treatment concentrations (Table 10). Both compounds are also known to be toxic to penaeid nauplii below treatment concentrations (Castille and Lawrence 1986), indicating that potential exists for malachite green and formalin to affect adversely the pond ecosystem and, through effluent discharge, external waters. Both compounds are potentially harmful to humans and particular care is recommended in use (NCC 1990).

Antibiotics

Antibiotics are commonly used in the treatment of shrimp disease in hatcheries and growout ponds and several are currently in use (Table 10). Treatments are applied either by immersion or through

Table 10. Common chemicals used in penaeid shrimp hatcheries and growout ponds (from Apud et al. 1989; Minsalan and Chiu 1989; Saclauso 1989; Main and Fulks 1990; authors' personal observations).

Lime

Calcium hydroxide, calcium carbonate, calcium oxide (neutralization of acidity, sterilization of pond bottoms)

Piscicides and molluscides (ponds)

Tea seed cake (saponin, widespread use), calcium hypochlorite (*P. japonicus* ponds in Japan), derris root extract (rotenone, widespread use), tobacco dust (nicotine, widespread), endrin, DDT, aldrin, thiodan (not recommended but still used in ponds in some countries), organotins (molluscicide in ponds)

Disinfectants, water and soil treatment

EDTA (hatchery, 5-10 $mg \cdot l^{-1}$, reducing metal toxicity), sodium hypochlorite (hatchery, 5 $mg \cdot l^{-1}$, common disinfectant), benzalkonium chloride (pond, hatchery, disinfectant), zeolite (ion exchange resin, intensive ponds), potassium permanganate (ponds, occasional use), calcium carbide (sterilization of pond sediments)

Chemotherapeutants

Copper sulfate (ponds, algicide and antimicrobial), malachite green (hatchery and ponds, fungal infections, in widespread use, 1-2 $mg \cdot l^{-1}$), formalin (hatchery and ponds, antimicrobial, parasitic treatments, 25-250 $mg \cdot l^{-1}$), methylene blue (protozoa infections, 8 $mg \cdot l^{-1}$)

Antibiotics for control of bacterial disease, including chloromycetin, chloramphenicol (hatcheries, widespread use, although now banned in some countries), oxytetracycline (ponds, widespread use), oxolinic acid (rare), furazolidone, streptomycin, tetracycline, nitrofurazone

incorporation of antibiotic in feed. There is also no doubt that antibiotics are used prophylactically to prevent outbreaks of disease, a practice which increases the exposure of pathogens, shrimp and the environment to the antibiotic.

The use of antibiotics in shrimp culture raises several issues of concern in relation to human health, product quality and the environment (Brown 1989; reviewed in Austin, this vol.). The potential environmental impact depends on the amount of antibiotic or antibiotic-exposed microorganisms reaching natural waters, which will be related to the treatment methods, amounts, frequency of treatments and physicochemical factors (Rasmussen 1988). The environmental impact of these compounds in tropical and subtropical environments is unknown. However, in Asia and Latin America, overuse of antibiotics has resulted in development of drug resistant shrimp pathogens and there is concern that transfer of resistance to human pathogens could result in development of resistance in human pathogens. Chloramphenicol, a widely used chemical in shrimp hatcheries, is of particular concern because of its importance in control of human *Salmonella typhae* infections.

Antibiotics use poses major problems for the shrimp culture industry. Widespread use of oxytetracycline in Taiwan, Thailand and the Philippines has resulted in development of resistant strains of *Vibrio*, which has made treatment of *Vibrio* infections, the most common growout disease in many Southeast Asian countries, extremely problematical. It is also thought that overexposure of larvae to antibiotics in Taiwanese hatcheries has been a major factor responsible for the very poor survival of *P. monodon* in growout ponds (Chen 1990). In the Philippines, drug resistant strains of the luminous *Vibrios* are thought to have developed as a result of widespread use of antibiotics (Sorgeloos 1990). This spread of resistant strains in Southeast Asia has probably been made easier by the frequent intermixing of effluent and influent water in many highly congested culture areas. Such examples highlight the potential damage to the industry caused through inappropriate antibiotic use. The wider effects on the environment outside of the farms remain to be explored.

Introductions and Transfers of Shrimps

Shrimps have been introduced (transported and released outside of the present species range) and transferred (transported and released within the present species range) for aquaculture purposes.

There are many examples of species being introduced to Asia from Latin America and *vice versa*. For example, *Penaeus monodon*, an Indo-West Pacific native, has been introduced to the Dominican Republic, Panama, Hawaii and Mexico from Taiwan (Welcomme 1988; Lightner et al. 1989) and *Penaeus vannamei*, an East Pacific native, has been introduced to the Dominican Republic and Venezuela (Welcomme 1988). *Penaeus stylirostris* (East Pacific native) and *P. vannamei* have been introduced into the Philippines for aquaculture purposes (Juliano et al. 1989). *Penaeus stylirostris, P. vannamei, P. brasiliensis* and *P. schmitti* have also been introduced at various times into Taiwan from Latin America (Liao and Liu 1989). Chiba et al. (1989) report the introduction of *Penaeus chinensis* from China to Japan. Within Asia, *Penaeus japonicus*, a native of China, Korea and Japan, has been introduced to Singapore from Japan (Chou and Lam 1989).

There has also been considerable transfer of shrimp within native ranges in Asia and Latin America, either as postlarvae or as mature broodstock. There are published reports of transfers of *Penaeus monodon* from the Philippines and Malaysia to Taiwan (Liao and Liu 1989) but it is well known that in both continents there is widespread transfer of shrimps within and between countries.

Experiences with aquatic species in general indicate that a number of problems may arise following the introduction of new species. Habitat changes, disruption of host community through competition, predation and stunting, genetic changes, introduction of diseases and parasites and socio-economic effects have all been documented (Welcomme 1988). There is some evidence that non-native shrimp are escaping from shrimp culture facilities into natural waters (Anon. 1990e), but there is little information on the impact of introductions and transfers on either cultured or wild shrimp communities.

There is evidence, however, that shrimp pathogens have been widely disseminated through introductions and transfers of shrimp. The infectious hypodermal hematopoietic necrosis virus (IHHNV), a serious pathogen of cultured shrimps, is suspected of being spread from its natural range on the Pacific coast of Latin America to the Middle East and Asia through transhipment of infected shrimps (Lightner et al. 1989). There is also anecdotal evidence in Southeast Asia that monodonbacilovirus (MBV) has been widely disseminated through movement of shrimp stock. Lightner et al. (1989) found MBV in *P. monodon* stocks introduced to Hawaii, Mexico and Tahiti from Southeast Asia and Taiwan. These pathogens present serious problems for the shrimp culturist, but information on impacts on wild stocks is non-existent, although concern over transfer from farm to wild stocks is growing. The experience with other crustaceans (e.g., Alderman et al. 1984) and finfish (NCC 1989) that disease organisms may be highly pathogenic to new species and the instances of severe and debilitating epizootics following translocation of pathogens to new hosts indicate the need for effective regulatory systems for prevention and control of the spread of pathogens to minimize risks to cultured and wild stocks.

Shrimp Disease, Production Loss and Environment

There is growing evidence that environmental impacts related to shrimp culture play a significant role in outbreaks of disease now affecting shrimp ponds in Asia and Latin America. Serious production losses from disease outbreaks have been reported from Taiwan (production losses of 42,000 t in 1987-88, Lin 1989; Chen 1990), Sri Lanka, Thailand (estimated US$27.2 million in 1989), China (*Vibrio* caused losses 10-30% of production; Main and Fulks 1990), Philippines and Ecuador (Anon. 1989d) amongst others. These examples indicate that disease in shrimp culture is of considerable economic significance (Liao 1990).

The causes of disease and production losses are in many cases difficult to ascertain, but several studies link outbreaks of disease to environmental factors. In Taiwan, deterioration of water quality and sediments in Taiwanese shrimp ponds have been closely linked to the severe 1987-88 production losses (Chen 1990; Lin 1989). In the central and eastern region of Thailand, many shrimp farms have suffered from mass mortalities after 2-3 crops (C. Kwei Lin, unpubl. data). Although specific data are lacking, these problems are being blamed on overloading of the inshore area and/or pond sediments with shrimp farm effluent. On the west coast of Sri Lanka, major mortalities in the shrimp culture industry are being blamed on self pollution of the main water supply canal, the 'Dutch' canal (J.M.P.K. Jayasinghe, pers. comm.). In Thailand, the production decline in the northern Gulf of Thailand has been so severe that shrimp culturists have abandoned 90% of the ponds in some provinces and farmers have moved to the southern part of the country. These problems, although poorly quantified, suggest that shrimp culture has exceeded the capacity of the ponds and the coastal environment to sustain shrimp culture.

Potential acid sulfate soils are widespread in Southeast Asia and parts of Latin America, particularly in mangrove forest areas (Macintosh 1982). The use of acid sulfate soils presents serious problems for shrimp culturists, as liming, seawater flushing and other management practices (Simpson et al. 1983) are required to avoid the low productivity, acidity and heavy metal toxicity associated with aquaculture on such soils. Several shrimp disease syndromes and production losses are linked to acid sulfate soils, including soft-shell syndrome, red disease and blue shrimps commonly reported from the Philippines, Thailand, Indonesia and Malaysia (Nash et al. 1988; Baticados et al. 1990). In some countries in Southeast Asia, production losses due to acid sulfate soil problems have again been so severe that farms have been abandoned, many of these on former mangrove forests, a very wasteful use of the mangrove resource.

Discussion

Environmental impact of shrimp culture has become a major issue in many tropical and subtropical countries with developing shrimp culture industries (Ong 1982; Chua et al. 1989a; Pullin 1989; Saclauso 1989) and the present paper affirms that there are significant environmental impacts associated with the explosive development of shrimp farming in Asia and Latin America. Unfortunately, there is a lack of quantitative data on the environmental impact of shrimp culture, something that needs to be rectified if present and future impacts are to be mitigated and shrimp culture is to be integrated successfully into the coastal zone.

Many of the problems affecting both shrimp culturists and the wider environment could be avoided by better site selection and improved management. The fact that ponds have been abandoned or are suffering significant disease outbreaks indicates considerable improvements in site selection procedures are necessary. The environmental problems that have arisen indicate that site selection should include more emphasis on proper environmental impact assessment. It is interesting to note that most shrimp farming manuals do not consider the wider environmental implications of shrimp culture development. The present criteria for site selection need to be reviewed and consideration given to the long-term capacity of areas to sustain shrimp culture developments.

The impact of shrimp culture on mangroves is a major issue and there is a growing realization that aquaculture development should not proceed indiscriminately in mangrove areas (Csavas 1988). The traditional extensive culture methods 'consume' the largest area of mangrove, with very low productivity in return, and as such the development of extensive culture systems seems difficult to justify as a long-term sustainable use of coastal resources. Intensification of pond culture can reduce the need for more conversion of mangrove area, whilst making better use of areas already converted (Kapetsky 1986). However, this option may be restricted by lack of technology, credit availability for upgrading of ponds by small-scale farmers and the unsuitability of many of the original extensive sites (Kapetsky 1987; Csavas 1988). Intensive forms of shrimp culture may also result in significant problems with effluent and disease.

Several authors have advocated semi-intensive shrimp culture as a means of avoiding unproductive use of large areas of land, whilst avoiding the worst problems of disease and effluent associated with intensive culture (Primavera 1989). Hirasawa (1988) has also demonstrated that semi-intensive shrimp culture may be the only economically viable culture method for farmers to survive fluctuations in cost and market price, because of the lower production costs. There are also thought to be benefits to semi-intensive culture in terms of reduced nutrient discharge (Edwards, this vol.). Therefore semi-intensive culture, combined with suitable selection of sites outside of mangrove areas could be encouraged as a more effective use of land and water resources, avoiding some of the environmental impacts of both intensive and extensive culture.

Several disease-related issues need to be addressed as a matter of urgency, because of their potential impacts on the shrimp culture industry and wider environment. The use of chemicals in intensive and, to a lesser extent, in semi-intensive culture needs to be examined in more detail, with practical guidelines developed, because of the significant environmental, public health and product quality concerns about the use of some compounds. The potential for translocation of pathogens to new hosts also indicates the need for more effective mechanisms for prevention and control of the spread of pathogens, to minimize risks to both cultured and wild stocks. There is also a need to understand better the relations between shrimp disease and environment, both as a means of preventing economically serious disease outbreaks and of indirectly promoting the need for more effective environmental management (e.g., reduced use of chemicals, improved pond water and soil quality).

Treatment of effluent offers some scope for reducing impacts on the water quality in the external environment. There has been a growing trend for fish farmers in temperate regions to treat effluent to reduce impacts on receiving waters. A major problem of aquaculture wastewater treatment is the dilute but high volume nature of the effluent in comparison with traditional forms of wastewater discharge (Muir 1982). Cost-effective technology is now available in several temperate countries to reduce loads of biochemical oxygen demand, suspended solids, nitrogen and phosphorus (Mäkinen et al. 1988; Bromage et al. 1989). There have been few attempts so far to treat effluent from intensive shrimp ponds on a commercial scale. Experience with temperate aquaculture suggests that pressure from legislators will be necessary to force farmers to adopt effluent control, but it is likely that the growing realization of negative impacts of shrimp pond effluent will result in treatment becoming important in the future.

So far, techniques for treatment of shrimp farm effluent are in the early stages of development. The wastewater during the period of culture is in most cases a portion of well-mixed pond water which, being rich in nutrients and microorganisms, could be suitable for culturing finfish, molluscs and seaweed. In Thailand, experiments have been conducted using the seaweed *Gracilaria* to remove nutrients and molluscs *(Perna viridis)* to remove solid matter from effluent water (C. Kwei Lin, pers. obs.) although such practices to remove solid matter have not yet been taken up on any commercial scale. In Hawaii, molluscs have also been used to improve quality of effluent from shrimp ponds. Such integrated approaches offer some scope for improving effluent quality and producing an additional cash crop. The quality of effluent discharged during and following harvesting is so poor that 'biological' methods, such as that mentioned above, are unlikely to be successful, and there may be a need to develop alternative strategies, for example, the use of settlement ponds to remove this potentially very damaging effluent.

In temperate aquaculture, there has also been a trend towards the use of highly digestible, 'low-pollution' diets, which are designed to reduce aquaculture pollutant loads (NCC 1990). The development of such diets, plus effective management of the pond environment to reduce food wastage and improve food conversion ratios, could also assist in reducing pollutant loads to the coastal environment, with long-term benefits to the shrimp farmer and the wider coastal environment.

This review demonstrates that some of the environmental problems, particularly water quality deterioration from shrimp pond effluent and poor site selection on acid sulfate soils, have also adversely affected the sustainability of the shrimp culture industry itself in several countries. In some, such as Thailand, these problems have been reduced by relocating farms to new areas, an option that broadens the impact on coastal resources and one which will become less available in future, as suitable sites for shrimp culture become used up, and pressure on the coastal zone from other users increases (Chua et al. 1989b). The fact that environmental problems are now affecting the shrimp farmer, other users of the coastal zone and the wider environment indicates an urgent need for effective planning, considering the long-term requirements of both the shrimp culturist and the wider environmental impacts, including the needs of other coastal resource users. In many countries, such an approach may require new legislation or more effective implementation of existing legislation (Van Houtte et al. 1989).

There is a need for economic principles to be applied to the use of coastal resources for shrimp culture, which consider the long-term value of shrimp culture versus the economic benefits of other uses, a notoriously difficult problem, but one which is beginning to be addressed (e.g., Kapetsky 1986; Dixon 1989). Such information will probably be necessary to provide policymakers with a basis for informed and balanced decisions over coastal resource utilization. Chua et al. (1989b) give an example from Malaysia where an annual fee of US$48/ha is required for conversion of mangrove to shrimp ponds. The value of undisturbed mangrove may significantly exceed this figure, with a fisheries value alone of up to US$2,777/ha (Kapetsky

1985), a value which ignores the economic value of other mangrove products, economic implications of increased coastal erosion and loss of livelihood for coastal communities, all of which need to be considered in a balanced management framework (Chua et al. 1989b).

The needs for effective planning, site selection and management with due consideration for the capacity of the environment to sustain shrimp culture, balanced against the needs of other users of coastal resources, are probably the main lessons to be learnt from the development of shrimp culture up to now. Shrimp culture can contribute significantly to the economies of many developing countries. However, there is no doubt from experiences so far that continued development of shrimp culture will require a more effective approach to environmental planning which integrates shrimp culture into the coastal environment in a sustainable manner.

References

Aiken, D. 1990. Shrimp farming in Ecuador. An aquaculture success story. World Aquacult. 21(1):7-16.

Alabaster, J.S. and J. Lloyd. 1982. Water quality criteria for freshwater fish. Butterworths, London.

Alderman, D.J., J.L. Polglase, M. Fragling and J. Hagger. 1984. Crayfish plague in Britain. J. Fish Dis. 7:401-405.

Anon. 1989. World shrimp farming 1988. Aquaculture Digest, San Diego, California.

Anon. 1990a. World shrimp farming 1989. Aquaculture Digest, San Diego, California.

Anon. 1990b. Out of Malaysia. World Shrimp Farming 15(9):1-4.

Anon. 1990c. Philippine update. World Shrimp Farming 14(7):3.

Anon. 1990d. Ecuador hit by shrimp pond deaths. Fish Farming Int. 17:7.

Anon. 1990e. Viruses concern state shrimpers. The Evening Post (Charleston, SC). 3 May 1990.

Apud, F., J.H. Primavera and P.L. Tores, Jr. 1989. Farming of prawns and shrimps. Aquaculture Extension Manual No. 5, Third Edition. Southeast Asian Fisheries Development Center, Aquaculture Department, Tigbauan, Iloilo, Philippines.

Arbhabhirama, A., D. Phantumvanit, J. Elkington and P. Ingkasuwan. 1988. Thailand. Natural resources profile. Oxford University Press, Singapore.

Bailey, C. 1988. The social consequences of tropical shrimp mariculture development. Ocean Shoreline Manage. 11:31-44.

Baticados, M.C.L., R.M. Coloso and R.C. Duremdez. 1986. Studies on the chronic soft-shell syndrome in the tiger prawn, *Penaeus monodon* Fabricius, from brackishwater ponds. Aquaculture 56:271-285.

Baticados, M.C.L., E.R. Cruz-Lacierda, M.C. de la Cruz, R.C. Duremdez-Fernandez, R.Q. Gacutan, C.R. Lavilla-Pitogo and G.D. Lio-Po. 1990. Diseases of penaeid shrimp in the Philippines. Aquaculture Extension Manual No. 16. Aquaculture Department, Southeast Asian Fisheries Development Center, Tigbauan, Iloilo, Philippines.

Bergheim, A., H. Hustveit and A.R. Selmer-Olsen. 1984. Estimated pollution loadings from Norwegian fish farms. II. Investigations 1980-1981. Aquaculture 36:157-168.

Beveridge, M.C.M. 1984. Cage and pen fish farming. Carrying capacity models and environmental impact. FAO Fish. Tech. Pap. 255, 133 p. FAO, Rome.

Boyd, C.E. 1978. Effluent from catfish ponds during fish harvest. J. Environ. Qual. 7:59-62.

Boyd, C.E. 1989. Water quality management and aeration in shrimp farming. Fisheries and Allied Aquacultures Departmental Series No. 2, 183 p. Auburn University, Alabama.

Bromage, N., J.P. Henderson and R. Watret. 1989. How to design a settlement pond. Fish Farmer 12(3):41.

Brown, J. 1989. Antibiotics: their use and abuse in aquaculture. World Aquacult. 20(2):34-35, 38-39, 42-43.

Carter, J. 1959. Mangrove succession and coastal changes in S.W. Malaya. Trans. Inst. Br. Geog. 26:79-88.

Castille, F.L. and A.L. Lawrence 1986. The toxicity of erythromycin, minocycline, malachite green and formalin to nauplii of the shrimp *Penaeus stylirostris*. J. World Aquacult. Soc. 17(1-4):13-18.

Chen, S.N. 1990. Collapse and remedy for the shrimp culture industry in Taiwan. Department of Zoology, National Taiwan University, Taiwan.

Chiang, H.C. and J.C. Lee. 1986. Study of treatment and reuse of aquacultural wastewater in Taiwan. Aquacult. Eng. 5:301-312.

Chiba, K., Y. Taki, K. Sakai and Y. Oozeki. 1989. Present status of aquatic organisms introduced into Japan, p. 63-70. In S.S. De Silva (ed.) Exotic aquatic organisms in Asia. Asian Fisheries Society Spec. Publ. 3, 154 p. Asian Fisheries Society, Manila.

Chien, Y.H., I.C. Liao and C.M. Yang. 1988. The evolution of prawn growout systems and their management in Taiwan, p. 143-168. *In* K. Murray (ed.) Aquaculture engineering technologies for the future. Institute of Chemical Engineers Symposium

Series. No. 111. Hemisphere Publishing Corporation, New York.

Cholik, F. and A. Poernomo. 1986. Development of aquaculture in mangrove areas and its relationship to the mangrove ecosystem, p. 93-104. *In* R.H. Mepham and T. Petr (eds.) Papers contributed to the Workshop on Strategies for Management of Fisheries and Aquaculture in Mangrove Ecosystems, 23-25 June 1986, Bangkok, Thailand and Country status reports on inland fisheries presented at the Third session of the Indo-Pacific Fishery Commission Working Party of Experts on Inland Fisheries, 19-27 June 1986, Bangkok, Thailand. FAO Fish. Rep. 370 (Suppl.), 248 p.

Chou, L.M. and T.J. Lam. 1989. Introduction of exotic aquatic species in Singapore, p. 91-97. *In* S.S. De Silva (ed.) Exotic aquatic organisms in Asia. Asian Fisheries Society Spec. Publ. 3, 154 p. Asian Fisheries Society, Manila.

Christensen, B. 1978. Biomass and primary production of *Rhizophora apiculata* BL. in a mangrove in southern Thailand. Aquat. Bot. 4:43-52.

Chua, T.E., J.N. Paw and E. Tech. 1989a. Coastal aquaculture development in ASEAN: the need for planning and environmental management, p 57-70. *In* T.E. Chua and D. Pauly (eds.) Coastal area management in Southeast Asia: policies, management strategies and case studies. ICLARM Conf. Proc. 19, 254 p.

Chua, T.E., J.N. Paw and F. Y. Guarin. 1989b. The environmental impact of aquaculture and the effects of pollution on coastal aquaculture development in Asia. Mar. Pollut. Bull. 20(7):335-343.

Civera, R. and J. Guillaume. 1989. Effect of sodium phytate on growth and tissue mineralisation of *Penaeus japonicus* and *Penaeus vannamei* juveniles. Aquaculture 77:145-156.

Csavas, I. 1988. Shrimp farming development in Asia, p 63-92. *In* Shrimp '88. Bangkok, Thailand, 26-28 January 1988. Infofish, Kuala Lumpur, Malaysia.

Dall, W. and D.M. Smith. 1986. Oxygen consumption and ammonia-N excretion in fed and starved tiger prawns, *Penaeus esculentus* Haswell. Aquaculture 55:23-33.

Dixon, J.A. 1989. Valuation of mangroves. Trop. Coast. Area Manage. 4(3):1-6.

Doumenge, F. 1989. Aquaculture in Ecuador, p. 1083-1094. *In* G. Barnabe (ed.) Aquaculture. Vol. 2. Ellis Horwood Series in Aquaculture and Fisheries Support. Ellis Horwood, New York.

FAO. 1989. Aquaculture production (1985-1988). FAO Fish. Circ. 815 (Rev. 2).

Ferdouse, F. 1990. Asian shrimp situation. Infofish Marketing Digest 1/90:32-38.

Fortes, M.D. 1988. Mangrove and seagrass beds of East Asia: habitats under stress. Ambio 17(3):207-213.

Hepher, B. 1985. Aquaculture intensification under land and water limitations. Geojournal 10: 253-259.

Hirasawa, Y. 1988. Shrimp culture economics in Asia. Paper presented at the Seventh Session of the IPFC Working Party of Experts on Aquaculture, 1-6 August 1988, Bangkok. Indo-Pacific Fisheries Commission. WPA/WP 16.

Hirono, Y. and S. van Eys. 1990. The Ecuadorean shrimp culture industry. Infofish Marketing Digest 4/90:53-58

Jayasinghe, J.M.P.K. and J.A. De Silva. 1990. Impact of prawn culture development on the present land use pattern in the coastal areas of Sri Lanka. Paper presented at the Symposium on Ecology and Landscape Management of Sri Lanka, June 1990, Colombo, Sri Lanka. 23 p.

Juliano, R.O., R. Guerrero III and I. Ronquillo. 1989. The introduction of exotic aquatic species in the Philippines, p. 83-90. *In* S.S. De Silva (ed.) Exotic aquatic organisms in Asia. Asian Fisheries Society Spec. Publ. 3, 154 p. Asian Fisheries Society, Manila.

Kapetsky, J.M. 1985. Mangroves, fisheries and aquaculture. FAO Fish. Rep. 338 Suppl.:17-36.

Kapetsky, J.M. 1986. Conversion of mangroves for pond aquaculture: some short-term and long-term remedies, p. 129-141. *In* R.H. Mepham and T. Petr (eds.) Papers contributed to the Workshop on Strategies for Management of Fisheries and Aquaculture in Mangrove Ecosystems, 23-25 June 1986, Bangkok, Thailand and Country status reports on inland fisheries presented at the Third session of the Indo-Pacific Fishery Commission Working Party of Experts on Inland Fisheries, 19-27 June 1986, Bangkok, Thailand. FAO Fish. Rep. 370 (Suppl.), 248 p.

Krom, M.D. and A. Neori. 1989. A total nutrient budget for an experimental intensive fishpond with circularly moving seawater. Aquaculture 83:345-358.

Kungvankij, P., L.B. Tiro, Jr., B.J. Pudadera, Jr., I.O. Potestas, K.G. Corre, E. Borlongan, G.A. Talean, L.F. Bustilo, E.T. Tech, A. Unggui and T.E. Chua. 1986. Shrimp hatchery design, operation and management. NACA Training Manual Series No. 1, 88 p. Network of Aquaculture Centres in Asia, Bangkok, Thailand.

Liao, I.C. 1990. The world's marine prawn culture industries: today and tomorrow, p. 11-27. *In* R. Hirano and I. Hanyu (eds.) The Second Asian Fisheries Forum. Asian Fisheries Society, Manila.

Liao, I.C. and H.C. Liu. 1989. Exotic aquatic species in Taiwan, p. 101-118. *In* S.S. De Silva (ed.) Exotic aquatic organisms in Asia. Asian Fisheries Society Spec. Publ. 3, 154 p. Asian Fisheries Society, Manila.

Licop, Ma. S. R. 1988. Sodium-EDTA effects on survival and metamorphosis of *Penaeus monodon* larvae. Aquaculture 74:239-247.

Lightner, D.V., R.M. Redman, T.A. Bell and R.B. Thurman. 1989. Geographic dispersion of the viruses IHHN, MBV and HPV as a consequence of transfers and introductions of penaeid shrimp to new regions for aquaculture purposes. J. Shellfish Res. 7:554-555.

Lin, C. Kwei. 1989. Prawn culture in Taiwan: what went wrong? World Aquacult. 20: 19-20.

Liu, J.Y. 1990. Present status and future prospects for shrimp mariculture in China, p. 16-28. *In* K.L. Main and W. Fulks (eds.) The culture of cold-tolerant shrimp: Proceedings of an Asian-U.S. Workshop on Shrimp Culture. Sponsored by the National Oceanic and Atmospheric Administration, United States Department of Commerce. The Oceanic Institute, Honolulu, Hawaii.

Macintosh, D.J. 1982. Fisheries and aquaculture significance of mangrove swamps, with special reference to the Indo-West Pacific region, p. 3-85. *In* J.F. Muir and R.J. Roberts (eds.) Recent advances in aquaculture. Vol. 1. Croom Helm, London.

Mahmood, N. 1986. Effect of shrimp farming and other impacts on mangroves of Bangladesh, p. 46-66. *In* R.H. Mepham and T. Petr (eds.) Papers contributed to the Workshop on Strategies for Management of Fisheries and Aquaculture in Mangrove Ecosystems, 23-25 June 1986, Bangkok, Thailand and Country status reports on inland fisheries presented at the Third session of the Indo-Pacific Fishery Commission Working Party of Experts on Inland Fisheries, 19-27 June 1986, Bangkok, Thailand. FAO Fish. Rep. 370 (Suppl.), 248 p.

Main, K.L. and W. Fulks, Editors. 1990. The culture of cold-tolerant shrimp: proceedings of an Asian-U.S. Workshop on Shrimp Culture. Sponsored by the National Oceanic and Atmospheric Administration, United States Department of Commerce. The Oceanic Institute, Honolulu, Hawaii.

Mäkinen, T., S. Lindgren and P. Eskelinen. 1988. Sieving as an effluent treatment for aquaculture. Aquacult. Eng. 7:367-377.

Martosubroto, P. and N. Naamin. 1977. Relationship between tidal forests (mangroves) and commercial shrimp production in Indonesia. Mar. Res. Indones. 18:81-88.

Matthes, H. and J.M. Kapetsky. 1988. Worldwide compendium of mangrove-associated aquatic species of economic importance. FAO Fish Circ. 814, 236 p.

Minsalan, C.L.O. and Y.N. Chiu. 1989. Effects of teaseed cake on selective elimination of finfish in shrimp ponds, p. 79-82. *In* J.L. Maclean, L.B. Dizon and L.V. Hosillos (eds.) The First Asian Fisheries Forum. Asian Fisheries Society, Manila.

Mohanty, A.N., P. Ravichandran, T. Rajyalakshmi and S.M. Pillai. 1989. Rate of ammonia excretion in *Penaeus monodon* (Fabricius). J. Aquacult. Trop. 4:93-95.

Moriarty, D.J.W. 1986. Bacterial productivity in ponds used for culture of penaeid prawns, Gelang Patah, Malaysia. Microb. Ecol. 12:259-269.

Muir, J.F. 1981. Management in *Clarias* culture, Thailand. Programme for the development of pond management techniques and disease control (DoF-UNDP/FAO THA/75/012) Thailand. National Inland Fisheries Institute, Bangkok, Thailand. 30 p.

Muir, J.F. 1982. Economic aspects of waste water treatment in fish culture, p. 123-135. *In* J.S. Alabaster (ed.) Report of the EIFAC workshop on fish-farming effluent. EIFAC Tech. Pap. 41, 166 p. Silkeborg, Denmark.

Naamin, N. 1987. Conversion of mangrove areas to tambaks aquaculture in Indonesia, p. 56-71. *In* Report of the Workshop on the Conversion of Mangrove Areas to Aquaculture. Iloilo City, Visayas, Philippines, 24-26 April 1986. UNDP/UNESCO Regional project on research and its application to the management of the mangroves of Asia and the Pacific, New Delhi. 175 p.

Nash, G., I.G. Anderson and M. Shariff. 1988. Pathological changes in the tiger prawn, *Penaeus monodon* Fabricius, associated with culture in brackishwater ponds developed from potentially acid sulfate soils. J. Fish Dis. 11:113-123.

NCC. 1989. Fish farming and the safeguard of the natural marine environment of Scotland. Nature Conservancy Council, Scottish Headquarters, Edinburgh.

NCC. 1990. Fish farming and the Scottish freshwater environment. A report prepared for the Nature Conservancy Council by the Institute of Aquaculture (University of Stirling). Nature Conservancy Council, Edinburgh. 285 p.

New, M. and U.N. Wijkström. 1990. Feed for thought. Some observations on aquaculture feed production in Asia. World Aquacult. 21:17-23.

Ong, J.E. 1982. Mangroves and aquaculture in Malaysia. Ambio 11:253-257.

Phillips, M.J., M.C.M. Beveridge and R.M. Clarke. 1991. Impact of aquaculture on water resources, p. 568-591. *In* D.E. Brune and J.R. Tomasso (eds.) Advances in aquaculture. Vol. 3. World Aquaculture Society, Baton Rouge, Louisiana. 606 p.

Poernomo, A. and V.P. Singh 1982. The problems, field identification and practical solutions of acid sulfate soils for brackishwater fish ponds, p. 49-61. *In* Report of Consultation/Seminar on Coastal Fish Pond Engineering, 4-12 August 1982, Surabaya, Indonesia. SCS/GEN/82/42. South China Sea Fisheries Development and Coordinating Programme, Manila.

Primavera, J.H. 1989. The social, ecological and economic implications of intensive prawn farming. Asian Aquacult. 11(1):1-6.

Pullin, R.S.V. 1989. Third world aquaculture and the environment. Naga, ICLARM Q. 12(1):10-13.

Rasmussen, F. 1988. Therapeutics used in fish production: pharmokinetics, residues and withdrawal periods. EIFAC/XV/88Inf. 13, 22 p. FAO, Rome.

Saclauso, C.A. 1989. Brackishwater aquaculture: threat to the environment? Naga, ICLARM Q. 12(3): 6-8.

Saenger, E.J., E.J. Hegerl and J.D.S. Davie. 1983. Global status of mangrove ecosystems. International Union for the Conservation of Nature and Natural Resources, Commission on Ecology Papers 3, 88 p.

Schuster, W.H. 1952. Brackish water ponds. Berita Perikanan-Jawatan Perikanan, Jakarta.

SEAFDEC 1989. Prawn industry impact worse. SEAFDEC Newsl. 12 (3):6.

Sedgwick, R.W. 1979. Effect of ration size and feeding frequency on the growth and food conversion of juvenile *Penaeus merguiensis* de Man. Aquaculture 16:279-298

Silas, E.G. 1987. Significance of the mangrove ecosystem in the recruitment of fry and larvae of finfishes and crustaceans along the east coast of India, particularly the Sundarbans, p. 19-34. *In* Report of the Workshop on the Conversion of Mangrove Areas to Aquaculture, 24-26 April 1986, Iloilo City, Visayas, Philippines. UNDP/UNESCO Regional project on research and its application to the management of the mangroves of Asia and the Pacific, New Delhi. 175 p.

Simpson, H.J., H.W. Ducklow, B. Deck and H.L. Cook. 1983. Brackishwater aquaculture in pyrite-bearing tropical soils. Aquaculture 34:333-350.

Snedaker, S.C. and C.D. Getter. 1985. Coastal resources management guidelines. Renewable resources information service. Coastal Management Publ. No. 2, 205 p. Research Planning Institute, Inc., Colombia.

Snedaker, S.C., J.C. Dickinson III, M.S. Brown and E.J. Lahmann 1986. Shrimp pond siting and management alternatives in mangrove ecosystems in Ecuador. Report for USAID. Office of Science Advisor, Washington, DC.

Solbe, J.F. de L.G. 1987. Report of the European Inland Fisheries Advisory Commission Working Party on Fish-Farm Effluents. Water Research Centre, Medmenham, UK. 20 p.

Sorgeloos, P. 1990. Controlling luminous bacteria in shrimp hatcheries. Larviculture and Artemia Newsl. 15:17.

Terchunian, A., V. Klemas, A. Segovia, A. Alvarez, B. Vasconez and L. Guerrero. 1986. Mangrove mapping in Ecuador: the impact of shrimp pond construction. Environ. Manage. 10:345-350.

Tseng, W.Y. 1987. Shrimp mariculture. A practical manual. Chien Cheng Kaohsiung, Taiwan.

Turner, R.E. 1977. Intertidal vegetation and commercial yields of penaeid shrimp. Trans. Am. Fish. Soc. 106:411-416.

Van Houtte, A.R., N. Bonucci and W.R. Edeson. 1989. A preliminary review of selected legislation governing aquaculture. Aquaculture Development and Coordination Programme. ADCP/REP/89/42. FAO, Rome.

Wajsbrot, N., M.D. Krom, A. Gasith and T. Samocha. 1989. Ammonia excretion of green tiger prawn *Penaeus semisulcatus* as a possible limit on the biomass density in shrimp ponds. Bamidgeh 41:159-164.

Walford, J. and T.J. Lam. 1987. Floating hatchery and net cage culture of *Penaeus indicus* in the Straits of Johore, Singapore. Aquaculture 67:11-32.

Warrer-Hansen, I. 1982. Evaluation of matter discharged from trout farming in Denmark, p. 57-63. *In* J.S. Alabaster (ed.) Report of the EIFAC Workshop on Fish-Farming Effluent. EIFAC Tech. Pap. 41, 166 p. Silkeborg, Denmark.

Welcomme, R.L. 1988. International introductions of inland aquatic species. FAO Fish. Tech. Pap. 294, 318 p.

Wickins, J.F. 1985. Ammonia production and oxidation during the culture of marine prawns and lobsters in laboratory recirculation systems. Aquacult. Eng. 4(3):155-174.

Wickins, J.F. 1986. Prawn farming today: opportunities, techniques and developments. Outlook on Agriculture 15:52-60.

Williamson, M.R. 1989. Development of a silt pump for aquacultural ponds. Aquacult. Eng. 8:95-108.

Discussion

BILIO: I would like to warn against taking and using easily available data, especially those provided by governments in certain countries. They may try to camouflage bad situations and the truth often lies between the picture that they give and that which is feared by ecologists.

M. PHILLIPS: Well, it is extremely difficult to get reliable data. For example, there is quite a variation in the literature concerning the area of mangroves in Ecuador that has been converted to shrimp culture. I agree with your request for a balanced view. Some of the data on mangrove use in Ecuador came from a LANDSAT study, under Ecuador-USA collaboration in the mid-1980s - so these are good data.

CHUA: I think that we are sometimes too sensitive about the amount of mangroves that have been taken off. We should consider in more detail the significance of a mangrove area that has been removed. These issues have been discussed for more than 25 years. Most studies simply state the total areas of mangrove that have been cleared. For aquaculture it is normally the riverine and river basin types of mangrove that are cleared. The margin mangroves are used for human settlements, etc., but statistics include all parts of the mangroves that have been cleared. So a statement that a certain clearance of mangrove has a given impact on the environment may not be very useful. Flying over the aquaculture areas in Ecuador you can see that the *estuarine* mangroves have been totally removed for shrimp farming. The recognition of this and its strong environmental impact are more important. We should discuss these more.

BILIO: I am not advocating absolute protection of all mangroves. The problem that remains is the indiscriminate attitude towards mangrove clearance. We must of course look at the different types of mangroves. There is also the old controversy about to what extent mangroves are important as nurseries for adjacent fisheries. This is bound to be different in different situations. Mangroves and coastline configurations are very diverse. They can be of little or of major importance as nursery areas. Moreover, if mangroves on sand or peat are destroyed, they will not easily re-establish themselves for perhaps hundreds of years whereas on silt and clay re-establishment is easy and rapid. It is also important to distinguish between degraded areas (which can perhaps best be used for aquaculture) and those areas that merit conservation because they are in good condition and/ or have a major function in coastal protection. *Then*, a discriminative choice can be made on how best to use and sustain mangrove productivity. The use of mangrove productivity in semi-intensive aquaculture needs much more study.

The most important thing is to avoid the view that mangroves can just be destroyed to make way for aquaculture. This is not a reasonable attitude - neither is absolute protection for all mangroves. Finally, we have to consider that market situations will also change. Perhaps shrimp culture will level off and decline as is the case with salmon culture and some operations in other areas. Then, new criteria come in.

RUDDLE: Yes. If shrimp culture does decline, what are the opportunity costs of having cleared mangroves for shrimp culture? What has been lost in terms of alternative uses? Is there any information on this aspect of mangrove economics?

BILIO: I do not know of any work, but this point is most important.

CSAVAS: There may be other considerations too. The truth is that mangroves are sub-optimal sites for shrimp culture. There is an inverse relationship between shrimp culture and mangrove area in Asia. Mangrove areas typically have acid sulfate soils that are not good for aquaculture. Shrimp culture expanded into mangrove areas because there were already traditional trapping and growing ponds there, often hundreds of years old. The best shrimp ponds are in fact inland, beyond the mangroves, where also construction costs are 40-60% cheaper than in the mangrove areas and yields may be 40-60% higher because of less acidity problems.

Shrimp culture development *is* already levelling off. The trend is towards increasing intensively of production in the existing extensive/semi-intensive ponds and decreasing the intensity of some highly intensive systems in Taiwan. The exploitation of new areas will slow down. The problems have come in those countries that had very small mangrove areas and have converted a high percentage of them to shrimp ponds: only Thailand, Philippines and Vietnam in Asia. In other Southeast Asian countries, including Indonesia, the impact of aquaculture on mangrove areas has been marginal.

BILIO: Concerning shrimp reproduction, for all species except *Penaeus japonicus* we can still only control one breeding cycle. Therefore there is still nearly total reliance on natural resources (gravid females or larvae) for seed supply. This limits seed supply in some areas. Therefore there is still a need for basic research to develop methods for continuously controlled reproduction over multiple generations of penaeids. Otherwise, the pressure on the natural resources will be too great. The same applies to many species of marine finfish that can still not be bred in captivity under controlled conditions as in land animal husbandry.

CSAVAS: In Asia, shrimp culture now relies entirely on seed produced in hatcheries from captured breeders. The pace of hatchery development in the last five years has been amazing. In Asia, there is no longer any reliance on wild-caught postlarvae. The same trend is occurring in Latin America. But I see the question here as an economic one, not a task for basic biological research. Rematuration and continuous reproduction can be achieved with all cultured penaeids, not just *P. japonicus*, but hatchery operators find it cheaper to purchase wild-caught gravid females. When these are scarce, they will shift to using cultured breeders, but this is more expensive. I agree, however, that the *methods* for using cultured breeders need to be improved and that this is a matter for research. The present methods are crude - not comparable to, say, induction of spawning in finfish by hormone injections.

Discussion on Latin American Shrimp Culture

BURAS: Is there any use of wastewater in this operation? Also, which environmental problems give most cause for concern?

ARELLANO: There is no wastewater use. The operation uses natural seawater. The problem is that the used seawater is dumped back into the sea. This may contain antibiotics - especially from hatcheries. We are also concerned about the future effects of the petroleum drilling and extraction operations along our coasts on the shrimp operations. We are also seeing elevation of some heavy metal levels, for example copper, in coastal waters where shrimp are raised. The sources of this are not known. We are also concerned about the construction of dams on the watersheds above the shrimp farms and possible run-off of pesticides into the shrimp farms.

BURAS: Have you tested for pesticides and herbicides in farmed shrimps?

ARELLANO: Yes. So far we are fortunate. They are not detectable, but there are detectable heavy metals.

BILIO: In the context of use of mangroves for aquaculture operations, the example of shrimp culture in Ecuador, on which Mr. Arellano has concentrated, may not be such a good example to follow as some people think. A lot of mangrove areas were destroyed, despite promises that the development would take place above the mangrove areas.

Environmental Management of Coastal Aquaculture Development

CHUA THIA-ENG
International Center for Living Aquatic Resources Management
MCPO Box 2631, 0718 Makati
Metro Manila, Philippines

CHUA, T.E. 1993. Environmental management of coastal aquaculture development, p. 199-212. *In* R.S.V. Pullin, H. Rosenthal and J.L. Maclean (eds.) Environment and aquaculture in developing countries. ICLARM Conf. Proc. 31, 359 p.

Abstract

This paper briefly outlines the trends of coastal aquaculture development worldwide and reviews its documented environmental impact including habitat destruction, impoverishment of biodiversity, nutrient and organic enrichment of coastal waters, harmful algal blooms, antibiotic resistance, change in macrobenthic communities and deterioration of coastal water quality. Policy guidelines and management strategies are discussed within the concept of sustainable development. Coastal aquaculture development and management programs at the national and local levels should be formulated within the framework of coastal area management. Specific guidelines pertaining to aquaculture operation and measures to mitigate adverse environmental impacts are explored. Regular monitoring is necessary and adequate regulatory measures in preventing or minimizing further deterioration of the coastal environment due to coastal aquaculture practices are suggested.

Introduction

For the purpose of this paper, the Food and Agriculture Organization's (FAO 1988) definition of the term "aquaculture" is adopted. The geographical coverage of the term "coastal," as defined by Sorensen et al. (1984), includes the shoreland influenced by the sea, the water column and the seabed extending to the edge of the continental shelf. Coastal aquaculture, as used in this paper, therefore, covers all brackishwater aquaculture practices operating landwards and marine aquaculture practices operating off the shoreline. Coastal aquaculture practices include seaweed farming and the culture of brackishwater and marine species of finfish, molluscs and crustaceans.

Fish production through aquaculture is increasing worldwide. At present, aquaculture constitutes about 12% of world fish production, but the trend is expected to accelerate in some countries. Total world production through aquaculture is expected to attain 22 million t by the turn of the century (FAO 1989).

FAO (1990) estimated that 56% of the 14.47 million t of aquaculture production in 1988 came from the brackishwater and marine environment. Over 90% of molluscs, crustaceans and seaweeds were derived from coastal aquaculture practices. Ninety-five per cent of the 1988 aquaculture production came from 20 nations, mostly Asian, earning a total of US$20.76 billion.

The levels and patterns of aquaculture practices vary significantly depending on the species cultured, sites and methods used. In finfish and crustacean culture, feeds (trash fish or formulated feeds) are extensively applied in cages, ponds, raceways or tanks. Molluscs and seaweeds are also densely farmed using floating rafts, long lines or stakes located in either the intertidal zone or further offshore. In milkfish (*Chanos chanos*) and seaweed (*Caulerpa* and *Eucheuma* spp.) farming, inorganic fertilizers are used. In Southeast Asia and some East Asian nations, organic manure is traditionally used to fertilize ponds and increase primary productivity either through direct application or through integrated fish and livestock culture. In Asia, most coastal aquaculture operations are small in size, extensive in operation, but densely aggregated, covering extensive areas of coastal land and water.

The need for food, employment and foreign exchange has influenced decisionmakers to opt for short-term economic benefits despite socioeconomic and environmental losses in the long run. Such is the case with coastal aquaculture which has developed rapidly in the last two decades and some countries have already experienced its negative impacts on the environment. In most developed nations, people have a high level of environmental awareness, and strict regulations are imposed on coastal aquaculture operations. This is not the case in most developing nations where environmental laws are inefficiently enforced or inadequate environmental regulations give rise to the widespread, uncoordinated development of aquaculture.

The destructive environmental impact of coastal aquaculture is gaining international attention. The International Council for the Exploration of the Sea (ICES) has established a Working Group on the Environmental Impacts of Mariculture for the North Atlantic region. In the tropics, various research efforts have assessed the level of impacts arising from large coastal aquaculture operations (Kapetsky 1982; Chua et al. 1989). Recently, the Joint Working Group of the Scientific Aspect of Marine Pollution (GESAMP) established a Working Group on Environmental Impacts of Coastal Aquaculture.

This paper explores various management options and policy guidelines for negating or minimizing the adverse environmental impact of coastal aquaculture practices and their future expansion.

Environmental Impact of Coastal Aquaculture

Coastal aquaculture has given rise to major environmental concerns, particularly by destroying coastal habitats, altering or impoverishing local biodiversity, affecting coastal communities and, in some cases, posing serious threats to human health. Table 1 summarizes the various environmental concerns brought about by coastal aquaculture. There have been efforts to appraise the environmental impact of various aquaculture practices (Chua et al. 1989; Folke and Kautsky 1989; Ackefors and Enell 1990). The ICES Working Group on Environmental Impacts of Mariculture has undertaken periodic reviews on these aspects and its appraisals have appeared in various reports (for example, ICES 1988, 1989; Rosenthal et al. 1988).

Table 1. Main environmental concerns caused by the activities of and farm discharges from coastal aquaculture practices.

Aquaculture systems	Activities and farm discharges										Environmental concerns						
	Wetland conversion	Introduced species	Wastes and fecal matters	Uneaten food	Organic fertilizers	Inorganic fertilizer	Drugs	Disinfectants	Pesticides	Antifoulants	Nutrient and organic enrichment	Toxicity to marine life	Antibiotic resistance in pathogens	Impoverishment of biodiversity	Human health hazards	Habitat destruction	Sediment accumulation
Pond	x	x	x	x	x	x	x	-	x	-	+	+	+	+	+	+	+
Tank	-	x	x	x	-	-	x	x	-	-	+	-	-	-	-	-	-
Raceway	-	x	x	x	-	-	x	x	-	-	+	-	-	-	-	-	-
Cage	-	x	x	x	-	-	x	-	-	x	+	+	+	+	-	-	+
Pen	-	x	x	x	-	-	-	-	-	-	+	-	+	+	-	+	+
Raft	-	x	x	x	-	-	-	-	-	-	+	-	-	+	-	-	+
Longline	-	x	x	x	-	-	-	-	-	-	+	-	-	+	-	-	+
Stake	-	x	x	x	-	-	-	-	-	-	+	-	-	+	-	+	+
Bottom	-	-	x	x	-	-	-	-	-	-	+	-	-	+	-	-	-
Hatcheries	-	x	x	x	-	x	x	x	-	-	+	-	-	-	-	-	-

x direct contribution to coastal water
\- none
\+ contribution to stated environmental concern

Destruction of Wetlands

Large areas of pristine tropical wetlands, predominantly mangrove swamps, have been cleared and converted to brackishwater fishponds. The boom of the shrimp farming industry in the 1970s and 1980s further contributed to rapid and unregulated clearing of mangroves. Over the last few decades, the Philippines has lost 75% of its mangroves, mostly to coastal aquaculture; while Ecuador, Indonesia, Malaysia, Thailand, and many other tropical developing nations have lost substantial areas (Snedaker et al. 1986; Terchunian et al. 1986; Olsen and Arriaga 1989; Paw and Chua 1991).

The process of converting mangrove areas to fish or shrimp ponds often releases water with low pH resulting from the oxidation of acid sulfate soils into the vicinity due to tidal flushings or leaching from newly constructed mud ponds during heavy rainfall (Poernomo and Singh 1982; Simpson et al. 1983). In their analysis of the environmental impact of mangrove conversion to fish or shrimp ponds, Chua et al. (1989) stated that, "... fish kills and the creation of bitter tasting river water have been attributed to the dissolution of sulfuric materials from areas with extensive acid sulphate soils... Although acute effects are localized and of short duration, chronic exposure leads to stress due to ionic imbalance in fish and possibly soft-shelling in shrimps resulting in low resistance to diseases and parasites ...."

Large-scale mangrove clearing endangers the survival of some endemic species; it destroys nursery grounds and other critical habitats; and affects the foodchain of the estuarine ecosystem and the habitats of mangrove terrestrial wildlife (UNESCO 1986; Paw and Chua 1991). Mangrove conversion to fishponds, if done at the shorefront, may also severely erode the coastline and increase the sediment load of coastal waters.

Increased Nutrient Load

Coastal aquaculture practices contribute to the enrichment of the coastal waters where farms are located. Uneaten food and residued organic or inorganic fertilizers are constantly added to the marine environment.

In intensive and semi-intensive shrimp farms, 10-30% of the pond water is exchanged each day either through tidal energy or pumping. A substantial amount of loose sludge, which is the most concentrated source of nutrient contaminants from shrimp and milkfish ponds, is discharged to coastal waters during harvests. These organic or inorganic substances enter neighboring waters or are deposited as the bottom sediments near the farming units. Unfortunately, there is very little quantitative data to assess accurately the environmental impact of nutrients released from coastal aquaculture installations, especially in the developing nations. Some estimates have been made in Sweden where the overall nutrient load from aquaculture is negligible compared to other nutrient sources, but local effects may be significant (Ackefors and Enell 1990).

Increased nutrient loads in the coastal environment may cause a significant change in the nutrient cycle (Folke and Kautsky 1989). Studies in Europe on cage culture of salmonids demonstrated a high sedimentation rate below the cage farms which often led to the build-up of anoxic sediments, increased levels of hydrogen sulfide, depletion of oxygen at the bottom and increased bacterial populations which affected macrobenthic organisms (Rosenthal et al. 1988; Folke and Kautsky 1989). The sedimentation rate from raft culture of mussels is comparatively lower primarily because of the filter-feeding habits of the cultured organisms and the absence of artificial feeds. But anoxic sediments do accumulate below large mussel farms with a substantial increase of sulfur bacteria (Folke and Kautsky 1989).

Soluble waste in the water column alters the natural composition of macro- and micronutrients, especially when large amounts of waste are discharged from intensive and semi-intensive fish and shrimp farms. Nutrients and organic wastes from aquaculture, together with other sources of eutrophication, may change the composition and abundance of phytoplankton populations (Takahashi and Fukazawa 1982; Folke and Kautsky 1989) and may stimulate or enhance blooms of harmful dinoflagellates or other phytoplankton species. Toxic algal blooms are associated with mass fish kills in many regions of the world (Doyle 1988; Maclean, this vol.). Ensuing changes in the number and type of the bacteria populations may increase the risk of disease outbreaks in cultured and wild organisms. The outbreak of Hitra disease in salmon farms in Norway, caused by *Vibrio salmonicida*, has been attributed to the self-polluted nutrient-enriched fishfarm environment (Folke and Kautsky 1989). Intensive shrimp farming in Thailand and Taiwan has caused serious disease outbreaks in recent years (C. Kwei Lin, pers. comm.; Liao 1990).

Toxicity to Cultured and Wild Organisms

Many chemicals are used in aquaculture for disease chemotherapy, disinfection, anesthesia, growth promotion and pest control (ICES 1988, 1989; Bailey and Jeffrey 1989; Meyer and Schnick 1989). Some, if abused or improperly applied, can be toxic to cultured and wild organisms.

Recent studies of tributyl tin (TBT), an antifoulant used in net-cages and small boats, have shown a high toxicity to crustacean larvae. Salmonid broodstock in treated cages showed a high incidence of reproductive failures. TBT is also persistent in marine sediments (ICES 1988). TBT was used extensively in cage farms in Europe but is now banned because of its toxicity. In most developing nations, antifoulants are not popularly used in aquaculture due to high costs and so it has not become an environmental issue. But TBT is still extensively used in ports and harbors in many parts of the world.

Pesticides are presently used in fishponds and net-pens to eradicate predators and other unwanted species. Some of these chemicals, such as Neguvon (trichlorfon) and Nuvan (dichlorvos), as presently employed in salmon net-pen culture, can kill crustaceans living in the general vicinity of the farm site. Plant-based biodegradable pesticides, such as tobacco dust (nicotine), tea seed cake (saponin) and Derris root extract (rotenone) and organic pesticides such as Gusathion (organophosphate), Brestan and Aquatin (organo-tin), are popularly used in most Asian nations in both fish and shrimp farms. The success of application of these pesticides depends on the experience and knowledge of fish farmers. While there are occasional reports of localized fish kills due to the untimely discharge of treated pond waters, there are no documented reports of major fish kills caused by the application of pesticides in brackishwater ponds. Most of the documented pesticide poisonings of aquatic organisms have been attributed to the impacts of agricultural runoff (Chua et al. 1989).

Antibiotic Resistance

Antibiotics are extensively used in aquaculture, especially in hatcheries and in semiintensive and intensive aquaculture operations involving high-market value commodities. Most are incorporated into feeds. In Norway alone, salmon culturists used close to 19.5 t of antibiotics in producing 45,000 t of salmon (Rosenthal et al. 1989). There are no statistics as to the amount and type of antibiotics used in coastal aquaculture practices elsewhere, but the quantity must be very large in view of intensive farm operations for salmon, eel, seabreams, seabass, yellowtail, tunas, mackerel, groupers and shrimp in various parts of the world. While there is insufficient knowledge to quantify the environmental impact of antibiotics, the environmental concerns are:

- the continued use of antibiotics and/ or their persistence in sediments could lead to the proliferation of antibiotic-resistant pathogens and this may complicate disease treatments (ICES 1988);
- antibiotics are transferred to wild fish in the vicinity of the cage/pen farms using medicated feeds. Shellfish may accumulate antibiotics in their tissues; and
- the presence of antibiotics in bottom sediments may affect natural bacterial decomposition and hence influence the ecological structure of the benthic microbial communities.

Impoverishment of Biodiversity

The transfer of species for aquaculture may alter or impoverish the biodiversity of the marine ecosystem through competition and interbreeding. Introduced species could escape to the wild and establish a self-sustaining reproducing population, thereby competing with and threatening some indigenous species. There are reports of salmonids which escaped from cages, negatively affecting indigenous fish species through predation, competition and habitat destruction (Folke and Kautsky 1989).

The introduction of disease-causing microorganisms from imported species can be harmful. Salmon smolts from Sweden introduced the trematode parasite to Norway, seriously infecting the natural salmon stocks in several rivers in Norway. Mussels in large-scale farming in the coastal lagoons, bays or inlets may compete heavily with other native filter feeders for planktonic food organisms and thus seriously affect their recruitment (Kaspar et al. 1985).

Socioeconomic and Health Implications

Coastal aquaculture may have serious socioeconomic and health implications. Large-scale mangrove conversion into fish and shrimp farms has displaced rural coastal communities which traditionally depended on mangrove resources for their livelihood.

About 30% of the shrimp ponds in Ecuador are either abandoned or not utilized. Over 90% of shrimp ponds in the Upper Gulf of Thailand, most of which are converted from mangrove areas, were abandoned after two growout seasons (C. Kwei Lin, pers. comm.). National governments have extensive programs to replant cleared mangrove areas. The irony is that some of these nations obtained loans from multinational banking institutions to convert their mangrove wetlands into shrimp farms and now they are borrowing again from the same banking institutions to replant the mangroves.

Economic losses from disease outbreaks caused by self-pollution in a nutrient-enriched fish farm environment can be considerable. The Hitra disease outbreak in Norway cost the growers 20% of the annual economic value of the Norwegian salmon farming industry.

The economic disaster of the shrimp industry in Taiwan in 1988 and 1989, in which shrimp production plunged from 90,000 t in 1987 to 50,000 t in 1988 and

20,000 t in 1989, shows the magnitude and severity of the social and economic implications of self-pollution from shrimp farming operations. While diseases were identified as the primary causative agent, excessive stocking densities, poor feeding practices, inadequate farm management as well as the indiscriminate use of drugs and other chemicals in pond preparation and disease treatments, all contributed to the stress on the culture organisms (Liao 1990). This created an environment conducive to the spread of pathogens. Millions of dollars were lost and recovery will take a considerable time. In Ecuador, climatic change and deterioration of water quality contributed to the decline of shrimp production from 70,000 t in 1988 to 45,000 t in 1989, although urban sewage, agricultural pesticides, fertilizers and other causes have been blamed.

Tropical shrimp culture has been found to have largely negative social consequences because its benefits in terms of profits or protein supply do not favor the coastal residents, most of whom are unskilled laborers (Bailey 1988). The costs of disrupting the coastal ecosystem, according to Bailey, include coastal erosion, saltwater intrusion into groundwater and agricultural fields, acidification and a reduction in supply of a wide range of valuable goods produced from the resources available in mangrove forests.

Coastal aquaculture's nutrient and organic loads added to the already nutrient-rich coastal waters in the tropics could have contributed to the frequency and intensity of algal blooms. While red tides may be caused by a variety of factors, nutrient enrichment may be one of the major causes (Doyle 1988; Maclean, this vol.). Paralytic shellfish poisoning (PSP) has been reported in areas affected by red tides. In Southeast Asia, close to 60 reported fatalities and more than 1,359 people have suffered from PSP due to outbreaks of *Pyrodinium* red tides (Hallegraeff and Maclean 1989). The economic impact of red tide outbreaks is also severe, particularly on the shellfish industry. A ban is usually imposed on marketing shellfish and fish from red tide-infested areas. Economic losses due to chronic or acute red tide outbreaks have been reported in various nations and these amount to millions of dollars (Maclean 1989). For example, in 1972, *Chattonella* red tides destroyed over 71 billion yen worth of farmed yellowtail; in 1971, red tide-related losses to the tourist industry in Florida, USA, reached US$18 million. An estimated loss of US$500,000 was reported in the Philippines during a red tide outbreak in 1989 (Maclean 1989).

Aquaculture products from culture sites with high *Escherichia coli* counts are not suitable for human consumption. Oyster, cockles and mussels cultured in highly polluted bays and coastal waters often cause serious gastrointestinal diseases resulting in the rejection of shellfish for exports. Newspaper reports on the presence of hepatitis virus in cockles have badly affected the consumption of cockles (*Anadara granosa*) in Malaysia and Singapore, thus, seriously inhibiting the expansion of the cockle culture industry.

There are yet no documented reports of serious economic consequences arising from the extensive use of drugs and other chemicals in coastal aquaculture. However, several chemicals used in aquaculture have been found to be potentially toxic or carcinogenic and thus pose health hazards to consumers. While the USA and a number of developed nations impose stringent regulatory measures on the use of chemicals in aquaculture (Meyer and Schnick 1989), many developing nations have none or only limited regulatory control. Hence, several chemicals banned for agriculture and aquaculture use in the USA, such as chloramphenicol, nitofurazone and TBT, are still being used in developing countries.

Policy Guidelines for Environmental Management of Coastal Aquaculture Practices and Their Development

Sustainable Development of Aquaculture

Aquaculture should be developed in the context of wise resource use and the environment. The ultimate goal of sustainable development is to develop natural resources in a manner that ensures a sustained increase in the level of societal and individual welfare (Dixon and Fallon 1989). As the Brundtland Commission (WCED 1987) put it, "Sustainable development implies the ability to meet the needs of the present without compromising the ability of the future generations to meet their own needs." Natural resources available for coastal aquaculture development are best utilized in a way that provides sustained or increased benefits to the community or individual who owns them. In the case of privately owned resources, the choice of aquaculture is usually driven by profit incentives and this economic activity may be changed depending on market forces.

In planning the best use of common property resources, the main beneficiaries should first be identified and their desired benefits determined when allocating the natural resources for aquaculture development in line with national interest. In other words, if a certain natural resource is found to be best developed for aquaculture which could provide economic and social benefits to the targeted community, appropriate development opportunities should be created to allow members of the targeted community to develop fully the resource under their control for their own welfare.

The long-term objectives of aquaculture development can be achieved through: (a) adequate planning and efficient implementation of aquaculture development and management programs based on the concept of sustainable development; (b) effective management of aquaculture practices; and (c) enforcement of mitigating and monitoring measures for any adverse environmental impact.

In formulating aquaculture development and management programs, just as in any natural resource development program, it is important to recognize limitations imposed by the present state of technology, social organizations, natural resources and by the ability of the biosphere to absorb the effects of human activities as emphasized in the report of the Brundtland Commission.

Development of Coastal Aquaculture Within the Framework of Coastal Area Planning and Management

Coastal aquaculture is one of the activities within the coastal environment and, therefore, its operations have direct or indirect effects on the marine ecosystem and other resource-based economic sectors operating in the area. As far as possible, coastal aquaculture practices should be carried out with minimal adverse impacts on the coastal and marine environment through adequate monitoring and mitigating measures as well as minimal conflicts in multiple-resource use. However, when conflicts arise, it may become necessary to consider tradeoffs associated with development alternatives. As such, coastal aquaculture should be developed within the overall framework of coastal area management. The coastal area is fragile and has an economically important resource system which is subject to pressures from increasing population and intense economic activities by various sectors. Therefore, the development of the coastal area requires a multisectoral, integrated and holistic approach in order to ensure wise and judicial use of the scarce common property resources.

National Policy and Aquaculture Planning and Management Program

A government's priorities for aquaculture development should be indicated in its national policy, but more importantly, the government should have a comprehensive, nationwide planning and management program for aquaculture development. Such a program should allocate specific areas or zones for aquaculture development taking into consideration socioeconomic viability and environmental sustainability and possible mitigating measures to reduce environmental impacts. Such a program should provide sufficient institutional supervision and monitoring of aquaculture operation procedures, the application of environmental impact assessment (EIA) and the implementation of regulatory measures.

Many developing countries currently promote aquaculture development, but lack comprehensive aquaculture development planning and management programs. Coastal aquaculture development in many countries appear largely driven by short-term economic incentives as exemplified by the large-scale development of shrimp farming.

To guide aquaculturists in the conduct of aquaculture operations and to safeguard the health of consumers, specific guidelines are needed on the use of chemicals, site selection, mangrove and other land use conversion, transfer of exotic species, intensity of farming operations, product quality control, etc. Such guidelines require constant review, refinement and updating.

Similarly, donor agencies must be aware of the possible environmental consequences of coastal aquaculture if it is not properly planned and managed. Financial intervention by external sources has been a major factor in determining aquaculture development in developing nations, especially as a source of foreign exchange.

Management Guidelines and Strategies

The purpose of management guidelines and strategies is to minimize the adverse environmental effects of coastal aquaculture either through preventive measures in the planning stage or through mitigating and regulatory measures when the culture practices are already in operation.

Individual, small-scale aquaculture practices may have little or no significant impact on the ecosystems. Traditional shrimp ponds operating within mangrove swamps using enclosed mud bunds and sluice gates as seen in Malaysia, Singapore and Thailand before the 1950s are good examples. Aquaculture practices have become an environmental concern only as the result of their expansion and intensification of culture operations or inputs (Chua and Paw 1987).

The main environmental concerns caused by the activities of the farm discharge from extensive coastal aquaculture practices and intensive scale of operation are summarized in Table 1. Most of the environmental issues are associated with pond and cage culture systems. Other culture systems such as tanks, raceways and hatcheries may also contaminate the coastal water if their discharges are not regulated. Because intensive culture systems involve treatment of water, farm discharge can be easily controlled. Sea-based culture practices include the farming of filter-feeding organisms and seaweeds using raft, long line, stake or bottom method. The main environmental concern is associated with sedimentation, particularly in the case of oyster and mussel culture.

Site selection is critical in ensuring a healthy culture environment. It is always important to determine the environmental conditions of any given site for aquaculture so that the intensity of farms and their level of operation can be determined. This is especially important in large-scale or extensive cultivation of filter-feeding molluscs because of the dependency of the cultured organisms on the primary productivity of the culture site. Intensive systems with a high degree of water management should incorporate waste treatment processes.

The choice of species for culture is normally based on economic, technical and social criteria. Whenever possible, polyculture of seaweeds with carnivorous finfish or shrimps or integrating the culture of filter-feeders with carnivorous species in cages or pens may help reduce sedimentation rate and nutrient loads. Folke and Kautsky (1989) argue that the net effect of the mussel culture cycle will always be that nutrients are removed from the environment at harvest. Mussel culture therefore counteracts eutrophication. The massive cultivation of mussels, oysters and cockles in the Ban Don Bay of Southern Thailand, in fact, might have contributed to nutrient reduction and reduced the primary productivity of the Bay which receives a considerable amount of organic loads from nearby urban centers. Folke and Kautsky (1989) further proposed to integrate mussel culture with cage farming. The mussels could filter the uneaten feed fragments from cage culture as well as other microorganisms filtered through the same process. The converted mussel meat can then be recirculated as fish feed.

Wetland Conversion

The use of mangrove wetlands for brackishwater finfish and shrimp culture should be regulated because huge expenses are incurred in wetland conversion and in remedial measures to mitigate coastal erosion, loss of habitats and other ecological damage. Shorefront and basin mangroves should be protected and preserved not only for its ecological value but to control erosion. Intensive shrimp farming can be sited in nonmangrove lowlands and use marine pumps to draw water from nearby estuaries. Except for the relatively low lease charges, mangrove swamps are not ideal sites for shrimp culture because of their acid sulfate soils and high levels of sedimentation.

If mangrove lands were assessed for their true value, many shrimp farmers would operate outside mangrove areas. A number of tropical countries where mangroves abound have established a National Mangrove Committee to determine the wise use of this scarce wetland resource. Although they function as advisories, the National Mangrove Committees have contributed immensely by guiding their respective governments in the judicious use of mangrove wetlands. The guidelines for mangrove management prepared by the International Union for the Conservation of Nature (IUCN), East-West Center at the University of Hawaii and the United Nations Educational, Scientific and Cultural Organization (UNESCO) (Hamilton and Snedaker 1984) should be considered in allocating mangrove resources for coastal aquaculture development.

The Use of Chemicals

Most information about the environmental effects of chemicals used in aquaculture comes from studies undertaken in North Atlantic region and is based on studies of cage culture of salmon in Canada, Norway, Sweden and other western European nations. The ICES Working Group on the Environmental Impacts of Mariculture has already begun comprehensive documentation of the chemicals and quantities used in fishfarming in the ICES nations. The United States Food and Drug Administration (FDA) has published a list of chemicals registered or approved for use in food fish culture (Schnick 1991). Some drugs used in many developing nations are not registered or banned from use by the FDA. In most developing nations where controls for drugs used in fish culture are less stringent, abuses are frequent.

Species Transfers for Aquaculture

Fish transfers have been well discussed by the ICES/EIFAC Working Group on the Introduction of Exotic Species. The "Code of Practice on the Introduction and Transfer of Non-indigenous Species," which is being adopted by the EIFAC nations, can be adopted (modified as necessary) worldwide.

The spread of diseases through species transfers poses great environmental risk, the consequences of which may be widespread and irreversible (Rosenthal et al. 1988). Strict regulatory control is necessary with respect to species transfers, especially the introduction of new exotic species, including evaluation of the potential effects of the introduction of diseases. The procedures outlined under the Code of Practice, though laborious and time-consuming, are necessary. However, the main problem is the inability of most nations to enforce screening measures.

Mitigating Measures

1. Replanting mangroves - In some nations, such as the Philippines, Thailand and Malaysia, efforts are being made to replant mangroves along shorefronts and river basins. While it is impossible to replace the original pristine mangroves, the replanted mangroves will replace much of the lost mangrove ecosystem and its essential functions.
2. Changing culture sites - One viable way of avoiding or reducing self-pollution impacts in cage culture is to resite the farm after operating for several years. This is possible with floating cage farms. In some bays where cage farms are densely distributed, site rotation becomes more difficult and moving to deeper waters would incur high costs. Technically, site rotation and operating in deeper waters using deepwater cages, rotating or submerged cages are possible. This culture method helps reduce nutrient loads while promoting growth due to cleaner water and healthier conditions. Moving floating cage farms or rafts away from red tide-infested areas may be necessary to avoid mass mortality or contamination.
3. Reducing the sinking rate of feeds and manipulating the feeding regime - A substantial amount of feeds used in cages is lost and attributed to sinking, inaccurate estimation of the fish biomass in the cages or overfeeding. The amount of feeds lost could be reduced by formulating feeds with slower sinking rates and by developing appropriate methods to assess fish biomass accurately so that the right amount of feed is given. An understanding of the feeding habits and digestion rate of the cultured species is also useful to determine accurately the feeding rate for optimal growth and conversion.
4. Removing organic sediments - This measure has been used in Europe to remove mud underneath the cages using siphons operated by divers. However, the removed sludge is often released into the nearby marine environment. In shrimp or milkfish ponds, the

loose sludge is pumped and discharged into nearby estuaries, thus distributing the sediments and organic substances into the nearby waterbodies. The sludge should be deposited on land, for example, on mud bunds of the pond.

5. Reducing the nutrient load - Seaweeds such as *Gracilaria* and *Caulerpa* can be polycultured with shrimp or milkfish in ponds or cultivated in the pond's exit canal to reduce the nutrient load of the water entering the sea.
6. Removing suspended solids from ponds - Uneaten food particles, fecal matter and other solids suspended in the water column in fish and shrimp ponds can be removed by precipitation in settling tanks or by culturing filter-feeding organisms, such as oysters or mussels. This form of biological filter could be fairly effective in reducing sediment loads in ponds or tanks.

Regulation and Monitoring

Regulatory measures are needed to license aquaculture farms. This is to enable the government authorities to regulate the number and size of farms and their distribution in a particular location. Licenses are usually valid for one year but renewable upon application.

In the ICES member countries, the farmers themselves regularly monitor environmental parameters and farm operations to determine the trends of environmental changes over time. This is especially important for cage and pen culture wherein the technical and economic feasibility of the culture site must be assessed for continued operation. A government program should be established to monitor regularly the spatial and temporal changes of environmental parameters on a regional basis. The types of parameters to be monitored depend on the culture system used, but the essential ones include: biological oxygen demand (BOD), nitrogen and phosphorus content, suspended solids, hydrogen sulfide, grain size of solids, water temperature, salinity, dissolved oxygen, primary productivity (chlorophyll content, phytoplankton biomass) and coliform bacteria count.

Pollution indicators should be used to monitor specific signs of organic pollution. For floating raft or longline culture, benthic macrofauna underneath the culture systems should also be monitored to detect any significant structural changes. The monitoring program should also include sampling time, frequency and number of samples to be taken, standardized sampling and analytical procedures.

Education and Training

Not all fish farmers understand the value of environmental management, that it is one way to ensure continued smooth and sustainable operation. The level of technical knowledge of fish farmers needs upgrading and improvement, especially with regard to the application of modern technologies. Both education and training should form an integral part of national aquaculture development and management programs.

Research

Scientific research should be an inseparable component of an environmental management program. Accurate and up-to-date information is needed to provide the scientific basis for environmental management. Research has been carried out in the ICES member countries. The main focus, however, has been on elucidating the effects of aquaculture operations on the seabed and nutrient input in the water mainly through cage culture. Other major studies cover the environmental consequences of chemical usage in coastal aquaculture, particularly, antibiotic resistance and the accumulation of antibiotics in nontarget organisms. These data will also be useful for developing management guidelines for coastal aquaculture. The scope of research, however, should be expanded to cover the environment absorptive and carrying capacity of coastal waterbodies and culture system, respectively; the strengthening of existing knowledge to quantify the effects of aquaculture inputs on the environment and to determine the socioeconomic and environmental costs of coastal aquaculture. Research is also needed to determine the environmental impact of coastal aquaculture in both the developed and developing nations.

References

Ackefors, H. and M. Enell. 1990. Discharge of nutrients from Swedish fish farming to adjacent sea areas. Ambio 19: 28-35.

Bailey, C. 1988. The social consequences of tropical shrimp mariculture development. Ocean Shoreline Manage. 11:31-44.

Bailey, T.A. and S.M. Jeffrey. 1989. Evaluation of 215 candidate fungicides for use in fish culture, p. 1-10. *In* Investigations in fish control. US Fish and Wildlife Services, National Fisheries Research Center, Washington.

Chua, T.E. and J.N Paw. 1987. Aquaculture development and coastal zone management in Southeast Asia: conflicts and complementarity, p. 2007-2021. *In* O.T. Magoon, H. Converse, D. Miner, L.T. Tobin, D. Clark, G. Domurat (eds.) Proceedings of the 5th Symposium on Coastal and Ocean Management. Vol.2. American Society of Civil Engineers, New York.

Chua, T.E., J.N. Paw and F.Y. Guarin. 1989. The environmental impact of aquaculture and the effects of pollution on coastal aquaculture development in Southeast Asia. Mar. Pollut. Bull. 20: 335-343.

Dixon, J.A. and L.A. Fallon. 1989. The concept of sustainability: origins, extensions and usefulness for policy. Soc. Nat. Resour. 2: 73-84.

Doyle, J. 1988. Health risks in aquaculture from marine toxins (red tides), p. 167-175. *In* E. Grimaldi and H. Rosenthal (eds.) Efficiency in aquaculture production: disease control. Proceedings of the Third International Conference in Aquafarming "Acquacoltura 86," 9-10 October 1986. Verona, Italy.

FAO (Food and Agriculture Organization). 1988. Definition of aquaculture. IPFE/WPA/WP2, 3 p.

FAO (Food and Agriculture Organization). 1989. Review of developments in aquaculture. Proceedings of the 18th Session of the Committee on Fisheries, 10-14 April 1989, Rome. COFI/89/Inf. 5, 15 p.

FAO (Food and Agriculture Organization). 1990. Aquacult. Min. No. 8, 4 p.

Folke, C. and N. Kautsky. 1989. The role of ecosystems for a sustainable development of aquaculture. Ambio 18: 234-243.

Hallegraeff, G.M. and J.L. Maclean, Editors. 1989. Biology, epidemiology and management of *Pyrodinium* red tides. ICLARM Conf. Proc. 21, 286 p.

Hamilton, L.S. and S.C. Snedaker, Editors. 1984. Handbook for mangrove area management. United Nations Educational, Scientific and Cultural Organization, Paris, International Union for the Conservation of Nature, Geneva and East-West Center, Honololu, Hawaii. 123 p.

ICES. 1988. Report of the Working Group on environmental impacts of mariculture. International Council for the Exploration of the Sea. CM 1988/F:32, 91 p.

ICES. 1989. Report of the Working Group on environmental impacts of mariculture. International Council for the Exploration of the Sea. CM 1989/F:11, 70 p.

Kapetsky, J.M. 1982. Some potential environmental effects of coastal aquaculture with implications for site selection and aquaculture engineering. SCS/82/CFE/CP-23, p. 76-82. FAO/UNDP-SCSP Consultation/Seminar on Coastal Fishpond Engineering, 4-12 August 1982, Surabaya, Indonesia.

Kaspar, H.F., P.A. Gillespie, I.C. Boyer and A.L. MacKenzie. 1985. Effects of mussel aquaculture on the nitrogen cycle and benthic communities in Kenepuru Sound, Marlborough Sound, New Zealand. Mar. Biol. 85: 127-156.

Liao, I.C. 1990. Aquaculture in Taiwan, p. 345-369. *In* M.M. Joseph (ed.) Aquaculture in Asia. Asian Fisheries Society, Indian Branch, Mangalore.

Maclean, J.L. 1989. Economic aspects of *Pyrodinium* red tides in the Western Pacific, p. 179-185. *In* G.M. Hallegraeff and J.L. Maclean (eds.) Biology, epidemiology and management of *Pyrodinium* red tides. ICLARM Conf. Proc. 21, 286 p.

Meyer, F.P. and R.A. Schnick. 1989. A review of chemicals used for the control of fish diseases. Aquat. Sci. 1: 623-710.

Olsen, S. and L. Arraiga. 1989. A sustainable shrimp mariculture industry for Ecuador. International Coastal Resources Management Project Tech. Rep. TR-E-6, 276 p. University of Rhode Island, Rhode Island.

Paw, J.N. and T.E. Chua. 1991. An assessment of the ecological and economic impact of mangrove conversion in Southeast Asia, p. 201-212. *In* L.M. Chou, T.E. Chua. H.W. Khoo, P.E. Lim, J.N. Paw, G.T. Silvestre, M.J. Valencia, A.T. White and P.K. Wong (eds.) Towards an integrated management of tropical coastal resources. ICLARM Conf. Proc. 22, 455 p.

Poernomo, A. and V.P. Singh. 1982. The problems, field identification and practical solutions of acid sulfate soils for brackishwater fishponds, p. 49-61. *In* Report of Consultation/Seminar on Coastal Fishpond Engineering, 4-12 August 1982, Surabaya, Indonesia. South China Sea Fisheries Development and Coordinating Programme, Manila, Philippines, SCS/GEN/82/42.

Rosenthal, H., D.P. Weston, R.J. Gowen and E.A. Black. 1988. Report of the ad hoc study group on environmental impact of mariculture. ICES (International Council for the Exploration of the Sea) Coop. Res. Rep. No. 154, 81 p.

Schnick, R.A. 1991. Chemicals for worldwide aquaculture, p. 441-467. *In* Fish health management in Asia-Pacific. Report on a regional study and workshop on fish disease and fish health management (ADB/NACA). ADB Agricult. Dept. Rep. Ser. No. 1. Bangkok.

Simpson, H.J., H.W. Ducklow, B. Deck and H.L. Cook. 1983. Brackishwater aquaculture in pyrite-bearing tropical soils. Aquaculture 34: 333-350.

Snedaker, S.C., J.C. Dickinson III, M.S. Brown and E.J. Lahmann. 1986. Shrimp pond siting and management alternatives in mangrove ecosystems in Ecuador. Prepared for the US Agency for International Development, Miami, Florida. 77 p.

Sorensen, J.C., S.T. McCreary and M.J. Hershman. 1984. Institutional arrangements for management of coastal resources. Prepared by Research Planning Institute, Inc., Columbia, South Carolina, for National Park Service, US Department of the Interior, Washington, DC. 165 p.

Takahashi, M. and N. Fukazawa. 1982. A mechanism of "red-tide" formation. II-Effect of selective nutrient stimulation on the growth of different phytoplankton species in natural water. Mar. Biol. 70: 267-273.

Terchunian, A., V. Klemas, A. Segovia, A. Alvarez, B. Vasconez and L. Guerrero. 1986. Mangrove mapping in Ecuador: the impact of shrimp pond construction. Environ. Manage. 10: 345-350.

UNESCO (United Nations Educational, Scientific and Cultural Organization). 1986. Mangroves of Asia and the Pacific: status and management. UNDP/UNESCO Research and Training Pilot Program on Mangrove Ecosystems in Asia and the Pacific Tech. Rep. Natural Resources Management Center and National Mangrove Committee, Ministry of Natural Resources, Manila. 538 p.

WCED (World Commission on Environment and Development). 1987. Our common future. Oxford University Press, Oxford.

Environmental Impact of Tropical Inland Aquaculture

M.C.M. Beveridge
M.J. Phillips*
Institute of Aquaculture
University of Stirling
Stirling FK9 4LA
Scotland, UK

BEVERIDGE, M.C.M. and M.J. PHILLIPS. 1993. Environmental impact of tropical inland aquaculture, p. 213-236. *In* R.S.V. Pullin, H. Rosenthal and J.L. Maclean (eds.) Environment and aquaculture in developing countries. ICLARM Conf. Proc. 31, 359 p.

Abstract

The environmental impact of tropical freshwater aquaculture is examined in terms of demand for resources (land, water, feed, construction materials) and waste outputs (fecal and urinary wastes, uneaten food, therapeutants and other chemicals, pathogenic microorganisms and parasites, feral animals). Sadly, there have been few studies of relevance carried out.

Problems of resource use appear to be highly localized and, in many cases, reversible. Although intensive production without water reuse or treatment can result in eutrophication of receiving waters, there are as yet few problems. Future legislation should be directed towards protection of receiving waters rather than effluent waste concentrations. Misuse of chemicals and therapeutants, accidental and otherwise, is fairly widespread, particularly among more profitable sectors of the industry.

It is recommended that guidelines be produced to help decisionmakers assess the risks to the environment posed by unrestricted movement of species and transgenic organisms.

Introduction

This paper sets out to review the environmental impact of tropical inland aquaculture. That aquaculture might possibly have adverse impacts in tropical regions has only begun to be recognized, let alone be the subject of research, and inevitably, the review draws heavily on results from studies carried out in northern Europe and North America. The emphasis is deliberately on the technical rather than on economic, social or planning aspects and dwells but briefly on subjects such as integration of aquaculture with agriculture and disease as these are covered elsewhere in this volume.

Tropical Inland Water Production

In Table 1, available production data on principal tropical freshwater species are shown. The most important group is the carps, accounting for approximately 3.4 million t·year^{-1}. Although production is growing rapidly, tilapias still account for less than 300,000 t·year^{-1} whilst only around 30,000 tonnes of tropical catfishes (i.e.,

*Present address: Division of Agricultural and Food Engineering, Asian Institute of Technology, GPO Box 2754, Bangkok 10501, Thailand

Table 1. Production of important tropical freshwater species in 1987 (Source: FAO 1988).

Species	Production (tonnes)
Carps	
Common carp	927,735
Grass carp	535,691
Silver carp	1,340,718
Bighead carp	631,435
Total	3,435,579
Tilapias	246,399
Catfishes	30,238
Giant river prawn	15,103

excluding *Siluris glanis* and *Ictalurus punctatus*) are grown. Farmed giant prawn (*Macrobrachium rosenbergii*) production totals some 15,000 tonnes per annum.

Although FAO has begun to record statistics on production by system and environment, no reliable statistics are as yet available. Most freshwater fish culture in the tropics, however, is undoubtedly pond-based. Lake-based cages and pens probably account for about 10%, whilst tanks and raceways account for 5% or so of production.

Extensive, semi-intensive and intensive methods can be employed to produce fish. In intensive aquaculture, no supplementary feeds are used and the fish must rely on prevailing, natural productivity. This method is not recommended for ponds, although, by necessity, it is still apparent in the poorest parts of the tropics (e.g., Bangladesh). By contrast, extensive methods have proved economically viable in eutrophic lakes in the Philippines and elsewhere (Beveridge 1987). Most tropical inland aquaculture, however, is semi-intensive, relying on the use of low-protein, generally plant-based, agricultural by-products as feeds. Feeding rates are highly variable and generally empirically derived. Intensive aquaculture is dependent upon the use of high protein, formulated feeds which supply all or virtually all of the animal's nutritional needs. Aeration and automatic feeders are also often employed and as a result production costs are high, limiting culture to high value and carnivorous species (see Shepherd and Bromage 1988 for examples). In some parts of the tropics and subtropics - in areas where water is in short supply or where the growing season is restricted and/or where there is a market for high value products (Japan, Taiwan, Israel) - some intensive inland aquaculture occurs. Although there is a trend towards intensification, the majority of inland aquaculture in the tropics is concerned with low value finfish species and intensive methods, as defined above, are therefore unlikely to become widespread.

Aquaculture and Environmental Impact

All aquaculture developments have an impact on the environment. Prior to assessing impacts it is necessary to be able to identify and quantify inputs and outputs. Aquaculture, like all other forms of production, including agriculture, uses resources; water, land, labor, materials for construction, feedstuffs are all necessary. There are also outputs: not only the fish that are being farmed, but also uneaten food, feces and other waste products as well as excess therapeutants and other chemicals.

Resource Use

Land Use and Space

Few countries have specific legislation dealing with use of land for aquaculture purposes. However, the laws regarding private and public property rights, agricultural development policies and land reform acts, environmental protection policies and national security interests largely control access and use of land for aquaculture (van Houtte et al. 1989).

Not all types of land are suitable. Most freshwater pond farms in the tropics are built on arable land, although marginal areas may be used, particularly where land is in short supply. In some communities where fish farming is firmly established the areas of arable land involved can be substantial. Among rice growers in Central Luzon, Philippines, for example, an average of 20% of farm area is devoted to rice-fish culture (Tagarino 1985)[1].

In the Philippines and Indonesia and elsewhere, productive ricefields have been converted to fishponds. In Taiwan, it is estimated that the land devoted to aquaculture production doubled between 1965 and 1985 to 70,000 ha (Lee 1987). This has had a major impact on land utilization. Whilst in several prefectures - Kaohsiung and Pingtung - much of the land was either marginal or unsuitable for agriculture, some change in the use of prime agricultural land has undoubtedly occurred.

Aquaculture has been involved in land use conflict issues in several parts of the tropics, particularly in highly urbanized or overcrowded areas. In Singapore and Hong Kong, shortages of land have adversely affected production of fish for the aquarium trade (Tan and Siow 1989) whilst in China, shortages of land near reservoirs has led to the development of water-based "cove culture" for fish fingerlings for restocking (Lu 1986).

Water-based aquaculture does not, of course, utilize land. However, it does occupy areas of lakes and rivers, and this can result in competition for resources that is similar in many respects to the competition for land. The best-known example is that of the pen culture industry in Laguna Lake, Philippines, in which rapid and uncontrollable expansion during the late 1970s and early 1980s resulted in an area of some 35,000 ha, equivalent to more than one-third of its surface area, being occupied by milkfish (*Chanos chanos*) pens (Beveridge 1984). Access to fishing grounds and navigation routes used by lakeshore communities was severely disrupted, thus causing a great deal of social tension. The presence of the fishpens in the shallow inshore areas was also believed to have disrupted fish spawning and nursery grounds, thereby depressing fisheries yields.

[1]This is now very much an overestimate - **Eds.**

Water Use

In theory, freshwater aquaculture may use water at almost any stage in the hydrological cycle; in practice, surface waters, which comprise only 0.3% of the earth's water resources, are most commonly used. Much of the water is abstracted from rivers or streams although some may already have been used for irrigation purposes. Groundwater resources are considerably more abundant but, though often locally important, appear to be used to a lesser extent on a global scale (Phillips et al. 1991). Studies of land-based tilapia farms in the Philippines have shown that the majority of ponds rely on irrigation water (39%) although some use pumped groundwater (32%) or surface runoff and springs (29%) (Sevilleja 1985). Industrial or municipal wastewaters appear to be rarely used.

Aquaculture is a net consumer of water and most forms require the use of considerable quantities (Muir and Beveridge 1987; Phillips et al. 1991). Although, in much of the tropics, water use for aquaculture is regulated either specifically or as part of the right to use water for fishing or agricultural purposes (van Houtte et al. 1989), in practice water is often available with few constraints. In areas where water is scarce, conflicts can arise. In Israel, where water has to be paid for, for example, it has been established that in many cases it is more profitable to use the water for crop irrigation than for aquaculture. Thus, many fishponds have been destroyed or deepened to form irrigation or dual purpose fish culture/irrigation reservoirs in which fish culture is restricted to only part of the year (Hepher 1985; Milstein et al. 1989; Sarig 1989a, 1989b).

In Table 2, water uses for various intensities and types of tropical aquaculture are given. Note that the quantities used per tonne of production vary enormously since they are determined not only by seepage and evaporative losses but also by intensity of production (i.e., stocking density and the use of feeds and fertilizers and management). Krom et al. (1989) attempted to categorize production systems on the basis of flow regimes. Low flow, extensive or semi-intensive types, defined as systems in which water is added simply to counteract evaporative and seepage losses, predominate in the tropics. Seepage from ponds varies by at least a factor of 10 up to 2.5 $cm \cdot day^{-1}$, depending upon soil types and pond surface area (Boyd 1979). Evaporative losses may be as great as 2.5 $cm \cdot day^{-1}$, although in the subtropics it is more typically around 0.5 $cm \cdot day^{-1}$ (Huet 1972; Hepher and Pruginin 1981; Teichert-Coddington et al. 1988). If one assumes total losses from evaporation and seepage of 1-2 $cm \cdot day^{-1}$ as typical of tropical conditions, then each ha of fishpond will "consume" 100-200 m^3 water per day. Total water requirements for ponds have been estimated by Hepher and Pruginin (1981) to vary between 35 and 60,000 $m^3 \cdot ha^{-1} \cdot year^{-1}$ in order to maintain an average water depth of 1.5 m throughout the growing season (240 days) and counteract losses estimated at between 1 and 2 $cm \cdot day^{-1}$.

Intensive production of some species tolerant of exceptionally poor water quality (e.g., catfish) can occur with little increased water use (Colman and Jacobson 1991). In most instances, however, intensification of production requires increased water use in order to maintain water quality (Krom et al. 1989). Intensive, high flow-rate freshwater systems do occur in the tropics, although they are rare. Such systems can be open (i.e., flow-through) or closed (some recycling occurs). Most of these systems occur in urban areas or parts of the world in which water is scarce (Table 2) (Hepher 1985; van Rjin et al. 1986; Krom et al. 1989).

The uncontrolled abstraction of water may adversely affect conditions in surface waters by:

- changing channel shape and patterns of sedimentation, affecting siltation and water movement;

Table 2. Water requirements in tropical freshwater fish farming. The estimated values for extensive, semi-intensive and intensive pond culture assume mean pond depth of 1 m during growing season, water losses of 1-2 $cm \cdot day^{-1}$ and 350 days production per year (see text).

Species and system	Production ($t \cdot ha^{-1} \cdot year^{-1}$)	Water ($m^3 \cdot t^{-1}$)
Ponds		
extensive	0.3 - 0.8	44,000 - 233,000
semi-intensive	1.0 - 5.0	7,000 - 70,000
intensive	3.0 - 8.0	4,000 - 23,000
[1]sewage-fed tilapia, Thailand	6.8	
[2]intensive catfish, Thailand		50 - 100
[3]intensive tilapia, Taiwan	17.4	
[4]polyculture, Israel		5,000 - 12,000
Tanks		
intensive tilapia, Kenya		50,000 - 60,000
Raceways		
[4]intensive carp, Japan	1,443	

Source: [1]Edwards et al. (1987); [2]Muir (1981); [3,4]Hepher (1985); [4]Sarig (1989b).

- reducing spawning or nursery areas for fish stocks;
- causing barriers to migratory fishes;
- altering thermal regimes; and
- altering biological communities through loss of dilution capacity between inflow and outflow.

There is no evidence that such problems have occurred in the tropics. By contrast, excessive removal of groundwater for aquaculture purposes has caused problems. In Taiwan, use of groundwater on a large scale has led to salination and subsidence (Chien et al. 1988).

Seed

The requirements for seed can be enormous. Most inland aquaculture, however, relies on hatchery-produced fry and fingerlings and, by contrast with tropical marine fish and shrimp culture, there are few reports of excessive overfishing of wild stocks. In Bangladesh, the collection of wild carp fry for stocking fishponds has undoubtedly contributed to the decline of fish stocks.

Perhaps the biggest concern posed by the requirements for fish seed, however, is disease (see below).

Feed

Demand for feedstuffs is recognized as a constraint to aquaculture growth in a number of countries including China (Guan and Chen 1989). Although aquaculture is likely to account for 20-25% of world fishmeal supplies by the year 2000 (Pike et al. 1990), it is argued that fishmeal supplies are sufficient, despite criticisms from some quarters that a number of stocks exploited for reduction to fishmeal are already overfished. However, inland tropical aquaculture relies on diets made largely from locally available materials of agricultural origin. Fishmeal is used only to a limited extent and demand from this quarter is unlikely to exacerbate fishmeal supplies.

Extensive aquaculture relies on natural food production. The culture of tilapias and milkfish in cages and pens by extensive means has been popular in the Philippines and elsewhere in Asia but has tended to follow a pattern in which there has been overexploitation of the food resource, resulting in falling production and increased reliance on supplemental feeds (Beveridge 1984). Extensive aquaculture production is related to primary productivity (Almazan and Boyd 1978; Aquino and Nielsen 1983) (Fig. 1), although the exact nature of the relationship remains poorly understood. Beveridge (1984) proposed the use of an empirical model to help predict carrying capacity although to date it remains to be tested.

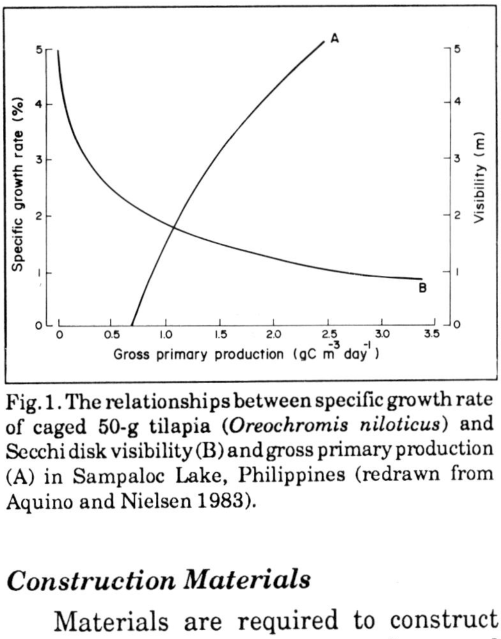

Fig. 1. The relationships between specific growth rate of caged 50-g tilapia (*Oreochromis niloticus*) and Secchi disk visibility (B) and gross primary production (A) in Sampaloc Lake, Philippines (redrawn from Aquino and Nielsen 1983).

Construction Materials

Materials are required to construct holding facilities such as cages and pens and demand for wood and bamboo in particular can be enormous. It has been estimated that to construct a 1-ha pen requires several thousand bamboos and a hundred or so hardwood trunks (Beveridge 1984). In Laguna Lake in the Philippines, the demand for construction materials is tremendous. If one assumes a lifespan of three years for bamboo and a fishpen area of 15,000 ha, then several million culms per year are required. Demand now greatly exceeds supply (Palijon 1986) and bamboos reportedly have to be imported from distant parts of the country.

Outputs from Intensive Aquaculture

There are three principal sources of aquaculture wastes: uneaten food, excreta, and chemicals and therapeutants. Mortalities and escaped (feral) fish may also be considered as wastes.

Chemicals and Drugs

There is little published data on the quantities of chemicals and drugs in western aquaculture (Beveridge et al. 1991) and even less is known about the types or quantities used in the tropics. Chemicals include compounds deliberately employed in the hatchery (to induce spawning, control stickiness of eggs, determine sex), to improve productivity (lime, fertilizers), control pests (pesticides, insecticides, herbicides, molluscicides, piscicides), treat or control diseases (fungicides, parasiticides, disinfectants), control fouling, and pacify fish during handling (anesthetics). Others, particularly materials used in construction of tanks or raceways, may inadvertently leach into the aquatic environment. Few antifoulants are used in freshwater cage or net pen culture, especially in tropical freshwater environments (Beveridge 1987).

LIME AND FERTILIZERS

The most important in terms of quantities used are lime and fertilizers. Lime increases the pH of pond soil and water, increases alkalinity and hardness, reduces humic acid content in the water, and generally improves benthic and phytoplankton production (Boyd 1979). Quantities used can be large - up to 2-3 $t \cdot ha^{-1}$ during the production cycle. Adverse effects of liming on the environment, however, are likely to be minimal, even when ponds are unduly leaky.

Most pond farmers in the tropics use fertilizers, types and amounts being determined by cost, availability, species and intensity of culture. Fertilization plays a major role in determining water quality (Boyd 1979).

THERAPEUTANTS

Although a very large number of therapeutants have some application in aquaculture (Herwig 1979; Meyer and Schnick 1989), farmers tend to rely on a fairly small number of compounds (see, for example, Solbe 1982), in part because of ignorance of what is available but also partly because of costs and, in western Europe and North America at least, legislation (Michel 1986). Evidence from recent tropical fish diseases workshops (Davy and Chouinard 1983; Shariff 1989) suggests that the number of therapeutants used in the tropics may be more limited than in temperate countries, although there are some important differences, particularly where antibiotics and antimicrobials are concerned. Again, the range and quantities used in tropical inland culture differ from those used in mariculture (Brown 1989). Whilst traditional, plant-based compounds have long been employed in the treatment of fish diseases in the tropics (Pantulu 1979), the most common therapeutants in use today are formalin, potassium permanganate, Dipterex and malachite green. Typical concentrations used are given in Table 3. Repeated treatments are necessary in many types of infection. Chemical treatment may be by bath, direct application (injection) or administration in food and what is not absorbed/metabolized by the fish is released into the environment. Bath treatment is the most common method for dealing with parasites, and although large, valuable fish such as broodstock, on occasion may be treated in tanks, chemicals are usually added directly to ponds (Sarig 1971). There have been no studies on the effect of these compounds on the pond ecosystem although it is known that most are toxic to aquatic life in general and often at lower concentrations than those used to treat fish. For example, formalin is toxic to algae at around 0.7-1.2 $mg \cdot l^{-1}$ and to zooplankton at around 5 $mg \cdot l^{-1}$ (Bringmann and Khun 1980; Corstjens and Monnikendam 1973 *in* Nature Conservancy Council 1991), whilst the concentrations used in ponds are typically 25-100 $mg \cdot l^{-1}$ (Table 3). Although it is generally assumed that the chemicals used break down rapidly in the pond environment, there is little specific information on decomposition rates or breakdown products, on the quantities absorbed by pond sediments, or the possibilities for transfer of chemicals to crops if pond sediments are removed. There have been some studies of the organophosphate insecticide Dipterex (2,2,2-trichloro-1-hydroxyethyl phosphate) which has been widely used in aquaculture since 1961 to control ectoparasites and predatory insects and to reduce the numbers of large zooplankton in fry ponds (Ellis 1974). Although this compound is known to be highly toxic to fish (Flores-Nava and Vizcarra-Quiroz 1988; Haque and Barua 1988), it is believed to be rapidly hydrolyzed under tropical pond conditions (pH > 7.5; 25-30°C) (Ellis 1974).

In cage and pen culture, treatment is usually carried out *in situ* (Beveridge 1987) with the resultant release of all of the chemical into the environment. Although treatment of caged fish seems to be rare, the open nature of cage systems should be stressed.

Antibiotics are not commonly employed in tropical freshwater aquaculture although they are used on occasion to treat broodstock, particularly in Asia. They are rarely added to feed but tend to be administered by injection. The uptake, distribution and elimination of therapeutants are determined by species, the formulation and physicochemical characteristics of the therapeutant, feed composition and environmental factors (Rasmussen 1988). Elimination can take anything from a few days to a few months. There have been few studies relevant to tropical species or culture conditions (Millar 1984; Nouws et al. 1989). In a comparison of antibiotic accumulation and elimination in tilapia and rainbow trout tissues, Millar (1984) found that levels are

Table 3. Concentrations of chemicals used in the control of tropical fish diseases.

Agent	Concentration (ppm)	Use
Formalin	25 - 100	Widely used for control of Protozoa and ectoparasites
$KMnO_4$	15 - 30	Bacteria, Protozoa, ectoparasites
Malachite green	[a]0.1 - 0.15	Widely used for control of fungi, Protozoa and ectoparasites
Dipterex	0.25 - 2.5	Control of ectoparasites (e.g., *Dactylogyrus*, *Gyrodactylus*)

[a]Unless bath treatment for fungi, such as *Saprolegnia*, when concentrations of 65-70 ppm are more typically employed.

highest in liver and lowest in muscle tissues. Among the tilapia all residues were eliminated within 96 hours whilst, by comparison, levels in rainbow trout tissues were still detectable 144 hours after administration. Differences were largely attributed to differences in temperature (27°C for tilapia, 6-8°C for trout). Information on elimination times for tropical species are urgently required since ingestion of residues of many antibiotics can have serious side effects (Table 4).

Losses of therapeutant administered in feed occur both as a result of leaching and as a result of feed losses. Leaching of oxytetracycline from food has been shown to be dependent upon temperature, pH and pellet surface area: volume ratio, with up to 20% being leached in 15 minutes.

Studies of hormones commonly used in tropical aquaculture, such as methyltestosterone, have shown that they are completely eliminated from the animal's body, although it takes several weeks (Johnstone et al. 1983). This is of some concern considering recent experimental work on their use as growth promoters (see Joseph (1988) for examples).

Little is known about the fate of unused antibiotics or their effect on the environment. Obviously, the potential exists to affect adversely natural bacterial communities and studies of freshwater salmonid farms by Austin (1985) have shown decreases in bacterial numbers in effluents during chemotherapy. Moreover, it can take many weeks for compounds such as oxytetracycline to break down, depending upon temperature, oxygen and light levels (Jacobsen and Berglind 1988; Samuelsen 1989). It has been clearly established from studies in Japan and elsewhere that the use of antibiotics in aquaculture has caused an increase in drug resistance among many of the important groups of pathogenic fish bacteria (Austin 1985; Michel 1986; Aoki 1988). In many bacteria, drug resistance is carried on R plasmids, circular pieces of DNA containing both a replication and a transmission gene. The R plasmid is transferred to daughter cells during cell division but may also be transferred to other bacteria during conjugation. The R plasmids in *Vibrio* and *Pasteurella piscicida* isolated from cultured Japanese fishes have been found to harbor various combinations of resistance markers to chloramphenicol, streptomycin, tetracycline, ampicillin, sulfonamides and trimethoprim (Michel 1986; Aoki 1988).

No studies of drug resistance among fish pathogens have been carried out in the tropics. However, such strains undoubtedly exist and research in this field, particularly in Southeast Asia, is urgently needed. Whilst oxytetracycline is probably the most widely used antibiotic, anecdotal evidence suggests that virtually all antibiotics, including chloramphenicol, which is regarded as the ultimate protection against human enteric pathogens such as *Salmonella typhae*, are in use. There are also risks of transfers of resistance from fish pathogens to bacteria, such as *Escherichia coli*, harbored by man.

Uneaten Food, Fecal and Urinary Products

There have been a number of different approaches to assessment of waste production, including direct measurement and the use of indirect mass balance techniques.

UNEATEN FOODS

In both semi-intensive and intensive aquaculture, food is provided. A proportion remains uneaten because quantities and quality are often inappropriate and because aquaculture systems and their management tend to confound optimization of ingestion (Beveridge et al. 1991). Studies indicate that the proportion of uneaten food varies from around 1% to as much as 30% and confirm that system, type of feed and management are important determinants

Table 4. Summary of toxic effects and nutritional hazards associated with the use of antibiotics in fish (modified from Michel 1986).

Drug	Undesirable effects on fish			
	Acute toxicity	Other effects	Persistence of residues	Risks for human beings
Kanamycin	Liver and kidneys of adult trout	-	8-10 days	Allergies and neurosensory disorders (internal ear)
Chloramphenicol	-	Growth retardation after prolonged treatment	48-72 hours	Bone marrow aplasia
Oxytetracycline	Rare	Growth retardation Sterility Immunosuppression	15-20 days 60 days	Digestive disorders Hepatorenal disorders
Erythromycin	-	Reversible nephrotoxicity	48-72 hours	Rare allergies
Sulphonamides	-	Sterility and nephrotoxicity after prolonged use	8 days 15 days	Hepatorenal disorders Leukopenia, allergy
Trimethroprim or Ormetoprim	-	-	48 hours - 5 weeks	Hepatorenal disorders Leukopenia, allergy
Nifurpirinol	Rare	-	24-48 hours	Carcinogenic?
Furazolidone	Individual intolerance	-	10 days	Digestive disorders Allergy
Quinolones	Oxolinic acid has sub-acute toxicity for ayu	-	72 days 96 hours	-

of wastage (Beveridge et al. 1991). Available data suggest that feed losses associated with pelleted feeds are less than for trash fish (although if the dry matter content of feeds is taken into account, the significance of these differences is negligible) and that the proportion of uneaten food in cages is considerably greater than those from tanks or ponds (Beveridge 1984).

EXCRETA

The undigested fraction of feed, together with mucus, sloughed intestinal cells and bacteria, is voided as feces whilst the digested portion is absorbed and metabolized. Nutrients absorbed in excess to requirements may be excreted together with end-products (ammonia and urea) derived from the catabolic breakdown of protein for energy purposes.

It is possible to estimate the fecal production associated with a diet from digestibility data on the principal constituents. Although this approach necessarily ignores the effects of variables such as temperature, body size, health, feeding rate and the synergistic/antagonistic effects of one dietary component on the digestibility of another, such estimates have been found to approximate those obtained by direct measurement (Beveridge et al., 1991). In Table 5, data are presented for a variety of warmwater fish species. Fecal production is typically around 250 $g{\cdot}kg^{-1}$ food fed on a dry weight basis for carnivorous species and somewhat higher for omnivorous/herbivorous species.

Most of the nitrogen excreted is in the form of ammonia. Data on ammonia production for common carp and tilapia are presented in Table 6. There is a great deal of variability not only between but also within species, since excretion rate is principally dependent upon nitrogen consumption although modified to some extent by temperature (Paulson 1980).

There have been few studies of bacteria excretion although it is known that the guts of fish can act as proliferation sites for Enterobacteriaceae, *Aeromonas* and fecal streptococci which enter the culture systems through the influent water or feed (Trust and Money 1972; Niemi and Taipalinen 1982; Austin and Allen-Austin 1985).

Table 5. Estimated diet digestibilities (%) and fecal production (g dry wt·100 g^{-1} ingested) from data for dietary components and proximate analysis (% dry wt). Typical diets for farmed tropical freshwater species have been used (modified from Beveridge et al. 1991).

	Catfish		Carp		Tilapia	
Dietary	Diet (%)	Digestibility (%)	Diet (%)	Digestibility (%)	Diet (%)	Digestibility (%)
Protein	35	80	35	85	35	70
Lipid	7	97	15	70	6	90
Carbohydrate	48	50	40	70	50	60
Ash	10	50	10	50	9	50
Overall diet digestibility (%)	64		73		64	
Fecal production (g·100 g^{-1})	36		27		36	

Table 6. Ammonia excretion rates for carps and tilapias.

Species	Production rate
C. carpio	[a]110-581 $mgN \cdot kg^{-1} \cdot day^{-1}$
O. mossambicus	[b]1.72 $mgN \cdot kg^{-1} \cdot hour^{-1}$
O. niloticus	[c]1.7-9.4 $mgN \cdot kg^{-1} \cdot hour^{-1}$

[a]Kaushik (1980); [b]Musisi (1984); [c]McKinney (pers. comm.).

MASS BALANCE ESTIMATES OF UNEATEN FOOD AND EXCRETA

An alternative to the use of data derived from physiological experiments is to employ a mass balance approach. It is possible to quantify uneaten food, feces and excreta, given data on feed qualities and quantities, FCR values, digestibilities and fecal composition, and to produce mass balance equations for various waste parameters, such as nitrogen, phosphorus or carbon, solids or biological oxygen demand (BOD). Consider nitrogen balance in tilapia, for example (Fig. 2). Assuming the nitrogen content of tilapia is 3% and that of a particular tilapia diet is around 8%, then for an FCR value of 1.6:1, approximately 98 kg of nitrogen is lost to the environment for every tonne of tilapia produced. By incorporating other data and a number of assumptions a more detailed picture can be assembled. If it is assumed that 20% of food is uneaten, then 102.4 kg of nitrogen is consumed but only 30 kg is retained per tonne of tilapia produced. It can be estimated from Table 5 that approximately 360 g of feces is produced per kg food ingested. If fecal N content is assumed to be around 4%, then 18.4 kg of nitrogen are voided as feces whilst the balance, 54 kg, is excreted as ammonia and urea.

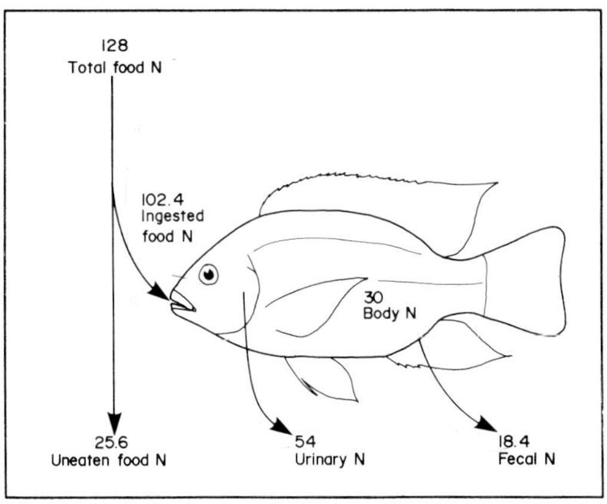

Fig. 2. Mass balance of nitrogen for 1-tonne tilapia production. Figures are in kg (see text).

Many such mass balance equations have been produced for fish (e.g., Braaten et al. 1983; Beveridge 1984; Gowan and Bradbury 1987; Heinsbroek 1988) and they can be reasonably accurate. None, however, directly accounts for food losses, and they cannot predict effluent composition in terms of suspended or settleable solids, nitrogen or phosphorus species, since both system and management play a major role in determining how much and in what forms the wastes might appear. The form in which wastes are discharged and temporal variability in effluent quality are important determinants of environmental impact. Most models also give no indication of diurnal or seasonal variability.

Unless removed or recovered, mortalities may be another source of wastes. Estimates from European aquaculture suggest that mortalities of between 1 and 5% typically occur during production. Whilst dead fish are readily detected and removed from tanks and raceways, this is not the case in ponds. In cage systems it is also difficult to observe or remove mortalities and a number of studies have pointed out that they may make a significant contribution to effluent nitrogen and phosphorus waste loadings (Penczak et al. 1982; Beveridge 1987).

WASTE PRODUCTION AND EFFLUENT

As discussed above, the quantities of uneaten food, feces and urinary products vary with species and type of food and are influenced by body size and season (i.e., metabolic rate). Waste production within a system increases with biomass and with food use. It is impossible, however, to predict farm effluent characteristics from waste production data alone. Alabaster (1982), in an analysis of data from European fish farms, pointed to strong correlations between parameters such as BOD and suspended solids (SS), and between NH_3 and chemical oxygen demand (COD), but also drew attention to the fact that effluent concentrations varied enormously even when standardized on a per unit production or biomass basis (see Beveridge et al. 1991). The reasons are that the system and its management play a large part in determining effluent characteristics. For example, water exchange rate influences the probability with which a food item is encountered and thus the proportion of food which remains uneaten, whilst flow characteristics will in part determine whether uneaten food and fecal particles remain intact and the proportion that settle within the system. Mode of discharge and the extent of treatment or dilution prior to discharge further complicate the picture.

Unfortunately, there has been little work on system-specific characterization of wastes. Nevertheless, it is possible to make some generalizations. Ponds, the most important system in tropical inland aquaculture, differ from most other systems in that with few exceptions (e.g., for carps in Japan; Suzuki 1986), water exchange is low, even in fairly intensive systems. Despite the fact that water quality in fishponds varies enormously depending on stocking density and management, the water quality of the day-to-day effluent is generally reasonable, even from intensively managed systems (Boyd 1978). Settleable solids levels, for example, are less than 3.3 $ml \cdot l^{-1}$. Problems can arise if ponds are drained prior to harvesting, when 95% of water may be discharged. Immediately following harvesting there is a marked increase in effluent phosphorus, ammonia, BOD and COD and settleable solids (Boyd 1978; Ellis et al. 1978; Solbe 1982; Bohl 1988; Butz 1988) (Table 8). Although there is a correlation between biomass harvested and effluent solids loading (Ellis et al. 1978), it is weak because of pond soil erosion.

Chemical budgets have been used to try and identify and quantify nutrient gains and losses and have been applied to intensively managed subtropical and tropical freshwater ponds and to subtropical brackishwater ponds (Avnimelech and Lacher 1979; Boyd 1985; Daniels and Boyd 1989). Data for channel catfish ponds are summarized in Table 7. These show that only 25-30% of nitrogen, phosphorus and organic matter applied in the form of feeds was harvested as fish. However, there was no accumulation of nitrogen and organic matter in sediments, the former as a result of denitrification and volatilization and the latter because of respiration. Most phosphorus not accumulated in the fish was trapped in the sediments and very little of any of the inputs was lost through seepage or overflow, even during pond drainage.

There is a more or less constant exchange of water through tank and raceway systems, determined by biomass and environmental conditions. Suspended solids and BOD effluent loadings in concrete raceways, expressed in terms of per cent food fed, have been shown to fall with increased water exchange (Butz 1988). Effluent characteristics can change dramatically as a result of routine operations. Studies carried out at the University of Stirling and elsewhere show that during cleaning, effluent loadings increase dramatically (Alabaster 1982; Butz 1988; Kadri 1988). Alabaster (1982) observed that 70% of the daily BOD, 75% of total-P and 10% of total N from tank-based, intensively managed systems could be discharged during a 30-minute cleaning period.

Cage and other water-based systems differ from land-based ponds, tanks and raceways in that no wastes are trapped within the system. Waste production from able 7. Average gains and losses (kilograms) for nitrogen, phosphorus, chemical oxygen demand (COD), and dissolved oxygen DO) in three channel catfish ponds (each 405 m²) on the Auburn University Fisheries Research Unit (modified from Boyd 1985). onds were stocked with 350 channel catfish (mean weight 16 g) and the growout period was 27 weeks.

tem	Nitrogen		Phosphorus		COD		DO	
ains								
Fish stock	0.12	(1%)	0.036	(2%)	1.6			
Feed	11.15	(92%)	1.682	(96%)	235.9	(36%)		
Nitrogen fixation	?							
Photosynthesis					413.1	(63%)	413.1	(77%)
Inflow from pipe	0.24	(2%)	0.035	(2%)	9.7	(1%)	6.3	(1%)
Rainfall	0.52	(4%)	0.001		0		2.8	(1%)
Runoff	0.05		0		0		0.3	
Aeration					0		9.9	(2%)
Net diffusion							104.8	(20%)
otal	12.08		1.754		660.3		537.2	
osses								
Fish harvest	2.99	(25%)	0.521	(30%)	60.1	(9%)		
Overflow and draining	0.89	(7%)	0.119	(7%)	34.4	(5%)	3.2	(1%)
Seepage	1.28	(11%)	0.145	(8%)	31.8	(5%)	0	
Respiration in water column					309.3	(47%)	309.3	(58%)
Benthic respiration					103.8	(16%)	103.8	(19%)
Fish respiration					120.9	(18%)	120.9	(23%)
Denitrification and diffusion of ammonia	6.92	(57%)						
Uptake by mud			0.969	(55%)				
Total	12.08		1.754		660.3		537.2	

Table 8. Quality of effluents from channel catfish ponds during the two phases of fish harvest (modified from Boyd 1978).

Variable	Harvest phase: Draining	Seining
Settleable matter (ml·l⁻¹)	0.08	28.5
BOD (mg·l⁻¹)	4.3	28.9
COD (mg·l⁻¹)	30	342
Filterable orthophosphate (μg·l⁻¹ as P)	16	59
Total phosphorus (mg·l⁻¹)	0.11	0.49
Total ammonia nitrogen (mg·l⁻¹)	0.98	2.34
Nitrate (mg·l⁻¹)	0.16	0.14

semi-intensive and intensive cage systems has been quantified using a mass balance approach in which inputs and outputs are estimated (Beveridge 1984, 1987) although, as discussed above, such an approach gives no information on solid or dissolved fractions. Measurements of the solid component of effluents from cage carp production in Indonesia have been made recently (Costa-Pierce and Roem 1990). Comparisons with solid effluent loads from temperate rainbow trout cage production (Merican and Phillips 1985) show that solids loadings and nutrient losses were considerably higher in the latter, probably as a result of higher feeding rates and feed losses and the higher nutrient concentrations in salmonid diets. Costa-Pierce and Roem (1990) estimated that losses of C, N and P in solid wastes from caged carp on average accounted for 5.4, 3.5 and 0.1%

respectively, of nutrient inputs. However, further work is required in order to quantify the dissolved fraction and the extent of leaching that might occur under tropical conditions.

ENVIRONMENTAL IMPACT OF UNEATEN FOOD AND EXCRETA

The impact of aquaculture effluent is determined both by the nature of the effluent and where and how it is discharged. Laws dealing with the discharge of water from aquaculture are to be found in most countries (van Houtte et al. 1989). Prior authorization for the discharge of any waste into a receiving water body is required, the authorization either being incorporated into the grant/license/permit to farm fish or granted separately. Aquaculture effluents characteristically have low levels of wastes. Nevertheless, even the waters drained prior to harvesting can have significantly higher levels of settleable matter, BOD, COD, total phosphorus, and total ammonia nitrogen than the receiving streams (Boyd 1978). Because it is important that the size and quality of the receiving waters be taken into account, legislation in some countries is directed at controlling the water quality of the receiving waters rather than at controlling effluent quality.

There have been no studies of the environmental impact of aquaculture in the tropics. Studies in temperate regions have shown that the discharge of wastes from farms can cause measurable changes in the water chemistry of the receiving bodies (Kilambi et al. 1976; Alabaster 1982; Bergheim et al. 1982; Beveridge 1984; Munro et al. 1985; Phillips et al. 1985, 1986). The principal effect on running waters is to increase ammoniacal nitrogen and phosphorus concentrations immediately downstream of the discharge. In addition, cage culture can cause long-term elevations of lake carbon, nitrogen and phophorus levels. Although carbon is usually present in excess in freshwater systems, increases in nitrogen and phosphorus, one or other of which tends to limit productivity in freshwaters (OECD 1982; Rast et al. 1989), will lead to eutrophication. A general increase in algal densities may thus result (Eley et al. 1972; Kilambi et al. 1976; Hays 1980). Cyanobacteria and other species tolerant of high P:N ratios are likely to become dominant (Stirling and Dey 1990; Nature Conservancy Council 1991) and such shifts in phytoplankton community structure will have implications not only for water quality but also for autotrophic food webs.

Changes in the benthic environment immediately under cages or downstream from land-based farms as a result of settlement of solid wastes have also been noted (Merican and Phillips 1985; Nature Conservancy Council 1991). The extent of waste settlement is dependent upon particle size and density, as governed by Stoke's Law, although in shallow, exposed lake or reservoir sites, resuspension and dispersion may reduce accumulation (Ackefors 1986). These sediments are richer in P, N and C levels than natural sediments and bacterial decomposition of organic matter can result in anaerobic conditions within a few millimeters of the sediment surface (Alabaster 1982; Enell and Löf 1983). Under completely anoxic conditions it is possible that H_2S could be evolved. Whilst this has not been reported from freshwater sites, enhanced release of ammonia and phosphorus has been observed (Enell and Löf 1983). Release of gases is enhanced by bioturbation caused by the high numbers of pollution-tolerant macroinvertebrates that can occur.

Eutrophication can also bring about increased production and changes in the macrophyte and natural fish communities' structure. Evidence to date shows that freshwater aquaculture activities can cause increased growth rates among the natural fish community, in part because of ingestion of uneaten pellets (Kilambi et al. 1976;

reported incidences from the tropics.

There are differences in perception of the problems posed by feral species. Indeed, as Welcomme (1988) has pointed out, an objective evaluation may be impossible given the many different views of natural ecosystems. From a strict conservation standpoint any change to the aquatic community is detrimental. Conversely, rural societies in the tropics may be more tolerant of introductions if they lead to improvements in food supply. As evidence, Welcomme (1988) reports that only two species introduced for aquaculture purposes in the tropics (*O. mossambicus* and *C. batrachus*) are widely perceived as pests, although a further five species are regarded with mixed feelings, whilst four are generally regarded as beneficial.

An increase in awareness has fuelled the lobbies against genetic dilution of wild fish stocks (FAO 1981; Meffe 1987; Welcomme 1988; Pullin 1990; Skaala et al. 1990), although few yet appreciate that it is important that they be conserved as future genetic resources for aquaculture (Pullin 1986, 1988, 1990).

Discussion

This review has been structured around a consideration of inputs (resource use) and outputs and as a result has ignored a number of important issues, particularly with respect to the creation of new aquatic habitats. Aquaculture production in the tropics is growing and whilst some of the growth has been achieved by intensification, much has involved an expansion in the area devoted to production. The creation of new freshwater environments may increase species diversity but at the same time may result in an increase in the incidence of water-borne diseases such as schistosomiasis, malaria and yellow fever (Pollard 1981; Tucker 1983; Canter 1985). The creation of ponds for aquaculture purposes poses less of a risk due to the combination of high predation pressure on snail and insect vectors, regular draining of ponds and drying out of the pond muds and the application of heavy doses of lime to pond sediments prior to stocking. It should also be borne in mind that some of the expansion in freshwater aquaculture production in Asia will be in cages and pens which can use the massive increase in reservoirs built for other purposes. It has also been speculated that integrated methods of aquaculture could be important in the development of influenza pandemics (Scholtissek and Naylor 1988). There is, however, no supporting evidence and there are a number of reasons why this is highly unlikely, not least being the difference in temperature tolerance of mammalian and piscine viruses.

The review has also ignored the impact of fish farms on the natural landscape. Land-based production systems tend to have little visual impact, particularly if sited in agricultural areas where they may actually enhance the landscape by increasing the diversity of views. Cages and pens, on the other hand, are often unsightly and development may well become an issue in areas of recreational or scenic importance, as has happened in Europe and North America.

The establishment of new fish farms may disrupt wildlife. It is known that cage and pen structures can act as Fish Attractant Devices (FADs) and that the high densities of farmed fish and waste food may attract predators and scavenging species which in turn can displace local species (Beveridge 1987, 1989; Nature Conservancy Council 1991). Routine farm operations can adversely affect the breeding success of some species. Such impacts have not yet been studied in the tropics.

However, it is important that the impacts of inland aquaculture in the tropics be put into some sort of context. Demand for land and water resources has caused problems, although largely of a local nature.

Nevertheless, it is possible to improve utilization and to minimize problems through better planning. There are a number of ways of reducing water demand including reduction of seepage losses, intensification of use, and reuse of water. Integration of agricultural water use with aquaculture can be implemented with benefit to both (Little and Muir 1987; Redding and Midlen 1991). The development of dual-purpose reservoirs in Israel (Leventer 1987), for example, has improved utilization and reduced the costs of water for aquaculture. More could also be done to reduce water loss from ponds by better siting and management. Losses can be further minimized by increasing pond depth. Sarig (1989a, 1989b) reports that water losses from ponds are 6-10 times higher (35-60,000 $m^3 \cdot ha^{-1} \cdot year^{-1}$) than those from the deeper dual-purpose reservoirs (6,000 $m^3 \cdot ha^{-1} \cdot year^{-1}$). Although there may be no costs involved, it is in the producers' interest to minimize seepage since production can be adversely affected because of the effect of high water exchange on alkalinity (Teichert-Coddington and Phelps 1989). Intensification of water use is also possible using the types of concrete-lined ponds developed in Taiwan and Israel (Hepher 1985) though waste output may be greater from such systems since there is little opportunity for biotransformation.

High water use undoubtedly occurs where water is plentiful and where there are no laws or costs to restrict use. In the EIFAC survey of European freshwater farms, for example, water consumption per tonne fish production was found to vary by a factor of more than 100 (Alabaster 1982). It is likely that where conservation of water is necessary or desirable, some sort of cost penalty must be enforced.

Nor should there be particular concern with regard to land use for freshwater aquaculture in the tropics. As stated above, much of the land that is used for ponds is agricultural and, unlike coastal aquaculture, there is little evidence of large areas of other types of land being converted to freshwater fish production. Moreover, intensification of production is a trend apparent in many parts of the tropics such as Taiwan, where production rose from an average of 1.4 $t \cdot ha^{-1}$ to 3.6 $t \cdot ha^{-1}$ between 1965 and 1985 (Lee 1987). It should also be borne in mind that changes in land use are not irreversible, being principally determined by economics (Hannig 1988). In Israel, for example, the area of land devoted to fish farming fell by 20% from 3,529 ha to 2,818 ha between 1979 and 1988 despite massive increases in production (Sarig 1989a). Cages and pens are of an even more temporary nature and can and have been readily dismantled as economic and political situations change (Beveridge 1984, 1987).

The production of wastes must also be set in context. With a few exceptions waste food, fecal and urinary products are likely to be a problem only during pond drainage and impacts are likely only where effluent waste concentrations are high and where the nutrient levels and flow rates of the receiving waters are low. In assessing the impact of such wastes it is therefore important that attention be focused on protection of the receiving waters rather than of effluent quality. In the few tropical countries practicing intensive pond-based aquaculture, the effluent is usually either drained into neighboring ponds or used for irrigation, for although there are as yet no supportive data it is undoubtedly of some value as a source of crop nutrients. However, it should also be pointed out that a number of the more recently established, intensive tilapia farms in Central America and the Caribbean are discharging effluent into relatively pristine rivers with unknown consequences.

Many studies and surveys have demonstrated strong relationships between the quantities and qualities of feed given and food conversion ratio with effluent characteristics such as suspended solids and BOD (e.g., Knosche 1972 *in* Rosenthal et al. 1987; Pursiainen 1988). Thus, it is possible to make some contribution to reducing effluent waste levels by improving food palatability and digestibility, although in view of the fact that the majority of tropical freshwater production is pond-based, it could be argued that there are perhaps better approaches. Land-based farms have a number of alternatives for reducing waste outputs. According to Tucker (1985), many catfish farms in the United States harvest without draining ponds thereby avoiding problems of effluent discharge although there is concern that this may result in long-term deterioration in pond water quality (Parker 1988). An alternative is to treat wastes prior to discharge and there are much data available from northern Europe and North America on the efficacy and costs of different design options (see Nature Conservancy Council 1991 for review). Wastes from water-based systems are much more difficult to contend with and cannot be treated in a cost-effective manner (Beveridge 1987; Nature Conservancy Council 1991). Models have been developed which attempt to predict the carrying capacity of tropical systems to aquaculture wastes (Beveridge 1984, 1987) although these await testing and verification. Moreover, it is known from studies in temperate climates that these models suffer from certain disadvantages and must be applied with care (Beveridge et al. 1991; Nature Conservancy Council 1991). They take no account, for example, of micronutrients.

Little is known about the impacts of the drugs and chemicals used or about residues and yet despite this it is clear that guidelines are frequently ignored, particularly in countries which have thriving industries and where profit margins are high. Dipterex, for example, is widely used as a piscicide in Asia at concentrations of up to 50 ppm, a 50-100 fold increase over its recommended use as a therapeutant for parasite control. New chemicals are being used with little comprehension of the risks involved. The insect fumigant phostoxin has recently become widely used in Bangladesh as a piscicide, any fish killed being netted and consumed. In water the chemical reacts to form phosphine gas (H_3P) which is highly toxic to humans when inhaled and it is not known whether any toxins accumulate in the flesh of the dead fish. Extracts from locally available plants such as *Derris elliptica* and *Croton tiglium* are increasingly being advocated (Guerrero 1989). This is largely to be commended: they are cheap and the toxicity of many is well known. However, it would be wrong to assume that because a compound is of natural origin it will have low toxicity to mammals and will break down rapidly in pond conditions (e.g., the piscicide *Diospyros cordifolia*; Sharma and Simlot 1971). Antibiotics are still widely used as a prophylactic in the transport of aquarium fishes despite the fact that it is clear that regular or systematic use exerts a major selective pressure on the development of resistance among bacterial populations and that scientists have campaigned against this practice for years.

Misuse of therapeutic agents in freshwaters is probably less than among other sectors of the industry since freshwater aquaculture is less profitable. Nevertheless, guidelines are urgently needed and the dangers of misuse need to be disseminated among farmers.

With one or two notable exceptions (e.g., the recent epizootic ulcerative syndrome outbreak in Asia and the aquarium fish trade), diseases among farmed tropical freshwater fish are not at present a major issue. A survey of the Proceedings from the International Symposia on Tilapias in Aquaculture (ISTA I and ISTA II), for example, reveals little discussion of the subject. However, as aquaculture spreads and intensifies, and unless fish movements are carefully regulated, disease problems will undoubtedly increase. This has long been recognized by scientists (e.g., Davy and Chouinard 1983) although governments apparently need further convincing. The trend of increased movements of fish for aquaculture purposes (Welcomme 1988) means that the other risks posed by introductions will also increase. Recent attempts by Central and Latin American countries in particular to examine the possibilities for culture of indigenous stocks are to be welcomed, although it would be naive to assume that many will prove superior to carps and tilapias. There are few sets of guidelines available which help decisionmakers evaluate the risks of introductions. The situation is becoming increasingly complex as programs for genetic improvement gain momentum and gene banks for commercially important fish species become established. A number of transgenic tilapias and carps have recently been experimentally produced (Brem et al. 1988; Indiq and Moav 1988; Kapuscinski and Hallerman 1990; Maclean and Penman 1990). A detailed consideration of the risks posed by such animals is urgently required before their use becomes widespread in aquaculture (Hallerman and Kapuscinski 1990).

Acknowledgements

We would like to thank the following who provided information for this review: Dr. Donald Baird, Dr. G. Nick Frerichs, Dr. Inocencio Higuera, Dr. Val Inglis, Dr. Kim Jauncey, Mr. Richard McKinney, Dr. Brendan MacAndrew, Mr. Stuart Millar, Ms. Theresa Redding, Professor Ron Roberts and Dr. Lindsay Ross, Institute of Aquaculture, Stirling; Dr. Anna Milstein, Dor, Israel; Dr. Imre Csavas, FAO Regional Office, Bangkok; Dr. Barry Costa-Pierce, ICLARM, Malaŵi; Dr. Kenneth Ruddle, Japan; Mr. Mike Baker, Commonwealth Development Corporation, Costa Rica; Mr. Uwe Barg and Dr. Manuel Martinez, FAO, Rome.

Dr. M.C.M. Beveridge is funded by the Overseas Development Administration of the UK Government.

References

Ackefors, H. 1986. The impact on the environment by cage fish farming in open water. J. Aquacult. Trop. 1: 25-33.

Alabaster, J.S. 1982. Report of the EIFAC workshop on fish-farm effluents. EIFAC Tech. Pap. 41. FAO, Rome.

Almazan, G. and C.E. Boyd. 1978. Plankton production and tilapia yield in ponds. Aquaculture 15: 75-77.

Anderson, I.G. and F. Shaharom-Harrison. 1986. *Sanguinicola armata* infection in bighead carp (*Aristichthys nobilis*) and grass carp (*Ctenopharyngodon idella*) imported into Malaysia, p. 247-250. *In* J.L. Maclean, L.B. Dizon and L.V. Hosillos (eds.) The First Asian Fisheries Forum. Asian Fisheries Society, Manila, Philippines.

Aoki, T. 1988. Drug-resistant plasmids from fish pathogens. Microb. Sci. 5:219-223.

Aquino, L.V. and B.H. Nielsen. 1983. Primary production and growth rate of *Tilapia nilotica* in cages in Sampaloc Lake, Philippines, p. 453-463. *In* L. Fishelson and Z. Yaron (comps.) Proceedings of the International Symposium on Tilapias in Aquaculture, Tel Aviv University, Israel.

Asian Institute of Technology. 1986. Buffalo/fish and duck/fish integrated systems for small-scale farmers at the family level. ODA Project No. R3919 final report. AIT Res. Rep. 198. 138 p. AIT, Bangkok.

Austin, B. 1985. Antibiotic pollution from fish farms: effects on aquatic microflora. Microbiol. Sci. 2: 113-117.

Austin, B. and D. Allen-Austin. 1985. Microbial quality of water in intensive fish rearing. J. Appl. Bacteriol. Symp. Supplement 1985: 207-226.

Avnimelech, Y. and M. Lacher. 1979. A tentative nutrient balance for intensive fish ponds. Isr. J. Aquacult. 31: 3-8.

Bergheim, A., H. Hustveit and A. Selmer-Olsen. 1982. Estimated pollution loadings from Norwegian fish farms. II Investigations 1980-1981. Aquaculture 36: 157-168.

Beveridge, M.C.M. 1984. Cage and pen fish farming. Carrying capacity models and environmental impact. FAO Fish. Tech. Pap. 255. 131 p.

Beveridge, M.C.M. 1987. Cage aquaculture. Fishing News Books Ltd., Farnham, Surrey.

Beveridge, M.C.M., M.J. Phillips and R.M. Clarke. 1991. A quantitative and qualitative assessment of wastes from aquatic animal production, p. 506-533. *In* D.E. Brune and J.R. Tomasso (eds.) Advances in world aquaculture. Vol. 3. World Aquaculture Society.

Bohl, M. 1988. National report for the Federal Republic

of Germany on suspended solids from land-based fish farms, p. 29-36. *In* M. Pursiainen (ed.) National contributions on suspended solids from land-based fish farms. Monistettuja Julkaisuja, Helsinki, Finland.

Boyd, C.E. 1978. Effluents from catfish ponds during fish harvest. J. Environ. Qual. 7: 59-62.

Boyd, C.E. 1979. Water quality in warmwater fishponds. Auburn University, Alabama.

Boyd, C.E. 1985. Chemical budgets for channel catfish ponds. Trans. Am. Fish. Soc. 114: 291-298.

Braaten, B., A. Ervik and E. Boje. 1983. Pollution problems on Norwegian fish farms. Aquaculture Ireland 14: 6-7.

Brem, G., G. Brenig, Horstgen-Schwarck and E.L. Winnacker. 1988. Gene transfer in tilapia (*Oreochromis niloticus*). Aquaculture 68: 209-219.

Bringmann, G. and R. Khun. 1980. Comparison of the toxicity thersholds of water pollutants to bacteria, algae and protozoa in the cell multiplication inhibition test. Water Res. 14: 231-241.

Brown, J. 1989. Antibiotics: their use and abuse in aquaculture. World Aquacult. 20(2): 34-35, 38-39, 42-43.

Bruton, M.N. 1986. The life history styles of invasive fishes in southern Africa, p. 201-208. *In* I.A.W. Macdonald, F.J. Kruger and A.A. Ferrar (eds.) The ecology and management of biological invasions in southern Africa. Oxford University Press, Cape Town.

Buras, N., L. Duek and S. Niv. 1985. Reactions of fish to microorganisms in wastewater. Appl. Environ. Microbiol. 50: 989-995.

Buras, N., L. Duek, S. Niv, B. Hepher and E. Sandbank. 1987. Microbiological aspects of fish grown in treated wastewater. Water Res. 21: 1-10.

Butz, I. 1988. Situation of fish farm effluents in Austria, p. 4-12. *In* M. Pursiainen (ed.) National contributions on suspended solids from land-based fish farms. Monistettuja Julkaisuja, Helsinki, Finland.

Canter, L.W. 1985. Environmental impact of water resources projects. Lewis Publishers, Chelsea, Michigan.

Chien, Y.H., I.C. Liao and C.M. Yang. 1988. The evolution of prawn grow-out systems and their management in Taiwan, p. 143-168. *In* K. Murray (ed.) Aquaculture technologies for the future. Institute of Chemical Engineers Symposium Series 111.

Cloete, T.E., D.F. Toerien and A.J.H. Pieterse. 1984. The bacteriological quality of water and fish of a pond system for the treatment of cattle feed effluent. Agr. Wastes 9:1-15.

Colman, J.A. and A.R. Jacobson. 1991. Review and development of aquaculture models for predicting and exploiting solute flux at the sediment-water interface, p. 460-488. *In* D.E. Brune and J.R. Tomasso (eds.). Advances in aquaculture. Vol. 3. World Aquaculture Society.

Costa-Pierce, B.A. and C.M. Roem. 1990. Waste production and efficiency of feed use in floating net cages in a eutrophic tropical reservoir, p. 257-271. *In* B.A. Costa-Pierce and O. Soemarwoto (eds.) Reservoir fisheries and aquaculture development for resettlement in Indonesia. ICLARM Tech. Rep. 23, 378 p.

Daniels, H.V. and C.E. Boyd. 1989. Chemical budgets for polythene-lined, brackishwater ponds. J. World Aquacult. Soc. 20(2):53-60.

Davy, F.B. and A. Chouinard, Editors. 1983. Fish quarantine and fish diseases in Southeast Asia: report of a workshop held in Jakarta, Indonesia, 7-10 December 1982. International Development Research Centre, Ottawa. 79 p.

Edwards, P., C. Polprasert and L.W. Kok. 1987. Resource recovery and health aspects of sanitation. Asian Institute of Technology Res. Rep. 205. 137 p. AIT, Bangkok, Thailand.

Eley, R.L., J.H. Carroll and D. De Woody. 1972. Effects of caged catfish culture on water quality and community metabolism. Proc. Okla. Acad. Sci. 52:10-15.

Ellis, J.E., D.L. Tackett and R.R. Carter. 1978. Discharge of solids from fish ponds. Prog. Fish-Cult. 40:165-166.

Ellis, J.T. 1974. A review of the literature on the use of Masoten in fisheries. US Fish and Wildlife Service, Washington, DC.

Enell, M. and J. Löf. 1983. Environmental impact of aquaculture sedimentation and nutrient loadings from fish cage culture farming. Vatten/Water 39 (4):364-375.

FAO. 1981. Conservation of genetic resources in fish: problems and recommendations. FAO Fish. Tech. Pap. 217.

FAO. 1988. Aquaculture production 1984-1987. FAO Fish. Circ. 815 (Rev. 1). 130 p.

Flores-Nava, A. and J.J. Vizcarra-Quiroz. 1988. Acute toxicity of Trichlorfon (Dipterex) to fry of *Cichlasoma urophthalmus* Gunther. Aquacult. Fish. Manage. 19:341-345.

Frerichs, G.N., S.D. Millar and M. Alexander. 1989. Rhabdovirus infection of ulcerated fish in Southeast Asia, p. 396-410. *In* W. Ahne and E. Kurstak (eds.) Viruses of lower vertebrates. Springer Verlag, Berlin.

Gowan, R. and N. Bradbury. 1987. The ecological impact of salmonid farming in coastal waters: a review. Oceanogr. Mar. Biol. Ann. Rev. 25:563-575.

Guerrero, Ma.R. 1989. Fish toxicants from local plants for better pond management. Greenfields 19(9):29.

Hallerman, E.M. and A.R. Kapuscinski. 1990. Transgenic fish and public policy: regulatory concerns. Fisheries 15(1):12-20.

Hannig, W. 1988. Towards a blue revolution: socioeconomic aspects of brackishwater pond cultivation in Java. Gadjah Mada University cultivation in Java. Gadjah Mada University Press, Yogyakarta, Indonesia. 404 p.

Haque, M.Z. and G. Barua. 1988. Toxic and sub-lethal effect of the Dipterex on *Oreochromis niloticus*. Bangladesh J. Fish. 11:75-79.

Haylor, G.S. Some aspects of the biology and culture of the African catfish *Clarias gariepinus* (Burchell 1822) with particular reference to developing African countries. *In* J.F. Muir and R.J. Roberts (eds.) Recent advances in aquaculture 4. (In press).

Hays, T. 1980. Impact of net pen culture on water quality and fish populations in Bull Shoals reservoir. Completion Report to the Arkansas Game and Fish Commission (AGFC Proj. 2-338-R-1). Arkansas Game and Fish Commission, Fayetteville, Arkansas. 10 p.

Heinsbroek, L.T.N. 1988. Waste production and discharge of intensive culture of African catfish, *Clarias gariepinus*, European eel, *Anguilla anguilla*, and rainbow trout *Salmo gairdneri*, in recirculation and flow through units in the Netherlands, p. 55-61. *In* M. Pursiainen (ed.) National contributions on suspended solids from land-based fish farms. Monistettuja Julkaisuja, Helsinki, Finland.

Hejkal, T., C.P. Gerba, S. Henderson and M. Freeze. 1983. Bacteriological, virological and chemical evaluation of a wastewater-aquaculture system. Water Res. 17:1749-1755.

Hepher, B. 1985. Aquaculture intensification under land and water limitations. GeoJournal 10(3):253-259.

Hepher, B. and Y. Pruginin. 1981. Commercial fish farming. John Wiley, New York.

Herwig, N. 1979. Handbook of drugs and chemicals used in the treatment of fish diseases: a manual of fish pharmacology and materia medica. Charles C. Thomas, Springfield, Illinois.

Hoffman, G.L. and G. Schubert. 1984. Some parasites of exotic fishes, p. 233-261. *In* W.R. Courtney Jr. and J.R. Stauffer Jr. (eds.) Distribution, biology and management of exotic fishes. John Hopkins University Press, Baltimore.

Huet, M. 1972. Textbook of fish culture. Fishing News Books Ltd., Farnham, Surrey.

Indiq, F.E. and B. Moav. 1988. A prokaryotic gene is expressed in fish cells and persists in tilapia embryos and adults following microinjection, p. 221-225. *In* Y. Zohar and B. Breton (eds.) Reproduction in fish: basic and applied aspects of endocrinology and genetics. INRA Press, Paris.

Jacobsen, P. and L. Berglind. 1988. Persistance of oxytetracycline in sediments from fish farms. Aquaculture 70:365-370.

Johnstone, R., D.J. Macintosh and R.S. Wright. 1983. Elimination of orally administered 17a-methyltestosterone by *Oreochromis mossambicus* (tilapia) and *Salmo gairdneri* (rainbow trout) juveniles. Aquaculture 35: 249-257.

Joseph, M.M., Editor. 1988. The First Indian Fisheries Forum, Proceedings. Asian Fisheries Society, Indian Branch, Mangalore, India.

Kadri, S. 1988. A study of a screen filter's performance in treating salmonid hatchery effluent. University of Stirling, Stirling, Scotland. M.Sc. thesis.

Kapuscinski, A.R. and E.M. Hallerman. 1990. Transgenic fish and public policy: anticipating environmental impacts of transgenic fish. Fisheries 15(1):2-11.

Kaushik, S.J. 1980. Influence of nutritional status on the daily patterns of nitrogen excretion in the carp (*Cyprinus carpio* L.) and the rainbow trout (*Salmo gairdneri* R.). Reprod. Nutr. Dev. 20:1751-1765.

Kilambi, R.V., C.E. Hoffman, A.V. Brown, J.C. Adams and W.A. Wickizer. 1976. Effects of cage culture fish production upon the biotic and abiotic environment of Crystal Lake, Arkansas. Final Report to the Arkansas Game and Fisheries Commission and the US Department of Commerce. Department of Zoology, University of Arkansas, Fayetteville, Arkansas.

Krom, M.D., A. Neori and J. van Rijn. 1989. Importance of water flow rate in controlling water quality processes in marine and freshwater fish ponds. Isr. J. Aquacult. 41(4):23-33.

Lee, K.K.F. 1987. The economic impact of technological innovation on aquaculture development: the innovation cycle in Taiwan, 1953-1985. University of Hawaii, Honolulu. 139 p. Ph.D. dissertation.

Leventer, H. 1987. The contribution of silver carp *Hypophthalmichthys molitrix* to the biological control of reservoirs. Mekoroth Water Co., Nazareth, Israel. 103 p.

Little, D. and J.F. Muir. 1987. Integrated aquaculture. Institute of Aquaculture, Stirling, UK.

Lu, X. 1986. A review of reservoir fisheries in China. FAO Fish. Circ. 803.

Maclean, N. and D. Penman. 1990. The application of gene manipulation to aquaculture. Aquaculture 85:1-20.

Meffe, G.K. 1987. Conserving fish genomes: philosophies and practices. Environ. Biol. Fish. 18:3-9.

Merican, Z.O. and M.J. Phillips. 1985. Solid waste production from rainbow trout *Salmo gairdneri* Richardson cage culture. Aquacult. Fish. Manage. 16: 55-70.

Meyer, F.P. and R.A. Schnick. 1989. A review of chemicals used for the control of fish diseases. Rev. Aquat. Sci. 1: 693-710.

Michel, C. 1986. Practical value, potential dangers and methods of using antibacterial drugs in fish. Rev. Sci. Tech. Off. Int. Epiz. 5: 659-675.

Millar, S. 1984. Antibiotic residue study in rainbow trout and tilapia. Napier College of Commerce and Technology, Edinburgh. 60 p. M.I. Biol. thesis.

Milstein, A., M. Feldlite, N. Mozes, and Y. Avinmelech. 1989. Limnology of reservoirs used for fish farming and crop irrigation, with integrated free and cage fish culture. Isr. J. Aquacult. 41: 12-22.

Muir, J.F. 1981. Management in *Clarias* culture, Thailand. UNDP/FAO Project Report THA/75/012. 18 p. National Inland Fisheries Institute, Bangkok, Thailand.

Muir, J.F. and M.C.M. Beveridge. 1987. Water resources and aquaculture development. Arch. Hydrobiol. 28: 321-324.

Munro, K.A., S.C. Samis and M.D. Nassichuk. 1985. The effects of hatchery effluents on water chemistry, periphyton and benthic invertebrates of selected British Columbia streams. Canadian Fisheries and Aquatic Sciences. MS Report 1830, 147 p.

Musisi, L. 1984. The nutrition, growth and energetics of tilapia, *Sarotherodon mossambicus*. University of London, UK. Ph.D. thesis.

Nature Conservancy Council. 1991. Fish farming and the Scottish freshwater environment. Report prepared for the Nature Conservancy Council by the Institute of Aquaculture, University of Stirling, Institute of Freshwater Ecology, Penicuik; and the Institute of Terrestial Ecology, Banchory. NCC, Edinburgh. 283 p.

Niemi, M. and I. Taipalinen. 1982. Faecal indicator bacteria at fish farms. Hydrobiol. 86:171-175.

Nouws, J.F.M., J.L. Grondel, A.R. Schutte and J. Laurensen. 1989. Pharmacokinetics of ciprofloxacin in carp, African catfish and trout. Tijdschr. Diergeneeskd. 114:682-687.

OECD. 1982. Eutrophication of waters, assessment and control. Organisation of Economic Cooperation and Development, Paris.

Palijon, A.M. 1986. There's a big demand for bamboo. Greenfields 16(7):12-16.

Pantulu, V.R. 1979. Floating cage culture of fish in the lower Meking River basin, p. 423-427. *In* T.V.R. Pillay and W.D. Dill (eds.) Advances in aquaculture. Fishing News Books Ltd., Farnham, Surrey.

Parker, N. 1988. Fish culture in the United States, p. 268-301. *In* J. Shepherd and N. Bromage (eds.) Intensive aquaculture. Blackwell Professional Books Ltd., Oxford.

Paulson, L.J. 1980. Models of ammonia excretion for brook trout (*Salvelinus fontinalis*) and rainbow trout (*Salmo gairdneri*). Can J. Fish. Aquat. Sci. 37:1421-1425.

Penczak, T., W.. Galicka, M. Molinski, E. Kutso and M. Zalewski. 1982. The enrichment of a mesotrophic lake by carbon, phosphorus, and nitrogen from the cage culture of rainbow trout *Salmo gairdneri*. J. Appl. Ecol. 19:371-393.

Phillips, M.J., M.C.M. Beveridge and L.G. Ross. 1985. The environmental impact of salmonid cage culture in inland fisheries: present status and future trends. J. Fish Biol. 27(a): 123-137.

Phillips, M.J., M.C.M. Beveridge and J.A. Stewart. 1986. The environmental impact of cage culture on Scottish freshwaters, p. 504-508. *In* J.F. de L.G. Solbe (ed.) Effects of land use on fresh waters. Ellis Horwood Ltd., Chichester.

Phillips, M.J., M.C.M. Beveridge and R.M. Clarke. 1991. Impact of aquaculture on water resources, p. 568-591. *In* D.E. Brune and J.R. Tomasso (eds.) Advances in world aquaculture. Vol. 3. World Aquaculture Society.

Pike, I.H., G. Andorsdottir and H. Mundheim. 1990. The role of fish meal in diets for salmonids. IAFFM Tech. Bull. 24. 72 p. International Institute of Fish Meal Manufacturers, Potters Bar, UK.

Pollard, N. 1981. The Gezira Scheme - study in failure. Ecologist 11(1): 21-31.

Polprasert, C. 1989. Organic waste recycling. John Wiley, Chichester.

Pullin, R.S.V. 1986. The worldwide status of carp culture, p. 21-34. *In* R. Billard and J. Marcel (eds.) Aquaculture of cyprinids. Institut National de la Recherche Agronomique, Paris. 502 p.

Pullin, R.S.V., Editor. 1988. Tilapia genetic resources for aquaculture. ICLARM Conf. Proc. 16, 108 p.

Pullin, R.S.V. 1990. Down-to-earth thoughts on conserving aquatic genetic diversity. Naga, ICLARM Q. 13(1): 5-8.

Pursiainen, M., Editor. 1988. National contributions on suspended solids from land-based fish farms. Monistettuja Julkaisuja, Helsinki, Finland.

Rasmussen, F. 1988. Therapeutants used in fish production: pharmacokinetics, residues and withdrawal periods. EIFAC/XV/88/Inf. 13. 25 p. FAO, Rome.

Rast, W., M. Holland and S.-O. Ryding. 1989. Eutrophication management framework for the policy-maker. MAB Digest 1. United Nations Educational, Scientific and Cultural Organisation, Paris. 83 p.

Redding, T.A. and A.B. Midlen. 1991. Fish production in irrigation canals: a review. FAO Fish. Tech. Pap. 317. 111 p. FAO, Rome.

Rosenthal, H., D. Weston, R. Gowen and E. Black, Editors. 1987. Report of the *ad hoc* Study Group on "Environmental Impacts of Mariculture". ICES CM 1987/F:2. 81 p.

Samuelsen, O.B. 1989. Degradation of oxytetracyline in seawater at two different temperatures and light intensities, and the persistence of oxytetracycline in the sediment from a fish farm. Aquaculture 83:7-16.

Sarig, S. 1971. The prevention and treatment of diseases of warmwater fishes under subtropical conditions, with special emphasis on intensive fish farming. T.F.H. Publications, Neptune City, New Jersey.

Sarig, S. 1989a. The fish culture industry in Israel in 1988. Isr. J. Aquacult./Bamidgeh 41:50-57.

Sarig, S. 1989b. Introduction and state of art of aquaculture, p. 1-19. *In* M. Shilo and S. Sarig (eds.) Fish culture in warm water systems: problems and trends. CRC Press, Boca Raton, Florida.

Scholtissek, C., and E. Naylor. 1988. Fish farming and influenza pandemics. Nature 331:215.

Sevilleja, R.C. 1985. Tilapia production in freshwater fishponds of Central Luzon, Philippines, p. 115-126. *In* I.R. Smith, E.B. Torres and E.O. Tan (eds.) Philippine tilapia economics. ICLARM Conf. Proc. 12, 261 p.

Shamsudin, M.N. 1986. Bacterial examination of fingerlings of bighead carp (*Aristichthys nobilis*) and grass carp (*Ctenopharyngodon idella*) imported into Malaysia, p. 235-240. *In* J.L. Maclean, L.B. Dizon and L.V. Hosillos (eds.) The First Asian Fisheries Forum. Asian Fisheries Society, Manila, Philippines.

Shariff, M. 1989 Diseases of aquarium fishes and their treatment, p. 67-81. *In* F.M. Yusoff and K.T. Siow (eds.) The Proceedings of Aquarium Fish Industry - A New Frontier. Malaysian Fisheries Society, Johore Bahru, Malaysia.

Sharma, K.P. and M.M. Simlot. 1971. Piscicidal effect of temru *Diospyros cordifolia* Roxb. J. Inland Fish. Soc. India 3:57-62.

Shepherd, J. and N. Bromage. 1988. Intensive aquaculture. Blackwell Professional Books Ltd., Oxford.

Shotts, E.B., Jr. and J.B. Gratzek. 1984. Bacteria, parasites and viruses of aquarium fish and their shipping waters, p. 302-312. *In* W.R. Courtney, Jr. and J.R. Stauffer, Jr. (eds.) Distribution, biology and management of exotic fishes. John Hopkins University Press, Baltimore.

Skaala, O., G. Dahle, K.E. Jorstad and G. Naevdal. 1990. Interactions between natural and farmed fish populations: information from genetic markers. J. Fish Biol. 36:449-460.

Solbe, J.F. de L.G. 1982. Fish farm effluents: a United Kingdom survey. EIFAC Tech. Pap. 41: 29-56. FAO, Rome.

Stirling, H.P. and T. Dey. 1990. Impact of intensive cage fish farming on the periphyton and phytoplankton of a Scottish freshwater loch. Hydrobiologia 190:193-214.

Suzuki, R. 1986. Intensive carp rearing in Japan, p. 327-333. *In* R. Billard and J. Marcel (eds.) Aquaculture of cyprinids. Institut National de la Recherche Agronomique, Paris.

Tagarino, R.N. 1985. Economics of rice-fish culture systems, Luzon, Philippines, p. 127-150. *In* I.R. Smith, E.B. Torres and E.O. Tan (eds.) Philippine tilapia economics. ICLARM Conf. Proc. 12, 261 p.

Tan, B.Y. and K.T. Siow. 1989. The aquarium fish trade and industry in Singapore, p. 23-26. *In* F.M. Yusoff and K.T. Siow (eds.) The Proceedings of Aquarium Fish Industry - A New Frontier. Malaysian Fisheries Society, Johore Bahru, Malaysia.

Teichert-Coddington, D.R. and R.P. Phelps. 1989. Effects of seepage on water quality and productivity on inorganically fertilised tropical ponds. J. Aquacult. Trop. 4:85-92.

Teichert-Coddington, D.R., N. Stone and R.P. Phelps. 1988. Hydrology of fish culture ponds in Gualaca, Panama. Aquacult. Eng. 7:309-320.

Trust, T.J. and V.G. Money. 1972. Bacterial population of diets for aquarium fishes. J. Fish. Res. Board Can. 29:429-433.

Tucker, C.S., Editor. 1985. Channel catfish culture. Elsevier, Amsterdam.

Tucker, J.B. 1983. Schistosomiasis and water projects: breaking the link. Environment 25(7):17-20.

van Houtte, A.R., N. Bonucci and W.R. Edeson. 1989. A preliminary review of selected legislation governing aquaculture. ADCP/REP/89/42. 81 p. FAO, Rome.

van Rijn, J., S.R. Stutz, S. Diab and M. Shilo. 1986. Chemical, physical and biological parameters of superintensive concrete fish ponds. Isr. J. Aquacult. 38(2):35-42.

Welcomme, R.L. 1988. International introductions of inland aquatic species. FAO Fish. Tech. Pap. 294.

Zale, A.V. 1987. Growth, survival and foraging abilities of early life history stages of blue tilapia, *Oreochromis aureus,* and largemouth bass, *Micropterus salmoides*. Environ. Biol. Fish. 20: 113-128.

Environmental Issues in the Control of Bacterial Diseases of Farmed Fish

B. Austin
Department of Biological Sciences
Heriot-Watt University
Riccarton, Edinburgh EH14 4AS
Scotland

AUSTIN, B. 1993. Environmental issues in the control of bacterial diseases of farmed fish, p. 237-251. *In* R.S.V. Pullin, H. Rosenthal and J.L. Maclean (eds.) Environment and aquaculture in developing countries. ICLARM Conf. Proc. 31, 359 p.

Abstract

Fish disease control strategies pose certain actual and perceived environmental problems. These problems center around the possible escape of pathogens into the receiving waters, the dissemination of living and dead vaccines, and the release of pharmaceutical compounds. Such compounds may accumulate in filter feeders such as mussels, pass into potable water supplies, or induce the development of antibiotic resistant microbial communities.

Introduction

There has been a steady increase in aquaculture production since the end of World War II. The total worldwide production in 1987 was 13.2 million tonnes (FAO 1989), mostly from Asia (see also Pullin and Csavas, this vol.). Commensurate with the increase in production, aquaculture has undergone a transition from an ancient, relatively primitive craft to a science relying on sophisticated technology, and incorporating recent advances in genetics, microbiology, nutrition, process technology and sewage treatment. In industrialized countries, aquatic species are farmed intensively and, in the case of fin fish, the production rates may exceed 75 $t \cdot ha^{-1}$ with water replenishment and aeration. However, there is a need to supply all feed, which must contain high levels of animal protein. This compares with the less intensive/extensive methods (usually practiced in developing countries), which are largely natural and involve only limited or no supplementary feeding of the stock or fertilization of the water. Here, the production rates range from 0.05-0.5 $t \cdot ha^{-1}$ (Purdom 1977; Ackefors and Rosen 1979).

In general, aquaculture requires less energy per unit of protein than farming land animals. However, fish excrete ammoniacal wastes, which contribute to the organic content of the receiving waters. Moreover, management practices are often inappropriate, especially in terms of feeding and cleaning regimes: uneaten food may be allowed to accumulate in the fish holding areas; and dead animals may remain in close association with other supposedly healthy stock for considerable periods. These practices favor the development of large microbial communities, comprising a diverse array of taxa (Austin and Allen-Austin 1985).

Some of these organisms may be instrumental in causing diseases, particularly when the stocking densities are too high and the animals become stressed. Some diseases have the potential to cause significant mortalities among stock. Problems associated with the use of disease control strategies form the basis of this paper.

Bacterial Diseases of Farmed Fish

To date, over 50 bacterial species have been associated with disease in freshwater and marine fish (Table 1). Worldwide, the greatest economic losses have been caused by *Aeromonas hydrophila*, *A. salmonicida*, *Cytophaga-Flavobacterium-Flexibacter*,

Table 1. Bacterial diseases of fish (after Austin and Austin 1987).

Bacterial pathogen	Disease	Host range	Geographical distribution
Acinetobacter sp.	Acinetobacter disease	Atlantic salmon, channel catfish	Norway, USA
Aeromonas hydrophila	haemorrhagic septicaemia	most freshwater fish	worldwide
A. salmonicida	furunculosis	salmonids	worldwide
A. sobria	septicaemia	salmonids	Europe
Clostridium botulinum	botulism	salmonids	Denmark, UK
Cytophaga aquatilis	gill disease	salmon	USA
Edwardsiella ictaluri	enteric septicaemia of catfish	channel catfish	USA
Edw. tarda	edwardsiellosis	eels, channel catfish	Japan, USA
Eubacterium tarantellus	eubacterial meningitis	striped mullet	USA
Flavobacterium piscicida	flavobacteriosis	marine fish	USA
Fla. branchiophila	gill disease	marine fish	Japan, USA
Flexibacter columnaris	columnaris	freshwater fish	worldwide
Fle. maritimus	gill disease	black-, red seabass	Europe, Japan
Fle. psychrophilus	coldwater disease	salmonids	Australia, Europe, USA
Lactobacillus spp.	pseudokidney disease	salmonids	Canada, UK, USA
Mycobacterium spp.	mycobacteriosis	most fish species	worldwide
Nocardia spp.	nocardiosis	most fish species	worldwide
Pasteurella piscicida	pseudotuberculosis	white perch, striped bass, yellowtail	USA
Planococcus sp.	planococcicoccis	rainbow trout	UK
Plesiomonas shigelloides	plesiomonas septicaemia	rainbow trout	Portugal, UK
Providencia rettgeri	septicaemia	silver carp	Israel
Pseudomonas anguilliseptica	red spot	eels	Japan, UK
Ps. chlororaphis	septicaemia	amago trout	Japan
Ps. fluorescens	septicaemia, fin rot	most fish species	worldwide
Renibacterium salmoninarum	bacterial kidney disease	salmonids	Europe, Japan, North and South America
Serratia liquefaciens	serratia septicaemia	Atlantic salmon	UK
Sporocytophaga sp.	saltwater columnaris	salmonids	UK, USA
Staphylococcus epidermidis	septicaemia	red seabream, yellowtail	Japan
Streptococcus spp.	streptococcicosis	many fish species	Japan, South Africa, USA
Streptoverticillium salmonis	streptomycosis	salmonids	USA
Vibrio alginolyticus	septicaemia	seabream	Israel
V. anguillarum	vibriosis	most marine fish	worldwide
V. carchariae	vasculitis	sharks	worldwide
V. cholerae (non-01)	vibriosis	ayu	Japan
V. damsela	vibriosis	damsel fish	USA
V. salmonicida	Hitra disease	Atlantic salmon	Norway, UK
V. vulnificus	vibriosis	eels	Japan, North America
Yersinia ruckeri	enteric red mouth	salmonids	Australia, Europe, North America

Edwardsiella tarda, *Pasteurella piscicida*, *Renibacterium salmoninarum*, *Streptococcus* spp., *Vibrio* spp. and *Yersinia ruckeri*. Often, the occurrence of disease stems from bad husbandry and adverse water quality conditions, which debilitate the animals leaving them vulnerable to opportunistic invaders. A typical example would be the occurrence of enteric redmouth disease, which is influenced by water quality factors (Austin and Austin 1987). In particular, crowded conditions cause high levels of ammonia and metabolic waste levels in the water and decrease the amount of dissolved oxygen, which precipitates epizootics (Bullock and Snieszko 1975). Similarly, finrot and vibriosis are exacerbated in polluted water (Bullock and Snieszko 1975; Ziskowski and Murchelano 1975; Minchew and Yarbrough 1977; Rfdsaether et al. 1977).

Disease Control Strategies

The methods used for controlling fish diseases are summarized in Table 2. Certainly, there is debate about what may be done to control diseases in wild stock other than improving water quality (by controlling pollutants) and preventing the release of pathogens and infected fish from aquaculture facilities. Yet, both of these possibilities lack appropriate supportive scientific data. However, much progress has been made in controlling diseases of farmed stock. Besides the more common reliance on vaccines and antimicrobial compounds, there is good evidence that improved husbandry and management practices, movement restrictions and use of genetically (disease) resistant fish strains contribute to overall disease control strategies.

The severe economic pressures facing some areas of aquaculture provide temptation to maximize production regimes, which introduce stress (an abstract term with no precise meaning) related problems into the stock. There may be further complications caused by adopting inadequate management practices in which aeration and water flow are insufficient, overfeeding occurs, and hygiene declines below the threshold level allowing disease to occur (Austin and Austin 1987). It is good practice to prevent the accumulation of organic material (feces and uneaten food) and biofouling communities, regularly remove and destroy dead or moribund animals, ensure the proper disinfection of items entering the fish-holding facilities, and ensure adequate aeration (Austin and Austin 1987).

The benefits of genetically resistant stock for disease control have been realized for many decades (e.g., Embody and Hayford 1925). Similarly, the role of dietary supplements in controlling disease outbreaks, e.g., the use of vitamin C for mediating infections caused by *Edwardsiella*

Table 2. Methods for controlling fish diseases.

Classification of fish stocks	Disease control measure
Wild fish	Control pollution (water quality)
	Control dissemination of pathogens and release of diseased animals from fish farms
Farmed fish	Adoption of suitable husbandry/management practices
	Restriction on movement of infected stock
	Use of genetically (disease) resistant fish strains
	Use of adequate diets or, where appropriate, dietary supplements
	Use of vaccines (prophylaxis)
	Use of antimicrobial compounds (chemoprophylaxis and chemotherapy)

ictaluri (Durve and Lovell 1982), have been documented. However, aquaculture has been slow in adopting the concepts. The policy of preventing the movement of infected stock and thereby reducing the risk of transferring disease, has usually met with strong resistance from aquaculturists. Clearly to be effective, such policies must have the support of the industry and ensure adequate compensation for diseased stock.

In human and veterinary medicine, vaccines may be considered a primary prophylactic tool. There is a general acceptance of the value of vaccines in aquaculture. However, effective formulations, i.e., formalin inactivated cultures, are available for only a few pathogens, namely *Aeromonas salmonicida*, *Vibrio anguillarum*, *V. ordalii*, *V. salmonicida* and *Yersinia ruckeri* (see Austin and Austin 1987). Many more formulations are necessary if vaccines are to have an overall beneficial effect on aquaculture production.

An attitude prevails that antimicrobial compounds are the mystical savior of all the woes of fish farming. This is a pity because, in the author's opinion, antimicrobial compounds should be regarded as the last line of defence against disease, to be used only when all other possibilities have been exhausted. Instead, there is a tendency to consider antimicrobial compounds as a primary (and in some cases, the only) means of disease control. It is conceded that antimicrobial compounds control many of the bacterial diseases which are troublesome in aquaculture (Table 3). However, there are many problems appertaining to the use of antimicrobial compounds in fisheries.

Table 3. Antimicrobial compounds commonly used to control bacterial diseases of fish (after Austin and Austin 1987).

Antimicrobial compound	Diseases controlled	Method of administration/dosage
Ampicillin	coldwater disease, streptococcicosis	immersion; 5-10 $mg{\cdot}l^{-1}$ of water for several hours
Benzalkonium chloride	fin rot, gill disease	immersion; 1-2 $mg{\cdot}l^{-1}$ of water for 1 hour
Chloramine B or T	fin rot, gill disease, mycobacteriosis	immersion; 18-20 $mg{\cdot}l^{-1}$ of water at pH 7.5-8.0 for up to 1 hour repeat for 2-3 days
Chloramphenicol	carp erythrodermatitis, columnaris, enteric redmouth, fin rot, furunculosis, septicaemias, pasteurellosis, ulcers, vibriosis	food additive; 50-75 $mg{\cdot}kg^{-1}$ of food daily for 5-10 days immersion; 10-50 $mg{\cdot}l^{-1}$ of water for up to 1 hour
Chlortetracycline	coldwater disease saltwater columnaris	food additive; 10-20 $mg{\cdot}kg^{-1}$ of food, feed for 1-14 days
Copper sulfate	columnaris, fin rot	immersion; 1-5 $mg{\cdot}l^{-1}$ of water for up to 1 hour
Doxycycline	mycobacteriosis, streptococcicosis	food additive; 500 $mg{\cdot}kg^{-1}$ of food

continued

Table 3 continued

Antimicrobial compound	Diseases controlled	Method of administration/dosage
Erythromycin	bacterial kidney disease (BKD)	food additive; 10-100 $mg{\cdot}kg^{-1}$ of fish daily for 10-21 days
Flumequine	furunculosis	food additive; 6 $mg{\cdot}kg^{-1}$ of fish daily for 6 days
Furanace	coldwater disease, columnaris, fin rot, gill disease, septicaemias, vibriosis	food additive; 2-4 $mg{\cdot}kg^{-1}$ of fish daily for 3-5 days immersion; 0.5-1.0 $mg{\cdot}l^{-1}$ of water for 5-10 minutes
Furazolidone	furunculosis, vibriosis	food additive; 25-75 $mg{\cdot}kg^{-1}$ of fish daily for 10-20 days
Iodophors	acinetobacter disease, BKD, flavobacteriosis, bacteriosis, furunculosis, septicaemias, mycobacteriosis	immersion; 50-200 mg of available iodine per liter of water for 10-15 minutes
Kanamycin	fin rot, mycobacteriosis, septicaemias, vibriosis	food additive; 50 $mg{\cdot}kg^{-1}$ of of fish daily for 7 days
Malachite green	columnaris, fin rot, gill disease, fungal infections	immersion; 1-5 $mg{\cdot}l^{-1}$ of water for up to 1 hour
Methylene blue	enteric redmouth	immersion; 1-3 $mg{\cdot}l^{-1}$ of water for 3-5 days
Nitrofurantoin	vibriosis	immersion; 50 $mg{\cdot}l^{-1}$ of water for 1 hour
Oxolinic acid	columnaris, enteric redmouth, furunculosis, septicaemias, vibriosis	food additive; 10 $mg{\cdot}kg^{-1}$ of fish daily for 10 days
Oxytetracycline	acinetobacter disease, coldwater disease, columnaris, edwardsiellosis, enteric redmouth, fin rot, furunculosis, gill disease, streptocociccosis	food additive; 75 $mg{\cdot}kg^{-1}$ of fish daily for 10 days
Penicillin G	BKD	food additive; 75 $mg{\cdot}kg^{-1}$ fish daily for 21 days
Quaternary ammonium compounds	coldwater disease, columnaris, gill disease	immersion; 1-2 $mg{\cdot}l^{-1}$ of water for 1 hour
Sodium nifurstyrenate	streptococcicosis	food additive; 50 $mg{\cdot}kg^{-1}$ of fish daily for 3-5 days

continued

Table 3 continued

Antimicrobial compound	Diseases controlled	Method of administration/dosage
Streptomycin	mycobacteriosis, septicaemias	food additive; 50-75 $mg{\cdot}kg^{-1}$ of fish daily for 5-10 days
Sulfisoxazole	BKD, coldwater disease, columnaris, furunculosis, mycobacteriosis	food additive; 200 $mg{\cdot}kg^{-1}$ of fish daily for 14 days
Sulfamerazine	BKD, coldwater disease, columnaris, enteric redmouth, furunculosis, septicaemias	food additive; 200 $mg{\cdot}kg^{-1}$ of fish daily for 14 days
Sulfamethazine	BKD, coldwater disease, columnaris, furunculosis, vibriosis	food additive; 100-200 $mg{\cdot}kg^{-1}$ of fish daily for 10-20 days
Sulfonamide-potentiated	enteric redmouth, furunculosis, plesiomonas septicaemia, vibriosis	food additive; 30 $mg{\cdot}kg^{-1}$ of fish daily for 5-10 days
Tetracycline	columnaris, furunculosis, streptococcicosis	food additive; 75-100 $mg{\cdot}kg^{-1}$ of fish daily for 10-14 days
Tiamulin	enteric redmouth	food additive; 5 $mg{\cdot}kg^{-1}$ of fish daily for 14 days

Environmental Problems Associated with Aquaculture

Even the most carefully managed aquaculture facility is capable of causing environmental problems relating to:

- the release of organic material, particularly feces and uneaten food;
- the dissemination of micro-organisms, including pathogens;
- the accidental or deliberate release of dead or live (genetically engineered or crippled) vaccines;
- the release of pharmaceutical or bioactive compounds; and
- the release of live (genetically modified) and living or dead diseased fish.

Although discussion will focus primarily on environmental problems associated with disease control strategies, it is pertinent to consider additional environmental issues in supposedly healthy farms.

The Release of Organic Material from Fish Farms

It is recognized that large quantities of organic material, in the form of feces and uneaten food, escape from fish-holding facilities (Austin and Allen-Austin 1985). Moreover, it may be assumed that these quantities are maximal during periods of tank cleaning. Their effect on the environment is largely unknown. Conceivably, the organic material would serve as an excellent nutrient source for native aquatic communities of both micro- and macroorganisms. Research is sorely needed to determine the significance of organic material on receiving waters and sediment.

The Dissemination of Microorganisms from Fish Farms

Research has shown that there is a net contribution of microorganisms, notably bacteria, from fish farms to the aquatic environment (Table 4), with the problem appearing to be more pronounced in marine rather than freshwater sites (Austin and Allen-Austin 1985). Data from a freshwater reservoir fishery in England indicated that bacterial populations in the effluent exceeded those of inflow waters on five occasions (out of a total of seven), albeit without apparent seasonality. Ironically, parallel data revealed that microbial populations in the fish-holding facilities consistently exceeded those of the effluent (Austin and Allen-Austin 1985). It was suggested that the higher counts in tank water reflected localized enhanced nutrient levels. Perhaps the microorganisms would be released in the effluent during periods of tank cleaning. Yet overall, there was no evidence of any major imbalance in the composition of the bacterial microflora leaving the reservoir fishery via the effluent (Allen et al. 1983).

In a parallel investigation, there was very limited seasonal fluctuation in the aerobic heterotrophic bacterial counts obtained from a freshwater fish farm in England (Table 4). As with the reservoir site, water from the fishponds contained more bacteria than inflow or effluent (Austin and Allen-Austin 1985). In 12 direct comparisons, the number of bacteria in effluent was higher than in inflow water on six occasions, i.e., 50% of the total (Table 4).

The difference between sizes of microbial populations in inflow and effluent waters was most pronounced in an English coastal marine fishery (Table 4). Here, the numbers of aerobic heterotrophic bacteria in effluent were

Table 4. The numbers of aerobic heterotrophic bacteria in the inflow and effluent waters of fish farms (data from Austin 1983; Austin and Allen-Austin 1985).

Type of fish farm	Period of sampling	Numbers of bacteria (ml^{-1}) Inflow	Effluent
Freshwater, reservoir, rearing rainbow trout	February	1.3×10^3	1.8×10^3
	April	1.03×10^6	8.6×10^7
	May	3.0×10^5	1.0×10^6
	June	7.8×10^7	6.52×10^5
	July	6.7×10^3	1.0×10^4
	August	6.1×10^5	6.3×10^5
	October	1.3×10^4	7.3×10^3
Freshwater, river, rearing rainbow trout	January	8.8×10^4	7.6×10^4
	February	2.73×10^5	3.9×10^4
	March	2.7×10^4	1.8×10^4
	April	1.9×10^4	2.1×10^4
	May	2.2×10^4	2.3×10^4
	June	2.1×10^4	1.4×10^4
	July	2.0×10^3	4.0×10^3
	August	1.5×10^4	7.0×10^3
	September	2.9×10^4	4.5×10^4
	October	1.6×10^4	3.9×10^4
	November	1.9×10^4	1.4×10^4
	December	1.1×10^4	3.2×10^4
Marine site, producing turbot	January	1.0×10^4	5.0×10^5
	April	2.5×10^6	4.0×10^7
	August	4.0×10^9	1.1×10^{10}
	November	4.0×10^8	8.0×10^9

generally at least ten fold higher than those from inflow water. Moreover, there was evidence of a seasonal fluctuation in the bacterial numbers, with minimum and maximum counts present in winter and summer, respectively (Table 4).

A comparison of the bacterial composition of the inflow and effluent waters of the marine fish farm revealed many changes. In particular, there was a decrease in the proportion of *Acinetobacter calcoaceticus*, *Hyphomicrobium*, *Hyphomonas*, *Micrococcus*, *Photobacterium*, *Pseudomonas*, *Staphylococcus* and *Vibrio*, with a commensurate increase in *Bacillus*, coryneforms, Enterobacteriaceae representatives and *Prosthecomicrobium* leaving the site via the effluent (Austin 1982, 1983). Although most of these taxa are of limited relevance to the receiving waters, the presence of Enterobacteriaceae representatives is undesirable insofar as the family contains human pathogens and, therefore, constitutes a potential public health risk.

The Release of Live Fish

Undoubtedly, live fish escape from aquaculture facilities but the effect on the environment is unclear. There are unpublished data that, in freshwater, salmonid escapees may outcompete some members of the wild fish populations to the annoyance of anglers. In some English river systems, rainbow trout which have presumably escaped from fish farms, have apparently outcompeted the native brown trout. However, the long-term environmental effects remain unclear. Moreover, it is unknown what damage may result from the accidental release of genetically modified fish.

Environmental Problems Associated with Disease Management Strategies

The presence of disease within farmed stock and the resulting disease control strategies will pose certain actual and perceived environmental problems. Such problems center on the possible escape of pathogens directly into the receiving waters or via diseased fish, the dissemination of living or dead vaccines, and the release of pharmaceutical compounds.

Escape of Pathogens

It is difficult to prove that aquaculture poses any problems to the environment in terms of the release of pathogens directly into the waterways or indirectly via dead or moribund (diseased) fish. In short, there is a dearth of published information. Yet, it seems probable that pathogens will indeed escape from farms, and this situation could explain the spread of diseases between neighboring farm sites and to wild fish stocks.

Problems with Vaccines

Currently, most vaccines are administered by bathing (see Austin 1984). It is the author's personal experience that, after use, the spent vaccine is often poured into the outflow of the farm. Thus, pathogens (albeit killed) would be released into the waterway. In ecological and epizootiological studies employing serological or DNA probe technology, the presence of such cells could give a false impression of the presence of potential pathogens in the aquatic environment. However, of greater importance is the possible use of live (nonvirulent) vaccines and the perceived risk of a return to a virulent state following dissemination in the aquatic environment. This possibility could be ecologically disastrous.

Pharmaceutical Compounds

There is a small but steadily increasing literature indicating environmental problems resulting from use of pharmaceutical compounds in aquaculture (e.g. Jacobsen 1988; Samuelsen 1989). These problems center on the unexpected longevity of bioactive substances in animal tissues, the fate of compounds in the environment, and the development and transfer of resistance in microbial communities.

TISSUE RESIDUES

The persistence of antimicrobial compounds in fish tissues has a tenuous link with environmental issues. However, for completion, the topic merits discussion here. Certainly, there is an increasing literature indicating that pharmaceutical compounds linger in fish tissues for greater periods than had hitherto been recognized. For example, McCracken et al. (1976) established that trimethoprim remained in rainbow trout muscle for 77 days after the cessation of treatment. After statistical modeling, Salte and Liestøl (1983) recommended that for rainbow trout maintained at a water temperature below 10°C, a withdrawal period of 60 days is necessary when using oxytetracycline and potentiated sulfonamides. This author's personal experience suggests that this period is much longer than normally practiced on fish farms.

DISCHARGE OF PHARMACEUTICAL COMPOUNDS IN THE AQUATIC ENVIRONMENT

The widespread use of antimicrobial compounds in aquaculture has generated fears about the potential release of the bioactive component into the aquatic environment. This could lead to uptake with potable water supplies and thus cause risks of allergic type reactions in human consumers. In addition, damage could occur in biological filtration systems. It is possible to make calculations on the maximum level of antimicrobial substances which could enter waterways from fish farm effluent (Table 5). Such calculations assume that the manufacturer's recommended dosage has been administered on a given occasion to all the stock on a farm and that all the bioactive ingredient has been voided immediately from the site. For these calculations, an estimated minimum water flow of 500 $m^3 \cdot day^{-1}$ for each tonne of fish (Stevenson 1980) has been assumed. These worst-case

Table 5. Theoretical maximum levels of antimicrobial compounds leaving aquaculture facilities following administration of chemotherapeutic regimes. These calculations assume that the manufacturer's recommended dosage has been administered, and that all the compound is voided immediately from the animals via effluent.

Antimicrobial compound	Theoretical maximum concentration leaving fish farm in effluent
Ampicillin	1:100
Benzalkonium chloride	1:500
Chloramine B or T	1:50
Chloramphenicol	1:6,666,666
Chlortetracycline	1:1,250,000,000
Copper sulfate	1:200
Doxycycline	1:1,000,000
Erythromycin	1:5,000,000
Flumequine	1:83,333,333
Furanace	1:125,000,000
Furazolidone	1:6,666,666
Iodophors	1:5
Kanamycin	1:10,000,000
Malachite green	1:200
Methylene blue	1:333
Nitrofurantoin	1:10,000,000
Oxolinic acid	1:50,000,000
Oxytetracycline	1:6,666,666
Penicillin G	1:6,666,666
Quaternary ammonium compounds	1:500
Sodium nifurstyrenate	1:10,000,000
Streptomycin	1:6,666,666
Sulphisoxazole	1:2,500,000
Sulphamerazine	1:2,500,000
Sulphamethazine	1:2,500,000
Sulphonamides - potentiated	1:16,666,666
Tetracycline	1:5,000,000
Tiamulin	1:100,000,000

scenarios indicate that use of oral medication leads to only negligible quantities of antimicrobial compounds leaving aquaculture facilities in the effluent. However, the use of baths may cause high concentrations of antimicrobial compounds to reach the aquatic environment. In the absence of any supportive experimental data, the calculations suggest that discharge of antimicrobial compounds does not constitute a realistic environmental problem. However, recent published data suggest that only 20-30% of antimicrobial compounds are actually taken up by fish from medicated food. Thus, approximately 70-80% reaches the environment (Jacobsen 1988; Samuelsen 1989), notably from uneaten food (Jacobsen and Berglind 1988). If this is correct, then of the 48.5 t of antibiotics used in Norwegian fish farms during 1987 (Gray 1990), 34-39 t reached the environment.

With oxytetracycline in seawater, it has been established that degradation proceeds rapidly. Experiments carried out with continual illumination indicated that at 5° and 15°C, degradation occurred within 128 and 168 hours, respectively. In darkness, these times increased to 390 and 234 hours (Samuelsen 1989). However, most oxytetracycline becomes bound to particulate matter and is deposited at the bottom of (or beneath) the fish-holding facilities, in the case of marine cage sites. Here, oxytetracycline may remain in concentrations capable of causing antibacterial effects for 12 weeks after the cessation of treatment (Jacobsen and Berglind 1988). In conditions where sedimentation is slow, the half life of oxytetracycline is 32 days. Yet if this antibiotic containing material becomes covered with 4 cm of sediment, degradation of the bioactive substance proceeds even more slowly. With this scenario, it has been reported that the half life is 64 days (Samuelsen 1989). Such antibiotic-containing sediment affects the fauna. For example, detectable levels of oxytetracycline have been found in blue mussels (*Mytilus edulis*) located 80 m from a fish farm using the same antibiotic (Møster 1986).

DEVELOPMENT OF ANTIBIOTIC-RESISTANT MICROBIAL COMMUNITIES

A very realistic problem is the development and spread of antibiotic resistance among members of the native aquatic microbial communities. It has been determined that the administration of medicated food has a dramatic effect on the microbial populations within the digestive tract. In one study, it was determined that use of oxolinic acid, oxytetracycline and sulfafurazole (which are used to combat infections caused by Gram-negative bacteria) caused a substantial increase in bacterial numbers throughout the gastrointestinal tract, with maximal populations in the lower intestine. The bacteria, comprising an essentially different range of taxa to those from untreated fish, were generally resistant to the antibiotics in use (Austin and Al-Zahrani 1988). In contrast, erythromycin and penicillin G, which are used to treat diseases caused by some Gram-positive bacterial fish pathogens, resulted in a rapid reduction in bacterial populations within the digestive tract. Again, the microorganisms were largely resistant to the antimicrobial compounds in use (Austin and Al-Zahrani 1988). Moreover, it was observed that antibiotic-resistant bacteria were shed, via the feces, into the surrounding water.

The effect of chemotherapeutants on bacterial populations within freshwater aquaculture facilities was addressed by Austin (1985a). Effectively at three sites which used antimicrobial compounds (oxolinic acid, oxytetracycline and potentiated sulfonamide) to combat serious bacterial diseases, there was a statistically significant influence on the bacterial communities in effluent (Table 6). Overall, there was a reduction in the number of

Table 6. Effect of chemotherapeutants on bacterial communities in fish farm effluent (based on Austin 1985a).

Freshwater site	Nature of chemotherapy regime	No. of aerobic heterotrophic bacteria (ml^{-1}) Inflow	Effluent
A	nil	3.0×10^2	3.7×10^3
B	7 days before therapy	2.4×10^3	5.6×10^3
	during therapy with oxolinic acid	3.47×10^3	2.26×10^3
	1 day after conclusion of therapy	4.04×10^3	4.0×10^3
C	1 day before therapy	2.13×10^3	4.0×10^3
	during therapy with potentiated sulfonamide	3.62×10^3	2.14×10^3
	7 days after conclusion of therapy	3.19×10^3	6.76×10^3
D	18 days before therapy	2.24×10^3	3.87×10^3
	during therapy with oxytetracycline	2.37×10^3	1.73×10^3
	9 days after conclusion of therapy	2.72×10^3	4.32×10^3

bacteria and range of taxa in effluent during periods of chemotherapy. Of course, this is not direct proof of inhibition by antimicrobial compounds and could reflect other changes in the fish farm environment, such as reduced quantities of organic material which may result from reduced feeding regimes during periods of diseases. However, the effluent communities increased in size within a few days of stopping treatment (Table 6). Of greater importance, it was realized that use of an antimicrobial compound led to an increased level of resistance to it among bacteria in the effluent. This increased level of resistance decreased soon after the cessation of therapy. In short, it was concluded that use of an antimicrobial compound resulted in an increased proportion of antibiotic-resistant bacteria leaving fish farms via effluent.

Plasmids, conferring antibiotic resistance properties, abound in fish pathogens (e.g., Toranzo et al. 1983; Aoki 1989) and native aquatic bacteria (Burton et al. 1982), particularly those in the vicinity of fish-holding facilities (e.g., Aoki and Egusa 1971; Aoki 1975; Aoki et al. 1977b, 1985). Workers have provided evidence of a widespread resistance to antimicrobial compounds (including numerous cases of multiple resistance; see Aoki 1989) among fish pathogens, notably *Aeromonas hydrophila* (Aoki et al. 1971a), *A. salmonicida* (Aoki et al. 1971b), *Edwardsiella tarda* (Aoki et al. 1977a), *Pasteurella piscicida* (Aoki and Kitao 1985), *Pseudomonas fluorescens* (Aoki et al. 1977b), *Streptococcus* spp. (Aoki et al. 1990), and *Vibrio anguillarium* (Aoki et al. 1974). For example, Aoki et al. (1985) reported 111 out of 139 strains (80% of the total) of *Vibrio anguillarum* were resistant to tetracyclines. Toranzo et al. (1984) found that 88% of bacteria isolated from cultured rainbow trout were antibiotic resistant, by means of plasmids. Takashima et al. (1985) indicated that 18/262 isolates (8% of the total) of *Pasteurella piscicida* were transferably resistant. It is conceivable that plasmid-mediated antibiotic resistance could be transferred from fish farm organisms to bacteria of human and/or veterinary significance. Initial work (Austin and Helon, unpubl.

data) has suggested that antibiotic resistance may indeed be transferred between related bacterial groups; in this pilot study, resistance was transferred from *Pseudomonas* to *Chromobacterium*.

Recommendations

It is clear that much needs to be done to reduce environmental problems associated with aquaculture. In particular, it is argued that the use of antimicrobial compounds and, perhaps, vaccines should be controlled by Codes of Practice. Already, such a Code of Practice has been proposed to regulate the use of antimicrobial compounds in aquaculture (Austin 1985b). It is worthwhile reiterating the salient points of this Code:

1. restriction in the availability of antimicrobial compounds to qualified personnel;
2. use of antimicrobial compounds to be strictly in accordance with the instructions;
3. storage of antimicrobial compounds to be in the prescribed manner for the effective life of the product;
4. access to pharmaceutical compounds to be denied to laymen and inexperienced personnel;
5. suitable withdrawal periods for chemotherapeutants to be used before animals are removed from the aquaculture facility;
6. release of pharmaceutical compounds into the aquatic environment to be prevented;
7. medically important compounds to be banned or restricted to use with certain specified important diseases;
8. rotation in the use of various categories of antimicrobial compounds;
9. unused antimicrobial compounds to be disposed of safely; and
10. a program of surveillance to be adopted to ensure that the Code of Practice is carried out.

The primary reasons for drafting this Code of Practice were to prevent/reduce the development of bacterial resistance and to protect the human consumer. Parallel schemes have been adopted in certain branches of human medicine (e.g., Modr 1982; Walton 1982). The obvious benefit stems from bringing an awareness of the problems involved to all interested parties. Although most of the points listed in the suggested Code of Practice are self-evident, it is necessary to explain the reasoning for others. In particular, medical compounds should be used prudently to reduce the risk of resistance developing among (medically) useful pharmaceutical compounds, and to avoid recrimination from the medical profession. However, it has been demonstrated for human pathogens that when the use of antimicrobial compounds is restricted or stopped, resistance to them declines (Forfar et al. 1966; Phillips and Cooke 1982). Therefore, rotating the use of antimicrobial compounds may be effective in preventing the development of a resistant microbial community (Walton 1982).

References

Ackefors, H. and C. Rosen. 1979. Farming aquatic animals. Ambio 8:132-142.

Allen, D.A., B. Austin and R.R. Colwell. 1983. Numerical taxonomy of bacterial isolates associated with a freshwater fishery. J. Gen. Microbiol. 129:2043-2062.

Aoki, T. 1975. Effects of chemotherapeutants on bacterial ecology in the waters of ponds and the intestinal tracts of cultured ayu (*Plecoglossus altivelis*). Japan J. Microbiol. 19:7-12.

Aoki, T. 1989. Ecology of antibiotic-resistant determinants of R plasmids from fish pathogenic bacteria, p. 571-576. *In* T. Hattori, Y. Ishida, Y. Maruyama, R.Y. Morita and A. Uchida (eds.) Recent advances in microbial ecology. Japan Scientific Societies Press, Tokyo.

Aoki, T. and S. Egusa. 1971. Drug sensitivity of *Aeromonas liquefaciens* isolated from freshwater fish. Bull. Japan. Soc. Sci. Fish. 37:176-185.

Aoki, T., S. Egusa, Y. Ogata and T. Watanabe. 1971a. Detection of resistance factors in the fish pathogen *Aeromonas liquefaciens*. J. Gen. Microbiol. 65:343-349.

Aoki, T., S. Egusa, T. Kimura and T. Watanabe. 1971b. Detection of R factors in naturally occurring *Aeromonas salmonicida* strains. Appl. Microbiol. 22:716-717.

Aoki, T., S. Egusa and T. Arai. 1974. Detection of R factors in naturally occurring *Vibrio anguillarum* strains. Antimicro. Ag. Chemother. 6:534-538.

Aoki, T., T. Arai and S. Egusa. 1977a. Detection of R plasmids in naturally occurring fish pathogenic bacteria, *Edwardsiella tarda*. Microbiol. Immunol. 21:77-83.

Aoki, T., T. Kitao and T. Arai. 1977b. R plasmids in fish pathogens, p. 39-45. *In* S. Mitsuhashi, L. Rosival and V. Krcméry (eds.) Plasmids, medical and theoretical aspects. Springer Verlag, Berlin.

Aoki, T. and T. Kitao. 1985. Detection of transferrable R plasmids in strains of the fish-pathogenic bacterium, *Pasteurella piscicida*. J. Fish Dis. 8:345-350.

Aoki, T., T. Kanazawi and T. Kitao. 1985. Epidemiological surveillance of drug resistant *Vibrio anguillarum* strains. Fish Pathol. 20:199-205.

Aoki, T., K. Takami and T. Kitao. 1990. Drug resistant non-hemolytic *Streptococcus* sp. isolated from cultured yellowtail, *Seriola quinqueradiata*, a marine fish. Dis. Aquat. Organ. 8:171-177.

Austin, B. 1982. Taxonomy of bacteria isolated from a costal marine, fish-rearing unit. J. Appl. Bacteriol. 53:253-268.

Austin, B. 1983. Bacterial microflora associated with a coastal, marine fish-rearing unit. J. Mar. Biol. Assoc. UK 63:585-592.

Austin, B. 1984. The future of bacterial fish vaccines. Vaccine 2:249-254.

Austin, B. 1985a. Antibiotic pollution from fish farms: effects on aquatic microflora. Microbiol. Sci. 2:113-117.

Austin, B. 1985b. Chemotherapy of bacterial fish diseases, p. 19-26. *In* A.E. Ellis (ed.) Fish and shellfish pathology. Academic Press, London.

Austin, B. and D. Allen-Austin. 1985. Microbial quality of water in intensive fish rearing. J. Appl. Bacteriol. Suppl. 207S-226S.

Austin, B. and D.A. Austin. 1987. Bacterial fish pathogens: diseases of farmed and wild fish. Ellis Horwood, Chichester.

Austin, B. and A.M.J. Al-Zahrani. 1988. The effect of antimicrobial compounds on the gastrointestinal microflora of rainbow trout, *Salmo gairdneri* Richardson. J. Fish Biol. 33:1-14.

Bullock, G.L. and S.F. Snieszko. 1975. Hagerman redmouth, a disease of the Enterobacteriaceae. Fishery leaflet 42, Department of the Interior, Fish and Wildlife Service, Washington, DC.

Burton, N.F., M.J. Day and A.T. Bull. 1982. Distribution of plasmids in clean and polluted sites in a South Wales river. Appl. Environ. Microbiol. 44:1,026-1,029.

Durve, V.S. and R.T. Lovell. 1982. Vitamin C and disease resistance in channel catfish (*Ictalurus punctatus*). Can. J. Fish. Aquat. Sci. 39:948-951.

Embody, G.C. and C.O. Hayford. 1925. The advantage of rearing brook trout fingerlings from selected breeders. Trans. Am. Fish. Soc. 55:135-142.

FAO. 1989. Aquaculture production (1984-1987). FAO Fish. Circ. 815. Rev. 1. FAO, Rome.

Forfar, J.O., A.J. Keay, A.F. MacCabe, J.C. Gould, and A.D. Bain. 1966. Liberal use of antibiotics and its effect in neonatal staphylococcal infection with particular reference to erythromycin. Lancet ii:295-300.

Gray, J. 1990. Antibiotic use in Norwegian fish farms. Mar. Pollut. Bull. 21:4.

Jacobsen, P. 1988. Legemiddelbrukens påvirkning av oppdrettsmiljfet. Norsk Fiskeoppdrett 1:38-39.

Jacobsen, P. and L. Berglind. 1988. Persistence of oxytetracycline in sediments from fish farms. Aquaculture 70:365-370.

McCracken, A., S. Fidgeon, J.J. O'Brien and D. Anderson. 1976. An investigation of antibiotic and drug residues in fish. J. Appl. Bacteriol. 40:61-66.

Minchew, C.D. and B.D. Yarbrough 1977. The occurrence of Fin rot in mullet (*Mugil cephalus*) associated with crude oil contamination of an estuarine pond-ecosystem. J. Fish Biol. 10:319-323.

Modr, Z. 1982. Statutory control of antibiotic use in man versus voluntary restriction, p. 211-226. *In* C.H. Stuart-Harris and D.M. Harris (eds.) The control of antibiotic resistant bacteria. Academic Press, London.

Møster, G. 1986. Bruk av antibiotika i fiskeoppdrett Sogn og Fjordane Distriktshøgskole, 5800 Sogndal, Norway. (In Norwegian)

Phillips, I. and D. Cooke. 1982. The control of antibiotic prescribing in a London teaching hospital, p. 201-209. *In* C.H. Stuart-Harris and D.M. Harris (eds.) The control of antibiotic resistant bacteria. Academic Press, London.

Purdom, C.E. 1977. Ministry of Agriculture, Fisheries and Food, Fish Cultivation Research. Lab. Leafl. 35:1-28.

Rødsaether, M.C., J. Olafsen, J. Raa, K. Myhre and J.B. Steen. 1977. Copper as an initiating factor of vibriosis (*Vibrio anguillarum*) in eel (*Anguilla anguilla*). J. Fish Biol. 10:17-21.

Salte, R. and K. Liestøl. 1983. Drug withdrawal from farmed fish: depletion of oxytetracycline, sulfadiazine, and trimethoprim from muscular tissue of rainbow trout (*Salmo gairdneri*). Acta Vet. Scand. 24:418-430.

Samuelsen, O.B. 1989. Degradation of oxytetracycline in seawater at two different temperatures and light intensities and the persistence of oxytetracycline in the sediment from a fish farm. Aquaculture 83:7-16.

Stevenson, J.P. 1980. Trout farming manual. Fishing News Books Ltd., Farnham, Surrey.

Takashima, N., T. Aoki and T. Kitao. 1985. Epidemiological surveillance of drug-resistant strains of *Pasteurella piscicida*. Fish Pathol. 20:209-217.

Toranzo, A.E., J.L. Barja, R.R. Colwell and F.M. Hetrick. 1983. Characterization of plasmids in bacterial fish pathogens. Infect. Immun. 39:184-192.

Toranzo, A.E., P. Combarro, M.L. Lemos and J.L. Barja. 1984. Plasmid coding for transferrable drug resistance in bacteria isolated from cultured rainbow trout. Appl. Environ. Microbiol. 8:872-877.

Walton, J.B. 1982. Changes in management systems to avoid the development of antibiotic resistance, p. 103-119. *In* C.H. Stuart-Harris and D.M. Harris (eds.) The control of antibiotic resistant bacteria. Academic Press, London.

Ziskowski, J. and R. Murchelano. 1975. Fin erosion in winter flounder (*Pseudopleuronectes americanus*) from the New York Bight. Mar. Pollut. Bull. 6:26-28.

Discussion

BOGEL: Your reference to Codes of Practice and control of antibiotic use could be extended to include husbandry practices that can replace procedures that involve antibiotics. Also, in warmblooded livestock production now there is the use of probiotics that influence the intestinal ecosystem in a positive way to reduce the impact of pathogens. Is there anything similar in fish that you know of?

AUSTIN: There is one pharmaceutical company, which I will not name, that is working on this and sees a good sales potential for its products.

BOGEL: What kind of compound is it? A carbohydrate? There seem to be some very simple sugars that can change intestinal ecosystems rapidly.

AUSTIN: I'm afraid I can't say.

PULLIN: Many of the finfish species cultured in developing countries have important intestinal flora. This applies to the mullets, tilapias and perhaps most or all of the carps. Antibiotic use in finfish culture in developing countries is very limited at present, but were it to expand, it may affect the intestinal flora of such fish, their general well-being and their culture performance.

AUSTIN: We have looked at the action of some antimicrobial compounds in invertebrates, where the intestinal flora is also alleged to contribute to digestion. Provided that the antimicrobial compound is used for the correct chemotherapeutic period and *not* continuously, then the adverse effect, if any, is extremely shortlived because the gut is very rapidly recolonized. Moreover, such antimicrobials do not sterilize the gut unless something is drastically wrong - such as a very wrong dosage. These tests were done on temperate species of penaeid shrimps and oysters.

KING: You have suggested a Code of Practice to restrict the use of antibiotics and other antimicrobials in aquaculture. To what extent do you see this being implemented?

AUSTIN: I have published a tentative Code of Practice giving the rationale for restrictions[1]. What I presented here is a summary of that Code of Practice. Codes of Practices only work when people actually want to adopt them and do not seek refuge in exceptions whenever it seems convenient. The problem with aquaculture is that fishfarmers can obtain compounds from many diverse sources. The quality of those compounds is often questionable. There should be restrictions on the sources of compounds used in aquaculture, to ensure quality control. Also, a fish farmer must know what he is doing when he uses a compound.

KING: My main concern is not the misuse of antimicrobials but rather that restrictions might prevent some of their benefits, with correct use, from being enjoyed in developing countries.

AUSTIN: Well, in the medical profession there used to be a view that if anything went wrong you prescribed an antibiotic. That view has changed in medicine, but a similar view persists in some branches of aquaculture. I think that, irrespective of restrictions, the overall usefulness of antibiotics in aquaculture will probably decline as we learn more.

EDWARDS: But even in medicine in some developing countries the rate of antibiotic use is very high.

BURAS: What are the alternatives to use of antibiotics in aquaculture? You mentioned using different compounds in sequence. Also, what effect do antibiotics have on the bacteria in fishponds? These bacteria are vital to pond production and management.

---

[1]Austin, B. 1984. Chemotherapy of bacterial fish diseases, p. 19-26. *In* A.E. Ellis (ed.) Fish and shellfish pathology. Academic Press, London.

AUSTIN: Over 50 pharmaceutical compounds, antibiotics or similar compounds, have been or are being used on fish farms. Take one disease example, furunculosis, which is caused by *Aeromonas salmonicida*. The classic treatments have been with potentiated sulfonamides oxytetracycline, and more recently oxolinic acid. The problem is that fish farmers and veterinarians (who are very much to blame here) will repeatedly use or prescribe the same compound. It is in these situations that resistance builds up. Suppose a farmer uses a drug once or twice ...

BURAS: In what period?

AUSTIN: A matter of months. Thereafter, if the farmer uses a different compound, resistance to the first will dwindle and it could be used again in the future. Otherwise, resistance will build up and its effectiveness will decline.

BURAS: What are the effects on bacteria in ponds?

AUSTIN: We have only looked at a limited range of organisms. Generally antibiotics cause a reduction in the range of taxa and an increase in antibiotic resistance. However, there is evidence that the steady state that existed before antibiotic treatment can return within a relatively short time. I can't comment, however, on all bacteria. For example, we don't know what happens with chemolithotrophs.

ARELLANO: The misuse of antibiotics and the development of resistance are frightening. We need to look more at the conditions that support or prevent disease: nutrition, management, etc. The emphasis should be on prevention, not cure by antibiotics. Also, conditions on fish farms vary very much with location. So do the effective doses of antibiotics. It is no good working out treatments in one country for application in another under a totally different set of conditions. With regard to disease prevention by good nutrition, it is useful to compare the nutritional status of wild and farmed fish - especially broodstock and early life history stages. The role of microbiologists in aquaculture should be less on developing antibiotics and kits and advising feedmills to incorporate compounds like oxytetracycline. This is wrong. When this advice comes from developed to developing countries, they are confused and it is difficult for them to question it. Yet, Dr. Austin mentioned the use of antibiotics in intensive developed-country aquaculture and resulting problems in sediments. The microbiologists should rather help developing countries with understanding how to manage their extensive and semi-intensive aquaculture systems better. This does not preclude the use of antibiotics but it does need a new philosophy of disease control.

AUSTIN: The samples that I have obtained from Ecuador indicate an increasing problem of antibiotic resistance in vibrios. However, we must realize whereas it is easy to debate and propose disease prevention strategies, in reality such protocols are broken by farmers as soon as they face an immediate disease problem. Their problems need urgent solutions. Most farmers do not even want to wait for diagnosis, because they fear that most of their stock will die while they wait. Therefore, when a disease situation occurs, there is usually a scramble for pharmaceutical compounds. I don't see that changing. Disease is the end result of an unsatisfactory process.

Developing-Country Aquaculture and Harmful Algal Blooms*

JOHN L. MACLEAN
International Center for Living Aquatic Resources Management
MCPO Box 2631, 0718 Makati
Metro Manila, Philippines

MACLEAN, J.L. 1993. Developing-country aquaculture and harmful algal blooms, p. 252-284. *In* R.S.V. Pullin, H. Rosenthal and J.L. Maclean (eds.) Environment and aquaculture in developing countries. ICLARM Conf. Proc. 31, 359 p.

Abstract

Toxic algal blooms began to have significant impacts on developing-country aquaculture in the 1970s, including toxic shellfish and mass mortalities of fish and shrimp. Based on the experiences of developed countries, the potential exists in the waters of developing countries for a wide range of presently unrecorded toxins as well as other effects of algal blooms. The implications of these hazards are discussed. Their economic impact, which extends beyond the aquatic sector, is also discussed. Management measures used in both developed and developing countries at the industry as well as the government level are described and assessed. Finally, evidence of a relationship between red tides and aquaculture is discussed.

Introduction

Many scientists believe that visible algal blooms or red tides are environmental indicators in as much as there is a strong correlation between the number of red tides and the degree of coastal pollution or use of coastal waters for aquaculture (see e.g., Anderson 1989; Lam and Ho 1989; Okaichi 1989; Seliger 1989; Smayda 1989).

Potentially nuisance blooms in the sea have been around for a long time. Captain Cook observed *Trichodesmium* blooms in the Coral Sea in 1770 and along with his crew suffered ciguatera poisoning in the New Hebrides in 1774 (Hallegraeff 1990). The first paralytic shellfish poisoning report on the Pacific coast of north America was in 1793 during explorations by Vancouver in British Columbia (Conte 1984). The naturalist Poeppig was the first to record a red tide in Chile, in 1827. Darwin saw the next one there in 1835 (Unesco 1982).

In recent years, there appears to have been a rapid global increase in red tides which is reflected, according to Anderson (1989), in the increase in countries represented at the international meetings on toxic dinoflagellates, from three at the first such meeting in 1974, 17 at the second in 1978, 22 in the third (1985), to 27 in the fourth (1989).** New occurrences have been reported in a variety of locations at each meeting. A new directory of experts in the fields includes over 390 persons from 42 countries (White 1990a).

Whether the blooms are increasing or not, there has been a rapid growth of relevant literature. The 163 references in the present review, for example, while by no means

*ICLARM Contribution No. 915

**The topic of apparent increases in blooms has been pursued in recent reviews by Hallegraeff (1992, 1993) - Eds.

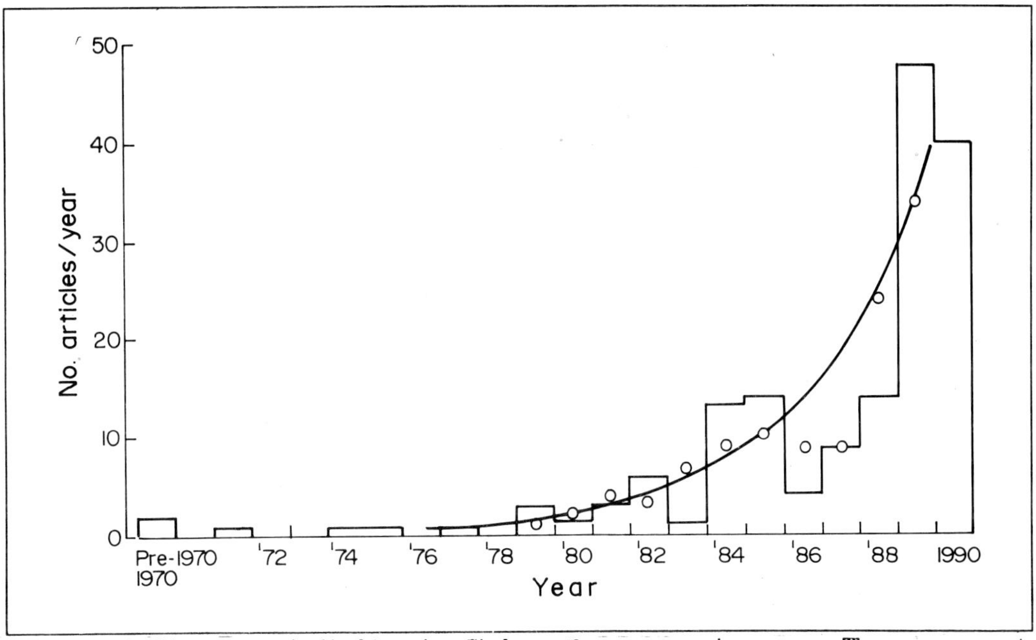

Fig. 1. Growth of literature as cited in this review. Circles are three-year running averages. The curve represents literature doubling every 2.25 years.

constituting a bibliography, exemplify the literature explosion. A plot of the publication dates (Fig. 1) shows that over 50% come from 1989 and 1990; the curve suggests a doubling of the annual literature output every 2-2.5 years over the past two decades. For comparison, aquaculture literature output doubled every five years from 1960 to 1980 (Maclean 1986).

One of the features of these algal blooms is a shift over the years from generally benign diatom blooms to flagellate blooms, possibly associated with decreased Si:P and N:P ratios, in part due to coastal enrichment from pollution and river-borne nutrients according to Smayda (1989). The flagellates include most of the toxic forms. Lam and Ho (1989) pointed out a clear shift from diatom- to dinoflagellate-dominated phytoplankton in Tolo Harbour, Hong Kong, as the waters became more polluted. Red tides there have increased dramatically also (Fig. 2). As Zou et al. (1985) pointed out, "A red tide can be looked upon both as a product and as a process in the eutrophication of estuarine and coastal ecosystems."

At the most recent international conference on toxic marine phytoplankton (Granéli et al. 1990), the influence of human activities was debated. In the conference overview, Taylor (1990) stated that "some blooms are plainly in response to eutrophication but others are equally plainly not." The following careful recommendation was approved and conveyed to the

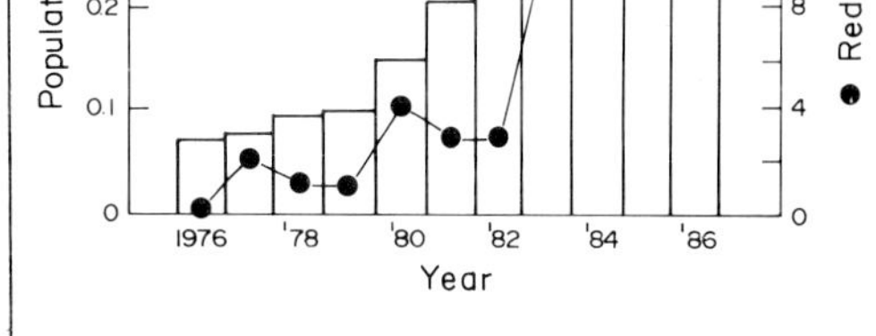

Fig. 2. Annual number of red-tides in Tolo Harbour, Hong Kong, and human population levels in Hong Kong from 1976 to 1986. From Lam and Ho (1989).

International Oceanographic Commission, Paris: "The conference participants reached a consensus that some human activities may be involved in increasing the intensity and global distribution of blooms and recommended that international research efforts be undertaken to evaluate the possibility of global expansion of algal blooms and man's involvement in this phenomenon" (Granéli et al. 1990, p. 517).

From the point of view of aquaculture, toxic algal blooms can be considered as part of the environment in which this sector is presently developing.

The coastal zone in many developing countries is becoming a major focus of attention to environmental managers in view of the largely uncontrolled developments taking place there, entailing massive destruction of natural habitats (Chua, this vol.) The Group of Experts on the Scientific Aspects of Marine Pollution (GESAMP) stated that "at the end of the 1980s, the major causes of immediate concern in the marine environment on a global basis are coastal development and the attendant destruction of habitats, eutrophication, microbial contamination of seafood and beaches..." (GESAMP 1990). Thus, it is the coastal environment of developing countries where we can expect to see dramatic increases in toxic algal blooms. The recent spreading of red tides in the Indo-Pacific (Maclean 1989a), for example, may be only a prelude to worsening problems ahead.

Meanwhile, that environment itself is in all probability undergoing changes such as atmospheric temperature increases and changing relative sea levels (Stewart et al. 1990), associated with global climatic changes which may mask, reinforce or negate some of the more local pollution/enrichment effects. We are dealing then with a growing problem in which there may be so many contributing factors - both primary and secondary - that it may be impossible or impractical to attempt to isolate any one cause or group of causes.

In the developing world, most of the interest in and workshops about harmful algal blooms (and, coincidentally over 80% of world aquaculture production) have been in Asia, and have happened fairly recently. A 1954 symposium on plankton in the Indo-Pacific made no mention of red tides or harmful algal species (FAO/Unesco 1954). Toxic blooms of *Pyrodinium* in Papua New Guinea in 1972, western Borneo in 1976 and 1980 and in the Philippines in 1983 (Maclean 1989b); fish kills in Hong Kong beginning in 1980 (Wong and Wu 1987); and an unprecedented red tide in Jinhae Bay, Korea (Park et al. 1989), raised enough concern for two regional meetings in 1984 (White et al. 1984; CSIRO 1985). The incidents in Hong Kong, Korea and the Philippines all involved aquaculture losses which continue to the present. The *Pyrodinium* situation was sufficiently alarming to warrant a special workshop in 1989 (Hallegraeff and Maclean 1989). In Latin America, a workshop on red tides on the Pacific coast was held in 1982 (Unesco 1982). There are no reports of similar activities in Africa. At the international level, two conferences in 1987 specifically addressed the impact of algal blooms on aquaculture (Dale et al. 1987; Jensen 1988).

In this review, I have attempted to consolidate information on the nature of the potential hazards, the extent of the algal bloom problem in developing-country aquaculture and, despite the uncertainty of future environments, some likely future scenarios.

Causative Organisms

The algae under consideration come from the Cyanophyta, cyanobacteria; Chrysophyta, chrysophytes; Pyrrhophyta, dinoflagellates; Raphidophyta, chloromonads, Bacillariophyta, diatoms; and Prymnesiophyta, prymnesioids - a mixture of primitive eukaryotes and prokaryotes (Taylor 1985). Some authors use classes rather than phyla to describe these organisms (e.g., Hargraves et al. 1989; Fukuyo et al. 1990) and the taxonomy of individual species as well as of groups remains under revision (e.g., several papers in Granéli et al. 1990).

Larger dinoflagellates commended most attention at earlier workshops but in recent years the importance of other groups - the raphidophytes, particularly *Chattonella*; chrysophytes, especially the brown tide *Aureococcus*; the prymnesiophyte *Chrysochromulina*; and the cyanobacteria - has been recognized. The picoplankton (0.2-2.0 mm or bacteria-sized cells), have been found to be a source of blooms and toxins; they include small flagellates and cyanobacteria. These organisms bloom seasonally in coastal waters at densities orders of magnitude higher than larger forms, viz up to $10^9 \cdot l^{-1}$ vs. $10^6 \cdot l^{-1}$ (Hargraves et al. 1989). Often the picoplankton are the dominant primary producers in the sea (Sieburth and Johnson 1989).

There is no comprehensive list of causative species and their number is certainly still growing. Taylor (1990) gave a current "minimum list" of well-established harmful marine phytoplankton species (i.e., excluding those which have been associated with oxygen depletion or gill clogging) as follows: dinoflagellates 27; chloromonads 5; chrysophytes 1; prymnesioids 5; silicoflagellates (chrysophytes) 1; and diatoms 4. The most recent list from Japan includes 300 freshwater and marine algal species including "not only causative organisms of red tide but also toxic species and organisms which are associated with other dominant species in red tides along Japanese and Southeast Asian coastal waters" (Fukuyo et al. 1990). Shumway (1990) lists about 44 toxic algae which affect shellfish. Some 12 genera of Cyanophyta have been implicated in producing acute lethal toxins (Carmichael et al. 1990).

Toxic Products

The various toxins produced by these algae (and by a variety of other marine organisms) have in common the property of modifying the functions of ion channels across cell membranes. On this basis, toxins can be grouped into three types: activators, stabilizers and occluders (Strichartz and Castle 1990).

Toxins affecting or potentially affecting aquaculture operations include activators such as the toxins causing neurotoxic shellfish poisoning, diarrheic shellfish poisoning, ciguatera and domoic acid, and occluders such as the paralytic shellfish poisons (saxitoxins) tetrodotoxin and anatoxins (Dale et al. 1987; Carmichael et al. 1990; Strichartz and Castle 1990). A list of the better known toxins is given in Table 1.

Neurotoxic Shellfish Poisons

Eight polyether toxins, collectively called brevetoxins, are presently known to be produced by the dinoflagellate *Gymnodinium breve*, the Florida red tide organism (Shimizu 1989). The toxins are potent fish killers and in aerosol form cause human respiratory irritation. They also cause a mild form of poisoning, similar to ciguatera, and of similar chemical structure to the ciguatoxins in humans ingesting contaminated shellfish. A bloom in North Carolina at the end of 1987 caused closure of 150,000 ha of shellfish grounds and lowered the crop value by $2 million (Tester and Fowler 1990). There seem to be no reports of hazards from this alga in other waters.

Diarrheic Shellfish Poisons

Diarrheic shellfish poisoning (DSP) in humans is caused by ingesting shellfish which have accumulated toxins from dinoflagellates of the genus *Dinophysis* or the benthic dinoflagellate *Prorocentrum lima*. Elever polyether toxins have been identified from shellfish in three groups - okadaic acid and derivatives (dinophysistoxins); pectenotoxins; and yessotoxin (Yasumoto and Murata 1990).

Table 1. Properties of various algal toxins.

Generic name	Number known	Molecular weight	Effect	LD_{50} mg/kg intraperitoneal mouse assay	Source
Brevetoxins	8	870-970	Potent	60 (rats)	Shimizu (1989)
			ichthyotoxins NSP		Poli et al. (1990)
Saxitoxins	17	~300	PSP	10	Hall et al. (1990)
Anatoxins	2	165, 252	Neurotoxic poisoning of animals, fish	20-200	Carmichael et al. (1990)
Hepatotoxins	8-10	800-1,000	Hepatoenteritis	30-1,000	Carmichael et al. (1990)
Diarrheic shellfish toxins	11		Nonlethal intestinal disorders	100-770 (LD_{99})	Yasumoto and Murata (199
Domoic acid	1	311	Gastrointestinal disorders, memory loss	-	Todd (1990)
Ciguatoxins	2	1,100	Gastrointestinal, cardiovascular and neurological disorders	0.45	Frelin et al. (1990)

Symptoms in humans are gastrointestinal and not fatal. DSP was only recognized in the mid-1970s and was probably previously confused with bacterial spoilage. However, recent reports (e.g., Dale et al. 1987; Shumway 1990) show that DSP is a significant problem in Europe and Japan, with incidences also in Australia, Chile, India and New Zealand. *Dinophysis* spp. are cosmopolitan phytoplankters, so the potential for problems elsewhere is high, and closures of shellfish farms may be longlasting. For instance, mussels in Sweden remained toxic almost continuously from October 1984 until summer 1986 (White 1988).

Ciguatoxins

Ciguatoxins are a group of disparate polyether toxins which cause the well-known illness in humans, ciguatera, from eating various tropical reef fish. The two major toxins are ciguatoxin (of which there are two "species") and maitotoxin. A third compound, scaritoxin, may be a form of ciguatoxin, while new "minor toxins" have recently been discovered (Legrand et al. 1990). Another different polyether toxin, palytoxin, has been found responsible for poisonings by triggerfish and xanthid crabs (Yasumoto and Murata 1990). The poisons are detected via mouse bioassay.

The benthic marine dinoflagellate *Gambierdiscus toxicus*, the major alga responsible for ciguatera, has now also been found to form toxic blooms on its macroalgal substrate. Population explosions of the alga in Tahiti after some 10 years of relative dormancy were closely associated with toxicity in the grazing surgeonfish *Ctenochaetus striatus* (Bagnis et al. 1990). *Gambierdiscus* showed clear seasonal population trends in Queensland, Australia

(Gillespie et al. 1985), but no such trend in Tahiti (Bagnis et al. 1985).

There are no records of ciguatera from farmed fish but the toxins could become important if ranching of tropical reef fish becomes practical and even for farming tropical carnivorous fish in marine enclosures, where they have access to small prey entering the enclosures.

Paralytic Shellfish Poisons

These poisons are produced by a number of marine dinoflagellates as well as by the predominantly freshwater cyanobacteria *Aphanizomenon flos-aquae*. *Aphanizomenon* produces aphantotoxins which have been found to be identical to saxitoxins (Shimizu et al. 1990). The best known dinoflagellate toxin producers are: *Gonyaulax* spp. now called *Alexandrium* spp. (Balech 1985), which have exhibited toxic blooms almost all around the world in tropical and temperate waters (e.g., Taylor and Seliger 1979; Anderson et al. 1985; Okaichi et al. 1989a); *Gymnodinium catenatum* which causes PSP in Australia, Europe, Japan and Venezuela (Hall et al. 1990); and *Pyrodinium bahamense* around Southeast Asia, the South Pacific (Maclean 1989d) and recently several countries along the Pacific coast of central America (F. Rosales-Loessener 1989, pers. comm.).

Eighteen closely-related "saxitoxins" have been discovered and a further six may possibly be found in future, based on the molecular structure of the group. The known "saxitoxins" include saxitoxin *per se*, neosaxitoxin and several gonyautoxins and decarbamoyl saxitoxins (Yasumoto et al. 1984; Hall et al. 1990). Several of these compounds in different proportions are found in the various algae and their consumers; each algal species exhibits a unique toxin "profile".

Saxitoxins accumulate mainly in filter-feeding bivalves, which in general are not lethally affected by them, and pass along the foodchain. The consumers include some gastropods, which prey on bivalves and become toxic, and also humans who are poisoned, sometimes fatally from paralytic shellfish poisoning (PSP). There are, of course, other pathways through zooplankton to fish and marine mammals, as well as to birds. Humans have contracted PSP by eating planktivorous fish (see below). The symptoms in humans are neurological, gastrointestinal and respiratory disorders (e.g., Pastor et al. 1989). Timely respiratory support prevents death. There is no proven antidote although the Philippine folk remedy of drinking coconut milk and brown sugar was reported to be effective in reducing toxicity of crude toxins (Gacutan 1986).

Amnesic Shellfish Poison

This form of intoxication in humans from eating bivalves was first reported in late 1987 in eastern Canada, when consumers of cultured mussels, *Mytilus edulis*, developed gastrointestinal disorders accompanied by short-term memory loss. There were some deaths. Unlike in other forms of algal poisoning, the neurological damage is permanent (Addison and Stewart 1989; Smith et al. 1990; Todd 1990).

The toxin was found to be domoic acid, a neurotoxic amino acid, detected by mouse bioassay and high-performance liquid chromatography (HPLC). The source was apparently a bloom of the diatom *Nitzschia pungens*. Agricultural run-off was suspected as the cause by Smith et al. (1990), whereas Addison and Stewart (1989) did not discount the possibility that the intensive aquaculture in the area may have been a cause of the bloom.

So far this toxin has not been found in developing countries. *N. pungens*, however, is a ubiquitous coastal species (Fukuyo et al. 1990).

Tetrodotoxin

Tetrodotoxin (TTX), best known for its dramatic effects on consumers of *fugu* puffer fish, is also found in other fish, frogs, newts, octopus, gastropods, starfish, crabs, flatworms, zooplankton, algae and bacteria (*Vibrio* spp. and several other genera) (Jeon et al. 1986; Narita et al. 1987; Tamplin 1990). As with saxitoxins, it is possible that bacteria are the primary producers of tetrodotoxin, which then accumulates up the foodchain (Narita et al. 1987; Tamplin 1990). Tetrodotoxin is not known to be associated with any dinoflagellate blooms, but is usually discussed in the same fora and has implications for aquaculture.

There are four tetrodotoxin toxins - tetrodotoxin itself being the best known. They are detectable by mouse bioassay and several more sophisticated biological as well as chemical techniques (Onoue et al. 1984; Tamplin 1990).

Pufferfish are farmed in Japan (and may be potential export crops in Asian developing countries). Saito et al. (1984) found that pufferfish from farms in several localities in Japan were all nontoxic. Toxicity was found only in wild-caught specimens. Not all pufferfish species are toxic. Those that are have very high resistance to interperitoneal administration of the toxins (minimum lethal dose of 300-750 mouse units (MU)/20-g body weight), compared to 1-20 MU for nontoxic pufferfish species (Saito et al. 1985b). The toxin seems to be a biological defense agent for the pufferfish (Saito et al. 1985c). Human poisonings are through ingestion of *fugu* or by being bitten, in the case of the Australian blue-ringed octopus.

Hepatotoxins

Anabaena and some other predominantly freshwater cyanobacteria - *Microcystis aeruginosa, M. viridis, Nodularia spumigena* and *Oscillatoria agardhii* - produce hepatotoxins, short peptide toxins which affect the liver and are also acutely lethal to animals (Carmichael et al. 1990). Six chemically related hepatotoxins are known which have about one-fifth the toxicity of saxitoxins and account for most of the cases of animal poisonings from freshwater cyanobacteria (Gorham and Carmichael 1988).

To date there are no confirmed instances of human death from these sources, but allergic and gastrointestinal problems are known and pets, livestock and wildlife, including fish, have been killed.

Toxicity problems from freshwater cyanobacteria have been recorded in Australia, Bangladesh, China, (12 countries of) Europe, India, Israel, Japan, Latin America, North America, South Africa, Thailand and the USSR (Carmichael et al. 1990). These countries are a fair cross-section of the globe and it is likely that in others, especially developing countries, the problem has not been fully recognized. Some of the species, e.g., *Anabaena flos-aquae, Microcystis aeruginosa* and *M. viridis*, are ubiquitous (Fukuyo et al. 1990). *Microcystis* is reported to dominate continuously warmer, shallow, eutrophic waters such as Indian temple ponds and Lake George in Uganda (Stirling and Dey 1990).

Anatoxins

Anatoxins are a group of neurotoxins produced by the freshwater cyanobacteria *Anabaena flos-aquae*. Six compounds have been isolated which have from one-twentieth to one-half the potency of saxitoxins (Carmichael et al. 1990). They can be detected by mouse bioassay and HPLC. The toxins are lethal. No antidotes are available (Gorham and Carmichael 1988).

Cytotoxins

Carmichael et al. (1990) noted that some freshwater cyanobacteria produce a variety of bioactive compounds of much lower toxicity to humans, collectively called cytotoxins, which cause dermatitis or irritation on contact. They can be lethal to mice, when administered interperitoneally. These authors pointed out that the various neurotoxic, hepatotoxic and dermatotoxic compounds produced by cyanobacteria are a direct and growing threat to animal and human water supplies.

Other Harmful Effects

Oxygen Depletion and Gill Clogging

Blooms of many algal species are not toxic in themselves but cause mortalities through oxygen depletion when blooms collapse. In theory this situation could apply to almost all algal bloom species. Some examples of these problems are given in Table 2.

Oxygen depletion in eutrophic ponds at night is a well known consequence of "excess" algal biomass, but similar problems are emerging in open marine waters. For example, in the southern Kattegat between Denmark and southern Sweden, "Oxygen deficits resulting from decomposition of algal matter, with the ensuing death of fish and other animals, consitute a regular environmental problem there" (Dahl et al. 1989).

Gill clogging due to mucus secretion was thought to be a major factor in fish kills caused by several flagellates (White 1988). However, species mentioned by White (1988) have been found to act through toxin production. One, the dinoflagellate *Gyrodinium aureolum* was found to kill seawater acclimatized rainbow trout (*Oncorhynchus mykiss*) by causing degeneration of gill tissue. Involvement of toxin(s) was presumed (Roberts et al. 1983). There is recent evidence of toxin production by the chloromonad *Heterosigma akashiwo* (R.J. Gowen, pers. comm.), also noted by White (1988) as causing only mucus clogging of gills.

Some larger dinoflagellates with spinous skeletons, such as *Chaetoceros convolutus, C. concavicornum* and *C. danicus*, can cause physical damage to gills, such that fish die of asphyxiation; this may become a serious problem for fish farmers (R.J. Gowen, pers. comm.).

The massive mortalities of farmed yellowtail (*Seriola quinqueradiata*) in the Seto Inland Sea of Japan by the chloromonad *Chattonella* are due to gill damage and production by the algae of highly unsaturated fatty acids which decrease the pH of the fish's blood, making gas exchange difficult (White 1988).

Nutrient Stripping

Uno and Sasaki (1989) report that diatom blooms affect nori (*Porphyra tenera*)

Table 2. Examples of algal species/situations causing anoxic/hypoxic conditions in aquaculture facilities in developing countries.

Species	Country	Commentary	Source
Cyanobacteria			
Trichodesmium erythraeum	Thailand (Gulf)	Extensive fish kills in farms	Suvapepun (1989)
Microcystis sp.	Philippines (Laguna de Bay)	Fish kills in fishpens	Ronquillo (1987)
Dinoflagellates			
Noctiluca sp.	Philippines (Manila Bay)	Deaths of fish in cages, mussels oysters, shrimp and crabs	Ronquillo (1988)
Noctiluca and other genera	Hong Kong	Fish kills in farms	Wong and Wu (1987); Lam and Yip (1990)

seaweed culture in Japan. Seasonal blooms of *Eucampia, Chaetoceros* and *Nitzschia* strip the seawater of nutrients, resulting in fading of dark brown color of the seaweed and lowering its commercial value. An index of photosynthetic activity per unit biomass has been developed which gives a 2-4 week warning to growers (Yamamoto and Fujisaki 1989).

Discoloration

A bloom of the ubiquitous ciliate *Mesodinium rubrum* was reported to have caused red discoloration of oysters (*Ostrea edulis*) in the Netherlands (Kat 1984). The author noted previous incidents of orange discoloration in oysters in Italy, pink oysters in Texas caused by a bacteria, blue-green oysters in France caused by a diatom and green oysters in Greece caused by coccolithophorid.

Tainting

Tainting of fish from algal blooms is another factor to be considered. Earthy tainting of freshwater fish flesh with geosmin was noted by Stirling and Dey (1990) as an effect of decomposing blooms. From personal experience, tainting is a seasonal problem in tilapias in Laguna de Bay, Philippines, attributed by some researchers to blooms of *Microcystis aeruginosa* but by others to actinomycetes (de Guzman 1990). A shipment of pond-grown penaeid shrimp from Ecuador was unmarketable as a result of intense musty flavor from geosmin, caused by cyanobacteria (Lovell and Broce 1985). A bloom of the diatom *Rhizosolenia chunii* in southeastern Australia in 1987 caused a strong bitter flavor in cultured and wild bivalve molluscs. Over a seven-month period, some 500 tonnes of mussels worth about $1 million were discarded (Parry et al. 1989).

In summary, many toxins are produced by algae which in blooms can render bivalve molluscs toxic to humans. Some toxins can be fatal to humans - the paralytic shellfish poisons (PSP) and amnesic shellfish poison - while others, diarrheic and neurotoxic shellfish poisoning, are milder but severe enough to close fish farms and fisheries for long periods. Human intoxication from fish containing PSP toxins has been known as well as from fish containing the more common ciguatera toxins. Fish and other organisms can be killed or even if they survive have their growth and fitness reduced by micro-algae via neurotoxins (anatoxins, brevetoxins, hepatotoxins and PSP toxins) as well as gill-clogging mechanisms and bloom-induced hypoxic or anoxic conditions. Finally, some algal bloom species can cause off-flavors and discoloration in various aquatic organisms.

Economic Impact

Economists seem rarely to have turned their attention to red tides. Yet the losses caused by blooms can be large, especially by developing-country standards. A list of estimated losses from blooms to mainly north American and Japanese shellfish industries given by Shumway (1990) shows figures ranging from $0.1 million to an astounding $430 million. These include fisheries as well as farms. Maclean (1989c) divided economic impact into three facets: occasional acute episodes; chronic situations; and permanent closures.

Acute Bloom Situations

In occasional acute situations, the public is taken by surprise and large numbers of people are affected by contaminated organisms. The number of persons affected ranges usually from 50 to 300, as in the first outbreaks of *Pyrodinium* in Sabah, Malaysia (Ting and Wong 1989), the Philippines (Gonzales 1989a), and Guatemala (Rosales-Loessener 1989); of domoic acid poisoning in eastern Canada (Todd 1990); and of neurotoxic shellfish poisoning in New Jersey, USA (Tester and Fowler 1990). In the Philippines, a second acute *Pyrodinium* episode occurred in 1987 (four years after the first) and resulted in over 200 cases of illness (Maclean 1989b). Such illness affects the resource rent from the fishery or aquaculture sector to some extent. The full impacts of blooms appear rarely to have been determined but are probably much larger than published figures imply.

A good example in the developing-country aquaculture context is the first outbreak of mussel (*Perna viridis*) poisoning in Manila Bay, Philippines, in 1988:

> In August and September 1988, the first outbreak in Manila Bay occurred. Thanks to the media, the whole seafood industry nearly ground to a halt, while mussel growers even tried to implicate freshwater products in an effort to offset the swing by consumers to tilapias and other freshwater organisms! All fish markets in Manila were depressed for over three months, similar to the case in San Francisco in 1980. Manila's seafood market handles 35% of the nation's landings. Thus, the losses were large, up to $300,000/day at the height of the scare. Japan and Singapore banned shrimp imports from the Philippines for an unknown period (although they were clean), which would have meant losses of $500,000/ day if the produce was not subsequently sold. Losses by mussel growers for a three-month period were more modest, about $950,000 in all" (Maclean 1989c).

Even the vinegar industry was affected, because mussels are usually eaten with a vinegar-based sauce. Quite probably, the prices of unaffected commodities such as poultry rose in response to the general scare on seafoods.

This "halo effect" (Shumway 1990) of red tide outbreaks into other sectors of the economy is not confined to developing countries; examples from developed countries are given by Maclean (1989c) and Shumway (1990). Misinformation in the media is a common cause.

As far as losses to aquaculturists themselves is concerned, there are few data from developing countries. Thai fish farmers lost some $1.16 million worth of fish in 1983 when a huge *Trichodesmium* bloom in the Gulf of Thailand collapsed, causing anoxic conditions along the coast (Suvapepun 1989). Mussel farmers in Manila Bay lost about $1 million worth of (condemned) produce during the 1988 *Pyrodinium* blooms there (Maclean 1989c). The single mussel farm in Brunei lost most of its crop which fell from the farm's longlines during a long ban on marketing in 1988 due to high PSP levels in shellfish. This involved total losses, including expenses in growing the crop, of about $0.1 million (Jaafar et al. 1989). A pond shrimp kill by an unknown dinoflagellate in Hupei and Shandong provinces of China in 1989 caused a financial loss of 300 million yuan (US$1=5.3 yuan) (Ger Guo Chang, pers. comm.).

Chronic Bloom Situations

In chronic red tide situations, large losses may still occur, a good example being the 71 billion yen loss of cultured yellowtail in the Seto Inland Sea of Japan in 1972 and previous and ongoing lesser but significant annual losses there (Fig. 3) (Okaichi 1989). However, in chronic situations, the public eventually recognizes that only particular aquatic products are at risk, thus reducing the "halo effect". Costs of regular monitoring become part of the annual loss of resource

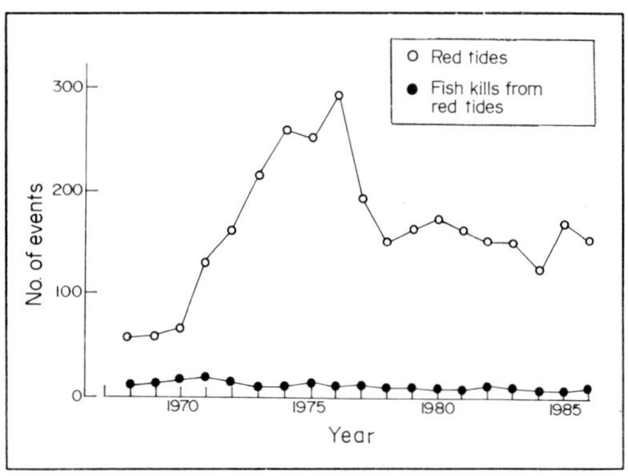

Fig. 3. Red tides in the Seto Inland Sea, 1968-1986. After Okaichi (1989).

rent. These costs may be $10,000-50,359 per year per affected area based on some recent examples (Maclean 1989c). A Canada-wide program of collection, testing, enforcement, management and information dissemination for the shellfish industry costs over $1 million per year (Pirquet 1988 cited in Shumway 1990).

Where red tides have been recognized as becoming chronic, interested parties in both government and the private sector can act more decisively. For example, in one of the earlier salmon farming fish kills in Norway, Tangen (1982) reported losses of 4 million Kroner from two farms due to *Gyrodinium aureolum* in 1982. In 1988, Norwegian farmers averted losses of $200 million by towing 120 farms out of threatened areas (Anon. 1988). Such strategies are further discussed below.

Permanent Closures

Permanent closures mean an indefinite loss of potentially high resource rents as well as losses of protein. There are examples of permanent and "near-continuous" closures in Canada (Taylor and Seliger 1979; White 1982). In developing countries, such economic sacrifices may appear to be untenable but could become necessary to avoid loss of life. The west coast of Sabah, Malaysia, is an area where there have been several deaths from PSP in most years since 1976, although red tides have only been found on a few occasions (Ting and Wong 1989). Not only does the prospect of developing a mollusc culture industry there appear bleak but the Sabah government recommends that people eat only small quantities of shellfish and that children should not eat them at all, even during "safe" periods (Wong and Ting 1989). Similar situations are likely to develop in other *Pyrodinium*-affected areas, which include many countries on both sides of the tropical Pacific (see Table 6, p. 266).

A summary of the main actual and potential "costs" associated with the above conditions - acute, chronic and closed areas - is given in Table 3. Evaluating many of them and even prioritizing them in terms of severity or importance is a difficult task but one which governments must face in making management decisions. Only a few of the cost factors lie within the aquaculture sector, but most have a bearing on its development.

Table 3. Actual and potential costs and losses associated with *Pyrodinium* red tides. Numbers represent relative severity of each factor from least (1) to most (5) important. (Source: Maclean and Ross 1989)

5	Loss of life
5	Cost of resettlement in a chronic red tide situation (transmigration)
4	Reduced price of uncontaminated seafoods
4	Cost of maintaining public awareness
3	Reduced price of suspect seafood
3	Loss of confidence by consumers
3	Loss of income by fishers, fishfarmers
3	Cost of monitoring and research (personnel, equipment, supplies)
2	Loss of condemned seafood (by fishers, fishmongers or agents)
2	Cost of publicizing and enforcing bans
2	Potential loss of business opportunities (aquaculture, fisheries)
2	Potential loss of resource use
1	Loss of wages of hired labor (fishers, fishmongers, drivers, etc.)
1	Loss of income by victims, in medicine, time off work, hospitalization (socioeconomic costs)
1	Loss of tourism income (coral reef resources)
1	Loss of income by seafood restaurants
1	Loss of foreign exchange earnings (private and government) through need to import and loss of export

Implications for Marine and Brackishwater Aquaculture

Estuarine and coastal environments contain noxious algal species which have affected virtually every type of aquaculture operation. Bivalves react to the presence of toxic dinoflagellates in a variety of ways as summarized by Gainey and Shumway (1988) and Shumway et al. (1990). These include shell valve closure; reduced filtration rate; food selection; inhibition of byssus production; change in oxygen consumption; change in heart rate; neurophysiological effects; and mortality. Presence and extent of the responses depend on the algal and bivalve species as well as the presence/absence of prior exposure of the bivalves to the dinoflagellates (Gainey and Shumway 1988). Some recent examples: the brown tide chrysophyte *Aureococcus anophageffrens* apparently caused starvation of scallops and reduced feeding in mussels (various papers in Cosper et al. 1989.); the diatom *Rhizosolenia chunii* caused high mortality of shellfish in Port Phillip Bay, Australia (Parry et al. 1989).

The implication of these data is that bivalves are affected during algal blooms to a varying extent that may include starvation and death. Kodama (1990) has taken the further step of suggesting that bivalves may become toxic by ingesting toxin-producing bacteria directly in the absence of flagellate blooms. Effects on the bivalves in this case, if any, have yet to be determined.

Intramuscular injections of tetrodotoxin and paralytic shellfish toxins into a variety of marine and freshwater clams showed the clams to be highly resistant with LD_{100}/20-g weight of over 300 MU for both types of poison (Hwang et al. 1990), indicating that these molluscs can accumulate high levels of the toxins.

Toxic dinoflagellates which cause PSP also affect fish. Fish larvae died rapidly on a diet of *Gonyaulax excavata* (*Alexandrium tamarense*) or of copepods which had been eating the dinoflagellates (White et al. 1989). The authors calculated that a first-feeding fish larva (in this case red seabream, *Pagrus major*) would need to eat only 6-11 *G. excavata* cells for a lethal dose. Zooplankton are probably much more resistant; for example, the dose of *A. tamarense* to kill brine shrimp (*Artemia salina*) is 10 times that required to kill mice on a weight-for-weight basis (Betz and Blogoslawski 1982).

Toxicity of PSP to adult marine and freshwater fish was investigated experimentally by Saito et al. (1985a). For most fish, lethal doses, administered interperitoneally, were 1-10 times that for the mouse [a 20-g mouse is the standard bioassay animal (Horwitz 1990)]. However, by oral administration, with a PSP solution absorbed onto a commercial feed, the lethal dose was much higher than by the interperitoneal route. Three aquaculture species were tested using both routes (Table 4).

Marine fish sampled had similar responses to these freshwater species. The authors concluded that fish were very resistant to ingestion of PSP. However, the toxin given in feed would probably have eluted rapidly, such that the oral tests above might not be valid (S. Hall, pers. comm.).

Table 4. Mimimum lethal doses (LD_{100}) of PSP in three aquaculture finfish species. Data from Saito et al. (1985a)

Species	LD_{100} MU/20-g body weight Interperitoneal	Oral
Cyprinus carpio	2-5	120
Salmo gairdneri (*Oncorhynchus mykiss*)	1-8	320-340
Oreochromis niloticus	2-5	>400

Hwang et al. (1990) subjected a large number of fish, crustaceans and molluscs to interperitoneal injections of tetrodotoxin and saxitoxins. Minimum lethal doses (LD_{100}/20-g weight) for fish were similar for the two groups of toxins and to values found by Saito et al. (1985a). The various crabs and shrimps tested were in many cases even more susceptible with LD_{100} values of 0.5-5.0 MU; shrimp were the most susceptible (0.5 MU).

In situ, adult fish may be killed by paralytic shellfish toxins during blooms. Estimated doses of toxins in various fish kills in Europe and north America (White 1984) were of the same order of magnitude as the experimental values in Table 4. White (1984) also found similar interperitoneal lethal doses to those of Saito et al. (1985a) and Hwang et al. (1990). He concluded that fish are unable to tolerate even small amounts of toxins in their bodies and hence are unable to accumulate them.

Nevertheless, there have been cases of PSP from eating planktivorous fish in Sabah, Malaysia (Ting and Wong 1989); Papua New Guinea (Maclean 1979) and the Philippines (Estudillo and Gonzales 1984). Samples tested shortly after capture in Brunei Darussalam showed that the toxin was confined to the digestive tract; the intestines were full of *Pyrodinium* cells (De Silva et al. 1989). It may be of interest that when Oshima (1989) re-examined the material after 18 months of storage, the intestines showed low toxicity, while the dorsal musculature was quite toxic.

Fish kills by other toxic algae, nearly all in fish farms, are widespread. Apart from *Chattonella*, the raphidophyte *Heterosigma akashiwo* has caused major losses of caged fish in Japan and also in Europe since the early 1970s (White 1988) and most recently (1989) a NZ$ 12 million loss of salmon in New Zealand (Hallegraeff 1990). Nontoxic algae cause similar damage by creating anoxic conditions (Table 2). *Pyrodinium* has caused extensive mortality of marine life during a decomposing bloom in Sabah, Malaysia (Maclean 1989a). This Sabah incident was around sheltered reefs while reefs exposed to currents were undamaged (E. Wood, unpubl. data).

A summary of recorded effects on existing aquaculture facilities in developing countries is given in Table 5. Although the list is not long, it shows that there must have been serious economic problems for farmers - major losses of shrimp in China and on a smaller scale in Malaysia; fish kills in Hong Kong, Singapore and Thailand; and public health problems - *Pyrodinium* has caused hundreds of illnesses and usually several deaths during each outbreak.

It is worthwhile drawing attention to the shrimp farming industry which despite boom and bust cycles due to disease and market idiosyncrasies, is a priority export-oriented industry for many developing-country governments in Asia (Bangladesh, China, India, Indonesia, Malaysia, the Philippines, Thailand, Vietnam) and Latin America (Brazil, Ecuador, Mexico, Panama, Peru). Cultured shrimp rose from 2% to 22% of total shrimp production between 1981 and 1988 (Liao 1990). As Table 5 shows, there has already been a large loss in shrimp farms from algal blooms in China and perhaps moderate losses in Malaysia. The Philippines has experienced shrimp export losses due to the "halo effect" as mentioned above, while development of shrimp farming in Brunei Darussalam and Malaysia (Sabah) is also threatened by the "halo effect" (Maclean 1989c).

A list of potential hazards in developing countries can be made by including (i) algal species in areas where they have exhibited toxicity to marine organisms which are not cultured but which suggest similar toxicity to or in potential aquaculture crops; and (ii) the "sleepers", planktonic algal species known to be present in a locality but not yet causing toxicity problems. Absence of problems may be due to absence of blooms, absence of aquaculture, absence of phytoplankton surveys, absence of toxin

Table 5. Incidence of toxic algal problems in developing-country aquaculture.

Country (site)	Algal species	Effect	Status	Relationship to aquaculture	Source
Brunei Darussalam	*P. bahamense* var. *compressum*	PSP	Chronic since 1976	Losses to mussel farmer	Jaafar et al. (1989)
China (Hupei and Shandong Prov.)	Unknown	Shrimp kill	First record, 1989	80% shrimp in 2,000 ha of ponds killed	Ger Guo-Chang (pers. comm.)
India (Tamil Nadu and Karnataka)	Unknown	PSP	First record, 1988	PSP, DSP found in shellfish harvesting beds, Karnataka	Baht (1981); Karunasagar et al. (1984, 1989)
Rep. of Korea (mainly Jinhae Bay)	26 species	Various	Chronic, increasing	"Severe damage" to farms	Park et al. (1989)
Malaysia (peninsular)	*Chattonella marina*	Shrimp kill	First report, 1985	Heavy losses in shrimp farms	Khoo (1985)
Malaysia (Sabah)	*P. bahamense* var. *compressum*	PSP	Chronic since 1976	Experimental cultured oysters toxic	Ting and Wong (1989)
Papua New Guinea	*P. bahamense* var. *compressum*	PSP	Maybe cyclic	Experimental cultured oysters toxic	Maclean (1989d)
Philippines	*P. bahamense* var. *compressum*	PSP	Increasing since first record, 1983	Caused mainly by eating cultured mussels	Gonzales (1989a)
Singapore	*Cochlodinium* *Chattonella* *Heterosigma*	Fish kills	Occasional	Mortality of groupers in cages	Lim (1989)
Taiwan (Tungkang)	Probably *Alexandrium tamarense*	PSP	First record, 1986	Caused by eating cultured clams (*Soletellina*)	Hwang et al. (1989); Su et al. (1989)
Thailand (Gulf)	Various	Fish kills	Chronic, increasing		Suvapepun (1989)
Venezuela	*Gymnodinium catenatum*	PSP		Stifled developing mussel culture industry	S. Hall (pers. comm.)

bioassays, or lack of recognition of a problem. The list given in Table 6 is no doubt incomplete.

Implications for Freshwater Aquacuture

The potential impact of toxic freshwater algae on aquaculture in developing countries, particularly in the tropics, depends to a large extent on resolving the paradox that: (a) lethal cyanobacterial blooms are common in temperate freshwaters; (b) resulting losses of fish crops and human illnesses are rare; and (c) the same algal species do not appear to be toxic in the tropics. It is worthwhile reviewing cyanobacteria toxicity before assessing its possible impact on aquaculture.

The prevalence of toxic cyanobacterial blooms can be gauged by a recent survey in Finland (Sivonen et al. 1990) in which 44% of 188 water bloom samples in fresh and brackishwater during 1985 and 1986 were found to be lethally toxic, containing either hepato- or neurotoxins. The authors noted that blooms were more common in the south, which was expected in view of the higher

Table 6. Potential hazards from various toxic algae to farmed aquatic organisms in some developing countries.

Algal species	Country (Locality)	Hazard	Source/Comments
Aphanizomenon spp.	Many countries	PSP in freshwater clams	
Chattonella spp.	Many countries	Fish kills	e.g., Singapore (Lim 1989); Philippines (Manila Bay) (Y. Fukuyo, pers. comm); India (Malabar coast) (Subrahmanyan 1954)
Dinophysis spp.	Many countries	DSP in shellfish	e.g., several *Dinophysis* species in Papua New Guinea (Maclean 1989a); Indonesia (Adnan, unpubl. data); Côte d'Ivoire (Dandonneau 1971); Mediterranean Egypt (Dowidar 1974); Puerto Rico (Margalef 1957); Thailand (Suvapepun 1989)
Gonyaulax monilata	Ecuador Pakistan (Arabian Sea)	Fish kills	Unesco (1982); Rabbani et al. (1990)
Microcystis aeruginosa	Many countries	Hepatoenteritis in humans; fish kills	Inadequate data to determine risk
Protogonyaulax sp.	Mexico (Gulf)	PSP	Described from Mexico; toxic isolates
Pyrodinium bahamense var. *compressum*	Costa Rica (Pacific) El Salvador Guatemala (Pacific) Mexico (Pacific)	PSP in cultured shellfish	Extensive *Pyrodinium* red tides causing PSP illnesses on the Pacific coast of Central America, November-December 1989. Previous Nicaragua (Pacific) episodes; also in Guatemala (see text)
	Fiji	PSP in shellfish	Causative organism unknown; most probably *Pyrodinium*. Five cases between 1975 and 1983 (Raj 1983)
	Indonesia	PSP in fish and shellfish	*Pyrodinium* not confirmed, but presumed by Adnan (unpubl. data); deaths from eating planktivorous fish, *Sardinella* sp. and *Selaroides* sp. (Adnan 1984)
	Palau	PSP in *Pyrodinium*	Cultured by Harada et al. (1982)
Unknown sp.	India	PSP in shellfish	In coastal shellfish (Segar et al. 1988)
	Singapore	PSP in shellfish	PSP occurred in mussels (Tan and Lee 1986)
	Solomon Islands	PSP, ciguatera	PSP from oysters (*Crassostrea* spp.) affects 2 per 1,359 population; ciguatera "infrequent and unpredictable" (Eason and Harding 1987)
	Thailand (Gulf)	PSP	(Kodoma et al. 1988)

population, industry, agriculture and forestry there. Carmichael et al. (1990) similarly stated that as waterbodies become more eutrophic they support higher production of cyanobacteria. Cyanobacterial bloom samples from 15 British freshwaters over recent years have all been lethal although mortalities to animals drinking water were recorded on only two occasions (Codd and Poon 1988).

The mode of ingestion of these toxins is by drinking the contaminated water. However, saxitoxins have been found in freshwater clams (Y. Fukuyo, pers. comm.). Thus, paralytic shellfish poisoning from cultured freshwater bivalves is a distinct possibility.

Toxicity levels per unit biomass of cyanobacteria are known to vary widely from week to week and even within an individual bloom on a single occasion (Codd and Poon 1988).

Another confounding feature is that isolates of a single species may or may not contain toxins (Codd and Poon 1988), a feature shared with some dinoflagellates. There is mounting evidence that, for saxitoxins at least, the toxin source may be bacterial symbionts rather than endogenous (Kodama 1990; Tamplin 1990).

What triggers toxin production? Production of toxins in a marine cyanobacteria (*Synechococcus* sp.) varied with salinity, temperature, light intensity and most interestingly with growth phase; no toxins were produced in the exponential growth phase but appeared suddenly during the stationary phase (Mitsui et al. 1989). For several dinoflagellate species, toxin production in cultures during the exponential growth phase was found to increase at lower temperatures as a result of lower growth rate (Hall 1982; Ogata et al. 1989).

In warmer latitudes, cyanobacteria tend to dominate in summer and when waterbodies become more eutrophic; *Microcystis aeruginosa* is frequently dominant in wastefed fishponds (Colman and Edwards 1987). Yet there are no reports of fish mortality from cyanobacterial poisons in tropical waters such as Lake George (Uganda) and tropical wastefed fishponds (Colman and Edwards 1987).

There is limited evidence that higher temperatures may suppress toxin production. For *M. aeruginosa* the optimum temperature for toxin production is about 25°C. The effects of light intensity are ambiguous (Codd and Poon 1988). The strains used in experiments appear to be from temperate sources. Would tropical strains behave differently? The tropical blooming cyanobacteria *Cylindrospermopsis (Anabaenopsis) raciborskii* was recently found to be severely hepatotoxic, apparently causing a large outbreak of hepatoenteritis from a tropical Australian reservoir (Hawkins et al. 1985). The authors noted that hepatoenteritis is common in many tropical countries and that *C. raciborskii* is one possible cause. The species blooms in water warmer than 25°C, although the toxicity tests of Hawkins et al. (1985) were only at 25°C.

Testing at higher temperatures of tropical freshwater blooms for toxicity and effects on fish is needed to resolve these issues. New reliable assay methods are needed also (Codd and Poon 1988).

As far as growth of the cyanobacteria themselves is concerned, the optimum temperatures may be higher than those noted above. Overall maximum rates of photosynthesis and growth are 25°C or greater; 27.5-32°C for *M. aeruginosa* (Roberts and Zohary 1987).

Toxicity to Fish

Phillips et al. (1985b) exposed rainbow trout (*Oncorhynchus mykiss*) to a bacteria-free culture of *M. aeruginosa*. The algae proved to be harmless in this form, presumably because the trout would not ingest them, although the algal toxicity was proven by intraperitoneal injections of ultrasonically broken cells from the culture, which were lethal to the fish. The authors cited two 1948 reports of an instance where decomposing *Aphanizomenon flos-aquae* proved toxic to a range of tank-held fish species in a river in Wisconsin, USA; while Tabthipwon et al. (1988) suspected that toxins from *M. aeruginosa* retarded growth of experimental tilapia (see below).

Colman and Edwards (1987) cited experiments in which carp, probably *Cyprinus carpio*, exhibited erratic behavior and mortality when fed a mixture of live *Aphanizomenon* and *Microcystis*; and mortality of golden shiner *Notemigonus crysoleucas*, when toxin from *A. flos-aquae* was added to the water. As with the study by Phillips et al. (1985b), interperitoneal injections were toxic in both experiments.

Phillips et al. (1985a) summarized the adverse effects of algae on farmed (and wild) fish populations as related not only to the toxins but also to increased turbidity, diurnal fluctuations of dissolved oxygen and increased concentrations of harmful metabolites during bloom die-offs.

Other than cyanobacteria, the freshwater dinoflagellate *Peridinium polonium*, one of several *Peridinium* species which bloom in many artificial lakes in Japan, produces an ichthyotoxin comparable in strength to that of the Florida red tide organism *Ptychodiscus brevis (Gymnodinium breve)*, and causes occasional fish kills (Oshima et al. 1989).

Much future aquaculture development in freshwater in developing countries will take place in relatively stagnant waters, such as flooded ricefields and ponds, and in cages in lakes and reservoirs.

The riceland available for rice-fish farming is enormous: there are 77 million hectares of irrigated riceland, for example (IRRI 1988). It must be remembered that cyanobacteria, through their nitrogen-fixing ability, play a vital role in building soil fertility; in addition they also produce substances that promote the growth of rice plants (Roger and Kulasooriya 1980). Thus, farmers encourage blooms. If fish production further encouraged cyanobacterial blooms through increased eutrophication, one would expect that there would be potential for poisoning not only the fish but also livestock and perhaps even humans downstream. However, this has not proven to be the case: where rice-fish culture is practiced, there have been no reports of illnesses.

In fishponds, cyanobacteria may play a major role in nitrogen fixation (Lin et al. 1988) and as a direct feed for phytoplanktivorous fish like some of the tilapias, including the most promising species, *Oreochromis niloticus*. Recent work has shown that 90% of fish yield in organically manured ponds come from food webs originating with algae (Schroeder et al. 1990) Yet their potential toxicity has rarely been taken into account. *O. niloticus*, in fact, seemed to be retarded in its growth when *Microcystis aeruginosa* was a significant part of the diet and, indeed, liver deterioration of the fish was found at higher levels of *Microcystis* in the feed (Tabthipwon et al. 1988). Recall that *M. aeruginosa* produces lethally toxic hepatotoxins (Carmichael et al. 1990). Codd and Poon (1988) noted that the algal genera which are typically dominant are those with toxin-producing species.

The area of freshwater lakes and reservoirs in Asia alone is some 13-14 million ha of which reservoirs currently comprise 5.5 million ha (De Silva 1988). Asian reservoir area may increase to 20 million ha by the year 2000 (Costa-Pierce and Soemarwoto 1987). The potential for aquaculture development in these waterbodies is large.

Algal blooms occur in such waterbodies and can be cropped by fish. For example, in the Saguling Reservoir, Indonesia, there are constant blooms of *M. aeruginosa* and the introduction of phytoplanktivorous fish for culture and capture fisheries was suggested (Costa-Pierce and Soemarwoto 1990; Munro et al. 1990). Again, there appears to be potential for poisoning of fish or mammals drinking the water.

Beveridge (1984) provided a model to calculate the aquaculture carrying capacity of enclosed waterbodies based on phosphorus values as a measure of the maximum algal biomass permissible, that maximum depending on the other purposes for which the water is used. Stirling and Dey (1990) observed that such models, being based on annual nutrient loads, overlook the possibility of short-term deleterious blooms. However, they recommend an upper limit of fish production in shallow well-mixed lakes of 3-4 t·ha^{-1}.

An elegant method of determining whether aquaculture activities contribute significantly to eutrophication in tropical enclosed waterbodies was provided by Costa-Pierce (1990). Through multiple regression analysis of water quality at several sites in an Indonesian reservoir, he determined that aquaculture, in this case cage culture of carp (*Cyprinus carpio*), had negligible impact on plankton compared to the effect of other endogenous and exogenous nutrients.

Freshwater algae themselves are now being grown for human food. The cyanobacteria *Spirulina* is the best known genus, used as a food supplement. But other genera such as *Anabaena*, which includes toxic strains, are coming on the market, posing potential health hazards (Gorham and Carmichael 1988).

Tropical countries would seem to have advantages for growing algal foods: strong sunlight and extensive ponds and paddies from which the algae might be harvested. Research is needed to identify and assess any hazards posed before this form of aquaculture develops. Monitoring toxicity of cultured blooms may become a routine expense.

Management Measures

Management at the Industry Level

Strategies for aquaculturists to offset the impact of toxic blooms include a variety of physical and chemical techniques.

MECHANICAL/PHYSICAL MEASURES

In Singapore, where affected fish cages and toxic plankton blooms are in the upper two meters, suggested steps to minimize contact between algae and fish include: a PVC "skirt" at least 2 m deep to surround individual cages or the whole farm; transfer of fish to deeper nets, 4 m deep; an airlift or water pump to draw up deeper water to disperse the algae; thinning out of the stocking density of sensitive fish; and towing of cage assemblages to safe areas (Lim 1989).

A management method for mussel beds in Samar, Philippines, was hinted at by Ronquillo (1987) in his observation that the poles on which the mussels grew were too close together and led to pollution, which provided nutrients for *Pyrodinium* to bloom. Moving the poles apart might encourage water movement and less waste buildup.

Adjusting depths of bivalves in cages or on longlines is unlikely to be a safe option. Maclean (1975) found toxic bivalves at 10 m depth in Port Moresby, Papua New Guinea, below depths of *Pyrodinium* blooms. Scallops on longlines in Funka Bay, Japan, were found to contain little toxin in the uppermost and lowest parts of the longlines but were highly toxic in the middle sections (S. Hall, reporting a presentation by V. Nishihama, pers. comm.). Recent experiments in eastern Canada showed that mussel growers cannot avoid toxicity in cultured mussels by altering the depth of holding structures due to the changing distribution of PSP toxin (from *Alexandrium excavatum*) in the mussels with depth over time (Desbiens et al. 1990).

Jones and Gowen (1985) modeled flushing rates and phytoplankton growth in Scottish sea-lochs and suggested that considerations of exchange and source water hydrography be included in aquaculture site selection there.

Dale et al. (1987) went further. They considered that a hydrographic survey should be undertaken as part of an aquaculture site selection process, followed by a pilot phase in which environmental parameters continue to be monitored; finally an ecological model should be set up to predict the occurrence of blooms. They also recommended that the literature be scanned to identify potentially harmful algae in the area; that sediments be examined for dinoflagellate cysts; and that available remote sensing data be obtained. However, it is difficult to imagine such guidelines being implemented in most developing countries.

CHEMICAL DISPERSANTS

Hallegraeff (1987) cited an earlier report on spraying copper sulfate from aeroplanes, which killed *Gymnodinium breve* blooms but caused anoxia when cells decomposed. White (1988) said that application of ammonium sulfate to brackishwater fishponds kills incipient blooms of *Prymnesium parvum*.

Chemical removal of a bloom of *Chattonella antiqua* in Shido Bay, Japan, was accomplished using 50 ppm of sodium percarbonate (Okaichi et al. 1989b). The bloom disappeared within two hours of application; the algae apparently sank and encysted. The area was suffering mortality of cultured yellowtail due to the *Chattonella* bloom at the time.

There does not seem to have been any further progress on using such chemicals but they may have a future role as for dispersants used in certain oil spill situations.

OZONATION

Ozone treatment was earlier reported to be effective against PSP toxins and brevetoxin (reviewed in Rosenthal 1981). Dupree (1981), summarizing previous literature, felt that 1-5 minute contact time at 0.56-1.0 mg $0_3 \cdot l^{-1}$ was sufficient to kill most pathogens but that ozone had high toxicity in itself, especially to oyster larvae. Blogoslawski (1988) concluded that ozonized seawater could not detoxify clams that had ingested cysts or had toxin bound in their tissues over long periods of time, but could destroy PSP in motile dinoflagellate cells.

There has been one experiment on the effect of ozone on *Pyrodinium*-toxic mussels (*Perna viridis*), in the Philippines (Gacutan et al. 1984), which reported that ozone was effective in lowering toxin levels, although rather slowly under the experimental conditions. The authors reported that PVP-iodide was also effective.*

DEPURATION

Experiments using clean seawater for toxin reduction in mussels (*Mytilus edulis*) in Korea showed that filtered seawater in an open system caused toxicity to drop by 40-92% over five days (Chang et al. 1988). White (1982) cited earlier experiments on clams and mussels with similar results. For *Pyrodinium*, toxic oysters (*Crassostrea echinata*) took three weeks to lose toxicity in a filtered open seawater system (Maclean 1975).

In vivo, however, depuration takes considerably longer. Shumway (1990) provided an extensive list of retention times of various bivalves exposed to different toxic dinoflagellate sources. From the list, an average figure would be 4-8 weeks for toxicity to drop below quarantine or detection levels, depending on mollusc and algal species. Thermal shock is also reported to speed up depuration (Blogoslawski and Neve 1979).

HEATING/CANNING

PSP toxins are largely unaffected by ordinary cooking methods and their activity is potentiated by the use of acid (vinegar) in preparations common in countries such as the Philippines.

*New results on rapid destruction of *Gymnodinium breve* toxins by Schneider and Rodrick (in press) by both direct and indirect (pre-ozonated water) ozone treatment hold promise of safe, effective toxin removal. **Eds.**

Detoxification of PSP toxins through canning is a little explored area, although a standard canning process can reduce toxin levels in scallops from 400 $MU \cdot g^{-1}$ to less than 4 $MU \cdot g^{-1}$. Most of the toxin reduction is due to the process while the remainder takes place over 30 days storage, according to Noguchi et al. (1980). An earlier report by Schantz (1973, cited by Arafiles et al. 1984) indicated that canned and processed shellfish may still contain up to 50% of the toxin even if the product has been heated to 115°C.

There does not seem to have been much progress or interest in this field since the above papers. However, White (1982) suggested canning as a viable option for Bay of Fundy clams in the face of increasingly toxic *Gonyaulax* blooms.

There is no chemical explanation for toxin breakdown or deactivation at the temperatures involved and the phenomenon may be simply a redistribution of toxin between meat and liquid parts of the canned shellfish (S. Hall, pers. comm.). Oshima's (1989) observation, mentioned above, that PSP toxins became distributed throughout fish tissues after 18 months frozen storage is pertinent here.

Microcystis toxin (hepatotoxin) is heat labile, such that proper cooking of fish raised in *Microcystis*-dominated ponds should render any such toxin in the fish harmless to humans [B. Hepher, pers. comm. to Colman and Edwards (1987)].

For tetrodotoxin, a combination of cooking and washing was effective in reducing highly toxic pufferfish liver to safe levels (Tsubone et al. 1986).

Management at the National Level

Red tide action plans have been produced in some Southeast Asian countries. All involve monitoring programs and public awareness. Measures related to aquaculture are: in Brunei Darussalam, a ban on harvesting and/or marketing of suspect fish and shellfish (De Silva et al. 1989); in Sabah, Malaysia, to cease picking, selling or eating all types of shellfish and snails, and even in the safe season not to consume a large amount of shellfish in one meal - children are urged not to eat shellfish at all (Wong and Ting 1989); in the Philippines a ban on harvesting, marketing and transporting of all kinds of marine shellfish from contaminated waters and provision of emergency loans to affected fish farmers (Gonzales 1989b).

In Hong Kong, where the major impact of red tides is fish kills in fish farms, simple sampling kits are issued to aquaculture representatives and government staff. On-site tests are performed by a "mobile squad" and fish farmers in the area are alerted and advised of means to minimize losses (Wong and Wu 1987).

In El Salvador, food was distributed by the government to clam harvesters who were obliged to withdraw their product from sale during the second red tide episode along the western Central American coast in April 1990 (S. Hall, pers. comm.).

Virtually every report dealing with management aspects of algal blooms calls for continual monitoring of plankton and suspect organisms, usually shellfish. As mentioned, this can be a costly exercise. In the Philippines, good progress towards a blowfly bioassay, rather than the more expensive mouse bioassay, has been made (A. Mendigo, pers. comm.).

No consideration seems to have been given by governments in developing countries to zoning as a means of setting aside for aquaculture, sites where pollution levels can be kept low and red tide incidence is unlikely. Indeed, in Southeast Asia at least, red tide management plans are somewhat fatalistic, assuming permanent problems and neglecting (probably for economic reasons) the rezoning of aquaculture operations, reduction of pollution or engineering works to alter current patterns (e.g., Corrales and Gomez 1990).

Remedial management has been attempted elsewhere. For example, the number of red tides in Japan's Seto Inland Sea dropped markedly, beginning four years after the passing of the 1972 Seto Inland Sea Environment Conservation Law (see Fig. 3) (Okaichi 1989). The law reduced land reclamation and required treatment facilities to reduce the chemical oxygen demand (COD). Reduction of pollution lessened the number of red tides. Yet, the number of fish kill events has not been decreasing and the value of lost fish harvests has been increasing (Okaichi 1989).

For algae which are not clearly associated with eutrophication, the physical habitat itself would need to be altered to be less favorable for the occasional dominance of toxic species. In fact, the decimation of mangrove forests around Southeast Asia may be one reason why toxic blooms of *Pyrodinium* are not more common in the region. A bioluminescent lagoon in Puerto Rico, caused by a "resident" population of *Pyrodinium*, lost its glow when a new channel to the sea was created; the channel increased the flushing rate of the lagoon and the *Pyrodinium* could no longer maintain their presence (H.H. Seliger, pers. comm.).

Coastal engineering may be a solution to reduce eutrophication and associated blooms and anoxic conditions. Dredging operations, however, invoke the danger of seeding benthic dinoflagellate cysts into the water column (Hallegraeff 1987).

Table 7 presents some of the present measures used by governments and suggests some alternatives. Moving an affected industry to a "clean" location is an option that could be less expensive than extensive monitoring and attempts at pollution abatement; harvesting red tide organisms is not a cost-effective measure at present and would depend on profitable use being found for chemicals from the organisms.

Similarly, coastal engineering options have to be considered against the value of the threatened or potential aquaculture (or fisheries) harvests in the area, as well as other environmental impact aspects. As aquaculture expands, the economic and public health incentives to change the shape of a bay or move a river mouth, for example, may become comparable to those that lead to massive dam construction for agriculture, flood control or drinking water. So far the whole field of algal blooms has been dominated by biologists and toxicologists.

Table 7. Present management concepts and alternative approaches to dealing with algal blooms in aquaculture.

Present concept	Alternative approach
Routine monitoring	"Once only" movement of aquaculture industry
Move cages out of red tide event by towing	Remove red tide events by coastal engineering
Attempts to lower pollution levels	Rezone aquaculture away from pollution
Use chemical dispersants in aquaculture crises	Harvest red tides
Insure, make grants/loans to affected farmers	Abandon (legislate against) aquaculture in the affected area

Imaginative thinking on long-term and large-scale solutions is lacking.

Management at the International Level

Some form of regional or international cooperative activity is usually recommended at the various international fora on red tides, but until recently little had been done. Dale et al. (1987) recommended a "red tide brigade" of international experts. Something of the kind was subsequently formed by the Intergovernmental Oceanographic Commission's Ocean Science and Living Resources Group which recently included red tide as a major thrust with the formation of the (OSLR/IOC) Group of Experts on Harmful Algal Blooms. The Western Pacific Group (WESTPAC) of the IOC also has a red tide section sponsoring research and workshops, while the United Nations Environment Programme is setting up regional management projects of an institution-building nature. The International Council for the Exploration of the Sea (ICES) investigates algal blooms through a Working Group on Phytoplankton and the Management of their Effects.

An International Red Tide Information and Assistance Service (IRTIAS) is being formed at Woods Hole Oceanographic Institution to assist developing countries especially and which will include advice on preventive measures for aquaculturists (White 1990b).

Aquaculture and Red Tides: An Uneasy Association

Some authors feel that aquaculture contributes to conditions favoring red tides. Smayda (1989) stated that "shellfish and finfish aquacultural activities are often followed by blooms of both benign and toxic algal species." He only gave one example but noted that "aquacultural activities augment and modify the local grazing structure as do benthic and pelagic fishery operations" and that this may be a cause of blooms. Hallegraeff (1990) also said that "the effluent from fish farms stimulates algal blooms".

The evidence is circumstantial but sometimes persuasive. Anderson (1989) pointed out that PSP in one Philippine site (Balete Bay, Davao Oriental) was first recognized just one year after mussel (*Perna viridis*) culture was introduced there. Corrales and Gomez (1990) noted that red tide sites in the Philippines are mostly sites of mussel farming. Port Moresby in Papua New Guinea, where the first recorded outbreak of *Pyrodinium* in the Indo-Pacific took place, was also the site of a large pearl oyster farm at the time. However, the larger picture does not support an association between *Pyrodinium* and aquaculture (or fishery) operations. Neither activity was present in other outbreak sites in Papua New Guinea or elsewhere.

A much closer relationship exists between *Pyrodinium* and mangroves. The only bloom of *Pyrodinium* where mangroves have not been in the general vicinity is western Manila Bay, Philippines, where, however, there used to be mangroves (Maclean 1989d). Margalef (1957) had earlier placed the nontoxic variety of *Pyrodinium* (*P. bahamense* var. *bahamense*) as a member of the mangrove community. Mangrove areas are generally sheltered and cheap to lease for which reason most coastal aquaculture in the Indo-Pacific region takes place in these areas - the habitat of *Pyrodinium*.

Another interesting feature of *Pyrodinium* is that the fossil record of cyst distribution is much broader latitudinally than the present distribution of the species (Matsuoka 1989).

For other harmful microalgae, particularly those enhanced by forms of pollution (see below), it is difficult to reject the hypothesis that the increased nutrients around fish, shrimp or mollusc farms might stimulate algal blooms.

Dispersal of red tide organisms via coastal or ocean currents is a known, and perhaps the chief mechanism, a recent interesting example being that of a bloom of the Florida red tide organism *Gymnodinium breve* which was carried several hundred kilometers northward to North Carolina in the Gulf Stream (Tester et al. 1989). Seliger (1989) lists a number of areas where toxic blooms of *Gymnodinium* and *Alexandrium* first appeared in 1971/1972, as a result of water movements probably attributable to El Niño during that period. Further, there is the possibility of long-distance dispersal of red tide organisms in ballast water (Maclean 1977; Hallegraeff et al. 1990) which poses dangers for aquaculture. Such organisms have been proven capable of surviving long voyages in ships' holds and may be carried into important aquaculture areas; regulation of ballast water discharge is called for (Hallegraef et al. 1990). There is also the possibility that the organisms may be spread through spread of cultured aquatic animals, particularly bivalves, to new growing areas, as in the Philippines, for example.

Two Types of Bloom Organisms

There seem to be two major types of marine blooms - those in which nutrient additions to coastal systems are obviously implicated and those in which coastal enrichment is not an obvious factor. Organisms such as *Gymnodinium breve*, some *Alexandrium* (*Gonyaulax*) species and *Pyrodinium* fall into the second category according to these authors. Perhaps *Gambierdiscus* belongs here also. One environmental factor that triggers major blooms may be El Niño. As noted above, Seliger (1989) felt that this was so for *Gymnodinium* and *Alexandrium*. The case is even stronger for *Pyrodinium*, for which all major outbreaks in the western Pacific since 1972 correlate with El Niño - Southern Oscillation (ENSO) events (Fig. 4).

Coastal enrichment, attributable to human activities, is almost certainly involved for other species and situations, as in the Seto Inland Sea (*Chattonella* spp.) (Okaichi 1989), in Jinhae Bay, Korea (Park et al. 1989), and Hong Kong (Wong and Wu 1987). The last two areas host a cocktail of bloom species. In Hong Kong, the evidence of human impact is most compelling (see Fig. 2) (Lam and Ho 1989).

Where does this leave the relationship between aquaculture and algal blooms? If the two types of blooms - either obviously associated with coastal enrichment or not - are treated separately, some future scenarios can be drawn.

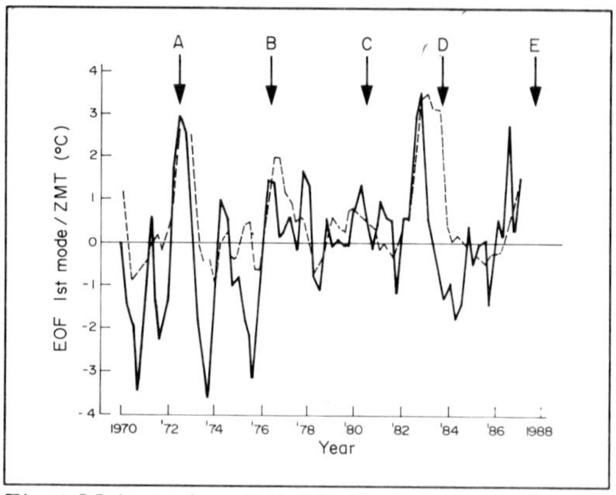

Fig. 4. Major toxic red tides in the western Pacific and ENSO events, 1972-1987. Graph shows empirical orthogonal functions (EOF, solid lines) of zonal wind anomalies, and zonal mean surface temperature anomalies (ZMT, broken line) over the near-equatorial eastern Indian and western Pacific Oceans. Strong positive anomalies are indicative of ENSO events and are seemingly correlated with *Pyrodinium* red tide events. Arrows show time of onset of red tides: A: Papua New Guinea, 1972; B: Borneo, 1976: C: Borneo, 1979-80; D: Philippines, 1983; E: Philippines, 1987. From Maclean (1989c).

Some Scenarios

What does the environmental future of developing-country aquaculture hold with respect to algal blooms? Obviously, either enrichment levels will be contained, increase or decrease.

In the unlikely event of decreasing levels of coastal enrichment, dominant algal species may revert to a mixture of more benign diatoms and of dinoflagellates that bloom with little (if any) enrichment, such as *Pyrodinium* in tropical waters. In this case, the affected countries can look forward to either regular or in the longer term cyclical toxic events. The short-term (but how short?) outlook is for chronic seasonal blooms that require monitoring and seasonal closures. The long-term perspective is for occasional acute events that may initially cause extensive losses of lives and investment (because monitoring activities will usually have lapsed between cycles).

If present levels of coastal enrichment can be contained, the outlook is for an ongoing uncertainty and lack of predictability about both the species and the severity of future red tides. Monitoring and insurance costs, occasional losses, public health concerns and uncertain investment returns may combine to make surviving coastal aquaculture products much more expensive.

Increased enrichment, if recent history is any guide, can only lead to a broader spectrum of toxic species or more or less continuous presence of a few dominant toxic species. Nuisance algal species which respond to coastal enrichment will increase in number and frequency of blooms as the level of enrichment increases. Most of these species cause fish kills through toxin production or by causing anoxic conditions. Aquaculture of finfish in such circumstances will become increasingly incompatible. Aquaculture operations themselves may contribute significantly to eutrophication, exacerbating the situation. If aquaculture continues, mollusc culture would be more suitable in these areas, since there is no additional nutrient input from feeding.

With regard to blooms that appear to be independent of coastal enrichment, the situation for aquaculture is unpredictable because the species appear to be spreading in geographic distribution. Most of them produce PSP toxins. Generally speaking finfish and crustacean culture would be more successful in such areas, because these organisms are neither susceptible to nor accumulate saxitoxins. Note that aquaculture activities in these areas may lead to eutrophication and perhaps a "succession" of bloom organisms towards those that obviously respond to enrichment.

K. Matsuoka (pers. comm.) suggested that a succession may be occurring in Manila Bay and nearby Zambales where *Pyrodinium* blooms have occurred in recent years. Noxious algal species such as *Chattonella* and *Alexandrium* were found in significant numbers there in a July 1990 survey in which he took part.

In enclosed freshwater waterbodies, given their multiple use, especially in tropical developing countries, cyanobacteria blooms are almost a certainty. Will these tropical blooms suddenly become toxic, as have other (marine) algal species? If the hypothesis is correct that bacterial symbionts (or parasites?) are responsible for toxicity in algae, then the "on" switch, i.e., an algal population assimilating toxic bacteria, may be easier than an "off" switch, i.e., an algal population losing its symbionts/parasites.

Finally, the prospect of abandoning aquaculture, e.g., through legislation against it in view of public health hazards that cannot be overcome reliably, is an alternative, which I found on sharing this manuscript, to have few supporters. For example, J.W. Hurst, Jr. (pers. comm.) wrote "I do not regard 'red tides' as an insurmountable barrier to the development of aquaculture. It must be understood that while the present methods for determining the safety of shellfish are expensive they are a necessary part of determining their safety."

My view is perhaps a jaded one in the light of the uncontrolled coastal developments taking place in most countries. There are so many sectors competing for use of inshore waters that the future of coastal aquaculture may be limited by them as much as by algal blooms. Red tides may be the "straw that breaks the camel's back" in this respect. However, I take heart from Dr. Hurst's (pers. comm.) conclusion that "Persons who are Third World nationals deserve high quality safe seafood and should be regarded as the most important customers of the aquaculture industry of that nation. This attitude will assure a high quality and safe product."

Acknowledgements

Thanks to Drs. W.W. Carmichael, R. Corrales, I.R. Falconer, Y. Fukuyo, S. Hall, G.M. Hallegraeff, K. Matsuoka, T. Okaichi, W. Pastor, A.D. Pongase, F. Rosales-Loessener, S. Shumway and A.W. White for timely advice and material for this review.

References

Addison, R.F. and J.W. Stewart. 1989. Domoic acid and the eastern Canadian molluscan shellfish industry. Aquaculture 77: 263-269.

Adnan, Q. 1984. Distribution of dinoflagellates at Jakarta Bay; Taman Jaya, Banten; and Benoa Bay, Bali: a report of an incident of fish poisoning at eastern Nusa Tenggara, p. 25-27. *In* A.W. White, M. Anraku and K.K. Hooi (eds.) Toxic red tides and shellfish toxicity in Southeast Asia. Southeast Asian Fisheries Development Center and International Development Research Centre, Singapore.

Anderson, D.M. 1989. Toxic algal blooms and red tides: a global perspective, p. 11-16. *In* T. Okaichi, D.M. Anderson and T. Nemoto (eds.) Red tides. Biology, environmental science, and toxicology. Elsevier, New York.

Anderson, D.M., A.W. White and D.G. Baden, Editors. 1985. Toxic dinoflagellates. Elsevier, New York.

Anon. 1988. Over 100 farms towed to avoid algal threat. Can. Aquacult. 4(5):43.

Arafiles, L.M., R. Hermes and J.B.T. Morales. 1984. Lethal effect of paralytic shellfish poison (PSP) from *Perna viridis* with notes on the distribution of *Pyrodinium bahamense* var. *compressa* during a red tide in the Philippines, p. 43-51. *In* A.W. White, M. Anraku and K.K. Hooi (eds.) Toxic red tides and shellfish toxicity in Southeast Asia. Southeast Asian Fisheries Development Center and International Development Research Centre, Singapore.

Bagnis, R., A.M. Legrand and A. Inoue. 1990. Follow-up of a bloom of the toxic dinoflagellate *Gambierdiscus toxicus* on a fringing reef of Tahiti, p. 98-103. *In* E. Granéli, B. Sundström, L. Edler and D.M. Anderson (eds.) Toxic marine phytoplankton. Elsevier, New York.

Bagnis, R., J. Bennett, C. Prieur and A.M. Legrand. 1985. The dynamics of three toxic benthic dinoflagellates and the toxicity of ciguateric surgeon fish in French Polynesia, p. 177-182. *In* D.M. Anderson, A.W. White and D.G. Baden (eds.) Toxic dinoflagellates. Elsevier, New York.

Baht, R.V. 1981. A report on an outbreak of mussel poisoning in coastal Tamil Nadu, India. National Institute of Nutrition. Indian Council of Medical Research, Hyderabad. 9 p.

Balech, E. 1985. The genus *Alexandrium* or *Gonyaulax* of the tamarensis group, p. 33-38. *In* D.M. Anderson, A.W. White and D.G. Baden (eds.) Toxic dinoflagellates. Elsevier, New York.

Betz, J.M. and W.J. Blogoslawski. 1982. Toxicity of *Gonyaulax tamarensis* var. *excavata* cells to the brine shrimp *Artemia salina* L. J. Pharm. Sci. 71:463-466.

Beveridge, M.C.M. 1984. Cage and pen fish farming. Carrying capacity models and environmental impact. FAO Fish. Tech. Pap. 255, 131 p.

Blogoslawski, W. and R. Neve, Rapporteurs. 1979. Detoxification of shellfish, p. 473. *In* D.L. Taylor and H.H. Seliger (eds.) Toxic dinoflagellate blooms. Elsevier North-Holland, New York.

Blogoslawski, W.J. 1988. Ozone depuration of bivalves containing PSP: pitfalls and possibilities. J. Shellfish Res. 7:702-705.

Carmichael, W.W., N.A. Mahmood and E.G. Hyde. 1990. Natural toxins from cyanobacteria (blue-green algae), p. 87-106. *In* S. Hall and G. Strichartz (eds.) Marine toxins. Origins, structure, and molecular pharmacology. American Chemical Society, Washington, DC.

Chang, D.S., I.S. Shin, H.Y. Goo, E.G. Oh, J.H. Pyun and Y.H. Park. 1988. Studies on the distribution, characterization and detoxification of shellfish toxin in Korea. 3. Detoxification of paralytic shellfish poison of sea mussel, *Mytilus edulis*. Bull. Korean Fish. Soc. 21:297-302. (Abstract only seen)

Codd, G.A. and G.K. Poon. 1988. Cyanobacterial toxins, p. 283-296. *In* L.J. Rogers and J.R. Gallon (eds.) Biochemistry of the algae and cyanobacteria. Oxford University Press, Oxford.

Colman, J.A. and P. Edwards. 1987. Feeding pathways and environmental constraints in waste-fed aquaculture: balance and optimization, p. 240-281. *In* D.J.W. Moriarty and R.S.V. Pullin (eds.) Detritus and microbial ecology in aquaculture. ICLARM Conf. Proc. 14, 420 p.

Conte, F.S. 1984. Economic impact of paralytic shellfish poison on the oyster industry in the Pacific United States. Aquaculture 39: 331-343.

Corrales, R.A. and E.D. Gomez. 1990. Red tide outbreaks and their management in the Philippines, p. 453-458. *In* E. Granéli, B. Sundström, L. Edler and D.M. Anderson (eds.) Toxic marine phytoplankton. Elsevier, New York.

Cosper, E.M., V.M. Bricelj and E.J. Carpenter, Editors.

1989. Novel phytoplankton blooms. Causes and impacts of recurrent brown tides and other unusual blooms. Springer-Verlag, Berlin.

Costa-Pierce, B.A. 1990. Multiple regression analysis of plankton and water quality relationships as affected by sewage inputs and cage aquaculture in a eutrophic tropical reservoir. Paper presented at the Second Asian Reservoir Fisheries Workshop, 15-19 October 1990, Hangzhou, China.

Costa-Pierce, B.A. and O. Soemarwoto. 1987. Proliferation of Asian reservoirs: the need for integrated management. Naga, ICLARM Q. 10(1):9-10.

Costa-Pierce, B.A. and O. Soemarwoto. 1990. Biotechnical feasibility studies on the importation of *Clupeichthys aesarnensis* Wongratana, 1983 from northern Thailand to the Saguling Reservoir, West Java, Indonesia, p. 329-363. *In* B.A. Costa-Pierce and O. Soemarwoto (eds.) Reservoir fisheries and aquaculture development for resettlement in Indonesia. ICLARM Tech. Rep. 23, 378 p.

CSIRO. 1985. Proceedings of the red tide workshop held at the CSIRO Marine Laboratories, 18-20 June 1984, Cronulla, N.S.W., Australia. Commonwealth Scientific and Industrial Research Organisation, Marine Laboratories, Hobart, Australia. 38 p.

Dahl, E., O. Lindahl, E. Paasche and J. Throndsen. 1989. The *Chrysochromulina polyepis* bloom in Scandinavian waters during Spring 1988, p. 383-405. *In* E.M. Cosper, V.M. Bricelj and E.J. Carpenter (eds.) Novel phytoplankton blooms. Causes and impacts of recurrent browntides and other unusual blooms. Springer-Verlag, Berlin.

Dale, B., D.G. Baden, B.M. Bary, L. Edler, S. Fraga, I.R. Jenkinson, G.M. Hallegraeff, T. Okaichi, K. Tangen, F.J.R. Taylor, A.W. White, C.M. Yentsch and C.S. Yentsch. 1987. Proceedings: International conference and workshop. The problems of toxic dinoflagellate blooms in aquaculture. Sherkin Island Marine Station, Cork, Ireland.

Dandonneau, Y. 1971. Étude du phytoplancton sur le plateau continental de Côte d'Ivoire. I. Groupes d'espèces associées. Cah. ORSTOM Sér. Oceanogr. IX: 247-265.

De Silva, M.W.R.N., A.H.M. Salleh, S.I. Mahali and S. Subramanian. 1989. Management of *Pyrodinium* red tides in Brunei Darussalam, p. 125-134. *In* G.M. Hallegraeff and J.L. Maclean (eds.) Biology, epidemiology and management of *Pyrodinium* red tides. ICLARM Conf. Proc. 21, 286 p.

De Silva, S.S. 1988. The reservoir fishery of Asia, p. 19-28. *In* S.S. De Silva (ed.) Reservoir fishery management and development in Asia. International Development Research Centre, Singapore.

Desbiens, M., F. Coulombe, J. Guadreault, A.D. Cembella and R. Larocque. 1990. PSP toxicity of wild and cultured blue mussels induced by *Alexandrium excavatum* in Gaspé Bay (Canada): implications for aquaculture, p. 459-462. *In* E. Granéli, B. Sundström, L. Edler and D.M. Anderson (eds.) Toxic marine phytoplankton. Elsevier, New York.

Dowidar, N.M. 1974. The phytoplankton of the Mediterranean waters of Egypt. I - a check list of the species recorded. Bull. Inst. Ocean. Fish. A.R.E. 4:321.

Dupree, H.K. 1981. An overview of the various techniques to control infectious diseases in water supplies and in water reuse aquacultural systems, p. 83-89. *In* L.J. Allen and E.C. Kinney (eds.) Proceedings of the bio-engineering symposium for fish culture. Fish Culture Section, American Fisheries Society, Bethesda, Maryland.

Eason, R.J. and E. Harding. 1987. Neurotoxic fish poisoning in the Solomon Islands. Papua New Guinea Med. J. 30:49-52.

Estudillo, R.A. and C.L. Gonzales. 1984. Red tides and paralytic shellfish poisoning in the Philippines, p. 52-79. *In* A.W. White, M. Anraku and K.K. Hooi (eds.) Toxic red tides and shellfish toxicity in Southeast Asia. Southeast Asian Fisheries Development Center and International Development Research Centre, Singapore.

FAO/Unesco. 1954. Proceedings of the Symposium on Marine and Freshwater Plankton in the Indo-Pacific, 25-26 January 1954, Bangkok, Thailand. FAO FI IndoPac. Fish. Counc. Rome, Italy.

Frelin, C., M. Durand-Clément, J.-N. Bidard and M. Lazdunski. 1990. The molecular basis of ciguatoxin action, p. 192-199. *In* S. Hall and G. Strichartz (eds.) Marine toxins. Origins, structure, and molecular pharmacology. American Chemical Society, Washington, DC.

Fukuyo, Y., H. Takano, M. Chihara and K. Matsuoka, Editors. 1990. Red tide organisms in Japan. An illustrated taxonomic guide. Uchida Rokakuho, Tokyo.

Gacutan, R.Q. 1986. Effects of coconut milk and brown sugar on crude toxins from mussels exposed to *Pyrodinium bahamense* var. *compressa*, p. 331-313. *In* J.L. Maclean, L.B. Dizon and L.V. Hosillos (eds.) The First Asian Fisheries Forum. Asian Fisheries Society, Manila, Philippines.

Gacutan, R.Q., M.Y. Tabbu, T. de Castro, A.B. Gallego, M. Bulalacao, L. Arafiles and F. Icatlo, Jr. 1984. Detoxification of *Pyrodinium*-generated paralytic shellfish poisoning toxin in *Perna viridis* from western Samar, Philippines, p. 80-85. *In* A.W. White, M. Anraku and K.K. Hooi (eds.) Toxic red tides and shellfish toxicity in Southeast Asia. Southeast Asian Fisheries Development Center and International Development Research Centre, Singapore.

Gainey Jr., L.F. and S.E. Shumway. 1988. A compendium of the responses of bivalve molluscs to toxic dinoflagellates. J. Shellfish Res. 7:623-628.

GESAMP. 1990. The state of the marine environment.

IMO/FAO/Unesco/WMO/WHO/IAEA/UN/UNEP Joint Group of Experts on Scientific Aspects of Marine Pollution (GESAMP). Rep. Stud. 39, 111 p.

Gillespie, N.C., M.J. Holmes, J.B. Burke and J. Doley. 1985. Distribution and periodicity of *Gambierdiscus toxicus* in Queensland, Australia, p. 183-188. *In* D.M. Anderson, A.W. White and D.G. Baden (eds.) Toxic dinoflagellates. Elsevier, New York.

Gonzales, C.L. 1989a. *Pyrodinium* blooms and paralytic shellfish poisoning in the Philippines, p. 39-47. *In* G.M. Hallegraeff and J.L. Maclean (eds.) Biology, epidemiology and management of Pyrodinium red tides. ICLARM Conf. Proc. 21, 286 p.

Gonzales, C.L. 1989b. Management of toxic red tides in the Philippines, p. 141-147. *In* G.M. Hallegraeff and J.L. Maclean (eds.) Biology, epidemiology and management of *Pyrodinium* red tides. ICLARM Conf. Proc. 21, 286 p.

Gorham, P.R. and W.W. Carmichael. 1988. Hazards of freshwater blue-green algae (cyanobacteria), p. 403-431. *In* C.A. Lembi and J.R. Waaland (eds.) Algae and human affairs. Cambridge University Press, Cambridge.

Granéli, E., B. Sundström, L. Edler and D.M. Anderson, Editors. 1990. Toxic marine phytoplankton. Elsevier, New York.

de Guzman, D.L. 1990. How to neutralize the off-flavor taste of tilapia. Currents (Philippine Council for Aquatic and Marine Research and Development) 2 May 1990. 2 p.

Hall, S. 1982. Toxins and toxicity of *Protogonyaulax* from the northeast Pacific. University of Alaska, Fairbanks. Ph.D. dissertation.

Hall, S., G. Strichartz, E. Moczydlowski, A. Ravindraw and P.B. Reichardt. 1990. The saxitoxins. Sources, chemistry, and pharmacology, p. 29-65. *In* S. Hall and G. Strichartz (eds.) Marine toxins. Origins, structure, and molecular pharmacology. American Chemical Society, Washington, DC.

Hallegraeff, G.M. 1987. Red tides in the Australian region. CSIRO Mar. Lab. Rep. 187, 14 p.

Hallegraeff, G.M. 1990. Status of harmful algal blooms in the Indo-West Pacific region. Paper presented to the OSLR/OIC meeting of the Group of Experts on Harmful Algal Blooms, 31 January - 2 February 1990, Paris. 10 p.

Hallegraeff, G.M. 1992. Harmful algal blooms in the Australian region. Mar. Pollut. Bull. 25:5-8.

Hallegraeff, G.M. 1993. A review of harmful algal blooms and their apparent global increase. Phycologia 32:79-99.

Hallegraeff, G.M. and J.L. Maclean, Editors. 1989. Biology, epidemiology and management of *Pyrodinium* red tides. ICLARM Conf. Proc. 21, 286 p.

Hallegraeff, G.M., C.J. Bolch, J. Bryan and B. Koerbin. 1990. Microalgal spores in ship's ballast water: a danger to aquaculture, p. 475-480. *In* E. Granéli, B. Sundström, L. Edler and D.M. Anderson (eds.) Toxic marine phytoplankton. Elsevier, New York.

Harada, T., Y. Oshima, H. Kamiya and T. Yasumoto. 1982. Confirmation of paralytic shellfish toxins in the dinoflagellate *Pyrodinium bahamense* var. *compressa* in Palau. Bull. Japan. Soc. Sci. Fish. 48:821-825.

Hargraves, P.E., R.D. Vaillancourt and G.A. Jolly. 1989. Autotrophic picoplankton in Narragansett Bay and their interaction with microplankton, p. 23-38. *In* E.M. Cosper, V.M. Bricelj and E.J. Carpenter (eds.) Novel phytoplankton blooms. Causes and impacts of recurrent brown tides and other unusual blooms. Springer-Verlag, Berlin.

Hawkins, P.R., M.T.C. Runnegar, A.R.B. Jackson and I.R. Falconer. 1985. Severe hepatotoxicity caused by the tropical cyanobacterium (blue-green alga) *Cylindrospermopsis raciborskii* (Woloszynska) Seenaya and Subba Raju isolated from a domestic water supply reservoir. Appl. Environ. Microbiol. 50:1291-1295.

Horwitz, W., Editor. 1990. Official methods of analysis. Association of Official Analytical Chemists, Washington, DC.

Hwang, D.F., C.H. Chueh and S.S. Jeng. 1990. Susceptability of fish, crustacean and mollusk to tetrodotoxin and paralytic shellfish poison. Nippon Suisan Gakkaishi 56:337-343.

Hwang, D.F., T. Noguchi, Y. Nagashima, I.C. Liao, S.S. Chou and K. Hashimoto. 1989. Paralytic shellfish poisoning in Taiwan, p. 419-422. *In* T. Okaichi, D. M. Anderson and T. Nemoto (eds.). Red tides. Biology, environmental science, and toxicology. Elsevier, New York.

IRRI. 1988. Rice facts 1988. International Rice Research Institute, Laguna, Philippines.

Jaafar, M.H., M.W.R.N. De Silva and P.H.Y. Sharifuddin. 1989. *Pyrodinium* red tide occurrences in Brunei Darussalam, p. 9-17. *In* G.M. Hallegraeff and J.L. Maclean (eds.) Biology, epidemiology and management of *Pyrodinium* red tides. ICLARM Conf. Proc. 21, 286 p.

Jensen, A., Moderator. 1988. The impact of toxic algae on mariculture. Proceedings of Aquanor 87. Norske Fiskeoppdretteres Forening, Trondheim, Norway.

Jeon, J.K., K. Miyazawa, T. Noguchi, H. Narita, S. Matsubara, M. Nara, K. Ito and K. Hashimoto. 1986. Occurrence of paralytic toxicity in marine flatworms. Bull. Japan. Soc. Sci. Fish. 52:1065-1069.

Jones, K.J. and R.J. Gowen. 1985. The influence of advective exchange on phytoplankton in Scottish fjordic sea lochs, p. 207-212. *In* D.M. Anderson, A.W. White and D.G. Baden (eds.) Toxic dinoflagellates. Elsevier, New York.

Karunasagar, I., H.S.V. Gowda, M. Subburaj, M.N. Venugopal and I. Karunasagar. 1984. Outbreak of paralytic shellfish poisoning in Mangalore, west coast of India. Curr. Sci. 53:247-249.

Karunasagar, I., K. Segar and I. Karunasagar. 1989. Potentially toxic dinoflagellates in shellfish harvesting areas along the coast of Karnataka State (India), p. 65-68. *In* T. Okaichi, D.M. Anderson and T. Nemoto (eds.) Red tides. Biology, environmental science, and toxicology. Elsevier, New York.

Kat, M. 1984. "Red" oysters (*Ostrea edulis* L.) caused by *Mesodinium rubrum* in Lake Grevelingen. Aquaculture 38: 375-377.

Khoo, E.W. 1985. Occurrences of "red tide" along Johore Straits, Malaysia, resulted in heavy mortality of shrimp. World Maricult. Soc. Newsl. 16(1):4.

Kodama, M. 1990. Possible links between bacteria and toxin production in algal blooms, p. 52-61. *In* E. Granéli, B. Sundström, L. Edler and D.M. Anderson (eds.) Toxic marine phytoplankton. Elsevier, New York.

Kodoma, K., T. Ogata, Y. Fukuyo, T. Ishimaru, S. Wissesang, K. Saitanu, V. Panichyakaran and T. Piyakarnchana. 1988. *Protogonyaulax cohorticula*, a toxic dinoflagellate found in the Gulf of Thailand. Toxicon 26:707-712.

Lam, C.W.Y. and K.C. Ho. 1989. Red tides in Tolo Harbour, Hongkong, p. 49-52. *In* T. Okaichi, D.M. Anderson and T. Nemoto (eds.) Red tides. Biology, environmental science, and toxicology. Elsevier, New York.

Lam, C.W.Y. and S.S.Y. Yip. 1990. A three-month red tide event in Hong Kong, p. 481-486. *In* E. Granéli, B. Sundström, L. Edler and D.M. Anderson (eds.) Toxic marine phytoplankton. Elsevier, New York.

Legrand, A.M., P. Cruchet, R. Bagnis, M. Murata, Y. Ishibashi and T. Yasumoto. 1990. Chromatogaphic and spectral evidence for the presence of multiple ciguatera toxins, p. 374-378. *In* E. Granéli, B. Sundström, L. Edler and D.M. Anderson (eds.) Toxic marine phytoplankton. Elsevier, New York.

Liao, I.C. 1990. The world's marine prawn culture industries: today and tomorrow, p. 11-27. *In* R. Hirano and I. Hanyu (eds.) The Second Asian Fisheries Forum. Asian Fisheries Society, Manila, Philippines.

Lin, C. Kwei, V. Tansakul and C. Apinhapath. 1988. Biological nitrogen fixation as a source of nitrogen input in fishponds, p. 53-58. *In* R.S.V. Pullin, T. Bhukaswan, K. Tonguthai and J.L. Maclean (eds.) The Second International Symposium on Tilapia in Aquaculture. ICLARM Conf. Proc. 15, 623 p.

Lim, L.C. 1989. Status of occurrences of phytoplankton bloom in Singapore. Paper presented to the Management and Training Workshop on *Pyrodinium* Red Tides, 23-30 May 1989, Brunei Darussalam. 8 p.

Lovell, R.T. and D. Broce. 1985. Cause of musty flavor in pond-cultured penaeid shrimp. Aquaculture 50:169-174.

Maclean, J.L. 1975. Paralytic shellfish poison in various bivalves, Port Moresby 1973. Pac. Sci. 29:349-352.

Maclean, J.L. 1977. Observations on *Pyrodinium bahamense* Plate, a toxic dinoflagellate in Papua New Guinea. Limnol. Oceanogr. 22:234-254.

Maclean, J.L. 1979. Indo-Pacific red tides, p. 173-178. *In* D.L. Taylor and H.H. Seliger (eds.) Toxic dinoflagellate blooms. Elsevier, New York.

Maclean, J.L. 1986. A database analysis kit for your information toolbox. Naga, ICLARM Q. 9(2):6-7.

Maclean, J.L. 1989a. Red tides in Papua New Guinea waters, p. 27-38. *In* G.M. Hallegraeff and J.L. Maclean (eds.) Biology, epidemiology and management of *Pyrodinium* red tides. ICLARM Conf. Proc. 21, 286 p.

Maclean, J.L. 1989b. Indo-Pacific red tides, 1985-1988. Mar. Pollut. Bull. 20:304-310.

Maclean, J.L. 1989c. Economic aspects of *Pyrodinium* red tides in the western Pacific, p. 179-185. *In* G.M. Hallegraeff and J.L. Maclean (eds.) Biology, epidemiology and management of *Pyrodinium* red tides. ICLARM Conf. Proc. 21, 286 p.

Maclean, J.L. 1989d. An overview of *Pyrodinium* red tides in the western Pacific, p. 1-8. *In* G.M. Hallegraeff and J.L. Maclean (eds.) Biology, epidemiology and management of *Pyrodinium* red tides. ICLARM Conf. Proc. 21, 286 p.

Maclean, J.L. and O. Ross, Rapporteurs. 1989. Discussion and recommendations on economic issues, p. 187-188. *In* G.M. Hallegraeff and J.L. Maclean (eds.) Biology, epidemiology and management of *Pyrodinium* red tides. ICLARM Conf. Proc. 21, 286 p.

Margalef, R. 1957. Fitoplancton de las costas de Puerto Rico. Inv. Pesq. VI:39-52.

Matsuoka, K. 1989. Morphological features of the cyst of *Pyrodinium bahamense* var. *compressum*, p. 219-229. *In* G.M. Hallegraeff and J.L. Maclean (eds.) Biology, epidemiology and management of *Pyrodinium* red tides. ICLARM Conf. Proc. 21, 286 p.

Mitsui, A., D. Rosner, A. Goodman, G. Reyes-Vasquez, T. Kusumi, T. Kodama and K. Nomoto. 1989. Hemolytic toxins in marine cyanobacterium *Synechococcus* sp., p. 367-370. *In* T. Okaichi, D.M. Anderson and T. Nemoto (eds.) Red tides. Biology, environmental science, and toxicology. Elsevier, New York.

Munro, J.L., Iskandar and B.A. Costa-Pierce. 1990. Fisheries of the Saguling Reservoir and a preliminary appraisal of management options, p. 285-328. *In* B.A. Costa-Pierce and O. Soemarwoto (eds.) Reservoir fisheries and aquaculture development for resettlement in Indonesia. ICLARM Tech. Rep. 23, 378 p.

Narita, H., S. Matsubara, N. Miwa, S. Akahane, M. Murakani, T. Goto, M. Nara, T. Noguchi, T. Saito, Y. Shida and K. Hashimoto. 1987. *Vibrio alginolyticus*, a TTX-producing bacterium isolated from the starfish *Astropecten polyacanthus*. Nippon Suisan Gakkaishi 53:617-621.

Noguchi, T., Y. Ueda, Y. Onoue, M. Kono, K. Koyama, K. Hashimoto, T. Takeuchi, Y. Seno and S. Mishima. 1980. Reduction in toxicity of highly PSP-infested scallops during canning process and storage. Bull. Japan. Soc. Sci. Fish. 46:1339-1344.

Ogata, T., M. Kodama and T. Ishimaru. 1989. Effect of water temperature and light intensity on growth rate and toxin production of toxic dinoflagellates, p. 423-426. *In* T. Okaichi, D.M. Anderson and T. Nemoto (eds.) Red tides. Biology, environmental science, and toxicology. Elsevier, New York.

Okaichi, T. 1989. Red tide problems in the Seto Inland Sea, Japan, p. 137-142. *In* T. Okaichi, D.M. Anderson and T. Nemoto (eds.) Red tides. Biology, environmental science, and toxicology. Elsevier, New York.

Okaichi, T., D.M. Anderson and T. Nemoto, Editors. 1989a. Red tides. Biology, environmental science, and toxicology. Elsevier, New York.

Okaichi, T., S. Montani, T. Lirdwitayaprasit, C. Ono and S. Yoshimatsu. 1989b. Countermeasures aimed at the *Chattonella* red tide which occurred in Shido Bay, Seto Inland Sea, in 1987. Paper presented at the Management and Training Workshop on *Pyrodinium* Red Tides, 23-30 May 1989, Brunei Darussalam.

Onoue, Y., T. Noguchi and K. Hashimoto. 1984. Tetrodotoxin determination methods, p. 345-355. *In* E.P. Ragelis (ed.) Seafood toxins. American Chemical Society, Washington, DC.

Oshima, Y. 1989. Toxins in *Pyrodinium bahamense* var. *compressum* and infested marine organisms, p. 73-79. *In* G.M. Hallegraeff and J.L. Maclean (eds.) Biology, epidemiology and management of *Pyrodinium* red tides. ICLARM Conf. Proc. 21, 286 p.

Oshima, Y., H. Minami, Y. Takano and T. Yasumoto. 1989. Ichthyotoxins in a freshwater dinoflagellate *Peridinium polonium*, p. 375-378. *In* T. Okaichi, D.M. Anderson and T. Nemoto (eds.) Red tides. Biology, environmental science, and toxicology. Elsevier, New York.

Park, J.S., H.G. Kim and S.G. Lee. 1989. Studies on red tide phenomena in Korean coastal waters, p. 37-40. *In* T. Okaichi, D.M. Anderson and T. Nemoto (eds.) Red tides. Biology, environmental science, and toxicology. Elsevier, New York.

Parry, G.D., J.S. Langdon and J.M. Huisman. 1989. Toxic effects of a bloom of the diatom *Rhizosolenia chunii* on shellfish in Port Phillip Bay, southeastern Australia. Mar. Biol. 102:25-41.

Pastor, N.I.S., I. Gopez, M.C. Quizon, N. Bautista, M. White and M. Dayrit. 1989. Epidemics of paralytic shellfish poisoning in the Philippines, 1988-1989, p. 165-171. *In* G.M. Hallegraeff and J.L. Maclean (eds.) Biology, epidemiology and management of *Pyrodinium* red tides. ICLARM Conf. Proc. 21, 286 p.

Phillips, M.J., M.C.M. Beveridge and L.G. Ross. 1985a. The environmental impact of salmonid cage culture on inland fisheries: present status and future trends. J. Fish Biol. 27(Supp.A):123-137.

Phillips, M.J., R.J. Roberts, J.A. Stewart and G.A. Codd. 1985b. The toxicity of the cyanobacterium *Microcystis aeruginosa* to rainbow trout *Salmo gairdneri* Richardson. J. Fish Dis. 8:339-344.

Poli, M.A., C.B. Templeton, J.G. Pace and H.B. Hines. 1990. Detection, metabolism, and pathophysiology of brevetoxins, p. 176-191. *In* S. Hall and G. Strichartz (eds.) Marine toxins. Origins, structure, and molecular pharmacology. American Chemical Society, Washington, DC.

Rabbani, M.M., Atiq-Ur-Rehman and C.E. Harms. 1990. Mass mortality of fishes caused by dinoflagellate bloom in Gwadar Bay, southwestern Pakistan, p. 209-214. *In* E. Granéli, B. Sundström, L. Edler and D.M. Anderson (eds.) Toxic marine phytoplankton. Elsevier, New York.

Raj, U. 1983. Ciguatera, clupeotoxism and other seafood poisoning in Fijian waters and their impact to the utilization of marine resources. Institute of Marine Resources, University of the South Pacific, Suva. mimeo. pag. var.

Roberts, R.J., A.M. Bullock, M. Turner, K. Jones and P. Tett. 1983. Mortalities of *Salmo gairdneri* exposed to cultured *Gyrodinium aureolum*. J. Mar. Biol. Assoc. U.K. 63:741-743.

Roberts, R.D. and T. Zohary. 1987. Temperature effects on photosynthetic capacity, respiration, and growth rates of bloom-forming cyanobacteria. N.Z. J. Mar. Freshw. Res. 21:391-399.

Roger, P.A. and S.A. Kulasooriya. 1980. Blue-green algae and rice. International Rice Research Institute, Laguna, Philippines.

Ronquillo, I.A. 1987. Fish kills in Philippine lakes and coastal waters. Paper presented at the Seminar-Workshop on Disseminating Information on Environmental Hazards, Mines and Geo-Science Bureau, 23-25 September 1987, Quezon City, Philippines. 14 p.

Ronquillo, I.A. 1988. Manila Bay shellfish fishery: how safe are the mussels and oysters in Manila Bay? mimeo. 8 p.

Rosales-Loessener, F. 1989. The Guatemalan experience with red tides and paralytic shellfish poisoning, p. 49-51. *In* G.M. Hallegraeff and J.L. Maclean (eds.) Biology, epidemiology and management of *Pyrodinium* red tides. ICLARM Conf. Proc. 21, 286 p.

Rosenthal, H. 1981. Ozonation and sterilization, p. 219-274. *In* K. Tiews (ed.) Aquaculture in heated effluents and recirculation systems. Vol. I. Heenemann Verlagsgesellschaft mbH, Berlin.

Saito, T., J. Maruyama, S. Kanoh, J.K. Jeon, T. Noguchi, T. Harada, O. Murata and K. Hashimoto. 1984. Toxicity of the cultured pufferfish *Fugu rubripes rubripes* along with their resistibility against tetrodotoxin. Bull. Japan. Soc. Sci. Fish. 50:1573-1575.

Saito, T., T. Noguchi, T. Takeuchi, S. Kamimura and K. Hashimoto. 1985a. Ichthyotoxicity of paralytic shellfish poison. Bull. Japan. Soc. Sci. Fish. 51:257-260.

Saito, T., T. Noguchi, T. Harada, O. Murata, T. Abe and K. Hashimoto. 1985b. Resistibility of toxic and nontoxic pufferfish against tetrodotoxin. Bull. Japan. Soc. Sci. Fish. 51:1371.

Saito, T., T. Noguchi, T. Harada, O. Murata and K. Hashimoto. 1985c. Tetrodotoxin as a biological defence agent for puffers. Bull. Japan. Soc. Sci. Fish. 51:1175-1180.

Schneider, K.R. and G.E. Rodrick. The use of ozone to degrade *Gymnodium breve* toxins. Aquat. Living Resour. (In press).

Schroeder, G.L., G. Wohlfarth, A. Aklon, A. Halevy and H. Krueger. 1990. The dominance of algal-based food webs in fish ponds receiving chemical fertilizers plus organic manures. Aquaculture 86:219-229.

Segar, K., I. Karunasagar and I. Karunasagar. 1988. Dinoflagellate toxins in shellfishes along the coast of Karnataka, p. 389-390. *In* M.M. Joseph (ed.) The First Indian Fisheries Forum, Proceedings. Asian Fisheries Society, Indian Branch, Mangalore.

Seliger, H.H. 1989. Mechanisms for red tides of *Pyrodinium bahamense* var. *compressum* in Papua New Guinea, Sabah and Brunei Darussalam, p. 53-71. *In* G.M. Hallegraeff and J.L. Maclean (eds.) Biology, epidemiology and management of *Pyrodinium* red tides. ICLARM Conf. Proc. 21, 286 p.

Shimizu, Y. 1989. Toxicology and pharmacology of red tides: an overview, p. 17-21. *In* T. Okaichi, D.M. Anderson and T. Nemoto (eds.) Red tides. Biology, environmental science, and toxicology. Elsevier, New York.

Shimizu, Y., S. Gupta and H-N. Chou. 1990. Biosynthesis of red tide toxins, p. 21-28. *In* S. Hall and G. Strichartz (eds.) Marine toxins. Origins, structure, and molecular pharmacology. American Chemical Society, Washington, DC.

Shumway, S.E. 1990. A review of the effects of algal blooms on shellfish and aquaculture. J. World Aquacult. Soc. 21:65-104.

Shumway, S.E., J. Barter and S. Sherman-Caswell. 1990. Auditing the impact of toxic algal blooms on oysters. Environ. Auditor 2:41-56.

Sieburth, J.M. and P.W. Johnson. 1989. Picoplankton ultrastructure: a decade of preparation for the brown tide alga *Aureococcus anophagefferens*, p. 1-21. *In* E.M. Cosper, V.M. Bricelj and E.J. Carpenter (eds.) Novel phytoplankton blooms. Causes and impacts of recurrent brown tides and other unusual blooms. Springer-Verlag, Berlin.

Sivonen, K., S.I. Niemelä, R.M. Niemi, L. Lepistö, T.H. Luoma and L.A. Räsänen. 1990. Toxic cyanobacteria (blue-green algae) in Finnish fresh and coastal waters. Hydrobiologia 190:267-275.

Smayda, T.J. 1989. Primary production and the global epidemic of phytoplankton blooms in the sea: a linkage?, p. 449-483. *In* E.M. Cosper, V.M. Bricelj and E.J. Carpenter (eds.) Novel phytoplankton blooms. Causes and impacts of recurrent brown tides and other unusual blooms. Springer-Verlag, Berlin.

Smith, J.C., R. Cormier, J. Worms, C.J. Bird, M.A. Quilliam, R. Pocklington, R. Angus and L. Hanic. 1990. Toxic blooms of the domoic acid containing diatom *Nitzschia pungens* in the Cardigan River, Prince Edward Island, in 1988, p. 227-232. *In* E. Granéli, B. Sundström, L. Edler and D.M. Anderson (eds.) Toxic marine phytoplankton. Elsevier, New York.

Stewart, R.W., B. Kjerfve, J. Milliman and S.N. Dwivedi. 1990. Relative sea-level change: a critical evaluation. Unesco Rep. Mar. Sci. 54, 22 p.

Stirling, H.P. and T. Dey. 1990. Impact of intensive cage fish farming on the plankton and periphyton of a Scottish freshwater loch. Hydrobiologia 190: 193-214.

Strichartz, G. and N. Castle. 1990. Pharmacology of marine toxins. Effects on membrane channels, p. 2-20. *In* S. Hall and G. Strichartz (eds.) Marine toxins. Origins, structure, and molecular pharmacology. American Chemical Society, Washington, DC.

Su, H.M., I.C. Liao and Y.M. Chiang. 1989. A toxic dinoflagellate first recorded in Taiwan, p. 85-88. *In* T. Okaichi, D.M. Anderson and T. Nemoto (eds.) Red tides. Biology, environmental science, and toxicology. Elsevier, New York.

Subrahmanyan, R. 1954. On the life-history and ecology of *Hornellia marina* gen. et. sp. nov. (Chloromonadineae), causing green discoloration of the sea and mortality among marine organisms off the Malabar coast. Indian J. Fish. 1:182-203.

Suvapepun, S. 1989. Occurrences of red tide in the Gulf of Thailand, p. 41-44. *In* T. Okaichi, D.M. Anderson and T. Nemoto (eds.) Red tides. Biology, environmental science, and toxicology. Elsevier, New York.

Tabthipwon, Prathak, Chawalit Orachunwong, J. Kugler and J. Moreau. 1988. Influence of *Chlorella vulgaris* and *Microcystis aeruginosa* mixed with commercial carp pellets on growth of Nile tilapia (*Oreochromis niloticus*), p. 387-392. *In* R.S.V. Pullin, T. Bhukaswan, K. Tonguthai and J.L. Maclean (eds.) The Second International Symposium on Tilapia in Aquaculture. ICLARM Conf. Proc. 15, 623 p.

Tamplin, M.L. 1990. A bacterial source of tetrodotoxins and saxitoxins, p. 78-86. *In* S. Hall and G. Strichartz (eds.) Marine toxins. Origins, structure, and molecular pharmacology. American Chemical Society, Washington, DC.

Tan, C.T.T. and E.J.D. Lee. 1986. Paralytic shellfish poisoning in Singapore. Annals Acad. Med. 15:77-79.

Tangen, K. 1982. [New case of fish mortalities caused by planktonic algae]. Norsk Fiskeoppdrett 7/8:5-8. Can. Transl. Fish. Aquat. Sci. 4953 (1983).

Taylor, D.L. and H.H. Seliger, Editors. 1979. Toxic dinoflagellate blooms. Elsevier North-Holland, New York.

Taylor, F.J.R. 1985. The taxonomy and relationships of red tide flagellates, p. 11-26. *In* D.M. Anderson, A.W. White and D.G. Baden (eds.) Toxic dinoflagellates. Elsevier, New York.

Taylor, F.J.R. 1990. Red tides, brown tides and other harmful algal blooms: the view into the 1990's, p. 527-533. *In* E. Granéli, B. Sundström, L. Edler and D.M. Anderson (eds.) Toxic marine phytoplankton. Elsevier, New York.

Tester, P.A. and P.K. Fowler. 1990. Brevetoxin contamination of *Mercenaria mercenaria* and *Crassostrea virginica*: a management issue, p. 499-503. *In* E. Granéli, B. Sundström, L. Edler and D.M. Anderson (eds.) Toxic marine phytoplankton. Elsevier, New York.

Tester, P.A., P.K. Fowler and J.T. Turner. 1989. Gulf stream transport of the toxic red tide dinoflagellate *Ptychodiscus brevis* from Florida to North Carolina, p. 349-358. *In* E.M. Cosper, V.M. Bricelj and E.J. Carpenter (eds.) Novel phytoplankton blooms. Causes and impacts of recurrent brown tides and other unusual blooms. Springer-Verlag, Berlin.

Ting, T.M. and J.T.S. Wong. 1989. Summary of red tide and paralytic shellfish poisoning in Sabah, Malaysia, p. 19-26. *In* G.M. Hallegraeff and J.L. Maclean (eds.) Biology, epidemiology and management of *Pyrodinium* red tides. ICLARM Conf. Proc. 21, 286 p.

Todd, E.C.D. 1990. Amnesic shellfish poisoning - a new seafood toxin syndrome, p. 504-508. *In* E. Granéli, B. Sundström, L. Edler and D.M. Anderson (eds.) Toxic marine phytoplankton. Elsevier, New York.

Tsubone, N., Y. Fuchi, S. Morisaki, T. Mizokoshi, M. Shuto, M. Fujii, K. Yamada and K. Hayashi. 1986. Removal of toxicity from pufferfish liver by cooking. J. Food Hyg. Soc. Japan 27:561-564. (Abstract only seen).

Unesco. 1982. Mareas rojas en el plancton del Pacífico Oriental. Informes de la Unesco sobre ciencias del mar 19, 47 p. (In Spanish).

Uno, S. and K. Sasaki. 1989. Volumetric considerations of some species of diatoms composing red tides in the nori culture area in Ariake Bay, p. 157-160. *In* T. Okaichi, D.M. Anderson and T. Nemoto (eds.) Red tides. Biology, environmental science, and toxicology. Elsevier, New York.

White, A.W. 1982. Intensification of *Gonyaulax* blooms and shellfish toxicity in the Bay of Fundy. Can. Tech. Rep. Fish. Aquat. Sci. 1064, 12 p.

White, A.W. 1984. Paralytic shellfish toxins in finfish, p. 171-180. *In* E.P. Ragelis (ed.) Seafood toxins. American Chemical Society, Washington, DC.

White, A.W. 1988. Blooms of toxic algae worldwide: their effects on fish farming and shellfish resources, p. 9-14. *In* A. Jensen (moderator) The impact of toxic algae on mariculture. Proceedings of Aquanor 87. Norske Fiskeoppdretteres Forening, Trondheim, Norway.

White, A.W. 1990a. Toxic algal blooms: an international directory of experts in toxic and harmful algal blooms and their effects on fisheries and public health. International Marine Science Cooperation Program, Woods Hole Oceanographic Institution, Woods Hole, Massachusetts.

White, A.W. 1990b. International Red Tide Information and Assistance Service (IRTIAS), p. 509-511. *In* E. Granéli, B. Sundström, L. Edler and D.M. Anderson (eds.) Toxic marine phytoplankton. Elsevier, New York.

White, A.W., M. Anraku and K.K. Hooi, Editors. 1984. Toxic red tides and shellfish toxicity in Southeast Asia. Southeast Asian Fisheries Development Center and International Development Research Centre, Singapore.

White, A.W., O. Fukuhara and M. Anraku. 1989. Mortality of fish larvae from eating toxic dinoflagellates or zooplankton containing dinoflagellate toxins, p. 395-398. *In* T. Okaichi, D.M. Anderson and T. Nemoto (eds.) Red tides. Biology, environmental science, and toxicology. Elsevier, New York.

Wong, J.T.S. and T.M. Ting. 1989. Management of red tides in Sabah, Malaysia, p. 135-139. *In* G.M. Hallegraeff and J.L. Maclean (eds.) Biology, epidemiology and management of *Pyrodinium* red tides. ICLARM Conf. Proc. 21, 286 p.

Wong, P.S. and R.S.S. Wu. 1987. Red tides in Hong Kong: problems and management strategy with special reference to the mariculture industry. J. Shoreline Manage. 3:1-21.

Yamamoto, T. and K. Fujisaki. 1989. Prompt information on red tides in the cultural grounds of nori (edible laver), p. 161-164. *In* T. Okaichi, D.M. Anderson and T. Nemoto (eds.) Red tides. Biology, environmental science, and toxicology. Elsevier, New York.

Yasumoto, T. and M. Murata. 1990. Polyether toxins involved in seafood poisoning, p. 120-132. *In* S. Hall and G. Strichartz (eds.) Marine toxins. Origins, structure, and molecular pharmacology. American Chemical Society, Washington, DC.

Yasumoto, T., U. Raj and R. Bagnis. 1984. Seafood poisonings in tropical regions. Faculty of Agriculture, Tohoku University, Japan.

Zuo Jingzhong, Dong Liping and Qin Baoping. 1985. Preliminary studies on eutrophication and red tide problems in Bohai Bay. Hydrobiologia 127:27-30.

Discussion

BILIO: To what extent could monitoring for red tides be done by regular observations from satellites?

MACLEAN: This has been promoted widely in the USA, but it is not a generally useful technique. It might work for *Gymnodinium breve* but, in developing countries, *Pyrodinium* is a major cause of red tides. It often forms a subsurface bloom that cannot be seen. It also occurs in turbid waters, such as lagoons, where it cannot be seen. Moreover, *Pyrodinium* and some other algae responsible for toxic molluscs such as *Dinophysis* spp. often occur in densities that are insufficient for a definite bloom structure to be seen. This applies to some of those responsible for diarrheal shellfish poisoning. Only a few hundred organisms per liter are necessary to affect human health.

D. PHILLIPS: Some of the methods for testing for toxins are longwinded and impractical. By the time samples have got to a laboratory and been analyzed, serious incidents might have occured. Do we need better, shorter, quicker tests? And if these were available would there be the personnel, facilities and the inclination in developing countries to carry them out? Are we seeing here a really serious constraint to the expansion of coastal aquaculture in developing countries? Some of the toxins involved, for example from *Pyrodinium* are extremely potent.

GOWEN: On satellite imagery, I agree that it is not a useful technique. Harmful events are often very localized and the resolution of satellites is insufficient to pick them up. Also, satellite imagery cannot say anything about whether a high concentration of a microalga is harmful or not. Therefore, information from satellite imagery can be inconclusive and confusing.

On testing for toxins, there is a lot of developmental work now in progress, especially in N. America, on developing new techniques using High Performance Liquid Chromotography (HPLC) and immunoassays so as to move away from the more usual techniques like mouse bioassays. These techniques are, however, still under development. One problem, for example with HPLC tests for diarrheal shellfish poisoning is that some of the toxins have not yet been characterized. The 'suite' of toxins can vary. Some can be missed by using, for example, specific HPLC techniques for okadaic acid. Therefore, we still need to use rat bioassays in conjunction with HPLC and a mouse bioassay for paralytic shellfish poisoning. The ICES Working Group will publish a report that will contain details of the current status of tests for phycotoxins.

ROSENTHAL: We should probably consider to move away from using the term 'bloom'. For example, the reported bloom of *Chrysochromulina* in the Kattegat and Skagerrak in 1988 was not really a bloom. It was 18-20% at the most, of the total standing biomass of phytoplankton.

MACLEAN: As far as I am aware, the only developing country (and it is a rich one - Ed.) that has anything other than a mouse bioassay test is Brunei Darussalam. Last year an HPLC set-up was acquired there and was running but probably hasn't been run again since. Like many sophisticated tests, it has to be standardized against standard toxins and kept running. In Japan, researchers now use HPLC tests fairly regularly for paralytic shellfish poisoning toxicity tests. The HPLC tests can probably be considered inappropriate technology for developing countries at present.

In the Philippines, mouse colonies are being established at strategic locations for bioassays. Each mouse costs about US$0.20: very expensive for a developing country. An alternative technique using blowflies is being developed and is almost up and running. The cost per blowfly is only about US$0.03 to 0.04. Such economies in the development of appropriate technologies are very important in developing countries.

PULLIN: I find one big mystery regarding toxins and this applies to the toxicity of freshwater cyanobacteria, like *Microcystis*. Many of us working on pond fertilization actually *want* to produce eutrophic waters in which cyanobacteria, like *Microcystis*, are the dominant organisms. This is because cyanobacteria, unlike many microalgae, are excellent foods for tilapia. *Microcystis* is the dominant food of tilapia in natural waterbodies like Lake George. As far as I know there are no records of anyone ever having suffered from eating fish from such natural waters or from fishponds in Southeast Asia with *Microcystis* blooms, and yet the literature is full of references to *M. aeruginosa* as having lethal toxins. What is the explanation for this? Should pond fertilization enthusiasts be worried that their cyanobacteria may in future switch to producing toxins that are harmful to humans. Why aren't they producing much toxins now? Or are they?

MACLEAN: Well, rice farmers also encourage the same type of blooms, so there is another few hundred million hectares of *Microcystis*-laden water, quite apart from fishponds! The marine dinoflagellates that produce toxins similar to, for example *Aphanizomenon* in freshwater, have toxic and non-toxic strains. The latest thinking is that the toxins are produced by symbiotic or parasitic bacteria within the dinoflagellate. This may also be the case with cyanobacteria having similar toxins. If so, these may also become toxic in time, if the bacteria get in.

M. PHILLIPS: You quoted some work on *M. aeruginosa* in reservoirs in the UK, where toxin-producing populations were found to be extremely common. A few years ago, we took *Microcystis aeruginosa* samples from fishponds at AIT, Bangkok back to Geoff Codd at Dundee University. The most toxic strain they now have in their collection comes from these Bangkok samples! There may of course be some selectivity in the way that various workers are culturing their samples, but I suspect that toxin-producing *Microcystis* populations are probably widespread in the tropics. These *Microcystis* are very toxic to mice, but rainbow trout are unaffected by their presence and Nile tilapia eat them without any apparent ill effects. Clearly there needs to be research on the fate of phycotoxins ingested by fish and whether these can get into the muscle, because it seems certain that tropical fishponds have toxin-producing *Microcystis* populations.

PULLIN: So the mice in bioassays get sick - but I've never heard of any sickness in humans that have fallen into tropical fishponds or reservoirs being ascribed to exposure to *Microcystis*.

M. PHILLIPS: In temperate waters there is some evidence that blooms of *Microcystis* in reservoirs used for domestic water supply can cause liver damage and sickness. There seems to be no information from developing countries.

EDWARDS: Well, people don't normally drink water out of fertilized fishponds, but some small-scale farmers in developing countries do regard their fishpond (a hole-in-the-ground) as a multipurpose source of water. The late Dr. Balfour Hepher told me that these toxins from cyanobacteria are heat-labile, so that even if they are present in the parts of fish to be eaten, cooking will destroy them.

BEVERIDGE: We have measured ingestion rates by tilapia *(Oreochromis niloticus)* of *Microcystis* supplied by Professor Geoff Codd of Dundee University. The fish absolutely gorge themselves on *Microcystis*.[1] We are going to do some followup work comparing ingestion of toxic and nontoxic strains, but nontoxic strains are rare. Professor Codd claims to have the only non-toxic strain available in the UK, and possibly the only one in Europe. We have also done some work with filter-feeding carps. The dataset is not too good, but there is some evidence that they select against ingestion of cyanobacteria. Work at AIT by Dr. Amararatne Yakupitiyage has shown that *in vitro* conditions simulating the highly acidic stomach of tilapia cause very rapid rupture of the *Microcystis* cell wall: in a matter of seconds. We can speculate that these conditions also rapidly denature the toxin.

MACLEAN: The only source of reference that we have on the heat-labile nature of these toxins is Balfour Hepher's comment to Peter Edwards. This question is so important. It may have helped the publication of an erroneous view by WHO that all such toxins are heat-labile. This is a very important area for future research.

PULLIN: Regarding Jay Maclean's comment on ricefields, much of the rice-farming in Asia would collapse without the nitrogen fixing efforts of cyanobacteria. What can be the fate in the natural environment of the presumably huge quantities of cyanobacterial toxins that are produced every rice-growing season? Are they biodegraded rapidly in ricefields (and fishponds) when the cyanobacteria die? Does anyone know?

M. PHILLIPS: Good question!

MACLEAN: The WHO report[2] distributed says quite blandly that all shellfish are safe to eat if cooked. This is not so. If people took that advice, some would become ill, perhaps fatally.

BOGEL: Well, many groups and specialists have worked on this. The statements you refer to clearly derive from some of this work. There seems to be some gaps in the knowledge of those responsible for the document. This reveals a lack of communication, which we must remedy.

---

[1]Northcott, M.E., M.C.M. Beveridge and L.G. Ross. 1991. A laboratory investigation of the filtration and ingestion rates of the tilapia, *Oreochromis niloticus*, feeding on two species of blue-green algae. Env. Biol. Fish. 31: 75-85.

[2]Report of the WHO Consultation on Public Health Aspects of Seafood-Borne Zoonotic Diseases, Hanover, Federal Republic of Germany, 14-16 November 1989. WHO/CDS/VPH/90.86, 62 p. World Health Organization, Geneva.

Microbial Safety of Produce from Wastewater-Fed Aquaculture

NETTY BURAS
Department of Civil Engineering
University of Arizona, Tucson
Arizona 85721, USA

BURAS, N. 1993. Microbial safety of produce from wastewater-fed aquaculture, p. 285-295. *In* R.S.V. Pullin, H. Rosenthal and J.L. Maclean (eds.) Environment and aquaculture in developing countries. ICLARM Conf. Proc. 31, 359 p.

Abstract

Fish culture in wastewater-fed ponds has been practiced for centuries. However, public health issues in aquaculture have rarely been considered. Water quality is a major determinant of the quality of aquatic produce. Therefore, wastewater should be treated prior to introduction into fishponds in order to produce fish with minimal risks for handlers and consumers. Stressed fish are susceptible to disease. To ensure the production of healthy fish, conditions in the pond environment should be such as not to exceed the stress limits of the fish - the homeostatic plateau.

Introduction

A fishpond is a complex ecosystem in which microbial components play a central role. Microbial populations decompose organic matter which enters the pond and make possible the development of foodwebs, in which the fish feed. In wastewater-fed fishponds, there are also pathogenic microorganisms. Their survival, numbers, infective capability and the danger they might present to fish farmers, fish handlers and consumers are hazards to be considered and dealt with. Fish culture has been practiced for centuries in some parts of the world but associated health issues have rarely been considered. Thus, contaminated fish reach the market and spread pathogenic organisms among those who handle and eviscerate them prior to cooking.

Since the microbial quality of fishpond water influences that of the fish, it is important to minimize the number of pathogenic microorganisms in wastewater-fed fishponds. Substantial knowledge has been accumulated on the dynamics of microbial populations in wastewater-fed ponds, on the pathogens present in human excreta and on the potential hazards that they present. Awareness that these problems exist and understanding the dynamics of pathogen removal are essential for the development of sustainable wastewater-fed aquaculture.

Microbial Water Quality, Fish Quality and Disease

In some developing countries, mostly in warm climates, fishponds are regularly fertilized with nightsoil, manure or other available organic wastes, some of which contain human pathogens and parasites, for example, *Cryptosporidium*, *Giardia*, Rotaviruses, *Salmonella* and *Shigella*. Some of the diseases caused by these organisms are not severe: a few days of diarrhea, cramps, gastroenteritis. However, during these illnesses many pathogens are excreted and reach excreta-fed fishponds, especially through addition of nightsoil. Because of poor hygiene, family members attending to such fishponds and handling the fish are at continuous risk of reinfection. The pathogens establish themselves and become endemic and the microbial quality of the fish produced in these ponds is at best questionable.

In experiments performed in our laboratory, tilapia *(Oreochromis aureus)*, common carp *(Cyprinus carpio)* and silver carp *(Hypophthalmichthys molitrix)* were exposed for their entire growing season to various pond water qualities. When the concentration of bacteria in the pond water was high (10^6-10^7 ml^{-1}), the number of bacteria recovered from the various organs and muscles of the fish was also high. When the concentration of bacteria in the water was lower (10^3-10^4 ml^{-1}), small numbers of bacteria were recovered from the kidneys, pronephros, spleen and liver and none was recovered from the muscles (Buras et al. 1987). When carp *(Cyprinus carpio)* were exposed to waters containing *Salmonella montevideo* in concentrations of 10^3, 10^4 and 10^5 ml^{-1} and *Salmonella* were recovered from various organs and muscles, the concentration of bacteria in the contents of the digestive tract after 8 to 9 days was equal to or higher than the concentration in the water (Table 1). Similar results were obtained when tilapia were exposed to water containing *Salmonella*.

In experimental ponds, when fish (tilapia, carp and silver carp) were exposed to high concentrations of bacteria originating from wastewater for an entire growing season (120 days), bacteria were recovered from all organs examined and from muscles.

Fish grown in excreta-fed ponds may become contaminated with viruses, bacteria and parasites. The pathogens, like other bacteria, are carried passively on the scales or gills. Some enter the digestive tract from where they penetrate the blood, intraperitoneal fluid, spleen, liver, kidneys and muscles.

When pathogens in the pond water reach high numbers, the risk of infection increases. The number of pathogens required to cause illness in humans is called the infective dose. The infective dose for a few common pathogens has been established in healthy adult volunteers as follows: *Shigella* d1, 10; *Shigella flexneri* 2a, 180; *Vibrio cholerae* biotype Ogawa, 1,000; enteric viruses, 1.0 PFU (plaque forming units) (Plotkin and Katz 1967). For protozoan and helminth parasites, the infective dose may be a single cyst (Feachem 1983).

Certain waterborne disease outbreaks, like salmonellosis, suggest that even lower levels than those indicated by volunteer feeding studies can cause illness (Bryan 1974). Infants, elderly persons and

Table 1. Recovery of *Salmonella* from various tissues and the content of the digestive tract (DTC) of tilapia (*Oreochromis aureus*) exposed to various concentrations of the bacteria in water for 8-9 days (Buras et al. 1987).

Concentration of *Salmonella* in water (ml^{-1})	Time of exposure days	Numbers of *Salmonella* recovered Blood (ml^{-1})	Muscle (g)	Kidney (g)	Liver (g^{-1})	Spleen (g^{-1})	DTC (ml^{-1})
2.5×10^3	8	0	0	0	0	0	1.2×10^3
2.8×10^4	9	0	0	2	1	3	1.8×10^4
4.0×10^5	9	0	2	23	20	31	6.0×10^5

malnourished persons with concomitant illnesses are usually more susceptible to infection. Strauss (1985) and WHO (1989) reviewed the literature on the survival of pathogens in and on fish and concluded that:

1. Invasion of fish muscle by bacteria is very likely to occur when fish are grown in ponds containing concentrations of fecal coliforms greater than 10^4 to $10^5 \cdot 100\ ml^{-1}$ and the potential for muscle invasion increases with the duration of exposure of the fish to the contaminated water.
2. There is little accumulation of enteric organisms and pathogens on or penetration into edible fish tissue when the fecal coliform concentration in the fishpond water is below $10^3 \cdot 100\ ml^{-1}$.
3. Even at lower contamination levels, high pathogen concentrations may be present in the digestive tract and the peritoneal fluid of the fish.

The Concept of a Threshold Concentration

Our field and laboratory studies have shown a pattern by which bacteria, bacteriophages and human enteric viruses can enter the organs and muscles of fish and have helped to define and establish a threshold concept and threshold values for various microorganisms usually found in domestic wastewater (Buras et al. 1985). Bacteria from the water, entering the digestive tract of fish, are usually retained there due to the phagocytic activity of cells lining the digestive tract. Bacteria which have not been engulfed there reach the blood and/or lymph where some are ingested by macrophages (Allison 1978). The remaining bacterial cells are carried to various organs where additional phagocytosis takes place (Avtalion and Sharabani 1975). If the number of bacteria entering the fish is high and surpasses the capability of the phagocytes, bacteria remain in the blood stream and eventually reach the muscles.

Mononuclear phagocytes are mainly located in the kidneys, spleen, liver and peritoneal fluid. Hence, high numbers of bacteria are usually detected in these organs. The physiological state and enzymatic activity of fixed and wandering phagocytes determine the potential primary defense of the fish and also its limits. The phagocytes in fish organs constitute the reticulo-endothelial (RE) system. The threshold values indicate the capacity and the limits of the RE system. Microorganisms, among them pathogens, will not be overcome if their numbers exceed the capacity of the RE system. Thus, threshold concentrations define the limits of the capability of the fish to cope with the high concentrations of bacteria and viruses in the pond environment. When this limit is surpassed, the RE system is no longer effective in retaining and destroying organisms reaching the blood stream and bacteria or viruses from the water can reach any fish tissues, including muscles (Buras et al. 1987).

The threshold concentrations also indicate the vulnerability of the fish to fish pathogens. When the defense system of the fish is affected by increased exposure to microorganisms, opportunistic bacteria which are not pathogenic under normal conditions but are present in the pond water can initiate and establish infections and produce disease outbreaks. Threshold concentrations reflect the physiological status of the fish, give the limits of the fish's capacity to cope with the bacteria and viruses in its environment, and indicate the point beyond which it is incapable of dealing with fish pathogens. In experiments at the University of Arizona, the threshold concentrations of bacteria for the tilapia were $1.0\text{-}2.0 \times 10^4\ ml^{-1}$.

The Homeostatic Plateau and Stress in Fish

A fish is a biological system and depends upon its external environment to provide its necessary energy (inputs) and to accept its outputs (wastes). A fish functions through a cybernetic system that regulates the activities of its various internal systems around a preferred level. The cybernetic system itself has certain limits within which it functions. These limits constitute the homeostatic plateau described by Sutton and Harman (1973). Fish function within certain ranges of well defined parameters (pH, dissolved oxygen, temperature, etc.) which define the homeostatic plateau. Beyond these limits, fish cannot function well, become more vulnerable and are closer to death.

It is important to acknowledge the existence of such limits for the benefit of the fish and for their implications for fish growers, handlers and consumers. A fish grown within its homeostatic limits is healthy and can better withstand some fluctuations of environmental parameters, including short exposure to potentially hazardous microorganisms. The homeostatic range of the fish defines the required quality of their environment, i.e., the pond water in fish culture.

Stress is defined as a stimulus acting on a biological system and the subsequent reaction of the system (Pickering 1981). The reactions or responses to stress in farmed fish can be (1) an immediate physiological response or (2) long-term changes in performance capacity; for example, changes in behavior, decreased growth rate and increased susceptibility to disease (Pickering 1981). Stress can be caused by, among other factors, changes in the quality of the pond water such as high or low temperatures, low dissolved oxygen, free ammonia, pH, high levels of heavy metals, and pesticides and mineral contents. The stocking density of the fish and the quality of feeds available, especially their protein content and high concentration of bacteria can also induce stress, as can handling procedures such as sorting, weighing and measuring. Fish are often subjected to more than one stress at any given time. Some stresses are acute whereas others are chronic (Donaldson 1981).

It is now recognized that stress can disturb the defense mechanisms of fish, with consequent suppression or exaggeration of responses which may disturb physiological balance and ultimately result in disease (Mazeaud and Mazeaud 1981). The reticulo-endothelial system is an important defense system both directly by its phagocytic capacity to destroy pathogenic microorganisms and indirectly by its role in antibody production by B lymphocytes. As a result of stress, the number of lymphocytes in lymphoid tissues decreases. Ellis (1981) suggested that stress induces the production of some substances that suppress the phagocytic activity of the macrophages in fish. Stress in fish has been shown to produce lymphocytopenia, monocytopenia and neutrophilia (Ellis 1981).

The environment in the wastewater-fed ponds can be stressful for fish. In addition to high concentrations of organic matter, they also contain high numbers of bacteria. These can saturate the cellular immune system and put the fish at risk of infections (Wedemeyer and McLeary 1981).

At the University of Arizona, tilapia exposed to stressful conditions in pond water harbored high concentrations of bacteria in their muscles, indicating a physiological state conducive to disease. Stress can lower the threshold values for penetration of bacteria into the muscles of fish and increase their susceptibility to disease. It can also increase the blood sugar levels.

The Microbial Safety of Farmed Fish and Their Environment

The microbial quality of farmed fish is largely determined by the quality of the water in which they are farmed. It is advisable that the concentrations of potentially harmful bacteria in wastewater-fed ponds should not exceed a critical concentration which can be defined as the threshold concentration for the fish. With adequate retention times and good management it is believed that these goals can be achieved and that fish grown in ponds with bacterial numbers between 1.0 x 10^4 and 5.0 x 10^4 x 10^4 ml^{-1} will not pose health risks to consumers.

This level of water quality was obtained in tertiary lagoons in a wastewater-fed aquaculture project at San Juan de Miraflores, Lima, Peru (Bartone et al. 1990). The water quality in such tertiary lagoons depends upon the quality of the incoming wastewater and the retention times at preceding treatment stages. What characterizes this tertiary stage and differentiates these lagoons from previous ones is the intense microbial activity which takes place there with consequent changes in the microbial populations from predominantly enteric bacteria, including some pathogens, to bacterial populations that are very active in degrading organic material and establishing complex foodwebs. These changes are usually expressed in a reduction in bacterial numbers and the appearance of a variety of other microorganisms, including protozoans and various invertebrates on which fish can feed.

As the quality of the water in which the fish are reared determines the quality of the fish produced, it is important to monitor and maintain the microbial and chemical water quality of tertiary lagoons used for fish culture, at levels compatible with the threshold values of the fish. Moreover, these waters should not contain levels of toxic chemicals or heavy metals, which can accumulate in the muscles and organs of the fish.

Very few protozoan and helminth parasites, if any, are usually recovered from tertiary lagoon waters despite their prevalence in many parts of the world and their presence in domestic wastewaters; for example, *Entamoeba coli, Endolimax nana, Giardia lamblia, Cryptosporidium parvum, Ascaris lumbricoides, Strongyloides stercoralis* and *Hymenolepis nana* (Feachem et al. 1983; C. Sterling, pers. comm.). Studies conducted by Dr. C. Sterling (University of Arizona) on the presence of *Giardia lamblia* and *Cryptosporidium parvum* in many parts of the world showed them to be common in wastewater treatment lagoons, including the tertiary lagoons at the Lima plant. Detection was by a monoclonal antibody method. Standard sedimentation-flotation methods failed to detect their presence. *Giardia* and *Cryptosporidium* cysts are small and light and do not settle but rather float on the surface of the lagoon where they concentrate. They are transported with the flow of the water from one lagoon to the other. They can survive in pond waters for six to eight weeks (C. Sterling, pers. comm.).

Toxic Chemicals

Various agricultural chemicals enter domestic wastewaters and wastewater-fed fishponds. Such chemicals are commonly taken up by aquatic organisms and, as their degradation is often slow, they may persist in various tissues. Their concentration increases up the foodchain and can culminate in high concentrations in humans that consume fish. Exposure of fish to pollutants such as heavy metals, polychlorinated biphenyls (PCBs) and chlorinated hydrocarbon pesticides are cause of great concern for human and fish health. Pippy and Hare (1969) found that normally well-tolerated concentrations of copper and zinc, when coupled with low water flow and warm temperatures, caused infection in tilapia by facultative pathogens like *Aeromonas hydrophila*. The outbreaks were of considerable proportions and it was demonstrated that they were not caused by temperature elevation alone. Exposure to nitrite and ammonia has also been associated with fungal infections *(Saprolegnia)* in rainbow trout (*Oncorhynchus mykiss*) (Carballo and Muñoz 1991; M. Carballo, pers. comm.).

Wastewater Treatment Prior to Stocking Fish

Wastewaters used for fish farming should be treated prior to stocking fish. A series of three to four oxidation ponds can produce a suitable water quality, if managed properly. A storage or 'maturation' pond containing water of acceptable quality, from which the fishponds are fed, is also advisable. The advantages of a storage pond are that fluctuations in water quality of the recipient fishponds can be minimized, and it provides suitable conditions to encourage the growth of acceptable microorganisms, such as protozoans and, as a result, helps to reduce enteric organisms and helminth parasites.

Public Health Considerations Concerning Fish Grown in Wastewater-Fed Ponds

In evaluating the public health aspects of fish reared in the presence of domestic wastewater, the following facts are emerging:

1. Common pathogens of warm-blooded animals do not cause diseases in fish (Guélin 1962).
2. Some bacteria like *Vibrio parahaemolyticus* (Janssen 1970), *Clostridium botulinum* type E, *Clostridium tetani, Staphylococcus aureus, Erysipelothrix musiopathiae, Streptococcus, Shigella* and various serotypes of *Salmonella* survive and sometimes multiply in the gut, mucus and various other tissues of fish (Lawton and Morse 1980).
3. *Escherichia coli*, the most common bacterium found in the feces of warm-blooded animals, has been isolated from the stomach and intestines of fish (Kehr and Butterfield 1943; Janssen 1970).
4. The bacteria mentioned above (2,3) do not constitute the normal flora of fish, and their presence indicates that the fish have been exposed to water containing these organisms.
5. The bacterial flora of a fish reflect the bacteriological quality of the waters that the fish inhabits (Geldrich and Clarke 1966).
6. Specific bacteriophages have been isolated from fish intestines *whenever* enteric bacteria are present in the water (Guélin 1962; Geldrich and Clarke 1966).
7. Bacteria found in the gastrointestinal tract of fish are of primary significance as a source of occupational diseases of fish handlers (Reichenback-Klinke 1973).
8. Human enteric viruses have been isolated from effluents after treatment (Berg 1975; Wellings et al. 1975; Buras 1976). As the infective dose for human pathogenic viruses is considered to be very low (Plotkin and Katz 1967), their presence in the fish, even in small numbers, may constitute a public health problem.
9. When growing fish in wastewater-fed ponds, human parasites present special problems because they are prevalent in many parts of the world and are hard to eradicate. They have to be given, therefore, special consideration and attention (Feachem 1983; C. Sterling, pers. comm.).
10. Bacteria, among them pathogens like *Salmonella*, have been detected in the intestinal tract content and

the peritoneal fluid of fish. During the cleaning and evisceration of fish, pathogens present there contaminate the hands of the fish handlers and cleaners. Such contamination happens before the fish are cooked. Contaminated fish can transmit pathogens from the pond waters to fish handlers. The handlers and cleaners constitute, in this case, the primary foci for the transmission of pathogens to their families and later, when an infection has ensued, to other people.

11. In case of human enteric viruses, as the infective dose is very low, a viral outbreak can be initiated by the route described above, even if the number of viral particles in the pond water and the fish intestine is not very high.

The public health aspects of wastewater-fed aquaculture should be assessed not only by the presence or absence of coliform or fecal coliform bacteria in the blood, the organs or the muscles of fish but rather by including considerations of the physiological state of the fish, the capability of its reticulo-endothelial system to handle stress from exposure to large concentrations of microorganisms in the fishpond, the influences on the fish of chemicals in the water, the ability of the fish to withstand fish pathogens and the threshold values of contaminants.

Conclusions

The successful implementation of such public health safeguards depends upon the understanding by fish farmers of why it is important to manage fishponds ecologically. In many cases, they neglect public health aspects because they are not aware of their relevance. It is important, therefore, to include, in basic training programs on wastewater-fed aquaculture, lectures and demonstrations on:

- wastewater as a source of human pathogens;
- the routes of infection of various pathogens present in wastewater and fishponds;
- various treatment methods for human excreta (nightsoil and wastewater) prior to use in fishponds; and
- basic environmental hygiene and sanitation.

If extension services had the capability to conduct routine, simple microbiological tests, this would help in creating more awareness of the potential public health problems connected with wastewater-fed aquaculture.

References

Allison, A.C. 1978. Macrophages and their products in immunoregulation (as effector cells and as targets for adjuvants), p. 181-186. *In* G.R. Werner and F. Flooh (eds.) Pharmacology of immunoregulation. Academic Press, London and New York.

Avtalion, R.R. and R. Sharabani. 1975. Studies on phagocytosis of fish. Immunology 29:1181-1187.

Bartone, C.R., J. Moscoso and H. Nava. 1990. Reuse of waste stabilization pond effluents for fish culture: productivity and sanitary quality results, p. 673-680. *In* C.R. O'Melia (ed.) Environmental Engineering Proceedings of the 1990 Specialty Conference. American Society of Civil Engineers, New York.

Berg, G. 1975. Detection, occurrence and removal of viruses. J. Water Pollut. Control Fed. 47:1587.

Bryan, F.I. 1974. Diseases transmitted by foods contaminated by wastewaters, p. 16-45. *In* Wastewater use in the production of food and fiber. Proceedings of the Conference, Oklahoma City. EPA-660/2-74-041. US Environmental Protection Agency, Washington, DC.

Buras, N. 1976. Concentration of enteric viruses in wastewater and effluent. Water Res. 10: 295-298.

Buras, N., L. Duek and S. Niv. 1985. Reaction of fish to microorganisms in wastewater. Appl. Environ. Microbiol. 50:989-995.

Buras, N., L. Duek, S. Niv, B. Hepher and E. Sandbank. 1987. Microbiological aspects of fish grown in treated wastewater. Water Res. 21(1):1-10.

Carballo, M. and M.J. Muñoz. 1991. Effect of sublethal

concentrations of four chemicals on susceptibility of juvenile rainbow trout (*Oncorhynchus mykiss*) to saprolegnosis. Appl. Environ. Microbiol. 57(6): 1813-1816.

Donaldson, E.M. 1981. The pituitary-interrenal axis as an indicator of stress in fish, p. 11-49. *In* A.D. Pickering (ed.) Stress and fish. Academic Press, London and New York.

Ellis, A.E. 1981. Stress and the modulation of defense mechanisms in fish, p. 147-169. *In* A.D. Pickering (ed.) Stress and fish. Academic Press, London and New York.

Feachem, R.G. 1983. Infections related to water and excreta: the health dimension of the decade, p. 25-46. *In* J.A. Dangerfield (ed.) Water practice manuals 3. Water supply and sanitation in developing countries. The Institute of Water Engineers and Scientists, London.

Feachem, R.G., D.J. Bradley, H. Garelick and D.D. Mara. 1983. Sanitation and disease: health aspects of excreta and wastewater management. John Wiley & Sons, Chichester, UK.

Geldrich, E.E. and N.A. Clarke. 1966. Bacterial pollution indicators in the intestinal tract of fresh water fish. Appl. Microbiol. 14:429-437.

Guélin, A. 1962. Polluted water and the contamination of fish, p. 480-502. *In* G. Borgstrom (ed.) Fish as food. Vol. 2. Nutrition, sanitation and utilization. Academic Press, London and New York.

Janssen, W.A. 1970. Fish potential vectors of human bacterial diseases of fishes and shellfishes. Am. Fish. Soc. Spec. Publ. 5:284-290.

Kehr, R.W. and C.T. Butterfield. 1943. Notes on the relationship between coliform and enteric pathogens. Public Health Rep. 58:589-607.

Lawton, R.L. and E.V. Morse. 1980. *Salmonella* survival in freshwater and experimental infection in goldfish *(Carassius auratus)*. J. Environ. Sci. Health (A)15 4:339-358.

Mazeaud, M.M. and F. Mazeaud. 1981. Adrenergic responses to stress in fish, p. 49. *In* A.D. Pickering (ed.) Stress and fish. Academic Press, London and New York.

Pickering, A.D. 1981. Introduction: the concept of biological stress, p. 2-9. *In* A.D. Pickering (ed.) Stress and fish. Academic Press, London and New York.

Pippy, J.H. and G.M. Hare. 1969. Relationship of river pollution to bacterial infection in salmon (*Salmo polar*) and suckers (*Catostomus commersoni*). Trans. Am. Fish. Soc. 98: 685-686.

Plotkin, S.A. and A. Katz. 1967. Minimal infective dose of viruses for man by oral route, p. 151-166. *In* G. Berg (ed.) Transmission of viruses by the water route. Wiley, New York.

Reichenback-Klinke, H.H. 1973. Fish pathology. TFH Publications, Neptune City, New York.

Strauss, M. 1985. Survival of excreted pathogens in excreta and faecal sludges. WHO International Reference Centre for Wastes Disposal (IRCWD) News 23:4-9.

Sutton, D.B. and N.P. Harman. 1973. Ecology: selected concepts. Wiley, New York.

Wedemeyer, G.A. and D.J. McLeary. 1981. Methods for determining the tolerance of fishes to environmental stressors, p. 247-275. *In* A.D. Pickering (ed.) Stress and fish. Academic Press, London and New York.

Wellings, I.M., A.L. Lewis, W. Carrol, C.W. Mountain and W. Pierce. 1975. Demonstration of virus groundwater after effluent discharge on to soil. Appl. Microbiol. 29:751-757.

WHO. 1989. Health guidelines for the use of wastewater in agriculture and aquaculture. WHO Tech. Rep. Ser. 778. World Health Organization, Geneva, Switzerland.

Discussion

BOGEL: We have seen many attempts to simplify complex situations with general schemes. Some are eventually disproven by scientific challenges but we know of others that have withstood such challenges for over 20 years. The 'threshold principle' that you mentioned is very interesting. We should aim for more collaboration in this area to get more data on fish species-specific threshold phenomena, relative to all the other factors that we know can influence their resistance. This would help us greatly to set general standards. There are of course difficulties. We can imagine that wastewater treatment by simple sedimentation will be possible even in developing countries, but the standards that have been mentioned here are rather high: for example for coliforms and enterobacteria, 10^4 or even 10^3/ml has been mentioned. These would totally preclude, say, the addition of nightsoil to fishponds.

PULLIN: Your paper, Dr. Buras, focuses on wastewater reuse in aquaculture. Many of the fish species that are candidates for this in developing countries actually *eat* bacteria. This applies particularly to species that feed on freshwater detrital aggregates (Bowen 1987)[1]. It also applies to filter-feeders in the marine environment. I foresee the *encouragement* of bacterial production in some developing-country aquaculture systems as a very important and cost-effective foodchain. Bacteria are of many kinds and can be a significant food resource.

ARELLANO: I agree with this. With or without biotechnology, bacteria can be developed as a source of feed. Right now in Ecuador, some farmers are adding sugar to their ponds. They know that it increases the growth of microorganisms as feeds. Therefore, recommendations on bacterial levels in ponds must differentiate between 'good' and 'bad' bacteria, otherwise developing-country aquaculture may lose the opportunity to grow fish in this way, especially when the recommendations come from important sources in the more powerful developed countries.

BOGEL: In poultry raising, adding certain sugars to the feeds suppresses the growth of *Salmonella*. The effects of such carbohydrates in livestock raising are not fully understood, but they hold great promise for the future. Probably the same applies to fish and fishponds.

EDWARDS: I agree with Dr. Pullin. Dr. Buras, you were able to formulate this threshold principle initially because of a treatment malfunction during which the levels of harmful bacteria in the ponds became greatly elevated. This would not apply to well-managed wastewater-fed fishponds. Fish grown in well-managed sewage-fed systems show very high growth rates - probably close to physiological maxima. They are certainly not very stressed. In work at AIT, in putting septage directly into fishponds, the initial concentration of fecal coliforms in septage was about 10^8/100 ml. The total bacterial concentrations in the ponds was 10^4-10^5/ml, but this fell rapidly by dilution to about 10^6/100 ml and then over a 24 hour period fell to about 10^3/100 ml because of die-off in the pond. We did examine fish muscle and found that there was some penetration but the levels were low and within your standards[2]. It is not feasible or desirable to have complete treatment of sewage before sewage-fed fish culture. The main question for the kinds of fish to be grown in developing countries is 'what is the minimum amount of pretreatment needed before safe reuse in aquaculture?'

BURAS: You started from raw septage and added that to the ponds?

EDWARDS: Yes.

BURAS: So there you had a well-established ecosystem, even without passage through a series of ponds. Great! The requirement is to have a *stable* system.

EDWARDS: Yes, it was a stable system, as are most sewage-fed aquaculture systems in Asia. The exception is the overhung-latrine pond system where fresh nightsoil goes directly to the fish from a latrine built over the pond. This system is dangerous.

BOGEL: Two categories of risks are mentioned in the paper: (1) the risks to housewives and other fish handlers who handle and eviscerate the fish and (2) risks due to pathogens in fish muscle. Considering food hygiene with respect to (1), the handlers could come into contact with pathogens from the fish gut. For (2), the levels of, for example, *Salmonella* will probably usually be low and the fish will normally be cooked. However, *Salmonella* is an important pathogen. As few as 10 *Salmonella* can cause an infection. So we need to aim for absence of *Salmonella* from farmed fish. This is not easy for quite a range of traditional fish farming systems. There are recommendations at hand to improve the situation. Time will tell to what extent they can be implemented over the next 20 years.

KING: We need a holistic approach to aquaculture and fisheries management, from stocking to harvesting, processing and selling. Developing countries cannot yet afford the purification systems that are used in developed countries. However, managers can and should promote proper hygiene during handling, etc. This should be a part of the whole task of fisheries management.

BOGEL: Do you mean a holistic approach to the whole chain of events 'from feed to food', including monitoring and hygiene throughout, plus the education of all concerned?

KING: I think that the most important area remains the postharvest chain of events.

BOGEL: Well, the consumers are the last in the chain. We usually assume that they are educated about these aspects, but that may not be so and they are the ones at risk. We cannot transfer responsibility to the last in the chain.

KING: I am not proposing that. The improvements have to be made by handlers, processors and marketing agents. Education, of course, is needed throughout.

EDWARDS: You are right, and this applies to all fish and fish products, not just those from sewage-fed aquaculture. By 2025, more than half of the world's population will live in urban areas. The fish that they will consume will come from waterbodies, including natural waters that are subjected to increasing pollution. I can even conceive that fish from well-managed sewage-fed systems could be safer than some of these. We should also recognize that much of the fish in developing countries is sold to consumers as whole fish. Therefore, the consumer does the evisceration and other preparation before cooking. Clearly there needs to be education for the consumers who are last in the chain.

BOGEL: FAO has had projects on education of villagers in developing-country fisheries projects. Housewives were educated in hygienic handling of fisheries products. To what extent these efforts changed the attitudes and practices of the people, I do not know.

AUSTIN: I was smiling a little during part of Dr. Buras presentation because I made a study on smoked fish products in Edinburgh[3], a city of 500,359 people. We sampled fish, smoked by a cold-smoking process, from the shops of all the fishmongers in Edinburgh. The population of bacteria in the fish muscle exceeded 10^8-10^9/g. There was no apparent health problem.

BOGEL: In WHO, we are very aware that consumer associations are very sensitive and resistant to others giving them orders about the treatment of food which contains pathogens, for example *Salmonella*. They say, no this should not be on our shoulders; we want to spend our money on *safe* food. The consumers feel that if they are part of the safety chain, then all others in the chain should discharge their responsibilities and not just pass on orders to consumers. We are trying to deal with this through a partnership program. The partners are all concerned with the food production, including those who supply feeds, pharmaceuticals, etc. This partnership philosophy will be increasingly important in the EEC countries. Recently, there have been joint meetings involving feed suppliers, farmers, slaughterhouse operators and consumers. Perhaps a similar approach could work in the fisheries sector, although I recognize that the circumstances in most developing countries are very different.

ARELLANO: Do we know anything about the long-term immunological effects of exposure to wastewater?

D. PHILLIPS: You mean effects on humans?

ARELLANO: No - on aquatic organisms.

D. PHILLIPS: There is some information. Freeman and co-workers[4] have published on stress effects, immunochemistry and steroid hormones in fish. He considered that steroid hormones in fish were critical to their health and were important indicators of all forms of stress.

MORIARTY: One suspects that there are effects, but the evidence is inconclusive.

[1]Bowen, S.H. 1987. Composition and nutritional value of detritus, p. 192-216. *In* D.J.W. Moriarty and R.S.V. Pullin (eds.) Detritus and microbial ecology in aquaculture. ICLARM Conf. Proc. 14, 420 p.

[2]Edwards, P., C. Pacharaprakiti, K. Kaewpaitoon, V.S. Rajput, P. Ruamthaveesub, S. Suthriawut, M. Yomjinda and C.H. Chao. 1984. Reuse of cesspool slurry and cellulose agricultural residues for fish culture. AIT Res. Rep. No. 166. Asian Institute of Technology, Bangkok.

[3]Alexander, B. and B. Austin. 1986. Bacterial microflora associated with a commercial fish smoker. FEMS Microbiol. Letter 34: 309-312.

[4]Freeman, H.C., J.F. Uthe and G. Sangalang. 1980. The use of steroid hormone metabolism studies in assessing the sublethal effects of marine pollution. Rapp. P.-V. Réun. CIEM 179: 16-22.

This additional debate on public health was held subsequently.

AUSTIN: Is there any evidence at all that aquaculture in developing countries has led to an increase in disease among fish farm workers themselves or populations in the immediate vicinity of fish farms?

PALMER: There is no clear evidence for this in any of the literature that I have seen.

AUSTIN: I was previously asked to review this for developed and developing countries. The information was very sparse. There was clear evidence for increasing incidence of leptospirosis and probably *Aeromonas hydrophila* infections around aquaculture facilities.

BOGEL: Of course, the mosquito population goes up where manmade lakes are built. There we have evidence from such large ecological changes that some diseases can increase.

AUSTIN: But surely the point is that this is not in association with aquaculture *per se*.

BOGEL: I quite agree, but we can speculate that increasing the waterspace may increase the vector population.

PALMER: That is also my concern - that introduction of fish farming will increase waterborne diseases.

PULLIN: In the humid tropics, there is a lot of waterspace quite apart from fishponds. In rural Africa, one could imagine that constructing a new fishpond close to a homestead could increase exposure to waterborne diseases. However, the root cause of all these diseases is *poverty*. Where it makes sense to include aquaculture in rural development, this must proceed with public health campaigns and the alleviation of poverty. The easiest point of attack in the life cycles of waterborne parasites and pathogens, especially those spread by snails and mosquitos, is in the final host - humans. Most attempts to control or eradicate vectors and intermediate hosts have been spectacular failures, often with adverse environmental impacts and this will probably not change.

In Asia, the benefits from rural aquaculture development are enjoyed to a much greater extent than anywhere else in the developing world and the problems of waterborne diseases associated with aquaculture have been largely overcome. This shows that it can be done. We must hope that it can be done in Africa as well. They key is to have better public health proceed in parallel with aquaculture development. To have one without the other means missed opportunities and greater risks.

BOGEL: I welcome this philosophy. We would not inhibit progress in one area because another problem is not yet solved. For example, in Cairo there is inadequate waste disposal: the dogs eat feces. What happens if we limit our objective to controlling the dog population? Clearly the objectives of better waste disposal and dog control have to be considered in parallel.

BILIO: There may be some possibility to control mosquito larvae and snails through the use of fish.

EDWARDS: But, as Dr. Pullin said, snail control programs using fish have failed. For the mosquito problem, we should distinguish between well-managed and unmanaged ponds rather than as has been suggested in some commentaries on African aquaculture - between newly created and abandoned ponds. For example, in World War II, there were reports of malaria associated with coastal ponds in Indonesia, when these ponds were unmanaged and became weedy. This is not the case now. Introduced tilapia cleared a lot of the weeds and mosquito larvae from ponds and the malaria problem went down. It is also interesting that sanitation engineers do not consider that weed-free oxidation ponds are a suitable breeding ground for mosquitos. Fish consume some mosquito larvae, but good pond management is the main factor.

PALMER: But there are many abandoned ponds.

EDWARDS: Yes, and they are obviously unmanaged and full of mosquitos!

BOGEL: We must also not forget that introduction of new farming systems brings with it new problem-solving measures, such as the use of insecticides. Is the control of mosquito larvae in ricefields mainly due to fish eating them or more to insecticides?

PALMER: Mosquitos have become resistant to pesticides in some areas.

PULLIN: In Malaŵi, there has been recent work on control of bilharzia snails in fishponds with compounds derived from natural vegetation[1]. The initial results are encouraging for small-scale ponds, but I doubt the applicability of this technique to all waterbodies. In addition, the pursuit of the objective of snail control can itself have deleterious environmental effects. As an example, the golden snail (*Pomacea* sp.) was introduced to the Philippines in the hope of using it in livelihood projects to export 'escargots'. This was a bad decision made on false impressions. This species now infests and damages over 400,000 ha of ricelands. Its control and if possible eradication are matters of great public concern (Acosta and Pullin 1991)[2]. The main control measure is to use molluscicides based on organo-tin compounds. Huge amounts of these compounds are now being applied to ricelands, and thus to adjacent wetland areas and watercourses on the mistaken premise that an aquatic snail like this can be controlled or eradicated by chemical means.

EDWARDS: I think that there are examples from Israel and China where the introduction of aquaculture actually decreased the incidence of diseases. I believe that the conversion of swamps to fishponds in Israel reduced the incidence of malaria, when aquaculture was introduced there before World War II. In China, building fishponds in weed-infested swampy areas reduced the habitats for the snail intermediate hosts of bilharzia and there is no bilharzia problem there now.

BURAS: So we can assume that there they controlled the snail and it worked, but in other places it may not work.

PALMER: Well, there are areas where snails are well-controlled and areas where they are not. The latter are usually in the poorer areas, where the people need most help.

---

[1]Msonthi, J.D. 1991. Molluscicidal compounds of plant origin. Aquabyte 4(1):6-7; Chiota, S.S., J.H. Seyani and E.C. Fabiano. 1990. Molluscicidal and piscicidal properties of indigenous plants, p. 22. *In* B.A. Costa-Pierce, C. Lightfoot, K. Ruddle and R.S.V. Pullin (eds.) Aquaculture research and development in rural Africa. ICLARM Conf. Proc. 27, 52 p. (Abstract only).

[2]Acosta, B.O. and R.S.V. Pullin, Editors. 1991. Environmental impact of the golden snail (*Pomacea* sp.) on rice farming systems in the Philippines. ICLARM Conf. Proc. 28, 34 p.

Developing-Country Aquaculture, Trace Chemical Contaminants, and Public Health Concerns

David J.H. Phillips
Acer Environmental, Howard Court
Manor Park, Daresbury, Nr. Runcorn
Cheshire WA7 1SJ, England, UK

PHILLIPS, D.J.H. 1993. Developing-country aquaculture, trace chemical contaminants, and public health concerns, p. 296-311. *In* R.S.V. Pullin, H. Rosenthal and J.L. Maclean (eds.) Environment and aquaculture in developing countries. ICLARM Conf. Proc. 31, 359 p.

Abstract

This paper reviews the present knowledge of the abundance of trace chemical contaminants (metals and organochlorines) in aquaculture products in developing countries, and the implications of this for public health. Existing data from poisoning episodes involving trace contaminants are discussed, and several trace metals, dioxins, dibenzofurans and coplanar PCBs are identified as contaminants of general concern with respect to public health. It is noted, however, that not all of these are abundant in aquaculture products, and of those which are present in appreciable concentration, not all constitute significant toxicological threats. The trace elements and sources thereof which are of great concern to public health are mercury (from finfish in particular) and cadmium (from the consumption of bivalve molluscs and crustaceans). It is conceivable that selenium also constitutes a significant health risk in unusual circumstances, i.e., where marine mammals are consumed as a staple food source. The available database for organochlorines is very poor compared to that for trace metals; however, the little information which exists suggests that significant contamination of aquaculture products is probable in certain instances. There is an urgent need for some data on trace contaminants in aquaculture (and capture fisheries) products, such that developing-country populations which rely heavily on such foods may be adequately protected. The analysis of hair and blood samples for trace metals and of breast milk samples for organochlorines is recommended as a measure of the exposure of developing-country populations to trace contaminants.

Introduction

The human population of the world reached one billion in about 1840, and surpassed two billion in 1930. Over the intervening sixty years since 1930, global population growth has accelerated markedly, to attain total numbers of about 5.1 billion at present. This population increase has recently been fuelled principally by the developing countries of Latin America, Africa and Asia; in most of the western nations, fertility control has created conditions close to zero population growth over the last decade or more.

The growth of human populations is inevitably accompanied by industrial development, and both domestic and industrial wastes give rise to problems with the contamination of aquatic environments. Estimates of the rates of mobilization of metals (Table 1) serve to emphasize the scale of these anthropogenic changes. Many trace elements are now being mined at a

Table 1. The effects of human activity on the global mobilization rates of metals. Two estimates of the latter are included, derived from different authorities and reflecting the uncertainties in this parameter (after Phillips 1980).

Element	Geological rate[a] (10^3 t·year^{-1})	Human-induced rate (mining) (10^3 t·year^{-1})		Total in oceans (10^6 t)
Iron	25,000	319,000[a]	395,000[b]	4,110
Manganese	440	1,600	8,150	2,740
Copper	375	4,460	6,000	4,110
Zinc	370	3,930	5,320	6,850
Nickel	300	358	481	2,740
Lead	180	2,330	3,200	41
Molybdenum	13	57	74	13,700
Silver	5	7	9	137
Mercury	3	7	10.5	68
Tin	1.5	166	227	14
Antimony	1.3	40	65	274
Cadmium	No data	No data	17	68

[a]Study of Critical Environmental Problems, MIT, 1970.
[b]US Bureau of Mines, Minerals Yearbook for 1970.

pace which exceeds their natural mobilization rates by greater than an order of magnitude, with clear implications for the pollution of aquatic resources, both on a global and local scale. The contributions of anthropogenically derived metals have been quantified for certain local environments, and are very considerable for many of the more toxic elements (Förstner 1980). In addition, rivers and coastal waters are presently exposed not only to increasing quantities of natural materials such as metals and nutrients, but also to cocktails of industrially derived contaminants, many of which exhibit significant persistence and capacities for bioaccumulation.

Over the last three decades, most western nations have introduced controls on sewage and industrial discharges to inland and coastal waters. By contrast, the developing nations possess far fewer controls of this sort in most instances (e.g., Phillips and Tanabe 1989), and those which do exist are often ignored (Beanlands and Si 1988). As a result, the global utilization patterns for many aquatic contaminants have altered in recent decades, utilization in developing countries gradually taking over from the western countries. Perhaps the best example of this "southward tilt" concerns the use of DDT and other organochlorine pesticides, now heavily dominated by developing countries (Goldberg 1975). Such patterns of increasing contamination of subtropical and tropical waters are, however, repeated for many pollutants (Phillips 1991).

A further fundamental consequence of population growth is an ever-increasing requirement for food products, and seafoods are an important protein-rich component of many diets, particularly in developing countries. However, the use of coastal waters as a convenient receptacle for domestic and industrial wastes threatens the quality of seafoods. The recent trend in some developing countries towards a greater reliance on aquaculture rather than capture fisheries tends to exacerbate this competition between resources, in that aquaculture activities are generally sited in inshore waters, where contaminants are more likely to be found in significant quantity.

In combination, the changes discussed above have given rise to significant concerns with respect to the contamination of freshwater and marine products in developing countries, and species subject to aquaculture are considered to be particularly at risk. The present paper addresses these concerns from two different perspectives. The initial portion of the paper reviews the existing knowledge of potential human health impacts from contaminants. Due to the paucity of data specific to the potential impacts of freshwater and marine products, information relevant to public health in general is cited; data from the western nations and developing countries are included. The second portion of the paper discusses available data relevant in either specific or general terms to contaminants in aquaculture species in developing countries, to attempt to synthesize a perspective of the magnitude of the problem. It should be noted that contamination by bacteria and viruses or by therapeutants to control such organisms is not covered in this review. Similarly, trace chemicals which pose a threat to aquatic resources but are not of significance to public health (e.g., tributyltin) are not discussed here.

Trace Contaminants and Public Health

Mercury

The first widely documented instance of public health impacts due to aquatic contamination by a trace metal occurred at Minimata Bay in Japan, commencing in the early 1950s. Minimata Bay lies on the eastern aspect of the Shiranui Sea, off southwestern Kyushu. The local waters are partially landlocked, exhibiting a poor assimilative capacity for pollutants. Concerns were raised initially by fish kills and bird mortalities in the area, and these were followed in 1953 by the outbreak of a "dancing disease" in the local cat population (Harada and Smith 1975; Takizawa 1979). The affected cats exhibited coordination problems and convulsion, death frequently resulting. Dogs and pigs were also affected, but in smaller numbers.

In 1956, a five year-old girl admitted to the Minimata factory hospital operated by the Chisso Chemical Corporation was diagnosed to be suffering from brain damage. The symptoms displayed included disturbances of gait and speech, and delirium. Additional cases from the same geographical area were discovered shortly thereafter, leading to a report to the Minimata Public Health Department in mid-1956 that an outbreak of an unclarified disease of the central nervous system had occurred in the Minimata area.

The disease was found to be noninfectious, and research began to focus upon a possible poisoning origin, perhaps involving the Chisso Chemical Corporation factory in Minimata. This factory commenced industrial activities in 1907 with the manufacture of fertilizers and carbide products, but later expanded into petrochemicals and plastics. Acetaldehyde production commenced in 1932, and ceased in 1968; vinyl chloride was produced between 1941 and 1971. The factory effluents were found to contain a variety of trace metals, including arsenic, copper, lead, manganese, selenium, and thallium. Each of these was investigated in the mid-1950s as the possible source of the Minimata symptoms, but all were eventually discounted. The incidence of the disease continued to increase, and the predominantly neurological symptoms were identified in both adults and congenital cases.

In late 1958, it was noted that the epidemiological symptoms exhibited by the Minimata victims were virtually identical to those reported by Hunter and Russell (1954) for occupational poisoning of workers in a factory in England producing methylmercury. Mercury was discovered to have been employed as a catalyst in the production of both acetaldehyde and vinyl chloride in the Chisso Chemical plant. Recent estimates suggest that about 80 tonnes of mercury were lost to the environment from the acetaldehyde manufacturing process, whereas only 0.2 tonnes were derived from the production of vinyl chloride (Taylor 1982). Studies in 1959 confirmed that the disease in cats could be reproduced by feeding the animals with methylmercury. Later investigations established that both methylmercury and inorganic mercury were present in the Chisso effluents, and that methylation of inorganic mercury could occur in sediments in the environment (Jernelov 1969; Fujiki 1972; Takeuchi 1972; Taylor 1982). The cause of Minimata disease was established to involve the contamination of local seafood products by methylmercury, and their consumption by the local populace.

By the early 1960s, more than 100 victims of the poisoning had been officially recognized and almost 50 deaths had occurred. The discharge of mercury from the factory continued until 1968, however, and by late 1974, some 107 deaths had occurred amongst some 798 officially verified patients, and 2,800 additional inhabitants of the area had applied for official verification.

Similar cases of mercury poisoning have occurred elsewhere, including those at Niigata in northwest Honshu in Japan (also due to the contamination of fish by industrial effluents containing mercury); in Sweden from organomercurial dressing used on grain as an anti-fungicide (affecting mostly wildlife rather than the human population); and in the USA and Canada, involving mercury derived from the chlor-alkali industry. Serious episodes involving human poisoning through the consumption of mercury-treated grain or maize were also reported from Guatemala, Pakistan, Iraq and Ghana (Bakir et al. 1973; Förstner and Wittmann 1983).

These events have clearly established the threat posed by mercury to human health and have driven increased legislation intended to protect populations from mercury poisoning. Most developing and developed nations now possess controls on mercury levels in aquatic biota, the favored limit being 0.5 $mg \cdot g^{-1}$ wet weight in the majority of cases (Nauen 1983).

Cadmium

Cadmium is included with mercury as a "List I" or "Black List" contaminant by the European Economic Commission (EEC 1976), and this reflects its high toxicity to mammals (see also Simpson 1981; Hutton 1982). While cadmium in seafoods is not known to have given rise to public health problems, significant concerns exist over this possibility. These are generated principally by the itai-itai syndrome, described below.

In 1947, an unusual disease was recognized in the Jintsu River basin in Japan (Yamagata and Shigematsu 1970; Kobayashi 1971; Friberg et al. 1974). A total of 44 cases was recognized initially, and the disease became known as itai-itai (literally translated as "ouch-ouch" disease, the name being derived from the cries of pain made by those suffering from the affliction). About 100 deaths occurred due to itai-itai disease among a total of about 200 patients prior to 1966 (Förstner and Wittmann 1983; Mance 1987). The most important symptoms of the disease included a softening of the skeletal bones, pseudo-fractures of these bones, and (in severe cases) skeletal deformation. In addition, albumin and other proteins were found in the urine of affected patients, due to kidney damage.

The primary cause of the itai-itai syndrome is now believed to be the contamination of the Jintsu River basin by cadmium, which was present in both sludges and effluents derived from a zinc mine operated by the Makioki Company in the upstream river catchment. The waters of the river were used downstream for irrigating ricefields, and the rice crop produced was severely contaminated.

Concentrations of cadmium in rice from the affected area approximated 0.7 $\mu g \cdot g^{-1}$ wet weight, greater by an order of magnitude than levels recorded for rice elsewhere in Japan (Friberg et al. 1974). Cadmium uptake from ingested rice was thought to be the principal cause of the disease in those affected. Controls on the discharge of cadmium from the mine were introduced in the mid-1950s and the incidence of itai-itai diminished rapidly thereafter.

Cadmium nevertheless remains of considerable concern in terms of public health. International health authorities have promulgated a provisional tolerable week intake for cadmium in humans of 400-500 µg (FAO/WHO 1972). However, populations in many areas of the world have been documented to approach or exceed this level of intake (Friberg et al. 1974; Simpson 1981; Hutton 1982; Sherlock 1983). High-risk groups ingesting unusually large amounts of the element in particular types of seafoods or offal undoubtedly exist in some populations, and subclinical effects (mostly involving the liver and kidney cortex) may occur (CEC 1978; Hutton 1982).

Other Trace Metals

Lead is of considerable concern as an environmental contaminant, particularly with respect to its effects on the central nervous system and the significant sensitivity of children to the element. Despite conflicting evidence, a general global increase in lead concentrations is now accepted to have occurred since the late 1700s (e.g., see Grandjean 1981; Wolff 1990). Lead is mostly taken up by humans from aerial sources through the lungs, but significant accumulation of the metal also occurs in cases of occupational exposure; from the ingestion of pica (paint, ashes, etc.) by children; from glazes on crockery; and from the dissolution of the element in potable water reticulation systems (Förstner and Wittmann 1983; Nriagu 1988).

Arsenic is also famous for its toxicity to humans, being a popular agent for deliberate homicides. However, accidental cases of poisoning have also occurred, such as that in Japan in 1955, caused by the contamination of milk powder containing arsenic-contaminated sodium phosphate, employed as a stabilizer. In total, 12,000 cases of poisoning of children were reported, 130 of these being fatal. There is evidence from both this event and other cases of drinking water contamination or occupational poisoning by inorganic arsenic that the effects may linger for considerable periods (Tsuchiya 1977; Förstner and Wittmann 1983; Nriagu 1988). Both arsenic and lead have been also associated with poisoning of humans in the USA through the consumption of "moonshine" (illicit whiskey), the principal impacts of being neurological and nephrological (Gerhardt et al. 1980; Harrington 1980). It is important, however, to recognize that the toxicity of arsenic is related fundamentally to its chemical form. While the inorganic forms of the element are of very considerable toxicity to mammals, many organic forms of arsenic are of insignificant toxicity. These include a variety of chemical forms of arsenic documented to be present in aquatic products (see below and Phillips 1990).

Chromium is of considerable toxicity to humans, especially in its hexavalent form. Problems involving chromium have occurred due to both air pollution and the disposal of contaminated wastes, in Japan in particular (Förstner and Wittmann 1983; Nriagu 1988).

Public Health Regulations for Trace Metals

As a result of public health concerns generated by incidents such as those discussed above, many regional and national authorities have promulgated legislation intended to restrict the exposure of human populations to trace metals. Certain of these regulations deal specifically with trace metals in food products from fresh or marine waters, as in some cases, such species exhibit a propensity to accumulate relatively high concentrations of particular metals.

The most comprehensive review to date of public health regulations for trace metals in freshwater and marine fish and shellfish is that of Nauen (1983); data from this source are presented in Table 2. It is evident that standards are most common for mercury, lead, arsenic and cadmium, with fewer regulations promulgated for copper, tin, zinc and other elements.

Table 2. Median international standards and ranges in such standards (μg·g-1 wet weight) for trace elements in freshwater and marine fish and shellfish (after Nauen 1983).

Element	Median standards Fish	Shellfish	Range	Number of countries with standards
Antimony	1.0	1.0	1.0 to 1.5	3
Arsenic	1.5	1.4	0.1 to 5.0	11
Cadmium	0.3	1.0	0.05 to 2.0	10
Chromium	1.0	1.0	1.0	1
Copper	20	20	10 to 100	8
Fluoride	150	-	150	1
Fluorine	17.5	-	10 to 25	2
Lead	2.0	2.0	0.5 to 10	19
Mercury	0.5	0.5	0.1 to 1.0	28
Selenium	2.0	0.3	0.3 to 2.0	3
Tin	150.0	190.0	50 to 250	8
Zinc	45	70	40 to 100	6

Organochlorines

By comparison to their highly significant impacts on many aquatic and terrestrial life forms (e.g., Hunt and Bischoff 1960; Hermann et al. 1969; Reijnders 1986; Moriarty 1988; Spies and Rice 1988), most organochlorines exhibit relatively low toxicities to humans. However, a few groups of organochlorines are of considerable significance in terms of their potential toxicity to humans. These include the dioxins (especially the most potent dioxin, which is 2,3,7,8-tetrachloro dibenzo-*p*-dioxin, also known as TCDD), the dibenzofurans and certain of the polychlorinated biphenyl (PCV) congeners.

The first major human poisoning episode due to organochlorines occurred in 1968 in Japan. The cause of this was the contamination of rice oil produced by the Kanemi Rice Oil Company, and the incident became known as the Yusho (translated as rice oil) affair. The oil was inadvertently contaminated by a commercial PCB preparation used as heat exchange fluid in its production, and over 1,500 persons were affected through the consumption of the oil (Kuratsune 1980). Tanabe et al. (1989) have recently shown that the causative agent of the poisoning was most probably a dibenzofuran (2,3,4,7,8-pentachloro dibenzofuran) produced through heating the oil in the cooking process, rather than the PCBs originally present in the oil. However, certain coplanar PCBs and polychlorinated quaterphenyls may also have played a role in generating the toxic symptoms (Mochiike et al. 1986; Tanabe et al. 1989). A similar case of poisoning occurred in Taiwan in 1979, due to the contamination of rice-bran oil by PCBs (Chen and Hsu 1986).

In 1976, a major accident occurred at a chemical plant in Seveso in northern Italy, leading to the contamination of the surround-ing area by dioxins (Margerison et al. 1980; Strigini 1983). The ICMESA chemical plant at Seveso (owned by the Hoffmann-La Roche company) had produced trichlorophenol since about 1970, reportedly in breach of their industrial license. In July 1976, an exothermic reaction occurred in a chemical reactor producing trichlorophenol, causing a safety valve to break, thus releasing a cloud of contaminated material into the atmosphere. Significant quantities of dioxins were present in the material released, including 2,3,7,8-tetrachloro-dibenzo-*p*-dioxin, which is of extreme toxicity to mammals. Significant contamination was later found to extend over several hundred hectares, affecting a population in excess of 5,000 (over 1,000 of whom were evacuated from the area). All agricultural activities were ordered to cease in the area and all animals were destroyed. Decontamination activities were undertaken in the most significantly affected areas, involving the removal of soil, its replacement by clean material, and the removal of some of the building and their contents. All contaminated material was destroyed by incineration in Switzerland. Epidemiological studies were poorly controlled, however, and the effects on the local population are hard to estimate with confidence, although these are believed to include a high incidence of chloracne and possible birth defects (Strigini 1983).

Both dioxins and dibenzofurans are also produced through the pyrolysis of PCBs. Several cases of local contamination by these compounds have been recorded, due to either the burning of PCBs at low temperatures, or to fires involving older transformers which contain PCBs (e.g., USCDC 1985; Thompson et al. 1986). Subsequent attempts at decontamination have not always met with success.

Several other cases of pesticide contamination also merit mention here. The inadvertent substitution of a fire retardant for a cattle feed supplement in Michigan in 1973 gave rise to poisoning of cattle by polybrominated biphenyls (PBBs), which were also passed to the human population through both milk and meat products (Schnare et al. 1984). The long-term impacts of this remain unclear, but PBBs appear to affect the liver and sensory nerves of humans. Also in the mid-1970s, the James River was severely contaminated by the organochlorine pesticide Kepone, produced by the Life Sciences Products plant in Hopewell, Virginia. While instances of human poisoning from the latter event were restricted to occupational exposure, the effects of Kepone on the environment were significant (Huggett and Bender 1980).

Developing-Country Aquaculture and Trace Contaminants

Information concerning the concentrations of trace contaminants in freshwater and marine products of developing countries is sparse, and that relevant specifically to aquaculture species is particularly rare in the literature. Given the preceding background data on public health, however, certain preliminary conclusions may be reached concerning the potential problems faced from contaminants in aquaculture species in developing countries. These problems will again be considered here on a contaminant-by-contaminant basis.

Mercury

There can be little doubt that mercury in freshwater and marine foods constitutes a significant public health threat in certain instances. In areas of gross contamination, public health effects have been observed, and these are essentially irreversible (see discussion above). While the element is accumulated to the greatest extent by large, long-lived predatory species such as tuna and shark (e.g., Phillips 1980; Eisler 1981), it may nevertheless be taken up to a significant degree by smaller species such as those subject to aquaculture, at least in circumstances of excessive local contamination. It is this latter route of extreme contamination, leading to elevated mercury levels in many marine products, which has given rise to environmental poisoning episodes in the past.

In developing countries, there is evidence that specific sources of mercury exist which create the type of local contamination which may give rise to public health impacts through the consumption of contaminated freshwater or marine species. Thus, for example, Suckcharoen et al. (1978) reported that while background levels of mercury in freshwater fish from Thailand were reasonably low (0.002-0.3 $\mu g \cdot g^{-1}$ wet weight; see also Menasveta and Siriyong 1977), individuals of the predatory fish species *Ophiocephalus striatus (Channa striata)* from a river in Samut Prakarn Province exhibited considerable contamination by the element (0.32-3.6 $\mu g \cdot g^{-1}$ wet weight). The source of the mercury was documented as a Japanese-owned caustic soda factory.

The same type of source has also been shown to be a significant contributor of mercury to coastal waters in India. The Thana Creek which runs into Bombay Harbour is highly contaminated by mercury, as are the waters off Karwar and the coast between Mangalore and Calicut (Kureishy et al. 1979; Zingde and Desai 1981; Qasim and Sen Gupta 1988; Sanzgiry et al. 1988). Elevated mercury concentrations are found in local finfish as a result of this contamination, predatory species being particularly affected.

In Indonesia, an outbreak of Minimata disease apparently occurred in the fishing village of Muara Angke, due to the consumption of fish contaminated by mercury from industrial wastes. Only one brief report on this topic (Webb 1983) is known to the author. It clearly merits further work.

While many developing countries rely heavily on fish as a staple source of protein and their populations may therefore accumulate moderate amounts of mercury, certain communities are exposed to unusually high concentrations of the element by virtue of the precise seafoods they consume. A recent study by Hansen (1990) concluded that mercury ingested principally in whale and seal meat may pose a significant threat to the Eskimo populations of Greenland, at least in terms of fetal exposure. This conclusion was based in part on the analysis of mercury in blood and hair samples, which have also been employed as an index of exposure by authors in Asia and elsewhere (e.g., see Suckcharoen et al. 1978; Sivalingam and Sani 1980; Airey 1983). There is good evidence from such investigations that the consumption of fish is the predominant source of mercury uptake in most communities (Airey 1983). Given this level of understanding of the sources of the element to humans, it is most unfortunate that the toxicological impact of mercury is not better characterized. In particular, the sublethal effects of the element on humans are very poorly understood, although fetal exposure is considered to be critical in a toxicological sense (Hansen 1990).

It may be concluded that the risk to human health from the consumption of freshwater and marine products containing mercury is significant, at least in the more contaminated areas. While the database relating directly to fish and shellfish employed for aquaculture is extremely poor, there is sufficient evidence from studies of the contamination of sediments and of capture fisheries species that certain areas of developing countries are highly polluted by mercury from industrial effluents. The production of any species by aquaculture methods in these areas needs careful consideration, particularly in locations where fish and shellfish form a major proportion of the total diet. This constraint has also been noted in the capture fisheries industry, particularly in areas of the world where large, long-lived predatory fish species constitute a significant proportion of the fisheries resource. While there have been calls to increase the permissible concentration of mercury in marine products to protect such fisheries (e.g., Anon. 1980), the scientific consensus supports the present public health limits, which range around 0.5 $\mu g \cdot g^{-1}$ wet weight for mercury in seafoods (Nauen 1983).

Cadmium

Several authors have concluded that the available safety margin is very low between the average daily exposure of populations to cadmium and the amount of the element required over time to cause sublethal effects such as renal tubular dysfunction (e.g., Friberg et al. 1974; Hutton 1982). This is partially because cadmium is of exceptional persistence in humans, exhibiting net accumulation to attain maximum concentrations at an age of about 50 years in most individuals. Considerable portions of many populations therefore ingest amounts of cadmium which approximate or exceed the provisional tolerable weekly intake of the element of 400-500 µg (FAO/WHO 1972). On this basis, it may be argued that the ingestion of any freshwater or marine food product containing excessive amounts of cadmium should be avoided.

Concentrations of cadmium are relatively low in the axial muscle tissues of most finfish species, but the element is significantly accumulated by bivalve molluscs and certain species of crustaceans (e.g., MAFF 1973; Phillips 1980; Eisler 1981; Simpson 1981). It is these organisms which are most frequently found to exceed public health standards laid down for cadmium freshwater and marine products (Nauen 1983). Thus, by direct contrast to the problems associated with mercury in seafoods, cadmium is of significance principally in particular phyla or groups of organisms which exhibit unusually great capacities for its bioaccumulation. By extension of this argument, toxicological impacts are likely to vary widely in any given population, depending on individual preferences for particular types of foods, only some of which are cadmium-rich. This scenario is complicated somewhat by the fact that not all foods which are cadmium-rich are aquatic in origin; thus, in addition to bivalve molluscs and crustaceans, the offal of terrestrial animals (especially kidney and liver tissue) is a recognized source of cadmium in high concentration (as is the smoking of tobacco).

Very little information is available on the cadmium concentrations of crustaceans farmed in developing countries; however, the database relating to cadmium in bivalve molluscs is rather better. The following discussion emphasizes values observed to be in excess of 1.0 $\mu g \cdot g^{-1}$ wet weight, which is the median international standard for permissible levels of cadmium in shellfish (Nauen 1983; Table 2).

Phillips et al. (1982a) conducted a broad-ranging survey of metals in seafoods in Hong Kong, analyzing samples caught locally and those purchased from retail markets. They noted occasional high cadmium values in certain crustaceans (e.g., the prawn *Penaeus japonicus*), but low levels in finfish. However, consistently high concentrations of the element were found in Pacific oysters (*Crassostrea gigas*) from the area of Deep Bay, to the northwest of the Territory. More extensive followup studies (Phillips et al. 1982b, 1986) confirmed that *C. gigas* grown in Deep Bay itself exhibited moderate cadmium levels of 0.4-1.1 $\mu g \cdot g^{-1}$ wet weight, but imported oysters from China frequently contained greater amounts of the element (up to 5.4 $\mu g \cdot g^{-1}$ wet weight). Later unpublished investigations involving dietary surveys suggested the existence of a high-risk group among the local community. These individuals consumed oysters frequently and significantly exceeded the provisional tolerable weekly intake of cadmium as a result (FAO/WHO 1972).

The Pacific oyster *C. gigas* is particularly efficient as a bioaccumulator of cadmium, and other bivalve species tend to contain rather lower amounts of the element in most locations, although this clearly depends on the degree of local contamination (Phillips et al. 1982b). Several studies of metals in bivalves other than *C. gigas* have been completed in Asia and elsewhere (e.g., Huschenbeth and Harms 1975;

Kumaraguru and Ramamoorthi 1979; Sivalingam and Bhaskaran 1980; Brown and Holley 1982; Hungspreugs and Yuangthong1984; Hungspreugs et al. 1984, 1989; Sivalingam 1984; Patel et al. 1985; Phillips and Muttarasin 1985). Few of these have noted cadmium levels to be consistently elevated above public health limits, although certain bivalve species exhibit concentrations of the element close to 1.0 $\mu g \cdot g^{-1}$ wet weight in some locations. For example, Phillips and Muttarasin (1985) found that cadmium concentrations in bivalves of commercial importance in Thailand varied from about 0.1 $\mu g \cdot g^{-1}$ wet weight in clams (*Paphia undulata*), through 0.4-0.7 $\mu g \cdot g^{-1}$ wet weight in cockles and rock oysters (*Anadara granosa* and *Crassostrea commercialis*, respectively), to a maximum of about 1.3 $\mu g \cdot g^{-1}$ wet weight in green-lipped mussels (*Perna viridis*). The last of these species exhibited the greatest variability in cadmium levels between locations and has been proposed as a suitable biomonitor for metals in Southeast Asia (Phillips 1984, 1985; Phillips and Rainbow 1988).

Information concerning cadmium concentrations in crustaceans from developing countries is sparser than that for bivalve molluscs, especially with respect to species subject to aquaculture. However, experience elsewhere has shown that prawns and shrimp are unlikely to be heavily contaminated, but that the brown meat of crabs may exhibit elevated concentrations of the element. There is a clear need for further studies of this topic, particularly in areas where *per caput* seafood ingestion is high and local industries exist which discharge cadmium.

It is concluded that cadmium in freshwater and marine products is of significance in terms of public health, but that specific problems are likely to arise only where species which heavily accumulate the element (e.g., oysters, certain crabs) are cultured or are subject to capture fisheries in regions which also support cadmium-producing industries (e.g., electroplaters; pigment and plastic industries; smelters; battery manufacturers). Any public health problems which arise will be likely to be due to long-term exposure to the element, and the health of specific populations could be monitored through the analysis of blood, hair or (preferably) kidneys at autopsy.

Other Trace Metals

While several other trace metals exert very significant impacts upon public health under certain circumstances of exposure, their accumulation through freshwater or marine food products is generally not considered to dominate exposure patterns for humans. This is the case, for example, for arsenic, chromium and lead.

Arsenic is present in freshwater organisms and marine biota in appreciable concentrations, and this has generated considerable public health-related concern over the last two decades in particular. Phillips (1990) has reviewed the present situation, noting in particular the dependence of the toxicological impacts of arsenic in aquatic species on the chemical speciation of the element. The arsenic found in most aquatic organisms is principally organic in nature, being present largely as methylated arsenicals and arseno-sugars in primary producers, and as arsenobetaine in higher organisms. The concentrations of inorganic arsenic present in most aquatic species are generally low (frequently below 1.0 $\mu g \cdot g^{-1}$ wet weight) and are thus of relatively little toxicological concern.

Concentrations of both chromium and lead in most marine products are also quite low (e.g., Phillips 1980; Eisler 1981). It is most unlikely that the consumption of freshwater or marine food products would constitute a highly significant source of either element compared to other routes of uptake, even in cases where the *per caput* consumption of such foods were high. While the accumulation by humans of any quantity of lead (by whatever route of absorption) may be considered to be undesirable, attempts to reduce lead concentrations in foods derived from fresh or marine waters would affect total uptake rates of the element very little. It may also be noted here that the very high values of lead reported by Menasveta and Cheevaparanapiwat (1981) for *Perna viridis* from the Gulf of Thailand are considered to be in error, presumably due to inadvertent contamination of the samples prior to, or during, their analysis (Phillips and Muttarasin 1985).

Among the other metals for which public health standards have been promulgated, copper, tin and zinc are of low toxicity to mammals, although they are of much greater toxic impact to aquatic organisms. Antimony is rarely reported to be present at significant concentrations in aquatic biota (Eisler 1981). Selenium is accumulated principally by predatory species and by long-lived fish and marine mammals, perhaps as a mechanism to detoxify bioaccumulated mercury. While it is conceivable that selenium in whale and seal meat is of toxicological importance to Eskimo populations (Hansen 1990), effects elsewhere in developing countries from the consumption of fish or invertebrates containing selenium are most unlikely.

Organochlorines

The database for organochlorines in aquatic species of developing countries is exceptionally poor. However, a few reports merit mention here, to provide an indication of the possibilities for contamination of such environments, and the consequent public health impacts.

Huschenbeth and Harms (1975) analyzed a wide range of invertebrates and fish from Thai waters for organochlorine pesticides and PCBs. No consistent regional pattern of contamination emerged from these studies, but residues of both pesticides and PCBs were detectable in certain species. Hungspreugs et al. (1989) have more recently concluded that pesticides are of importance on a local scale only in Thailand.

In Hong Kong, Phillips (1985) demonstrated that DDT and its metabolites, isomers of hexachlorocyclohexane (HCH), and PCBs were present in significant quantities in green-lipped mussels (*Perna viridis*), which were employed as a biomonitor of both trace metals and organochlorines. While the DDT and HCH groups of compounds were widely distributed, PCBs were present in specific hot-spot areas only, and one of these (Junk Bay, to the east of Victoria Harbour) was excessively contaminated. Later studies (Tanabe et al. 1987a; Kannan et al. 1989a, 1989b) documented the presence of coplanar PCBs in both *P. viridis* and sediments, and suggested that the marine foodchain would be significantly contaminated by these compounds, which are thought to be of appreciable toxic potential for humans (Tanabe et al. 1987b).

No data are known on the prevalence and distribution of dioxins or dibenzofurans in aquaculture products from developing countries. This is a significant gap in our present knowledge, and there is undoubtedly a case for the investigation of such extremely toxic contaminants in food products from fresh and marine waters in developing countries.

While the analysis of hair (and in some instances, blood or nails) is preferred to provide an indication of human exposure to trace metals (see above), the most appropriate indicators of contamination of human populations by organochlorines involve lipid-rich tissues, as these contaminants preferentially associate with such tissues. The best noninvasive technique concerns the use of breast milk, which is lipid-rich and therefore contains significant amounts of organochlorines. Studies of breast milk have been frequent in western nations (see review by Jensen 1983), and have generally indicated a decrease in the concentrations of DDT and other organochlorines in western populations, subsequent to the imposition of controls or bans on these compounds in the 1970s and 1980s. However, the southward tilt of the utilization of such pesticides (Goldberg 1975) has not been accompanied by increased analyses of the effects of such compounds on human populations in developing countries. Such data as are available suggest significant contamination of at least certain populations by organochlorines (see Winter et al. 1976; de Campos and Olszyna-Marzys 1979; Ip 1983; Al-Omar et al. 1986; Ip and Phillips 1989). The investigations of Ip and Phillips (1989) have indicated a link between the consumption of aquatic products and organochlorine contamination in ethnic Chinese in Hong Kong, and much further work is required to establish the extent and significance of this link in developing countries in general. It is relevant here that aquatic products form a very large part of the diet in many developing countries, and that breast feeding is common throughout the developing countries, often for long periods after birth.

References

Airey, D. 1983. Total mercury concentrations in human hair from 13 countries in relation to fish consumption and location. Sci. Total Environ. 31:157-180.

Al-Omar, M.A., F.H. Abdul-Jalil, N.H. Al-Ogaily, S.J. Tawfiq and M.A. Al-Bassomy. 1986. A follow-up study of maternal milk contamination with organochlorine insecticide residues. Environ. Pollut. Ser. A 42:79-91.

Anon. 1980. Discussions on mercury follow committee report. Aust. Fish. (October):2-10.

Bakir, F., S.F. Damluji, L. Amin-Zaki, M. Murtadha, A. Khalidi, N.Y. Al-Rawi, S. Tikriti, H.I. Dhahir, T.W. Clarkson, J.C. Smith and R.A. Doherty. 1973. Methylmercury poisoning in Iraq. An interuniversity report. Science 181:230-241.

Beanlands, G.E. and Z. Si. 1988. Coping with environmental impact assessment in the developing world, p. 12-17. *In* P. Hills, R. Keen, K.C. Lam, C.T. Leung, M.A. Oswell, M. Stokes and E. Turner (eds.) Pollution in the urban environment. Vol. 1. Vincent Blue Copy Co. Ltd., Hong Kong.

Brown, B.E. and M.C. Holley. 1982. Metal levels associated with tin dredging and smelting and their effect upon intertidal reef flats at Ko Phuket, Thailand. Coral Reefs 1:131-137.

CEC. 1978. Criteria (dose/effect relationships) for cadmium. Report of a Working Group of Experts prepared for the Commission of European Communities, Directorate-General for Social Affairs, Health and Safety Directorate. Pergamon Press, Oxford.

Chen, P.H. and S.-T. Hsu. 1986. PCB poisoning from toxic rice-bran oil in Taiwan, p. 22-37. *In* J.S. Waid (ed.) PCBs and the environment. Vol. III. CRC Press, Boca Raton, Florida.

de Campos, M. and A.E. Olszyna-Marzys. 1979. Contamination of human milk with chlorinated pesticides in Guatemala and in El Salvador. Arch. Environ. Contam. Toxicol. 8:43-58.

EEC. 1976. Council Directive of 4 May 1976 on pollution caused by certain substances discharged into the aquatic environment of the Community. Directive 76/464/EEC, European Economic Community, Brussels.

Eisler, R. 1981. Trace metal concentrations in marine organisms. Pergamon Press, New York.

FAO/WHO. 1972. Report of the Joint Committee on food additives. Food and Agriculture Organization and World Health Organization. WHO Tech. Rep. Ser. 505.

Förstner, U. 1980. Inorganic pollutants, particularly heavy metals in estuaries, p. 307-348. *In* E. Olausson and I. Cato (eds.) Chemistry and biogeochemistry of estuaries. John Wiley, New York.

Förstner, U. and G.T.W. Wittmann. 1983. Metal pollution in the aquatic environment. Second edition. Springer-Verlag, Berlin.

Friberg, L., M. Piscator, G.F. Nordberg and T. Kjellström, Editors. 1974. Cadmium in the environment. Second edition. CRC Press, Cleveland, Ohio.

Fujiki, M. 1972. The transitional condition of Minimata Bay and the neighbouring sea polluted by factory waste water containing mercury. *In* 6th International Conference on Water Pollution Research. Jerusalem.

Gerhardt, R.E., E.A. Crecelius and J.B. Hudson. 1980. Moonshine-related arsenic poisoning. Arch. Med. 140:211-213.

Goldberg, E.D. 1975. Synthetic organohalides in the sea. Proc. R. Soc. Lond. (B) 189:277-289.

Grandjean, P. 1981. Blood lead concentrations reconsidered. Nature (Lond.) 289:188.

Hansen, J.C. 1990. Human exposure to metals through consumption of marine foods: a case study of exceptionally high intake among Greenlanders, p. 227-243. *In* R.W. Furness and P.S. Rainbow (eds.) Heavy metals in the marine environment. CRC Press, Boca Raton, Florida.

Harada, M. and A.M. Smith. 1975. Minimata disease: a medical report, p. 180-192. *In* W.E. Smith and A.M. Smith (eds.) Minimata. Chatto & Windus Ltd., London.

Harrington, J.T. 1980. Alcohol, arsenic, and (rapidly) old kidneys. Arch. Int. Med. 140:167-168.

Hermann, S.G., R.L. Garrett and R.L. Rudd. 1969. Pesticides and the western grebe. A study of pesticide survival and trophic concentration at Clear Lake, Lake County, California, p. 24-53. *In* M.W. Miller and G.G. Berg (eds.) Chemical fallout: current research on persistent pesticides. Charles C. Thomas, Springfield, Illinois.

Huggett, R.J. and M.E. Bender. 1980. Kepone in the James River. Environ. Sci. Technol. 14:918-923.

Hungspreugs, M. and C. Yuangthong. 1984. The present levels of heavy metals in some molluscs of the Upper Gulf of Thailand. Water Air Soil Pollut. 22:395-402.

Hungspreugs, M., S. Silpipat, C. Tonapong, R.F. Lee, H.L. Windom and K.R. Tenore. 1984. Heavy metals and polycyclic hydrocarbon compounds in benthic organisms of the Upper Gulf of Thailand. Mar. Pollut. Bull. 15:213-218.

Hungspreugs, M., W. Utoomprurkporn, S. Dharmvanij and P. Sompongchaiyakul. 1989. The present status of the aquatic environment of Thailand. Mar. Pollut. Bull. 20:327-332.

Hunt, E.G. and A.I. Bischoff. 1960. Inimical effects on wildlife of periodic DDD application to Clear Lake. Calif. Fish Game 46:91-106.

Hunter, D. and D.S. Russell. 1954. Focal cerebral and cerebellar atrophy in a human subject due to organic mercury compounds. J. Neurol. Neurosurg. Psychiat. 17:235-241.

Huschenbeth, E. and U. Harms. 1975. On the accumulation of organochlorine pesticides, PCB and certain heavy metals in fish and shellfish from Thai coastal and inland waters. Arch. Fischereiwiss. 26:109-122.

Hutton, M. 1982. Cadmium in the European Community: a prospective assessment of sources, human exposure and environmental impact. Technical report prepared by the Monitoring and Assessment Research Centre, University of London, for the Commission of European Communities. Contract number 333-ENV UK.

Ip, H.M.H. 1983. Breast milk contaminants in Hong Kong. Bull. Hong Kong Med. Assoc. 35: 16 p.

Ip, H.M.H. and D.J.H. Phillips. 1989. Organochlorine chemicals in human breast milk in Hong Kong. Arch. Environ. Contam. Toxicol. 18:490-494.

Jensen, A.A. 1983. Chemical contaminants in human milk. Residue Rev. 89:1-128.

Jernelov, A. 1969. Conversion of mercury compounds, p. 68-74. *In* M.W. Miller and G.G. Berg (eds.) Chemical fallout: current research on persistent pesticides. Charles C. Thomas, Springfield, Illinois.

Kannan, N., S. Tanabe, R. Tatsukawa and D.J.H. Phillips. 1989a. Persistency of highly toxic coplanar PCBs in aquatic ecosystems: uptake and release kinetics of coplanar PCBs in green-lipped mussels (*Perna viridis* Linnaeus). Environ. Pollut. 56:65-76.

Kannan, N., S. Tanabe, T. Okamoto, R. Tatsukawa and D.J.H. Phillips. 1989b. Polychlorinated biphenyls (PCBs) in sediments in Hong Kong: a congener-specific approach to the study of coplanar PCBs in aquatic ecosystems. Environ. Pollut. 62:223-235.

Kobayashi, J. 1971. Relation between the 'itai-itai' disease and the pollution of river water by cadmium from a mine, p. 1-7. *In* Advances in water pollution research: Proceedings of the 5th International Conference. San Francisco.

Kumaraguru, A.K. and K. Ramamoorthi. 1979. Accumulation of copper in certain bivalves of Vellar Estuary, Porto Novo, S. India in natural and experimental conditions. Estuar. Coast. Mar. Sci. 9:467-475.

Kuratsune, M. 1980. The Yusho poisoning affair in Japan, p. 287-302. *In* R.D. Kimbrough (ed.) Halogenated biphenyls, terphenyls, naphthalenes, dibenzodioxins and related products. Elsevier/North Holland, Amsterdam.

Kureishy, T.W., M.D. George and R. Sen Gupta. 1979. Total mercury content in some marine fish from the Indian ocean. Mar. Pollut. Bull. 10:357-360.

MAFF. 1973. Survey of cadmium in food. Fourth Report of the Working Party on the Monitoring of Foodstuffs for Heavy Metals. Ministry of Agriculture, Fisheries and Food. Her Majesty's Stationery Office, London.

Mance, G. 1987. Pollution threat of heavy metals in aquatic environments. Elsevier Applied Science, New York.

Margerison, T., M. Wallace and D. Hallenstein. 1980. The superpoison. MacMillan, London.

Menasveta, P. and R. Siriyong. 1977. Mercury content of several predacious fish in the Andaman Sea. Mar. Pollut. Bull. 8:200-204.

Menasveta, P. and V. Cheevaparanapiwat. 1981. Heavy metals, organochlorine pesticides and PCBs in green mussels, mullets and sediments of river mouths in Thailand. Mar. Pollut. Bull. 12:19-25.

Mochiike, A., T. Matsuo, H. Kanamori, N. Hoshita and I. Sakamoto. 1986. Determination of PCQs by HPLC and its application to the analysis of Yusho patient blood and toxic rice oil and to the distribution of synthetic PCQs in mice. Chemosphere 15:599-606.

Moriarty, F. 1988. Ecotoxicology: the study of pollutants in ecosystems. Second edition. Academic Press, London.

Nauen, C.C. 1983. Compilation of legal limits for

hazardous substances in fish and fishery products. FAO Fish. Circ. 764, 102 p.
Nriagu, J.O. 1988. A silent epidemic of environmental metal poisoning? Environ. Pollut. 50:139-161.
Patel, B., V.S. Bangera, S. Patel and M.C. Balani. 1985. Heavy metals in the Bombay Harbour area. Mar. Pollut. Bull. 16:22-28.
Phillips, D.J.H. 1980. Quantitative aquatic biological indicator organisms - their use to monitor trace metal and organochlorine pollution. Applied Science Publishers Ltd., London.
Phillips, D.J.H. 1984. Regional monitoring of conservative pollutants in Asian coastal waters: the WESTPAC Bio-indicator Programme. Asian Mar. Biol. 1:1-15.
Phillips, D.J.H. 1985. Organochlorines and trace metals in green-lipped mussels, *Perna viridis* from Hong Kong waters: a test of indicator ability. Mar. Ecol. (Prog. Ser.) 21:251-258.
Phillips, D.J.H. 1990. Arsenic in aquatic organisms: a review, emphasizing chemical speciation. Aquatic Toxicol. 16:151-186.
Phillips, D.J.H. 1991. Selected trace elements and the use of biomonitors in subtropical and tropical marine ecosystems. Rev. Environ. Contam. Toxicol. 120:105-129.
Phillips, D.J.H. and K. Muttarasin. 1985. Trace metals in bivalve molluscs from Thailand. Mar. Environ. Res. 15:215-234.
Phillips, D.J.H. and P.S. Rainbow. 1988. Barnacles and mussels as biomonitors of trace elements: a comparative study. Mar. Ecol. (Prog. Ser.) 49:83-93.
Phillips, D.J.H. and S. Tanabe. 1989. Aquatic pollution in the Far East. Mar. Pollut. Bull. 20:297-303.
Phillips, D.J.H., G.B. Thompson, K.M. Gabuji and C.T. Ho. 1982a. Trace metals of toxicological significance to man in Hong Kong seafood. Environ. Pollut. (Ser. B) 3:27-45.
Phillips, D.J.H., C.T. Ho and L.H. Ng. 1982b. Trace elements in the Pacific oyster in Hong Kong. Arch. Environ. Contam. Toxicol. 11:533-537.
Phillips, D.J.H., E.Y.L. Chan, E.K.C. Pang and T.L. Tin. 1986. Trace metals in Pacific oysters (*Crassostrea gigas* Thunberg) marketed in Hong Kong. *In* J.L. Maclean, L.B. Dizon and L.V. Hosillos (eds.) The First Asian Fisheries Forum, Vol. 1. Asian Fisheries Society, Manila.
Qasim, S.Z. and R. Sen Gupta. 1988. Some problems of coastal pollution in India. Mar. Pollut. Bull. 19:100-106.
Reijnders, P.J.H. 1986. Reproductive failure in common seals feeding on fish from polluted coastal waters. Nature (Lond.) 324:486-487.
Sanzgiry, S., A. Mesquita and T.W. Kureishy. 1988. Total mercury in water, sediments, and animals along the Indian coast. Mar. Pollut. Bull. 19:339-343.
Schnare, D.W., M. Ben and M.G. Shields. 1984. Body burden reductions of PCBs, PBBs and chlorinated pesticides in human subjects. Ambio 13:378-380.
Sherlock, J.C. 1983. Dietary surveys on a population at Shipham, Somerset. Sci. Total Environ. 29:12-32.
Simpson, W.R. 1981. A critical review of cadmium in the marine environment. Prog. Oceanogr. 10:1-70.
Sivalingam, P.M. 1984. Ocean disposal and land reclamation problems of Penang, Malaysia. Conserv. Recycl. 7:85-98.
Sivalingam, P.M. and A.B. Sani. 1980. Mercury content in hair from fishing communities of the State of Penang, Malaysia. Mar. Pollut. Bull. 11:188-191.
Sivalingam, P.M. and B. Bhaskaran. 1980. Experimental insight of trace metal environmental pollution problems in mussel farming. Aquaculture 20:291-303.
Spies, R.B. and D.W. Rice. 1988. Effects of organic contaminants on reproduction of starry flounder, *Platichthys stellatus*, in San Francisco Bay, California, USA. II. Reproductive success of fish captured in San Francisco Bay and spawned in the laboratory. Mar. Biol. 98:191-200.
Strigini, P. 1983. The Italian chemical industry and the case of Seveso. UNEP Ind. Environ. (October-December):16-21.
Suckcharoen, S., P. Nuorteva and E. Hasanen. 1978. Alarming signs of mercury pollution in a freshwater area of Thailand. Ambio 7:113-116.
Takeuchi, T. 1972. Distribution of mercury in the environment of Minimata Bay and the Inland Ariake Sea, p. 79-81. *In* R. Hartung and B.D. Dinaman (eds.) Environmental mercury contamination. Ann Arbor Science, Ann Arbor, Michigan.
Takizawa, Y. 1979. Epidemiology of mercury poisoning, p. 325-365. *In* J.O. Nriagu (ed.) The biogeochemistry of mercury in the environment. Elsevier/North Holland Biomedical Press, Amsterdam.
Tanabe, S., R. Tatsukawa and D.J.H. Phillips. 1987a. Mussels as bioindicators of PCB pollution: a case study on uptake and release of PCB isomers and congeners in green-lipped mussels (*Perna viridis*) in Hong Kong waters. Environ. Pollut. 47:41-62.
Tanabe, S., N. Kannan, A. Subramanian, S. Watanabe and R. Tatsukawa. 1987b. Highly toxic coplanar PCBs: occurrence, source, persistency and toxic implications to wildlife and humans. Environ. Pollut. 47:147-163.
Tanabe, S., N. Kannan, T. Wakimoto, R. Tatsukawa, T. Okamoto and Y. Masuda. 1989. Isomer-specific determination and toxic evaluation of potentially hazardous coplanar PCBs, dibenzofurans and dioxins in the tissues of "Yusho" PCB poisoning victim and in the causal oil. Toxicol. Environ. Chem. 24:215-231.
Taylor, D. 1982. Minimata disease. Environ. Sci. Technol. 16:81A.
Thompson, H.C., D.C. Kendall, W.A. Korfmacher, K.L. Rowland, L.G. Rushing, J.J. Chen, J.R. Kominsky,

Rowland, L.G. Rushing, J.J. Chen, J.R. Kominsky, L.M. Smith and D.L. Stalling. 1986. Assessment of the contamination of a multibuilding facility by polychlorinated biphenyls, polychlorinated dibenzo-*p*-dioxins, and polychlorinated dibenzofurans. Environ. Sci. Technol. 20:597-603.

Tsuchiya, K. 1977. Various effects of arsenic in Japan depending on type of exposure. Environ. Health Perspect. 19:35-42.

USCDC. 1985. Morbidity and mortality week report. U.S. Centers for Disease Control 34(36).

Webb, K. 1983. Indonesia has Minimata disease. World Environ. Rep. (November):5.

Winter, M., M. Thomas, S. Wernick, S. Levin and M.T. Farvar. 1976. Analysis of pesticide residues in 290 samples of Guatemalan mothers' milk. Bull. Environ. Contam. Toxicol. 16:652-657.

Wolff, E. 1990. Evidence for historical changes in global metal pollution: snow and ice samples, p. 205-217. *In* R.W. Furness and P.S. Rainbow (eds.) Heavy metals in the marine environment. CRC Press, Boca Raton, Florida.

Yamagata, N. and I. Shigematsu. 1970. Cadmium pollution in perspective. Bull. Inst. Public Health, Tokyo 19:1-19.

Zingde, M.D. and B.N. Desai. 1981. Mercury in Thana Creek, Bombay Harbour. Mar. Pollut. Bull. 12:237-241.

Discussion

BOGEL: You have identified certain categories of risk: real risks that have been identified and quantified; theoretical risks, where the risk is deduced from other, terrestrial incidents and their data on tissue analysis, etc.; and also hypothetical risks where data are really lacking and we can only speculate. This seems to be a useful categorization for future discussion and recommendations. In particular, the importance of hypothetical risks must not be underestimated. Such risks may be real but as yet we do not have the machinery to investigate them. Heavy metals have been thoroughly investigated in German agriculture by sampling liver, kidneys, etc. from cows and analyzing for trace metals. The huge databank from this study was analyzed by computer and showed high lead levels near (say about 50 m either side) busy roads. So lead was correlated with proximity to traffic and traffic density. But the exact sites for sampling are very important. For example, some samples from lands adjacent to the Hanover-Dusseldorf highway showed lower levels. Samples to be compared *must* be taken at the same distance from the highway. Mercury was always correlated with human population density.

PHILLIPS: One thing that I omitted to mention about lead is that there is now good evidence for long-distance contamination through air masses for lead. Studies by Russ Flegal and co-workers in the USA[1,2], using isotopic ratios (particularly Pb^{206}/Pb^{207}) have shown, for example, that there is extremely good traceability of lead from Asia into the Pacific.

MUNRO: Concerning the use of agricultural chemicals with plantation crops (sugar, cocoa, etc.), can these affect farmed aquatic organisms? Are they detectable? Are they accumulative?

D. PHILLIPS: The chemicals used have changed over the last 10-15 years. However, in some developing countries, some of the organochlorine compounds that have been banned in developed countries are *still* in use. Even use of DDT is surprisingly high. One problem now is that a large number of chemicals have been developed for specific purposes and one would need data on each, to determine its properties in the environment. Some groups, like Professor Donald Mackay's[3], are trying to get around this problem by using generalized approaches - such as octanol water partition coefficients - a general description of whether a chemical is hydrophobic or hydrophilic.

There is a good paper by Clark and co-workers (1988)[4] dealing with the concept of fugacity. This is not a new concept but the way that the Mackay group uses it *is* new. Such generalized approaches will be used more and more. In developing countries, the lack of data on what is being used is a serious problem. It means that a lot of screening has to be done for different compounds that may be there. This is very costly. A lot of expenditure on analyses could be saved if reliable information on chemical usage was available.

PULLIN: Can you please define the term 'fugacity'?

D. PHILLIPS: In this context, basically it concerns the propensity of a chemical to enter the living component of the environment and to have an elongated lifespan in that component.

MUNRO: Are there any other specific chemicals which are liable to become accumulative, like DDT? Can dioxins accumulate in living organisms?

D. PHILLIPS: Oh yes, and organochlorines because they are hydrophobic and lipophilic and therefore sit in the lipid components of organs. Many of these chemicals have quite long half-lives in living organisms both in terms of their degradation *per se* and their rates of excretion. This applies to all organochlorines, some toxic trace metals and some herbicides.

MORIARTY: There are many misconceptions about the movement of pesticides along foodchains, especially for aquatic systems. DDT is often found in higher concentrations at higher levels in a foodchain, but this can be independent of position in the foodchain. Reinert did a beautiful experiment on this with an alga, a crustacean and fish in a tank, all exposed to dieldrin[5]. The alga acquired a high concentration, the crustacean higher and the fish highest of all. When the organisms were kept in separate tanks, he got the same results! The significant uptake by all was from the water.

D. PHILLIPS: Well, the weight of evidence shows that there is a dynamic equilibrium between the concentrations of hydrophobic chemicals in biota and those in their ambient surroundings. Whether this involves take up from the water or in food or both is largely irrelevant. What you get is what you get. At higher trophic levels, such as whales and other marine mammals there *is* clearly foodchain transfer of chemicals though maybe not always 'foodchain amplification'. In summary the concentration of say an organochlorine that one finds in an organism is simply the result of the dynamic partitioning of that chemical between the organism and its environment. I agree, however, with Dr. Moriarty that the concept of foodchain amplification of chemicals *per se* is almost entirely wrong.

EDWARDS: Some of the 100,000 or so manmade chemicals in current use are getting into sewage, which can be reused for aquaculture. Separation of industrial effluents from sewage is often impractical. This could pose more public health problems than even microbial diseases for consumers of sewage-raised fish. For example, the Calcutta wastewater-fed aquaculture systems are claimed to be free from industrial wastes, but this is clearly not true. Hundreds of factories there discharge wastes into the wastewater scheme. These are probably mostly small backyard industries, but they can be highly polluting. Preliminary studies indicate that heavy metal concentrations in Calcutta sewage are still relatively low because of limited industrial discharge, but this situation could change in the future.

D. PHILLIPS: I see the problem. In the developing countries, it would make sense to reuse the nutrients in sewage to raise fish but, on the other hand, industrial wastes should be kept away from the cultured organisms. Every sewage stream that I have ever looked at has some industrial effluents entering it. In rare cases, the concentrations of trace metals are too low to pose any problems for fish cultured downstream. In most cases, however, there are significant concentrations of organic and inorganic chemicals, derived from industry, which would pose major problems for some types of farmed aquatic organisms, particularly if there is little dilution. We could attempt to generalize as to which compounds would cause the most problems for certain finfish, molluscs and crustaceans, but note that good information about these compounds in sewage streams in developing countries is very scarce. Moreover, the concentrations vary greatly with time for each stream. There can be shock loadings.

EDWARDS: This is true. It probably means that sewage reuse in aquaculture must involve some waste treatment.

D. PHILLIPS: Not necessarily. It may mean this or perhaps just adequate dilution.

MUNRO: Surely this means that such reuse of these effluents in aquaculture will always be very risky. They could contain almost anything from industry. Rural effluents are presumably just of domestic origin. For example, the river entering Saguling reservoir, near Bandung, Indonesia, carries industrial wastes as well as domestic sewage. It is black in color. The fish cage units in the reservoir are therefore exposed to this.

RUDDLE: The distinction between urban and rural economies is becoming less and less clear. Effluents in the countryside are now rarely from domestic sewage alone.

BURAS: There is now increasing information on the retention time and biodegradation of chemicals. This can help in planning the reuse of effluents in fishponds so that the fish are exposed to biodegraded compounds, not the more toxic forms.

KING: Regarding international cooperation, could programs such as the regional seas program of UNEP, for example its South Pacific Environmental Programme (SPREP), be used for further studies?

D. PHILLIPS: I don't know the status of programs like SPREP regarding information on industrial discharges, but this should be explored.

---

[1]Flegal, A.R. and V.J. Stukas. 1987. Accuracy and precision of lead isotopic composition measurements in water. Mar. Chem. 22: 163-177.

[2]Flegal, A.R., K. Itoh, C.C. Patterson and C.S. Wong. 1986. Vertical profile of lead isotopic compositions in the north-east Pacific. Nature (London) 321: 689-690.

[3]Department of Chemical Engineering and Applied Chemistry, University of Toronto, Toronto, Ontario M5S 1A4, Canada.

[4]Clark, T., K. Clark, S. Paterson, D. Mackay and R.S. Norstrom. 1988. Wildlife monitoring, modeling, and fugacity. Environ. Sci. Technol. 22: 120-127.

[5]Reinert, R.E. 1972. Accumulation of dieldrin in an alga *(Scenedesmus obliquus), Daphnia magna,* and the guppy *(Poecilia reticulata).* J. Fish. Res. Board Can. 29: 1413-1418.

Discussion and Recommendations on Aquaculture and the Environment in Developing Countries*

COMPILED BY R.S.V. PULLIN
International Center for Living Aquatic Resources Management
MCPO Box 2631, 0718 Makati
Metro Manila, Philippines

PULLIN, R.S.V., Compiler. 1993. Discussion and recommendations on aquaculture and the environment in developing countries, p. 312-338. *In* R.S.V. Pullin, H. Rosenthal and J.L. Maclean (eds.) Environment and aquaculture in developing countries. ICLARM Conf. Proc. 31, 359 p.

Abstract

This is a compilation of information from tape recordings of a discussion on aquaculture and the environment in developing countries. The main topics discussed were the public health issues associated with aquaculture operations and aquatic produce, with particular emphasis on safe levels of contaminants; special problems associated with bivalve molluscs and crustaceans; safe reuse of wastewater in aquaculture; interactions between aquaculture and the environment, particularly in mangrove and other coastal areas; the impact of aquaculture on natural habitats and their biota; aquatic diseases; mechanisms for future action; and guidelines for policymakers and projects. Recommendations for research and for aquaculture development are provided. A schema for decisionmaking on aquaculture development is presented.

Introduction

This discussion was compiled from tape recordings made during round table discussions at the Bellagio Conference on Aquaculture and the Environment in Developing Countries. The compiler worked on the premise of retaining all potentially useful information, whether thoroughly referenced or not. Moreover, the sources of information have not been attributed to individuals, unlike the format used for transcribing question and answer sessions following specific presentations at the conference. The result is a compilation that although not thoroughly referenced, strives at least to record all the information contributed. It is intended to serve as an adjunct to the main contributions to the conference. The order of topics is not intended as a priority ranking.

Public Health

General Considerations

It is essential to consider the public health aspects of aquaculture as part of the *total resource system* of which aquaculture, its policymakers and practitioners (effectors), its beneficiaries and other affected persons form parts: in other words, a holistic, ecological approach is essential.

*ICLARM Contribution No. 826.

Public health issues in aquaculture concern many different groups - farm workers, transporters, sellers and processors of farmed aquatic produce - not just consumers; for example, the peeling of shrimps is the process that affords the greatest risk of contamination with *Salmonella*. All affected groups depend on *services*, ranging from local to national, to guarantee the safety of farmed aquatic produce. Such services include the formulation and implementation of quality control regulations, monitoring, inspection and certification. Education about public health and aquatic produce is also important.

Aquaculture is a fast-growing food production sector in a few developing countries and many more would like it to be so but generally lack effective legislation and clear policies on aquaculture development, the safety of aquatic produce and safeguards for fish farm workers, fish handlers, processors, sellers and consumers. In many developing countries, aquaculture is regarded as a subsector of fisheries and responsibility for aquaculture is vested in a Fisheries Department or equivalent, usually lacking adequate resources for working towards effective quality control of aquaculture produce.

The International Labor Organization has produced a book on Occupational Hazards in Agriculture (ILO 1983). There is nothing similar for aquaculture but a few developed countries (e.g., Norway, UK) have begun to consider the safety needs of workers on marine cage farms.

Codes of practice for fish and fisheries products have been elaborated by the Codex Alimentarius Commission, joint FAO/WHO Food Standards Programme (FAO 1982). A draft Code of Hygienic Practice for the Products of Aquaculture has been prepared by the FAO Fish Utilisation Service and circulated to member-countries (FAO 1991). The consolidated draft will be considered by the Codex Alimentarius Commission on Fish and Fishery Products. There are many works on aquatic produce quality and general hygiene (e.g., Kramer and Liston 1987). All relevant codes of practice and regulations should strive to maintain a balanced attitude between strict safeguards and realistic, attainable standards that will not unreasonably constrain or prevent development - especially for those in developing countries who could benefit greatly from aquaculture. For export produce, the application of codes of practice for quality control can have political as well as technical implications. The rejection of shipments from developing countries can have catastrophic effects on producers.

Producers: Healthy Approaches to Aquaculture Development

The nutrition and health of fish farm workers are interdependent. Aquaculture development and innovations (and indeed interventions of any kind in the agrarian systems of developing countries) must not cause economic shifts or changes in access to resources that will disturb or threaten basic household nutritional requirements. There is, of course, potential to *improve* the nutrition of farming families through aquaculture: they can consume small or otherwise less marketable produce - as in agriculture. However, new technology is never neutral and policymakers and planners must assess the possibility of negative effects.

Many families in existing and potential fish farming areas in developing countries are extremely poor and are on the margin with respect to protein-energy malnutrition. The increasing acceptance of the 'Farmer First' approach (Chambers et al. 1989), rather than top-down technology transfer will lessen threats from new interventions to the nutrition and health of farming families. Farmers must be asked for *their* perspectives on food production for livelihood and for family food security. For example, where land holdings are very limited (a common situation in developing countries) it may be inappropriate to try to persuade farmers to build ponds and thereby eliminate or reduce the capacity to grow crops for the families' own security. There is little or no margin for error, especially where climatic and other uncertainties threaten family food security, as in subSaharan Africa. An interactive combination of on-station and on-farm research, emphasizing the involvement of farmers at all stages of conceptualization and testing, is the best way to ensure that new developments avoid negative effects.

The same 'Farmer First' approach should be used for innovations such as use (and potential abuse) of new chemicals. For example, 'phostoxin' a compound developed for fumigation in godowns and which releases phosgene gas (mustard gas) when wet, has been widely used in fishponds in Bangladesh as a piscicide. Two tablets are generally used per pond. Fish killed are safe to eat but this compound, although cheap and easy to use, may cause harm to people if mishandled. This is a similar situation to the use of cyanide in coastal and inland fisheries in some tropical developing countries, though there the operations are not only risky but also illegal.

Consumers and Product Quality

Consumer acceptance of aquaculture produce is important in the rural and urban areas in developing countries and also for the developed countries that import their produce. In most market situations in developing and developed countries, consumers are generally not aware of the sources of origin. They can usually judge 'freshness' but can rarely assess the other characteristics of produce that relate to its quality and safety. Moreover, in developing countries, consumers are generally uneducated about the aspects of contamination of aquatic produce to which the codes of practice, formulated by developed countries and international organizations, refer. This means that such codes of practice are at present unworkable in most developing countries.

In developed countries, consumers are more educated about the safety and quality of aquatic produce. Their education was started by government agencies, perhaps 30-40 years ago, and now the *companies* that ship and sell aquatic produce continue to inform their customers about product quality and safety, through advertising and instructions on packages, etc. Developing countries could follow the same educational route, but would need adequate resources for their governments to make a start and would be hampered by low literacy levels and other communication difficulties. Moreover, most companies join such educational efforts only when it helps them competitively.

In some developed countries, consumers' perceptions, often encouraged by the media, may become biased in favor of wild-harvested or 'free-range' produce over intensively farmed produce. Whether this has any real basis in respect of differences in organoleptic qualities or not, it can affect production, advertising and sales. It will also increasingly affect developing-country producers aim their produce at export markets. The environmental conditions under which they raise their produce will increasingly affect its image.

In the long term, fish farmers in developed and developing countries could turn this perceived difference between wild and farmed organisms into a marketing advantage by showing that they can develop farmed breeds that are demonstrably better than wild fish; for example, in shape, dressing weight, color, taste, keeping qualities, etc. In other words, the differences could be emphasized and the point made that farmed breeds can be changed to improve and ensure quality whereas wild fish cannot. It is likely that wild-caught and farmed aquatic organisms will always have different images for most consumers.

The relative preferences of consumers can change rapidly. For example, in Italy and some other parts of Europe, consumers used to prefer wild-caught to farmed eels. However, environmental pollution (mercury in the River Elbe, heavy metals in the Bight of Gdansk/Danzig oilspills, etc.) and moves to grow fish in the effluents of nuclear power stations have brought, over the last five years, a swing in consumer preference for eels farmed in well-managed, safe farming environments. These are now more expensive than wild-caught eels. Similar rapid swings, associated with image changes, can be expected where new exotic species and products are marketed in competition with traditional produce. Aquaculture should be able to supply produce that is of guaranteed high quality and is tailored to consumer preferences.

Coastal Aquaculture and Public Health

GENERAL APPROACHES TO SAFE LEVELS OF CONTAMINANTS

Site selection for coastal fish farms is extremely important. In many cases, it will determine whether or not the produce can meet quality control criteria, including those relevant to public health. Many coastal dwellers in developing countries will continue to consume farmed aquatic organisms that fail to meet such criteria. These consumers can build up some resistance to some of the pathogens present in farmed aquatic organisms, but remain at risk.

Local consumption of seafood from areas like Manila Bay, Philippines, Deep Bay, Hong Kong and parts of the Inner Gulf of Thailand undoubtedly exposes the consumers to high levels of contaminants—especially microorganisms. However, despite this and the general absence of epidemics, there is no convincing epidemiological evidence that infective doses of pathogens or contaminants are significantly different between developing- and developed-country nationals. Indeed the same public health and safety criteria *must* be applied to all. For example, there is no rationale for lowering standards for trace metal contamination of seafood destined for developing-country consumers. Moreover, even where there is local 'tolerance' to some lower quality produce, lowering standards would put foreign visitors and other transients at risk.

Conservative standards are in fact essential for marine farmed aquatic produce because such produce, particularly bivalves, *is* known to have been responsible for human sickness and deaths.

Back-end depuration of farmed bivalves in developing countries remains the exception rather than the rule. It is costly and can also cause weight losses and reduced shelf-life of products. Where there are possibilities of health risks, these should be considered. However, selection and protection of pollution-free sites for bivalve farming are really the only long-term, cost-effective options for most developing countries.

Some of the acceptable limits for contaminants in farmed aquatic produce in the developing countries may also need revision. For example, the World Health Organization (WHO) recommended limit for mercury in fish is 0.5 $\mu g \cdot g^{-1}$ wet weight. This is based on an average adult human weight of 70 kg. However, Asian adults (both sexes) weigh on average 55 kg and children everywhere weigh significantly less.

F.A.J. Armstrong studied Eskimo populations that were exposed to mercury intake in fish and marine mammals over a long time and concluded that the 0.5 $\mu g \cdot g^{-1}$ limit was too conservative and that it could be raised to 1.0 or even 2.0 $\mu g \cdot g^{-1}$. The majority opinion now, however, is that the 0.5 $\mu g \cdot g^{-1}$ wet weight limit is probably about right and allows some safety margin (Hansen 1990). If the aim of any limit is to protect all or nearly all of the population, then a safety margin is essential. Its breadth is a matter for risk-benefit analysis. For example, it would be unreasonable to set limits that would protect from mercury toxicity a person who had chosen to eat swordfish as his/her only protein source (for example, in an extreme attempt to diet) but that would also preclude others from consuming such fish in moderate quantities out of preference or economic necessity.

On a related matter, there is evidence that human paralytic shellfish poisoning symptoms occur below the current worldwide safety level of 80 µg/100 g in the case of the causative dinoflagellate *Pyrodinium* in the Philippines (N.I.S. Pastor, pers. comm.).

The overall problem remains that the range of harmful contaminants, from viruses and bacteria to phycotoxins, industrial chemicals and trace metals, is very wide. Moreover, many of them have interactive effects on consumers. A scheme that sets scientifically based and reliable safe limits for all contaminants would be very difficult and expensive to develop. Alternative and complementary approaches, such as probability matrix techniques - for example, see Piedrahita and Tchobanoglous (1987) - should be further developed and tested.

Some farmed produce may contain dangerous levels of contaminants such as phycotoxins, trace metals and some microorganisms. For example, the kidneys and livers of terrestrial animals (cows, pigs) sometimes contain very high cadmium levels above the permissible limits. For aquatic produce, it must be emphasized that cooking by consumers, particularly boiling, does *not* necessarily render it safe. Any public health schemes that rely only on education of consumers to cook aquatic produce, however thoroughly, will fail.

The pathways of contaminants, including microorganisms, from farming to human consumption and other uses of aquatic produce are not well known. Such pathways need to be studied for *all* farmed aquatic produce (seaweeds to molluscs, crustaceans, other invertebrates and finfish), for tissues that are eaten raw as well as for processed items.

Monitoring of the safety of farmed aquatic produce is expensive, but robust monitoring and inspection facilities should be established, especially for microbial and trace metal contaminants in bivalve molluscs destined for domestic consumption. Public health is too important for governments to neglect this responsibility or to retain an *ad hoc* approach as at present. The cost can and should be passed on to the consumers for some products but for others - for example, green mussel *(Perna viridis)* in Asia - the profit margins may be much too narrow to permit this. If so, the ultimate solution, where the safety of produce cannot be assured and paid for, will be to farm less risky organisms such as finfish. An alternative approach is the 'Polluter Pays' principle - which has great appeal and is in line with modern attitudes to recognizing the effects of such externalities in development (as for example in new projects promoted by the World Bank). However, it has yet to be seen to work well even in the developed countries.

MOLLUSCS

Molluscs, particularly bivalves, are a commodity group with special characteristics with respect to safety and quality of produce and consumer acceptance. Unlike finfish, which do not normally accumulate dangerous levels of contaminants in the parts of them (usually muscle) eaten by humans and for which the culture environment (especially fishponds) often provides some homeostasis and attenuation of contaminants, bivalve molluscs feed and grow by filter-feeding on microorganisms and are grown almost exclusively in waters the quality of which cannot be controlled. Bivalves selectively accumulate some microorganisms and chemical contaminants, including heavy metals.

Because of this and consumer awareness of the potential problems, the production of cultured bivalves has recently levelled off or declined in many countries. The advice given to bivalve consumers in developing countries, where episodic toxic algal blooms occur, is often of limited use; for example, in some parts of Southeast Asia, adults are advised not to eat more than "a small amount at any one time", and children not to eat "any" bivalves when red tide organisms are present.

There are some guidelines on quality standards for shellfish including molluscs, such as WHO (1990) but there is an urgent need for updating these. Procedures such as the depuration and irradiation of contaminated produce are simply not affordable in many areas of production in developing countries. Moreover, depuration may not reduce all contaminants to acceptable levels even if prolonged. For most developing-country situations, selection of pollution-free sites and their protection are the obvious answers.

The tragedy of the constraints that contaminants pose for bivalve farming is that here is a huge potential source of animal protein that could be grown cheaply, using natural productivity; but its safety, quality and consumer acceptance are very difficult to ensure.

Bivalve mollusc beds are also often 'self-polluting' because of fecal buildup and siltation. This reduces yields and growth rates of a range of species; for example, oysters and scallops in Japan (see Csavas, this vol.). Water quality is also important; for example, in *Crassostrea gigas* culture beds in Deep Bay, Hong Kong, the growth time to market size was extended to four to seven years, compared to two years for culture with good water quality. These environmental problems are significant in other bivalve culture areas (e.g., Arakawa et al. 1971; Coeurdacier et al. 1983; FAO 1988; Mito and Fukuhara 1988).

CRUSTACEANS

Farmed crustaceans (which at present means almost entirely penaeid shrimps and *Macrobrachium* spp.) are very different from bivalve molluscs in their uptake of contaminants that can threaten public health. The greatest risks come from the processing procedures, not from uptake of contaminants during growout. Some producers allow farmed crustaceans to void their guts before processing, but this is not common.

The processing procedures for shrimp produced in Ecuador for export to the USA follow the Food and Drug Administration (FDA) guidelines. The main product is frozen shrimp. In 1983, there was an El Niño and blooms of cyanobacteria. Most of the ponds had a salinity of only about 4 ppt. The algae caused an off-flavor in the shrimps and shipments were seriously affected. The additional fear now is that contamination with human pathogens, like *Vibrio parahaemolyticus*, might occur. The problem is that it can take about ten hours to transport harvested shrimp from farms to packing plants. The packing plants send trucks and the shrimp are transported to the plants on ice. Their guts are removed ('deveined') at the plant, an FDA requirement. The FDA has been strict on rejection of shrimp shipments said not to meet their standards: in the opinion of some amounting to an 'overkill' attitude. Producers and processors greatly fear rejection of shipments and their quality control procedures have been improving rapidly since the early 1980s, especially in Asia, to guard against this.

SEAWEEDS

Some of the problems outlined above for molluscs and crustaceans apply also to farmed seaweeds, but the pathways of contaminants from the farm environment through to final products are imperfectly known. Seaweeds are known to accumulate trace metals and there is scope for microbial contamination in farming, processing, shipping and product storage. More research on this is needed.

Recommendations on Public Health

- Aquaculture must not cause unacceptable risks to public health or harm that negates the benefits of the improved livelihood and nutrition that its adoption can bring. It should be subject to similar controls and safeguards as are applied to agriculture, irrigation and other aspects of food production and handling. This concerns all persons affected by and involved in aquaculture, and to other users of waters in which aquatic organisms are farmed or which are affected by aquaculture: the farm workers, handlers and processors, sellers and consumers of farmed aquatic produce.
- The health of aquaculture workers and other users of waters in which aquatic organisms are farmed or which are affected by aquaculture must be safeguarded. There must be effective measures against exposure to waterborne pathogens and parasites, to toxic chemicals used on aquatic farms, and against reduction of water quality for purposes such as domestic supply, watering crops and livestock, washing clothes and utensils, bathing, sports and recreation. Examples of possible health risks to such persons include exposure to waterborne diseases, such as bilharzia and leptospirosis, and to chemicals, such as trace metals, pesticides, disinfectants, antibiotics and hormones; and the risks of working in the aquatic medium, including exposure to harsh climatic conditions and drowning.
- All available information on health risks from aquaculture should be compiled as a statistical database with the involvement of appropriate international organizations, such as WHO and ILO, so as to evolve guidelines and codes of practice. This database should be regularly updated as aquaculture expands and experience is increased.
- In order to mitigate or avoid public health risks from aquaculture development including consumption of aquatic produce, more research is needed on the real or perceived role of aquaculture operations and their produce in human disease transmission and the risk of chemical and microbial contamination. In particular, population-wide studies are required, especially in developing countries, of exposure to mercury, cadmium, organochlorine pesticides, polychlorinated biphenyls (PCBs) and dioxins. Consumption of aquatic produce may be an important source of exposure. Monitoring should employ noninvasive techniques (analysis of hair, urine, or breast milk - as appropriate) in addition to autopsy-derived samples, when available.
- Research is also needed on better methods for aquatic disease prophylaxis and control as an alternative to massive use of chemotherapeutants.
- Regarding the safety of aquatic produce, monitoring programs for trace contaminants, bioactive compounds and microorganisms are needed. Research and development programs are required to improve the tests available for ensuring the safety of aquatic produce. The present tests for biotoxins, for example, are inadequate to afford an acceptable degree of protection, even when widely employed. Very few of the many trace contaminants which may be present in aquaculture produce are monitored on a regular basis.

Recommendations Concerning Harmful Algae

There is clear evidence that hypereutrophication of coastal waters increases primary productivity and the occurrence of algal blooms, but statistically significant data for a global increase in harmful algal events are lacking. Such events are almost impossible to predict. This greatly hinders risk assessment, the cost-effectiveness of monitoring programs and the evaluation of potential adverse impacts.

- In all shellfish culture areas where phycotoxin accumulation may be a threat to public health, routine detection and quantification of phycotoxin levels in farmed aquatic produce should be undertaken using well-established methods. Monitoring the presence of harmful algal species is also useful for public health purposes but must not replace the testing of produce for phycotoxins.
- The possibilities for the use of established and newly developed methods of detecting and measuring phycotoxins in the farmed aquatic produce of developing countries should be thoroughly assessed.
- For coastal aquaculture, farm sites should be chosen where harmful algal events are rare or absent. The possibility of relocating cage farms away from affected areas should also be considered.
- Finfish farm management strategies should include routine microscopic analysis of water samples for identifying harmful species in addition to gross measurements of water transparency or phytoplankton density by Secchi disk or chlorophyll measurements. Such monitoring need not be elaborate, time-consuming or expensive. When combined with actions designed to minimize losses in the presence of harmful algal events - such as cessation of feeding, towing cages to an unaffected area, and preemptive harvesting - it can be very cost-effective. This recommendation applies at present to coastal farms rather than inland (freshwater) farms where comparable harmful algal events have as yet not occurred.
- Research is urgently needed on the prevalence and ecotoxicology of cyanobacterial toxins in the freshwaters used for aquaculture in developing countries.
- Multidisciplinary research is needed to understand better the ecophysiology of harmful algal events in coastal areas used for aquaculture. This should include evaluation of relevance for application in developing countries of the monitoring protocols and management strategies developed by the ICES Working Group on Plankton and others who have considered approaches to understanding and mitigating the effects of these events in developed countries.

Excreta Reuse in Aquaculture

WASTEWATER-FED AQUACULTURE

With increasing urbanization, there is a need to consider how to reuse human excreta, particularly wastewater, in aquaculture. Important systems already exist in Asia (Edwards 1985; Edwards and Pullin 1990). Can such systems be operated so as to safeguard public health? This is likely to be possible in some developing-country locations, but not in all. For some resource systems, it will be possible to meet safety thresholds with respect to human pathogens and pollutants (such as trace metals) whereas in other locations this will be impossible for technical or economic reasons or both. The setting of threshold levels for microorganisms and chemical contaminants in farmed aquatic organisms and in the water in which they are farmed may be useful approaches. However, the entire management process on waste-fed fish farms must be geared to raising produce that is recognized as safe and accepted as such by consumers.

Produce from aquaculture usually enters the same marketing chains as produce from capture fisheries. This applies to produce from wastewater-fed systems, as in Calcutta, as well as to other aquaculture systems. Sometimes the origins of the produce are disguised.

THE PROBLEM OF ADMIXTURES OF INDUSTRIAL WASTES

The mixing of industrial and agroindustrial wastes with domestic wastewater poses a very serious constraint to the reuse of the latter in aquaculture. Ideally, industry, agroindustry and other polluting enterprises should be required to treat and render safe their own wastes, but this has not been and will not be possible everywhere. Where there are significant admixtures of industrial and agricultural chemicals to wastewater, its reuse in aquaculture may be precluded because of the risks of contamination of the systems and their produce with trace metals and industrial chemicals toxic to fish and humans. However, given that raw sewage *must* be treated before use in aquaculture, channelling pretreated effluents to fishponds could augment chemical biodegradation; for example, by the action of fishpond bacteria. Where any such mixing of industrial, agricultural and domestic wastes occurs in existing systems or would occur in a newly proposed system, thorough studies should be made to assess these risks. Moreover, the effect of industrial wastes on fish, the pathways of such wastes through aquatic systems and their metabolism by fish, all require much greater study.

FUTURE REQUIREMENTS

The lack of adequate information and proven risk assessment tools at present makes it very difficult to give a scientifically based answer to the question of whether or not a given source of wastewater is appropriate for safe reuse in aquaculture. Formulation of a code of practice for wastewater reuse in aquaculture is needed, but this will depend upon further research because only then can accurate *numbers* be specified rather than using words like "significant" quantities.

Such a code of practice must enable the understanding and quantification of the pathways of wastewater chemicals and biota from their *original* sources through aquaculture reuse systems and ultimately to the consumers of aquatic produce. This will require studies encompassing everything from the original processes that produce wastewater and other wastes mixed with them, through any treatment processes, to the aquaculture reuse systems themselves and ultimately to measuring contamination in fish in the marketplace and in the homes of consumers.

Where limited research resources are available, the greatest and most rapid understanding of potential health risks often comes from studies on the final product. This has been the general experience of WHO in aspects of food safety. Clearly not all farmed aquatic produce can be sampled or monitored. This would be incredibly expensive. Some regulatory authorities, such as the FDA in the USA, avoid this by making inferences about the safety of produce because of knowledge about the status of the 'environment' in which it is produced; for example, some coastal areas in the Republic of Korea are designated as 'clean' for shellfish production and are periodically reassessed. None of these approaches, however, helps the planning of and investment in *new* enterprises where data on the production environment and final product are *de facto* not available. Investment needs protection whether it is from the small-scale private farmer or from large corporate, urban or community waste treatment and reuse concerns. Therefore, information guidelines and a code of practice that will have a *predictive* function for new enterprises are as important as the monitoring, regulation and (where necessary) improvement of existing enterprises. These are all lacking at present. It is also highly relevant that many developing countries lack adequate wastewater treatment and disposal systems.

There is an ongoing UNDP-World Bank campaign to improve this situation and the option of reuse through aquaculture can help to focus more attention on the needs and opportunities associated with wastewater as a resource, including the potential profits from reuse through aquaculture. WHO has recommended that wastewater reuse options should be considered in *all* attempts to improve its treatment and disposal in developing countries.

Recommendations on Wastewater Reuse in Aquaculture

The term 'wastewater' is used here in a generic sense to mean human excreta: whether fresh ('nightsoil'); in the form of sewage or wastewaters in the narrow sense (excreta, with added water to facilitate waterborne transportation); or other partially treated forms such as septage.

- The WHO recommendation (WHO 1989) that wastewater reuse should always be considered in schemes to improve sanitation is endorsed.
- More research on the public health aspects of existing wastewater-fed aquaculture systems is essential before they are further developed or promoted elsewhere. Such studies should include assessments of the risks and benefits to producers and consumers in addition to pathogen removal. Pathogen removal may occur in pretreatment and treatment processes before reuse of wastewater in aquaculture and during such reuse. Thereafter, aquaculture produce may also be treated for pathogen removal.
- Given that the scientific basis of wastewater-fed aquaculture was developed in Germany from the late nineteenth to the early twentieth century, culminating in the establishment in the late 1920s of the Munich sewage-fed fishponds (which are still in operation today), the voluminous German literature on the development of wastewater-fed fish culture should be comprehensively reviewed and the review published in English to facilitate global awareness and access.
- In wastewater-fed fish culture systems, the presence of industrial effluents should be closely monitored and minimized as much as possible. This is particularly critical for certain contaminants which may accumulate to relatively high concentrations in aquaculture produce; for example, mercury and hydrophobic pesticides.
- Wastewater should never be used for aquaculture without pretreatment. Fish should not be stocked in a wastewater-fed pond until it has an established and relatively stable community of plankton. This is important for pathogen removal. For raw sewage, a minimum detention time of 8-10 days in an anaerobic pond is recommended to remove settleable pathogens.
- A tentative critical density of $10^5 \cdot ml^{-1}$ total bacteria should not be exceeded in wastewater-fed fishponds, except locally for a few hours during loading of pretreated wastewater.
- Loading of wastewater into fishponds should be suspended for two weeks prior to fish harvest to eliminate *Cryptosporidium*. After harvest from wastewater-fed systems, fish should be held for at least a few hours in a suitable, isolated clean waterbody to facilitate evacuation of their gut contents.
- The threshold concentration of total bacteria in the muscle of fish harvested for human consumption from wastewater systems should not exceed $50 \cdot g^{-1}$. *Salmonella* should be absent.

- Viable eggs of human trematode parasites should be absent from wastewater-fed fishponds. Nightsoil should be stored for two weeks to eliminate such eggs before reuse in fishponds. For bilharzia (*Schistosoma* spp.) control, a carefully designed package of chemotherapy, health education, improved sanitation and snail control in fishponds and other waters is recommended.
- Vegetation in wastewater-fed ponds should be kept to a minimum to reduce the breeding of insect and other disease vectors and intermediate hosts.
- The health and safety of farm workers in wastewater-fed aquaculture should be given particular attention.
- For consumers, good hygiene should be promoted at all stages of handling and processing fish (evisceration, washing, cooking) as these can be major sources of infection.
- The culture of molluscs in wastewater-fed systems is not advisable because of their propensity to accumulate large quantities of contaminants (for example, metals and pesticides) even when these are present only in trace amounts in the wastewaters employed. Finfish may be cultured under conditions where the culture of molluscs is inadvisable, due to the lower propensity of finfish muscle to accumulate certain contaminants. This should not, however, be accepted as a general rule. Each case should be considered on its merits.
- Culture of luxury aquatic produce, such as crustaceans and high value finfish, in wastewater-fed systems is not recommended as it may damage product image and cause marketing problems, especially for export produce. This could harm other existing or future aquaculture operations in developing countries.
- In societies in which the direct consumption by humans of fish cultured in wastewater-fed systems is socially unacceptable, wastewater reuse in aquaculture could be designed to produce fish or other organisms for use as high-protein components of feed for other animals, including carnivorous fish.

Recommendations on Reuse of Industrial and Agricultural By-Products in Aquaculture

By-products from industry, agriculture and agroindustrial processes are resources with potential for reuse in aquaculture. These include waste heat, effluents and solid residues.

- Few opportunities exist in developing countries to use industrial waste heat effluents for aquaculture. Since temperature is generally high in tropical and subtropical regions, the utility of waste heat to maintain temperatures slightly above ambient is usually of short duration and seasonal. If such use is considered to increase growth or in the management of broodstocks and hatcheries, these effluents need to be free from or contain acceptably low levels of contaminants. Moreover, the utilization in aquaculture of waste heat effluents from power stations may be difficult because this is generally not considered when a power station is planned and constructed.
- Organic residues or by-products (what is left after agricultural production on farms or following processing of agricultural produce in agroindustry) can be used directly, or after processing, as nutritional inputs to aquaculture systems. Such organic residues include the by-products of animal husbandry (manure and slaughterhouse wastes), fisheries (trash fish and fish processing wastes), and the effluents of some intensive aquaculture systems.
- Organic residues from agroindustrial processing plants, in solid form or as

effluents, may be used directly as feeds or as feed ingredients for farmed aquatic organisms or as fertilizers in fishponds. Their value as a nutritional input to aquaculture systems increases with their carbon to nitrogen ratio.

- Reuse of organic residues or by-products of agriculture, animal husbandry (and intensive aquaculture) in adjacent aquaculture systems comprises integrated farming. Agricultural and agroindustrial by-products are commonly traded as feeds or fertilizers for fish culture. Reuse of agroindustrial effluents as fishpond fertilizers is widely recommended but insufficient land to construct fishponds adjacent to factories (and intensive livestock and aquaculture systems) is a frequent constraint to the development of such integrated operations.
- Supplementation of agroindustrial effluents with inorganic fertilizers may be required to correct imbalanced C : N ratios and facilitate effluent reuse in fishponds, as a fertilizer to stimulate natural food production.

Interactions between Aquaculture and the Environment

An Ecological Approach

Ecologists define ecosystems as being essentially self-contained, apart from receiving solar energy. Many aquaculture systems are similar to this. Ecosystems also impinge upon and affect each other. The basic living components (biota) and the nonliving (abiotic) components (habitat) all interact.

For aquaculture ecosystems there are three main variables:

1. the major categories of habitat - saltwater or freshwater;
2. the intensity and/or sustainability of the production operations and the species used; and
3. the value judgements of all persons concerned.

An ecosystem approach allows consideration of the impact of aquaculture on the environment and vice-versa.

Aquaculture development can in fact be an aid to conservation of adjacent wetland habitats when it is well-managed in terms of water quality and coexistence with wildlife. A good example is the bird sanctuary afforded by the Munich sewage-fish ponds (Bayernwerk, n.d.).

Environmentally compatible aquaculture development should be an integral part of coastal and catchment zone management as a whole. It can *enhance* and help conserve habitats and biodiversity, as well as having the potential for negative impacts. Farmers and fisherfolk can have a strong political force in influencing governmental authorities to pursue policies and plans that will accentuate the *positive* environmental impacts and sustainability of aquaculture.

The indigenous knowledge of existing and potential farmers is very important here. For example, South Pacific islanders are well aware of the links between giant clam conservation and coral reef conservation in general. Similarly, groups of small-scale fishers in Ecuador know the value of mangroves for supporting their fisheries and will in future organize and campaign for the wise management of mangrove resources.

Another major factor in habitat change and habitat destruction in developing countries is the intervention of *government* in pursuit of self-sufficiency in food production or power generation. This often means the impoundment of inland waters. Impoundment has impacts not just on the inundated areas but on all areas downstream, including coastal habitats; and also on areas upstream, if the migrations of fish and other biota are disrupted.

In such circumstances, the concept of 'acceptable levels of change' is almost impossible to apply, because what may be acceptable or highly desired by government and by, for example, farmers who were formerly restricted to rainfed farming systems but have gained irrigated lands post-impoundment, may be strongly resisted by those for whom it has real or perceived negative consequences: displaced persons, upstream and downstream communities and conservationists.

Mangroves and Other Coastal Wetlands

Mangrove areas have been used for construction of shrimp ponds in tropical developing countries. In Brazil, mangrove cutting to develop shrimp ponds, (individual size as much as 200 to 300 ha) has changed the local climate so much that the aquaculture operations themselves have had to be terminated. In the dry season, there is insufficient freshwater to counteract rises in salinity (up to 45 ppt) from evaporation.

In some other countries, significant amounts of mangrove forests have been destroyed. However, this is not a simple issue. In some countries (for example, Thailand) most of the mangrove areas cleared for aquaculture were secondary mangroves (i.e., perhaps having good coverage, but long exploited for various purposes and not pristine primary mangroves). At least in Asia, shrimp ponds have recently been and are now being constructed either on former mangrove areas that were cleared or postcleared long ago (and can be considered degraded) or on more recently cleared areas for which the primary purpose of clearance was timber abstraction (logging, woodchip industries and charcoal production), or by adapting traditional trapping-ponds, or on areas inland from the mangrove belt.

In Thailand, Chaichavalit (1989) showed that only 32% of new shrimp farms had been built in mangrove areas, including nipa palm stands. Of the remainder, 10% were built in areas previously occupied by shrimp trapping or growing ponds and up to 28% on former rice, coconut or other farm lands. Assuming that 25-30% of the Thai shrimp pond area is outside the mangroves, then of the 90,000 ha of shrimp ponds reported in Thailand (Rosenberry 1989), about 65,000 ha occupy former mangroves. However, the total area of mangrove destruction in Thailand since 1961 has been reported as 172,000 ha (Chantadisai 1989), of which only up to 38% can be attributed to shrimp farming. Assuming that the same situation with 25-30% of shrimp ponds outside the mangrove belt also holds true for the Philippines, then Philippine shrimp farms account for not more than 17% of the total area of 310,000 ha of mangroves destroyed since 1920 (Zamora 1989). The total area of Philippine shrimp ponds is now about 200,000 ha (Rosenberry 1990). Therefore, in many of the areas for which shrimp culture has been blamed as a major culprit for mangrove destruction, it actually accounts for less than half of the total area of mangroves destroyed (see Csavas, this vol.).

Aquaculturists in Asia are therefore, more often than not, the *end-users* of already degraded or destroyed mangroves, rather than the primary culprits of mangrove destruction. Their activities have often enjoyed strong government support. Nevertheless, many of the areas that they have occupied could and should have been rehabilitated. There is usually an element of social conflict in mangrove use because the original users of most mangrove areas were relatively poor (small-scale fishers and farmers), whereas most shrimp farming and logging activities are run by big business.

The loss of mangrove habitats can bring associated losses of wildlife, from insects, such as fireflies, to proboscis monkeys. The main problems of loss of mangroves on the shore and in estuarine areas are, however, the reduced protection from erosion by wind and water. This increases coastal erosion and soil erosion.

Policy formulation regarding the further conversion of mangrove and other coastal areas for aquaculture is location-specific and cannot be generalized. For example, in Indonesia, it is possible and probably desirable for more mangrove areas to be converted for aquaculture, as part of integrated coastal zone management, whereas in Thailand it is probably not advisable to convert any more mangrove areas into coastal ponds because mangroves as a resource system are already scarce there.

The main requirement now, rather than apportioning blame for past destruction, is to plan and execute plans for the sustainable use of mangroves, including aquaculture as a component of this, in balance with nature conservation and exploitative uses of mangroves.

The same issues of degradation and loss and the need to plan for future sustainable use and conservation also apply to other coastal wetlands. For example, 96% of the original wetlands in the San Francisco Bay area have been lost. Recent legislation is now compelling developers to find ways of 'making good' or 'giving back' lost habitats. This is called 'mitigation'. The same principle could be applied to mangrove and other wetland reestablishment in developing countries. A few funding agencies have tried this but the results have been very limited.

It is now recognized that conservation of pristine and near-pristine mangrove stands and other coastal habitats is very important so that the biodiversity of their communities is maintained for use and study by future generations. This can best be achieved by realistic valuation of these resources and the various options for their use.

Coastal habitats that are managed for integrated enterprises (say aquaculture, fisheries and tourism) provide a better chance for conservation of marine species than habitats that are not thus managed. Such benefits, however, depend upon the scale and intensity of the aquaculture systems. Clearly there are greater nature conservation benefits with modest-scale, less intensive aquaculture systems.

Inland Waters

When planning new freshwater impoundments and reservoirs for water supply, irrigation and power generation purposes, their potential for aquaculture and enhanced fisheries should also be considered from the *outset* of the planning phase. This must involve the production of management plans and assessment of environmental impact: for example, the effects of introductions of exotic species, if such are to be used for stocking of enhanced fisheries; the socioeconomic aspects of management of such fisheries (access, ownership, sustainability, etc.); and the carrying and holding capacity of the waterbodies for more intensive aquaculture, particularly cage culture; Costa-Pierce and Soemarwoto (1990) have provided a useful example.

The state-of-the-art of planning and managing enhanced fisheries and avoiding their possible negative environmental impacts is poorly developed. For some Latin American reservoirs, hatcheries have been built for stocking purposes and to compensate for losses of seed material from natural production. However, clear guidelines were lacking and the results have been poor (Quirós 1989). The same applies to construction of 'fish passes' to allow the migration of fish around dams. Some were constructed around new dams; other dams were built without them. The results are only partially documented and the effectiveness of the passes remains controversial (de Godoy 1975; Delfino et al. 1986).

Mixed farming is generally an aid to biodiversity conservation, unlike industrial monocropping. The integration of inland aquaculture into mixed farming enterprises is an example.

Fisheries Enhancement in 'Circumscribed' Marine Habitats

Fisheries enhancement can be attempted in a wide range of 'circumscribed' marine habitats, such as bays, estuaries, lagoons and reefs, but experience has been very limited and much more research is needed. This is akin to ranching. It can involve releases not only of finfish but also of invertebrates. For example, *billions* of bivalve seed are released yearly by some countries, notably Japan.

The main (and often the only) environmental impact of the release of stocking material is that it can greatly change the composition (diversity, abundance and genetic characteristics) of the natural communities of marine organisms. There are some management tools associated with such ranching operations: for example, control of predators, habitat enhancement (provision of artificial refuges), systems for harvesting migrating organisms and a whole range of access options from totally open access to licensed individual, community or corporate ownership and access rights.

The future goal for enhanced fisheries in 'circumscribed' habitats is to develop these into *managed ecosystems* - rather like meadows, rangelands and forests - not merely stocking them but also managing the species assemblages and, where relevant, vegetation cover. This implies major environmental impacts because formerly relatively undisturbed or new (in the case of new reservoirs), and other ecosystems will be managed in entirely different ways.

Impacts of Aquaculture and Enhanced Fisheries on Nature

Farmed breeds of aquatic organisms and hatchery-reared stocks for release may be close to wildtypes in their genetic characteristics or genetically modified by selection, hybridization, polyploidy or gene transfer, etc. Moreover, the species used may be exotic or native. Despite farmers' best efforts, escapes from fish farms usually occur, especially from the semi-intensive and extensive aquaculture operations prevalent in developing countries. Releases for enhanced fisheries are like massive escapes from aquaculture. Escapees or fish released into one waterbody, river basin or coastal locality will often spread to others, sometimes across national boundaries. Therefore all such operations put farmed or hatchery breeds into the natural environment. The same applies to agriculture and forestry, but there the impacts are easier to see and therefore sometimes easier to control.

Assessing and predicting the impact of this in natural waters are very difficult. Good data are scarce and there is no generally applicable theory or paradigm that will allow reliable forecasting of effects of escapes and releases.

Most attention has been focused on the effects of exotic species.* Value judgements are frequently involved; for example, the common carp *(Cyprinus carpio)* is a highly prized species in Indonesia, but is regarded as a pest in Australia. It is exotic to both. Moreover, most information on the effects of exotic fish transfers is from reports written soon after the transfer was made. Early reports of no harmful effects may not mean that there will not be any. For example, the explosion of the lamprey population and its damage to fisheries in the Canadian Great Lakes came almost 100 years after its first access to these waters. Therefore, data which indicate that the vast majority of transfers have not had any harmful effects do not remove the need for extreme caution in

*[To set the subject of exotic transfers in context, at the Seventh International Ichthyology Congress, The Hague, The Netherlands, 26-31 August 1991, it was noted that from 1,354 exotic fish introductions, only 24 have become established and dominant (of which 17 have had neutral or positive effects and only seven harmful ecological effects) - **Eds.**].

transfers of exotic species and the formulation and implementation of strict international codes of practice (Turner 1988).

The overall principle should be that if there is any reasonable expectation that harmful consequences will ensue with no guarantee of commensurate and sustainable benefits that would justify the 'cost' for present and future generations, then a transfer of an exotic species or breed should *not* proceed.

It is therefore particularly important that research institutions and development projects act *responsibly* and do *not* transfer exotic species other than in compliance with international codes of practice and, over and above that, after thorough appraisal of the likely costs, benefits and alternative approaches. There are many examples of transfers of exotic tilapia species in Africa by research and development agencies that should not have been made. Most of their results have been negative in terms of development and they have left a permanent legacy of mixtures of native and exotic species in natural waters, sometimes beyond the original target country.

A good example of this erroneous and shortsighted approach is afforded by some of the attempts to find a tilapia with good growth and other culture characteristics for use in brackishwater aquaculture in west Africa. External advice, accepted by some national collaborators, was that native lagoon species like *Sarotherodon melanotheron* were worthless for farming and that exotics from elsewhere in Africa, like *O. aureus, O. mossambicus* and *O. spilurus niger* should be introduced. Alternative approaches such as acclimatization to saltwater of early life history stages and/or broodstock of native species or the long-established exotic *O. niloticus* were not explored, despite the existence of publications on the approach (Watanabe et al. 1984, 1985). Transfers of exotics were made and research has been started towards production of salt-tolerant hybrids. Quite apart from the difficult management requirements that an industry based on hybrids would place upon new entrant farmers, this approach of producing hybrids with introduced exotic species should have been the *last* approach resorted to, not the first and only one, because of the possibility that these exotics could escape and colonize the brackishwater coastal systems of adjacent west African countries.

This example illustrates two common fallacies when new aquaculture developments are considered: (1) that the native species and breeds are probably inferior and not worth developing for aquaculture (this view is common even when no rigorous quantitative assessments have been done and sometimes even when few or no fish farms exist in the target area and has led to developed-country universities shipping exotic tilapias from their campuses to Africa) and (2) that such 'short cuts' are justified often because shipments of exotic material lend prestige to projects, their executors and donors. Even worse examples are expected to come from marine aquaculture where transfer of finfish and invertebrates from one ocean to another is likely to be made on the whim that 'it works here, so let's try it there'. Here the private sector is expected to be the most difficult to influence and regulate.

The ICES/EIFAC codes of practice (Turner 1988) need to be strictly followed by the public and private sectors. They require the importing country to prove that the transfer will not cause harmful impacts either in that country or to neighboring countries and is justified because it will bring proven benefits. These codes of practice have worked very well recently in Europe. The problem with applying them in some developing countries is that their decision tree can be 'felled' by commercial pressure groups or even a single ministerial opinion based on perceptions, not scientific evidence.

Aquaculture and enhanced fisheries will impact on natural biota and will affect ecosystems and their biodiversity. The main question is, what levels of change should be regarded as acceptable? The only existing policy guidelines are really: (1) to attempt to conserve biodiversity, *in situ* and *ex situ*, as much as possible and (2) to regard *all* farmed and hatchery-raised breeds of aquatic organisms, exotic or native, as potentially disruptive and able to impact on natural biota - *not* singling out exotic species or genetically modified breeds (for example, transgenic fish) as being the only organisms for which impact studies are required. A cautious approach with long-term projections is essential because once aquatic organisms are established in natural waters they are virtually impossible to eradicate.

Transfers of aquatic organisms can have very major effects on ecosystems. For example, the shipments of the oyster *(Crassostrea virginica)* from the east coast of the USA to the San Francisco Bay area and other transfers introduced other exotic organisms as well (Carlton 1979). Currently, 95% of the benthic species and of the benthic biomass in the San Francisco Bay area are exotic. These species are all highly opportunistic - which is why they become established. This may or may not be a 'better' environment now than when the transfers began about 200 years ago - but it is certainly much changed! This is a subject for value judgments, as are all situations in which different successful species outcompete others as a result of human interventions.

There is a possibility that farmed breeds may be produced in future so as to have less invasive, opportunistic characteristics than wildtypes, thereby lessening the chances that they would became established and outcompete wildtypes. However, the converse is also possible. Breeds developed for rapid growth and feed conversion, bigger size, different life history characteristics such as later maturity, greater fecundity, bigger egg size, different color, etc. could also theoretically displace natural populations and greatly reduce biodiversity.

Much of the world's inland aquaculture uses carps and tilapias. The wild genetic resources of both are under threat. The same may be true for catfishes and other groups. Conservation of wild genetic diversity for the future requires a realistic approach combining *in situ* and *ex situ* methods (Pullin 1990). *In situ* conservation is already difficult, given the ways in which aquatic resources are being harnessed for development and transfers of exotic species are continuing. For example, even some of the locations for collections of pure undisturbed stocks of tilapias in Africa identified by a 1987 workshop (Pullin 1988) have subsequently been found to be suspect when actual collecting expeditions were mounted: some wild stocks of *Oreochromis niloticus* in northern Sénégal appeared to be of mixed origins, probably having interbred with escapees of exotic stocks introduced for an ill-conceived aquaculture development project which failed. This shows that the few remaining undisturbed populations merit special conservation measures and their habitats designated as sites of special scientific interest.

The best prospects for conserving such populations and sites are in game parks and nature reserves (Pullin 1990). The desirability of using *native* species wherever possible has in fact been well recognized for a long time by some international bodies. For example, ICES included this in the early formulation of a code of practice as far back as 1972. Its European member-countries have recently undertaken to implement this strictly with respect to marine species. However, such long delays between formulation and implementation should not dissuade developing countries from formulating and implementing such measures. Indeed it is essential that they do so to protect their aquatic biodiversity. International agencies, research institutions and the private sector, in developed and developing countries, should set examples that will facilitate this.

Introductions of Pathogens, Parasites and Predators

The introduction of pathogens, parasites and predators along with aquatic organisms transferred for aquaculture and enhanced fisheries is widely perceived as a serious risk, but there are relatively few well-documented examples as yet, compared to the numbers of transfers. For microbial pathogens, there is great anxiety that bacterial and other diseases will spread from introduced farm stocks to wild populations, but there is very little evidence for this at present and general thinking is that it is likely to occur only in exceptional cases, though even one such case could be disastrous for important wild populations. Most existing evidence for serious disease outbreaks transferred among stocks are in the opposite direction: *from* wild fish to farmed fish. Many examples of disease outbreaks and attempts at their prevention and control are given by Grimaldi and Rosenthal (1988) and Lillelund and Rosenthal (1989). In a further review (Rosenthal 1990), the spread of *Gyrodactylus* in cage farmed salmon in Sweden and from them to farmed and wild stocks in Norway when smolts were transferred, and the introduction of the eel swim bladder parasite *(Anguillicola crassa)*, probably from Taiwan to the then Federal Republic of Germany in 1980, were cited as important examples of the severe and widespread damage that can be done by transfers of fish parasites. There have also been several instances of ectoparasite and helminths being transferred along with exotic fish in Asia. For example, *Sinergasilus major* and *Bothriocephalus gowkongensis*, two host-specific parasites of grass carp, were introduced to Sri Lanka (ADB/NACA 1991).

Making a distinction between infection from the wild and from transfers may not be very useful since many of the stocks collected and transferred for aquaculture and enhanced fisheries are from the wild or are close to wildtypes - there being very few selected farmed breeds from wild populations and natural waters.

In Asia, the world's major region for aquaculture, the International Development Research Centre (IDRC) of Canada has sponsored work towards effective quarantine of transferred aquatic organisms (IDRC 1983) and an Asian Fish Disease Research Network. Others are following this lead. For example, when ICLARM and its collaborators introduced new founder stocks of *O. niloticus* from Africa to the Philippines for a selective breeding program, these stocks were subjected to up to seven months rigorous quarantine in an isolation unit, from which effluents were disinfected. This was done in collaboration with the Philippine members of the IDRC-supported network. However, such rigorous quarantine remains the exception rather than the rule. For most transfers among developing countries, disease certification and quarantine remains a 'cosmetic' exercise. Moreover, testing for viruses is almost impossible as these have yet to be well described for most tropical species and many developing countries lack the necessary facilities.

Apart from quarantine, there are other approaches to minimize the spread of pathogens, parasites and predators through transfer of aquatic organisms. First, the organisms should be transferred in their early life history stages, preferably as fertilized eggs or embryos raised as near as possible under sterile conditions. This avoids the acquisition of a burden of pathogens and parasites by later stages and the transport of extraneous material and associated biota on skin and gill mucus and on shells, etc.

There should be practical contingency plans to attempt to cope with disease outbreaks or the spread of unwanted species that are unintentionally introduced along with the transfers. There are no clear guidelines for these in developing countries as yet.

Because disease outbreaks depend upon environmental factors and stress levels in the affected organisms, not just the presence of the pathogens or parasites, good husbandry and good maintenance of environmental quality can help to reduce or eliminate epizootics and establishment of pathogens and parasites where they have been inadvertently introduced. This is difficult, however, for some situations, especially where the farmed organisms cannot be seen and their health assessed regularly. Sometimes, even mortalities are not easily seen. Fish farmers often wait for evidence of a significant kill before realizing that something is going wrong.

Pathogen Virulence and Drug Resistance

In developed-country aquaculture, there is now cause for concern that some new and very aggressive fish pathogens are emerging. Some of these pathogenic species have counterparts in developed countries. Their pathogenicity is due to plasmids. One example is a *Serratia marcescens* that is tentatively associated with pollution. It may be a new organism or may have been missed before. It spread in Scottish waters perhaps from one point source where fish were contaminated by sewage. It has a very powerful protease. If it enters fish flesh even in very low numbers (comparable to the minimum infective doses for causing Salmonellosis in humans), it can kill a fish virtually overnight. The body proteins are destroyed extremely rapidly. This pathogen can also thrive at 35°C. Dr. B. Austin and colleagues at the Centers for Disease Control, Atlanta, are examining it as a possible threat to public health (McIntosh and Austin 1990).

Also, over the last few years some strains of bacteria pathogenic to fish have been found that are resistant to the antibiotics in common use. These may also become more aggressive in finfish and shellfish. The vibrios are a good example. These problems of pathogen aggressiveness and resistance to antibiotics are expected to worsen.

Responsible use of antibiotics and other drugs can mitigate this but in aquatic organisms, as in humans, improved living conditions and health care will not eliminate the appearance of new pathogens and disease outbreaks. Plasmid-mediated resistance and changes in pathogenicity will remain a problem.

There is an added danger here in that abuse ('bucket chemistry') in aquaculture of some of the antibiotics that are important in controlling human pathogens (for example, massive use of chloramphenicol as an attempted bacteriostat or chemotherapeutic in hatcheries) could facilitate the development of resistance in human pathogens. There are published lists (Austin 1984) that recommend the prohibition or restriction of use of some drugs for aquaculture purposes, but it needs updating. Chloramphenicol is never used in aquaculture in the United Kingdom and there is zero evidence for chlorampenicol resistance in pathogens from studies on waters associated with hospital wastes. In some developing countries which do use chloramphenicol in aquaculture, resistance of 30-45% has been found in aquatic bacterial populations (Austin 1985).

These problems are likely to be more prevalent in intensive than in semi-intensive or extensive aquaculture systems. However, these terms should be used with care when referring to bivalve culture beds which are usually termed extensive by aquaculturists but which can suffer huge losses from disease as their *populations* are dense. The definition of intensity in aquaculture usually refers to input levels (see p. 142).

For oxytetracycline, which is regularly used in aquaculture, resistance is commonly around 70% (Austin 1985). Ultimately it is a matter for value judgments whether to allow the use of an important or exclusive 'weapon' against human disease to be used, and therefore possibly abused in aquaculture as well. Chloramphenicol is widely used in some countries (for example in the poultry industry in India) whereas such uses are largely prohibited in Europe. The global picture is not a simple split between developed and developing countries attitudes. For example, in the Sudan, chloramphenicol is retained for use against typhus and its use for other purposes is totally prohibited.

The key to managing resistance so that antibiotics remain useful weapons against disease for the far future, is to use different drugs in succession. Withdrawal of use of a particular antibiotic can lead to a massive reduction in resistance to it by aquatic pathogens. Multiple resistance to a number of drugs is also reduced by this. This also applies to resistance to pathogens in the bodies of fish farm workers. This is what most aquaculturists have to learn: to *rotate* drug usage. Most do not do this at present.

Recommendations on Disease Control in Farmed Aquatic Organisms

Disease can be a major constraint to aquaculture development by causing serious losses of farmed stocks. Treatment is generally by antimicrobial compounds and this has raised concerns regarding development of disease resistance and the ecotoxicology of these compounds. There is a general lack of information on the significance of diseases in developing-country aquaculture in terms of mortalities and reduced culture performance, on the range of available drugs, and on the extent of their use. These information gaps should be remedied, and clear, widely accessible publications prepared on the correct use of drugs in aquaculture.

- Widespread routine use of drugs should be discouraged. Medication (treatment) should be considered as the last tool in a management scheme when other measures have proven inadequate. Disease control strategies should be improved with a view to reducing drug use in aquaculture.
- To avoid the development of drug-resistant strains, the use of antimicrobial agents should be rotated and the persistent or frequent use of a single chemical discouraged.
- Training in the appropriate use of drugs should be implemented. Such training should include correct methods of application and dosages, approved methods for disposal of residues and proper storage, and pertinent standardized analytical techniques.
- The establishment of fish health services in developing countries should be encouraged. Such services should ensure that veterinarians promote preventive measures through good husbandry practices, thereby reducing the need for frequent treatment.
- Research is needed to determine the longevity and fate of drugs entering and leaving aquaculture facilities, with emphasis on microbial degradation and mobilization. The utilization rates of pesticides in integrated agriculture-aquaculture and their pathways and possible accumulation in local populations should be measured.

Recommendations on Conservation of Biodiversity and Habitats

- Aquaculture development must not cause loss of or deleterious changes to the wild genetic resources of living organisms or their natural habitats. It should proceed with due regard for nature conservation and in the context of international, regional, national and local conservation and environmental management strategies.
- The likely consequences of escape of organisms from aquatic farms, their establishment in the wild and their effects on other organisms (including

humans) and their habitats, should be thoroughly evaluated *before* any aquaculture development proceeds and especially before any transfers of aquatic organisms are made. This applies to all aquatic organisms, whether native or exotic species and whether close to wild types in genetic characteristics or developed for farming processes by any genetic modification. Genetic modification means selective breeding and genetic management technology (for example, hybridization, chromosome manipulations, sex control and genetic engineering). These requirements also apply to research and development efforts in which aquatic organisms are purposefully released in natural or manmade waterbodies for enhanced (culture-based) fisheries, including ranching.

- All transfers of aquatic organisms for aquaculture research and development should follow established codes of practice; for example, those of the International Council for the Exploration of the Sea (ICES) and European Inland Fisheries Advisory Commission (EIFAC). These codes of practice should be applied to avoid adverse ecological impacts including the spread of fish diseases among countries and ecological zones (such as river basins) and among farms and natural waters within the same zone. In intensive aquaculture operations in some developed countries, substantial losses have been caused by the introduction of diseases, parasites and predators. This has led to the enforcement of these codes of practice and other stringent measures to reduce this risk. Developing countries should implement similar strategies to safeguard their aquaculture operations as well as their wild stocks of aquatic organisms.

Aquaculture Development Projects

General Considerations

For aquaculture development projects, the key to future sustainability is that the project personnel and government officials listen to existing and potential fish farmers, avoid interventions that will damage their interests and link improvements in livelihood, nutrition, health care and education, rather than separating them as unisectoral responsibilities - as has been the usual case. Formal, governmental technical support and extension services are generally inadequate in developing countries because of a lack of resources. Most extension services still use an outdated, top-down approach: officials telling farmers what is good for them.

The process of change in these formal mechanisms is bound to be slow. Meanwhile, *informal* mechanisms (particularly successful farmers helping new entrants through acting as key farmers and organizing farmers' clubs, etc.) are extremely important as these make use of the indigenous knowledge of farmers: the traditional wisdom of the rural communities, which officials usually lack and ignore. This is an enormous resource of environmental wisdom that can be tapped.

Proposed new calendars of activities, involving aquaculture innovations, must not disrupt essential elements of the farmers' existing calendar of activities; for example, the requirements for production of cereal crops, such as rice. This applies to all aquaculture, not just to integrated farming such as rice-fish culture. Moreover, the tradeoffs between expenditure of time, labor and other resources on aquaculture and other activities must be thoroughly assessed. It is a horrifying statistic that poor farm family members in developing countries spend an average of about 40% of their time being sick and/or attending to sick persons. This means that there is little or no time for innovations that intensify farm enterprises. The theoretical advantage of pond aquaculture here is that it can be done with relatively modest inputs of resources as a part-time enterprise and can improve farm family nutrition and health. Further studies on this should be closely linked to studies on primary health care and education.

Another important requirement for aquaculture development projects to avoid adverse effects on farm families is that they anticipate, as far as is possible, *worst case* scenarios such as losses from drought, theft, disease, etc. This is particularly important for farmers who are new entrants. They must not be 'oversold' on aquaculture as a risk-free enterprise, whether full time or part time. There is a rule of thumb (based on intuition, *not* on quantitative data) that all farmers of aquatic organisms, irrespective of the species and systems used, should anticipate on average a *total* loss of produce (and therefore usually loss of all profits) once in ten years; as a worse case scenario twice in ten years; and as a worst case scenario with those two years of total loss back-to-back. This has been dubbed 'when the aquaculture roof falls in'. Aquaculture insurance companies have more useful and data-based predictive mechanisms, but probably little or nothing exists for predicting small-scale pond aquaculture losses in developing countries.

It is very important that the assumptions made by planners, developers and investors are regularly questioned and amended if found to be false. If new aquaculture technology is developed on false assumptions, projects will fail because of this, not necessarily because of any technical shortcomings. An example is the development of recycling systems for aquaculture in Europe. Some investors jumped into implementing the technology *without* testing prior assumptions and hence there were failures and now this kind of aquaculture has a damaged image which, despite the availability of better technology, may take 10 years to overcome (Rosenthal 1981a, 1981b; Dijkema, in press).

Risk Assessment

Those concerned with risk assessment in aquaculture development (for example, in terms of waterborne diseases, microbiological safety of produce, trace chemical contaminants, etc.) can learn a lot from the quantitative techniques used in risk assessment in other fields of human endeavor. There are many powerful techniques available from assessing risks for insurance purposes. A probability matrix technique has been suggested for risk assessment in wastewater-fed aquaculture (Piedrahita and Tchobanoglous 1987) but not yet evaluated or used for this purpose.

A risk assessment procedure for wastewater-fed aquaculture is also suggested by Buras (this vol.). Probably an array of risk assessment techniques is needed because the variety of risks, the range of pathogens and the categories of persons at risk are very wide and change with time. For example, a new health risk to tilapia hatchery workers in the Philippines has come with the advent of hapa(cage)-based breeding techniques: workers spend much of the day wading the ponds containing the hapas and suffer from a variety of foot infections, especially fungal infections.*

Risk assessment should be an integral part of aquaculture development projects executed by development banks, FAO and other agencies. It should be included in their feasibility studies. Moreover, the responsibility for action on this could be taken up by FAO regional bodies, such as COPESCAL and IPFC for discussion at the regional level and followup at the national level.

Policymaking

An essential background to the development of mechanisms (such as codes of practice) for dealing with environmental issues in aquaculture

*This type of hatchery operation is now becoming rare. - **Eds.**

development is the formulation of clear policies with respect to aquaculture. Only recently have developing countries began to make such policies. Formerly, aquaculture was permitted to 'develop' as a kind of 'good idea', largely promoted by external advisors and agencies, usually with superficial (or no) prior analysis of all the factors that should influence policymaking on aquaculture. For example, a future 'gap' in fish supply from capture fisheries to meet the demands of a growing population would frequently be taken simplistically as a clear need and opportunity for aquaculture development, without any analysis of the total implications of this for the people involved, available resource systems or other options to raise plant and animal produce.

This is now changing. For example in Malaŵi, the government has begun to evolve a clear policy to support aquaculture development, especially small-scale pond aquaculture on mixed farms. The Government of Malaŵi sees the need for aquaculture development, despite the fish supply from Lake Malaŵi (the waters of which cover about a quarter of the country) because fish supply from the lake at affordable prices cannot reach some other increasingly populous areas and because aquaculture activities on mixed farms bring better nutrition and livelihood and spread the risks for the farmers.

Overall, however, the formulation of clear national policies on aquaculture development is rare. This has probably been a major factor in the failure of many attempts to develop aquaculture over the past 30 years, especially in Africa and Latin America. Also, there is a persistent problem in that aquaculture development efforts by different donors within a single country or region are often very poorly coordinated. Host country national coordinating mechanisms usually cannot cope with many projects and donor attitudes and objectives are very diverse. Many projects are 'donor-driven'. ICLARM and the International Food Policy Research Institute (IFPRI) are planning to work on guidelines to help with this as ICLARM is about to join the Consultative Group on International Agricultural Research (CGIAR) of which IFPRI is also a member.

In aquaculture development policy formulation, it must be recognized that the individual projects and R & D efforts that try to initiate aquaculture are often relatively short-lived. When aquaculture takes off it can spread like a forest fire through entrepreneurial activities and informal extension that can be very difficult to monitor and control. Therefore, policymaking must include full consideration of possible future development scenarios, their social and environmental implications, sustainability and the needs for extension, investment, monitoring, regulation and future *coordinated* support from donors. Policy formulation must also include predictions of how aquaculture will develop, who the farmers will be, how intensive and extensive, large- and small-scale operations might have to coexist and change with time. There is no justification here for considering so-called 'commercial' and 'socially oriented' aquaculture in isolation.

The ICES Experience

Future application of the ICES/EIFAC codes of practice in Europe will, for EEC countries, be based upon delineation of river basin areas, coastal zones and disease-free or disease-affected areas, rather than national boundaries. The emphasis will be on *ecological* borders. For example, after 1992, the UK would be able to sue France for perhaps a 20-year period if French introductions of exotic species, like coho salmon were shown to have adverse effects on British salmonid populations. Clearly such existing codes of practice are idealistic and not fully implementable and would be even less so in developing countries. However, their principles, particularly the ecological approach and their call for transnational arrangements, are applicable worldwide and developing countries should begin to develop similar arrangements, even though the road will be long.

Some donors, for example the Danish International Development Agency (DANIDA), have begun to document their guidelines on environmental aspects of fisheries and aquaculture development (DANIDA 1989). Other donors, particularly the CEC, are beginning to coordinate their efforts for a common approach to this in the future.

A Schema for Making Decisions on New Aquaculture Developments

A schema (Fig. 1) was developed from a flowchart devised by the ICES Working Group on the Environmental Impacts of Mariculture (ICES 1989). It provides three major pathways for evaluating the potential impact of aquaculture development in terms of socioeconomic interactions, technical aspects (state-of-the-art and management requirements), and ecological interactions. The boxed subject areas in these three columns are indicative of major considerations and are not given in sequential order of appraisal or relative importance. These will vary from case to case. A stepwise approach is essential to evaluate interactions within and among the three columns *before* the establishment of any aquaculture operation can be approved. For example, the type of aquaculture system and its scale will certainly affect the expected interactions at the proposed site (or area) while competition for water resources may prohibit the sustainable development of aquaculture even when environmental concerns are negligible. *Conversely*, effective and affordable technology for improved waste handling or reuse may permit the enlargement of aquaculture operations at a site or an area which otherwise has very limited assimilative capacity.

Integrated systems may not only improve the cost-effectiveness of aquaculture activities but may also make them environmentally acceptable. Some ecological impacts have been studied in detail (mainly in industrialized countries) and mathematical models have been formulated which can be used, with caution, as tools to estimate the extent of impact in some situations.

The interactions among columns and the overall decision on acceptability is made where the three boxes labelled 'Acceptable?' are interlinked. It is hoped that quantitative tools and appropriate software will be developed for this.

If the development is technically feasible and the anticipated impacts (socioeconomic and/or environmental) are acceptable, one can proceed with an accompanying monitoring program (see loop). Monitoring cannot, however, be an exercise on its own. It should serve a purpose (e.g., environmental protection, farm management, research for the advancement of knowledge, etc.). The data obtained through monitoring can be used in an iterative evaluation process to identify changes and trends in environmental quality. The results can be used for regulatory purposes and may lead to the identification of environmental quality standards.

References

ADB/NACA. 1991. Fish health management in Asia-Pacific. Report on a Regional Study and Workshop on Fish Disease and Fish Health Management. Asian Development Bank Agriculture Department. Rep. Ser. No. 1, Asian Development Bank, Philippines and Network of Aquaculture Centres in Asia, Bangkok, Thailand.

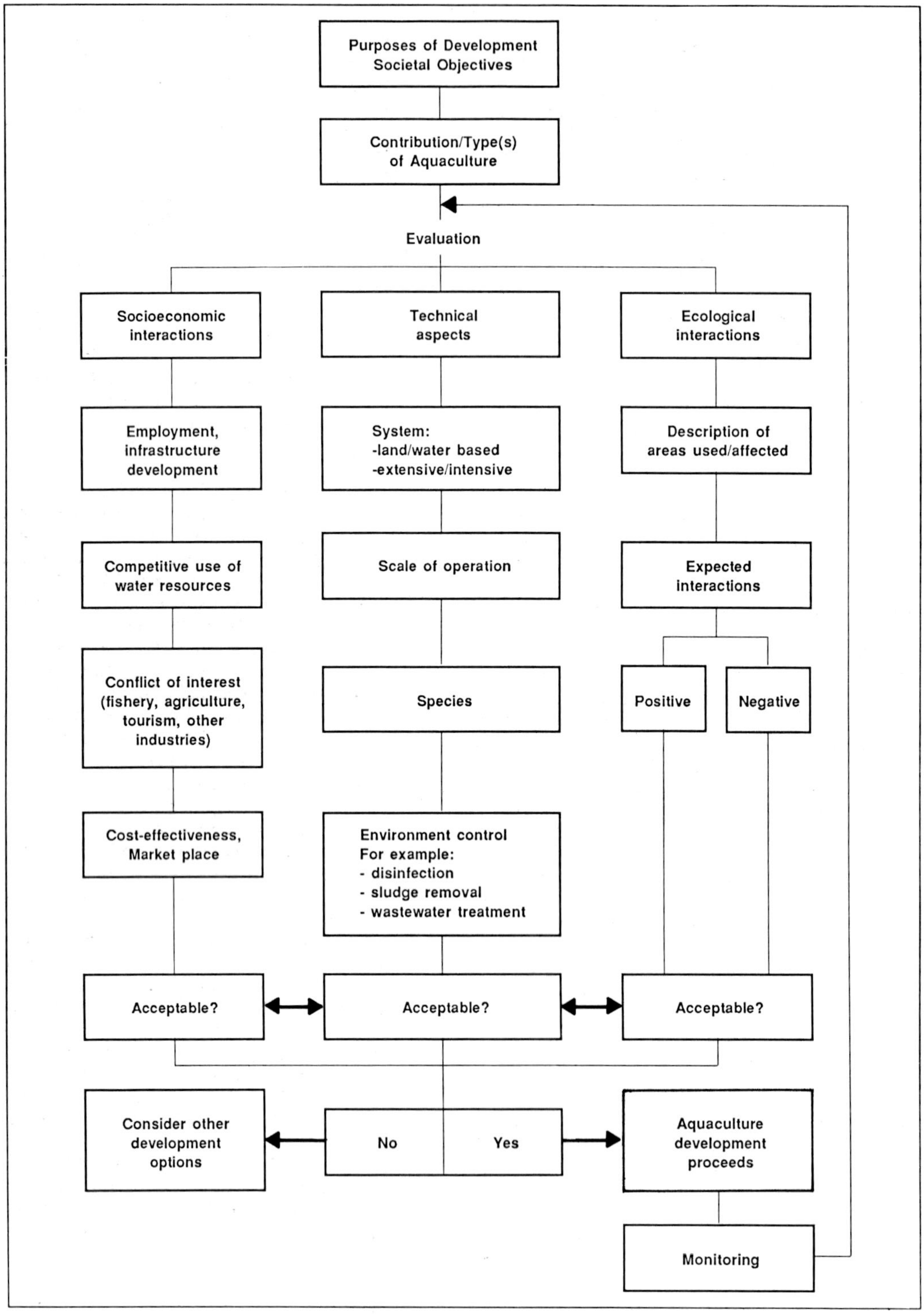

Fig. 1. A decisionmaking schema, devised during the conference, for considering proposed aquaculture developments and their environmental impacts. This was based on the flowchart published by ICES (1989).

Arakawa, K.Y., Y. Kusuki and M. Kamigaki. 1971. Studies on biodeposition in oyster beds. I. Economic density for oyster beds. Venus Jap. J. Malacol. 30(3):113-128. (In Japanese).

Austin, B. 1984. Chemotherapy of bacterial fish diseases, p. 19-26. *In* A.E. Ellis (ed.) Fish and shellfish pathology. Academic Press, London.

Austin, B. 1985. Antibiotic pollution from fish farms and its effect on the aquatic environment. Microbiol. Sci. 2:113-118.

Bayernwerk, A.G. n.d. Teichgut Birkenhof und Speichersee. Bayernwerk, Ascheim, Germany. (In German).

Carlton, J.T. 1979. Introduced invertebrates of San Francisco Bay, p. 427-444. *In* T.J. Conomos (ed.) San Francisco Bay: the urbanized estuary. American Association for the Advancement of Science, San Francisco.

Chaichavalit, S. 1989. An economical analysis of intensive giant tiger prawn culture: a case study of Samut Songkram. Thammasart University, Bangkok, Thailand. 92 p. M.E. thesis.

Chambers, R., A. Pacey and L.A. Thrupp, Editors. 1989. Farmer first: farmer innovation and agricultural research. Intermediate Technology Publications, London.

Chantadisai, T. 1989. Environmental impact of aquaculture on mangrove ecosystem. BIOTROP Spec. Publ. 37:215-223.

Coeurdacier, J.M., M. Héral, M. Liange, J. Moreau, J. Prou, D. Razet and P. Wilhot. 1983. Rejects d'élevage de poissons en eau de mer: consequences pour l'huitre *Crassostrea gigas*. Rev. Trav. Inst. Pêches Marit., Nantes 46(2):121-140 (In French).

Costa-Pierce, B.A. and O. Soemarwoto, Editors. 1990. Reservoir fisheries and aquaculture development for resettlement in Indonesia. ICLARM Tech. Rep. 23, 378 p.

DANIDA. 1989. A strategy for fisheries development: plan of action for integration of environmental considerations into Danish development assistance. Department of International Development Cooperation, Copenhagen.

Delfino, R., C. Baigún and R. Quirós. 1986. Esclusas para peces en la represa de Salto Grande. Consideraciones acerca de su functionamiento. Departamento de Aguas Continentales. Ser. Inf. Tec. Inst. Nac. Invest.-Desarr. Pesq. (Argent.) 3, 51 p. (In Spanish).

Dijkema, R. Developments in cultivating the African catfish *(Clarias gariepinus)* in the Netherlands: techniques, markets and perspectives. Proceedings of the 5th International Conference on Aqua-farming, Verona, Italy. (In press).

Edwards, P. 1985. Aquaculture: a component of low cost sanitation technology. World Bank Tech. Pap. 36, 45 p.

Edwards, P. and R.S.V. Pullin, Editors. 1990. Wastewater-fed aquaculture. Proceedings of the International Seminar on Wastewater Reclamation and Reuse for Aquaculture, 6-9 December 1988. Calcutta, India. Environmental Sanitation Information Center, Asian Institute of Technology, Bangkok, Thailand.

FAO. 1982. Reference manual to codes of practice for fish and fishery products. FAO Fish. Circ. 750, 257 p.

FAO. 1988. Report of the Seventh Session of the Indo-Pacific Fishery Commission, 1-6 August 1988. Bangkok, Thailand. FAO Fish. Rep. 411, 37 p.

FAO. 1991. Draft code of hygienic practice for the products of aquaculture. 33 p. FAO, Rome. (Manuscript).

Godoy, M.P. de. 1975. Peixes do Brasil, suborden Characoidei, Baia do Rio Mogi-Guassu. Editora Franciscana, Piraçicaba, Brazil. (In Portuguese).

Grimaldi, E. and H. Rosenthal, Editors. 1988. Efficiency in aquaculture production: disease control. Proceedings of the 3rd International Conference on Aqua-farming "Acquacultura 86", 9-10 October 1986. Verona, Italy. Edizioni del Sole 24 Ore, Milan, Italy. (In English and Italian).

Hansen, J.C. 1990. Human exposure to metals through consumption of marine foods: a case study of exceptionally high intake among Greenlanders, p. 227-243. *In* R.W. Furness and P.S. Rainbow (eds.) Heavy metals in the marine environment. CRC Press, Boca Raton, Florida.

ICES. 1989. Report of the Working Group on Environmental Impacts of Mariculture. Figure 1. Flowchart on principal procedures for the evaluation of environmental impact of mariculture, p. 15. Marine Environmental Committee. Mariculture Committee. International Council for the Exploration of the Sea. C.M. 1989/F. Copenhagen, Denmark.

IDRC. 1983. Fish quarantine and fish diseases in Southeast Asia: report of a workshop held in Jakarta, Indonesia, 7-10 December 1982. International Development Research Centre, Ottawa.

ILO. 1983. The ILO encyclopaedia of occupational health and safety. 3rd ed. 2 vols. International Labour Organization, Geneva.

Kramer, D.E. and J. Liston, Editors. 1987. Seafood quality determination. Proceedings of the International Symposium on Seafood Quality Determination, coordinated by the University of Alaska Sea Grant College Program, Anchorage, Alaska, 10-14 November 1986. Elsevier, Amsterdam.

Lillelund, K. and H. Rosenthal, Editors. 1989. Fish health protection strategies. Contributions to the Canadian-German Cooperation Programme. Federal Ministry for Research and Technology, Hamburg.

McIntosh, D. and B. Austin. 1990. Recovery of an extremely proteolytic form of *Serratia liquefaciens* as a pathogen of Atlantic salmon, *Salmo salar*, in Scotland. J. Fish Biol. 36:765-772.

Mito, S. and O. Fukuhara. 1988. Aquaculture development in Japan, p. 39-72. *In* J.V. Juario and L.V. Benitez (eds.) Perspectives in aquaculture development in Southeast Asia and Japan: contribution of the SEAFDEC Aquaculture Department. Southeast Fisheries Development Center, Iloilo, Philippines.

Piedrahita, R. and G. Tchobanoglous. 1987. The use of human wastes and sewage in aquaculture, p. 336-352. *In* D.W. Moriarty and R.S.V. Pullin (eds.) Detritus and microbial ecology. ICLARM Conf. Proc. 14, 420 p.

Pullin, R.S.V., Editor. 1988. Tilapia genetic resources for aquaculture. ICLARM Conf. Proc. 16, 108 p.

Pullin, R.S.V. 1990. Down-to-earth thoughts on conserving aquatic genetic diversity. Naga, ICLARM Q. 12(1):10-13.

Quirós, R. 1989. Structures assisting the migrations of non-salmonid fish: Latin America. COPESCAL Tech. Pap. 5, 41 p.

Rosenberry, B. 1989. World shrimp farming. Aquacult. Dig. 28 p.

Rosenberry, B. 1990. World shrimp farming 1989. Annual report of Aquaculture Digest, January 1990. San Diego, California. 28 p.

Rosenthal, H. 1981a. Recirculation systems in western Europe, p. 305-315. *In* K. Tiews (ed.) Aquaculture in heated effluents and recirculation systems. Vol. II. Proceedings of the World Symposium on Aquaculture in Heated Effluents and Recirculation Systems, Stavanger, Norway, 28-30 May 1980. Heeneman Verlagsgesellschaft mbH, Berlin.

Rosenthal, H. 1981b. Experiences with closed culture systems on a commercial scale. Mariculture Committee, International Council for the Exploration of the Sea, Copenhagen. C.M. 1981/F: 15, 15 p.

Rosenthal, H. 1990. Efficiency in aquaculture production: reflections on the 1986 Verona Conference, p. 8-14. *In* H. Rosenthal and E. Grimaldi (eds.) Efficiency in aquaculture production: technology improvements in farming systems. Proceedings of the 4th International Conference on Aquafarming "Acquacultura 88", 14-15 October 1988. Verona, Italy. Ente Fiere di Verona, Verona, Italy. (In English and Italian).

Turner, G.E., Editor. 1988. Codes of practice and manual of procedures for consideration of introductions and transfers of marine and freshwater organisms. EIFAC Occas. Pap. 23. FAO, Rome.

Watanabe, W.O., C.M. Kuo and M.C. Huang. 1984. Experimental rearing of Nile tilapia fry *(Oreochromis niloticus)* for saltwater culture. ICLARM Tech. Rep. 14, 28 p.

Watanabe, W.O., C.M. Kuo and M.C. Huang. 1985. Salinity tolerance of the tilapias *Oreochromis aureus, O. niloticus* and an *O. mossambicus* x *O. niloticus* hybrid. ICLARM Tech. Rep. 16, 22 p.

WHO. 1989. Guidelines for the use of wastewater in aquaculture. Report of a WHO Scientific Group. Tech. Ser. No. 778. World Health Organization, Geneva.

WHO. 1990. Report of WHO consultation on public health aspects of seafood-borne zoonotic diseases. WHO/COS/VPH/90.86. World Health Organization, Geneva.

Zamora, P.M. 1989. Philippine mangroves: their depletion, conversion and decreasing productivity. Wallaceana 58:1-5.

Author Index

Geographic Index

Species Index

Bellagio Environment Conference Participants. Seated (from left to right): F. Moriarty, R. Gowen, R.S.V. Pullin, N. Buras, V. Palmer, M. Bilio, H. King. Standing (middle row, from left to right): B. Austin, S. Cataudella, J.L. Maclean, T.E. Chua, H. Rosenthal, M. Martínez-Espinosa, D.J.H. Phillips, J.L. Munro, E. Arellano, K. Ruddle. Standing (back row, from left to right): M.J. Phillips, I. Csavas, P. Edwards, M.C.M. Beveridge.

List of Participants

Mr. E.M. Arellano*
Faculdad de Ing. Maritima y Marinas
Ciencias del Mar
Centro Nacional de Agricultura e Investigaciones Marinas
Rocafuerte y Loga
P.O. Box 5863
Guayaquil, Ecuador

Dr. B. Austin
Department of Biological Sciences
Heriot-Watt University
Riccarton, Edinburgh EH14 4AS
Scotland, UK

Dr. M.C.M. Beveridge
Institute of Aquaculture
University of Stirling
Stirling FK9 4LA
Scotland, UK

Dr. M. Bilio
Fisheries and Aquaculture
Division 422, GTZ, POB5180
D6236 Eschborn 1
Federal Republic of Germany

Dr. K. Bögel
Division of Communicable Diseases
World Health Organization
1211 Geneva 27
Switzerland

Dr. N. Buras
Department of Civil Engineering
University of Arizona
Tucson, Arizona 85721
USA

Dr. S. Cataudella
Dipartamento de Biologia
Università di Roma
"Tor Vergata"
Via E. Carnevale
00173, Rome, Italy

Dr. Chua Thia-Eng
ICLARM, MCPO Box 2631
0718 Makati, Metro Manila
Philippines

Mr. I. Csavas
FAO Regional Office for Asia and the Pacific
39 Phra Atit Road
Bangkok, 10200, Thailand

Dr. P. Edwards
Division of Agricultural and Food Engineering
Asian Institute of Technology
P.O. Box 2754
Bangkok 10501, Thailand

Dr. R. Gowen
Natural Environment Research Council
Dunstaffnage Marine Laboratory
P.O. Box 3, Oban
Argyll, Scotland
UK

Ms. Hannah King
7, Duchy Wood
Daisy Hill
Bradford, West Yorkshire BD9 6D2
UK

Mr. J.L. Maclean
ICLARM, MCPO Box 2631
0718 Makati, Metro Manila
Philippines

Dr. M. Martinez-Espinosa
Inland Water Resources and Aquaculture Service
Fishery Resources and Environment Division
Food and Agriculture Organization
Via delle Terme di Caracalla
00100 Rome, Italy

Dr. F. Moriarty
Queens Orchard, Little Eversden
Cambridge CB3 7HD
UK

Dr. J.L. Munro
ICLARM, MCPO Box 2631
0718 Makati, Metro Manila
Philippines

*Mr. Edgar Arellano was tragically killed in a motoring accident in 1991.

Ms. V. Palmer
Third World Medical
Research Foundation
New York, New York
USA

Dr. D.J.H. Phillips
Acer Environmental
Howard Court
Manor Park, Daresbury
Nr. Runcorn, Cheshire WA7 1SJ
England, UK

Dr. M.J. Phillips
Asian Institute of Technology
GPO Box 2754
Bangkok 10501, Thailand

Dr. R.S.V. Pullin
ICLARM, MCPO Box 2631
0718 Makati, Metro Manila
Philippines

Dr. H. Rosenthal
Fischereibiologische Abteilung
Institut für Meereskunde
an der Universität Kiel
Düsternbrooker Weg 20
D 2300 Kiel 1
Federal Republic of Germany

Dr. K. Ruddle
11-20 Matsugaoka-cho
Nishinomiya-shi
Hyogo-ken 662
Japan

Deutsche Gesellschaft für Technische Zusammenarbeit (GTZ), GmbH

The Deutsche Gesellschaft für Technische Zusammenarbeit (GTZ) GmbH is owned by the Government of the Federal Republic of Germany. It is commissioned to undertake specialist technical planning and implementation of measures for technical cooperation with developing countries. This takes place on the basis of a General Agreement with the Federal Government. Development policy is formulated by the Federal Ministry for Economic Cooperation (BMZ). GTZ draws on the facilities available in both the public and private sectors as far as this is expedient and cost effective.

ICLARM International Center for Living Aquatic Resources Management

The International Center for Living Aquatic Resources Management (ICLARM) is an autonomous, nongovernmental, nonprofit, international scientific and technical center which has been organized to conduct, stimulate and accelerate research on all aspects of fisheries and other living aquatic resources.

The Center was incorporated in Manila in March 1977. It became a member of the Consultative Group for International Agricultural Research (CGIAR) in May 1992.

ICLARM is an operational organization, not a granting entity. Its program of work is aimed to resolve critical, technical and socioeconomic constraints to increased production, improved resource management and equitable distribution of benefits in economically developing countries. The Center's work focuses in tropical developing countries on three resource systems - inland aquatic (mainly ponds and rice floodwaters), coastal and coral reef - in which research is carried out on their dynamics, on investigating alternative management schemes, and on improving the productivity of key species. The work includes cooperative research with institutions in developing countries, and supporting activities in information and training. The programs of ICLARM are supported by a number of private foundations and governments.

Policies are set by a Board of Trustees with members drawn from the international community. Direction of ICLARM, under the policies set by the Board, is the responsibility of the Director General.

MCPO Box 2631,
0718 Makati, Metro Manila
Philippines
Telephones 8175163, 8180466, 8189283, 8175255
Cable: ICLARM Manila
Telex: (ETPI) 64794 ICLARM PN
Fax: (63-2) 816-3183